Exploration with Short-Offset Grounded-Wire Transient Electromagnetic Method

Guoqiang Xue · Xin Wu · Weiying Chen · Nannan Zhou

Exploration with Short-Offset Grounded-Wire Transient Electromagnetic Method

Springer

Guoqiang Xue
Institute of Geology and Geophysics
Chinese Academy of Sciences
Beijing, China

Xin Wu
Institute of Geology and Geophysics
Chinese Academy of Sciences
Beijing, China

Weiying Chen
Institute of Geology and Geophysics
Chinese Academy of Sciences
Beijing, China

Nannan Zhou
Institute of Geology and Geophysics
Chinese Academy of Sciences
Beijing, China

ISBN 978-981-96-8845-6 ISBN 978-981-96-8846-3 (eBook)
https://doi.org/10.1007/978-981-96-8846-3

The original submitted manuscript has been translated into English. The translation was done using artificial intelligence. A subsequent revision was performed by the author(s) to further refine the work and to ensure that the translation is appropriate concerning content and scientific correctness. It may, however, read stylistically different from a conventional translation.

This Springer imprint is published by the registered company Springer Nature Singapore Pte Ltd.
The registered company address is: 152 Beach Road, #21-01/04 Gateway East, Singapore 189721, Singapore

If disposing of this product, please recycle the paper.

Preface

The artificial source electromagnetic method is an important means of deep resource detection. The key is to solve the problem of large depth and high precision detection through innovative methods and technologies, and further explore the detection performance of the artificial source excitation field. Based on the latest development of wide-area electromagnetic method and multi-channel transient electromagnetic method, my team and I have researched the Short-offset Transient. The characteristic of this method is that it has a small transmission and reception distance, stronger signals and larger bandwidth at short offset distances, which is more conducive to meeting the needs of deep and high-precision detection. With the support of peers at home and abroad, this method has made certain progress in recent years. This monograph collects the author's research results in the field of information extraction technology and detection application examples of short-offset transient electromagnetic method in recent years, hoping to contribute to the fine detection of the second depth space.

The key to solving the problems of deep and high-precision detection lies in innovative methods and technologies, and further exploring the detection performance of artificial source excitation fields. In 2013, the author led the team to propose short-offset transient electromagnetic measurement, which extended the measurement area of the grounded wire source from the far zone to the middle and near zone, forming the Short-offset Transient Electromagnetic (SOTEM). This method allows observation points to be arranged within a range of 0.3–2 times the detection depth from the field source. Its main feature is that it has a smaller transmission and reception distance, stronger signals and larger bandwidth under short offset distance observation, which can significantly improve the signal strength and post-processing quality of the data, and is more conducive to meeting the needs of deep and high-precision detection.

Compared with the loop source transient electromagnetic method, the short-offset transient electromagnetic method (SOTEM) can not only excite the horizontal component of the electric field, but also excite the vertical component of the electric field, making all components of the electromagnetic field related to the electrical structure of the formation, thus effectively detecting high-resistance and low-resistance areas. Because the observation point is closer to the source, the observation signal of this grounded source is stronger, has a higher signal-to-noise ratio, and thus significantly improves the detection accuracy.

SOTEM inherits the advantage of the small volume effect of the loop source transient electromagnetic method, while overcoming its limitation of exploration depth. In addition, SOTEM also retains the advantage of the strong signal of the grounded source long-offset transient electromagnetic method, and improves work efficiency and accuracy. This method effectively solves the scientific and technical problems of the current geophysical transient electromagnetic method in terms of large depth, high precision, and rapid detection, and also overcomes the difficulties of fine detection in areas with harsh surface terrain conditions and complex underground environments such as mountains and mines.

In recent years, the research team of the Institute of Geology and Geophysics of the Chinese Academy of Sciences has undertaken a number of national natural science projects, including "Research on Transient Electromagnetic Field Theory Based on Time-Varying Point Charge Current Element", "Key Issues of SOTEM Deep Detection", "Research on Fine Detection Technology of Ground-Well Transient Electromagnetic Method", "Research and Application of Multi-Line Source Transient Electromagnetic Transverse Magnetic Polarization Field Detection Method", "Research on Pseudo-Random Code Full Waveform Response Dimension Reduction Bayesian Inversion of Time Domain Electromagnetic Method", "Research on Theory and Method of Fine Detection of Short-offset Transient Electromagnetic Method", "Research on Fine Extraction Method of Multi-source Ground-Air Time Domain Electromagnetic Response Signal", and "Key Theory and Technology of Land-Sea Transient Electromagnetic Detection". At the same time, the research team has cooperated with many domestic units to achieve a number of application results, jointly researching and promoting the development of the short-offset transient electromagnetic method.

Focusing on the scientific problems and technical difficulties of SOTEM in deep and high-precision detection, more than ten doctoral students and postdocs have chosen related research topics and completed corresponding research. These topics include "Research on Transient Electromagnetic Theory of Point Charge Current Element", "Research and Application of Short-offset Transient Electromagnetic Method", "Research and Application of Key Technologies of Multi-channel Transient Electromagnetic Method", "SOTEM Depth Measurement Data Inversion Method", "Research on Multi-source Multi-component Detection Method of Short-offset Transient Electromagnetic Method", "2.5D Forward and Seismic Offset Imaging Research of Multi-channel Transient Electromagnetic Method", "Research on Depth Response of Electromagnetic Frequency Measurement Based on Harmonic Point Element", "Research on 3D Inversion of SOTEM Based on Particle Swarm-Pseudo Newton Hybrid Algorithm", and "Analysis and Information Extraction of SOTEM Response Characteristics with Excitation Effect", and have achieved phased research results.

Based on understanding the basic concepts of the transient electromagnetic method of the electrical source and the propagation mechanism of the transient electromagnetic field, the research team proposed a new model for calculating the electromagnetic field based on the assumption of time-varying point charge. This model replaces the traditional dipole source with a time-varying point charge element, considering the exact position vector of the source, eliminating the impact of the source size. At the same time, this method no longer goes through frequency-time transformation, but directly calculates the time-domain transient field based on the time-domain position function, thus avoiding the truncation error caused by frequency-time transformation. Based on this, the team proposed axial short-offset transient electromagnetic detection and semi-airborne transient electromagnetic method, which are helpful for detection in mountainous areas, mining areas and covered areas, and are suitable for our country's national conditions.

Furthermore, the research team introduced the processing concept of deep learning, designed a denoising autoencoder neural network, and achieved one-stop multi-source noise removal, which performs better than traditional methods. Drawing on advanced seismic technology, the team used a fully convolutional neural network to directly establish the mapping relationship between the electromagnetic diffusion field and the virtual wave field, thereby achieving high-precision interface structure imaging. At the same time, a dual-drive inversion method combining model-driven and data-driven was adopted, and a recurrent neural network was designed, which significantly improved the accuracy and precision of structural body position recognition.

This book introduces 9 application examples, covering multiple fields such as metal mines, coal mines, and engineering, specifically including:

Salt Mine Detection: Ye County in Henan is a world-class rock salt base. In the past, water dissolution method was used for mining, and the location of deep water-filled cavities was unknown. The main difficulties lie in the large thickness of the cover layer, deep burial depth, and the need to find low resistance in low resistance, which is difficult and the traditional detection effect is not good. A 1000-meter offset was used to detect a salt mine cavity buried at 1400 m. The cavity distribution delineated on the apparent resistivity-depth profile was revealed by drilling, verifying the detection capability of the SOTEM method. This method provides a new technical means for deep, high-resolution detection of underground mineral resources and is the first example of SOTEM electromagnetic fine detection in thickly covered areas in China.

Coal Mine Detection: Transient electromagnetic method is widely used in goaf surveys. The author compared the differences in response characteristics and detection capabilities of loop source and electrical source devices, and believed that the loop source transient electromagnetic signal is more sensitive to strata, while the electrical source transient electromagnetic method has a larger detection depth. The two devices were jointly applied to the investigation of stratum stability and surrounding goaf for a certain substation site selection, using loop source data to obtain shallow stratum information, and electrical source detection data to provide deep stratum information, achieving good results. The research results show that the electrical source transient electromagnetic method has the advantages of large detection depth, convenient construction, and high work efficiency, and is a deep detection electromagnetic method worth promoting.

Metal Mine Detection: The gold mines on the southern margin of the North China Craton are expected to become the second Jiaodong. Xiao Qinling and Xiong Ershan have both made breakthroughs in prospecting, but the Xiaoshan mining area has not yet made a breakthrough, mainly due to complex mineralization conditions. The previous detection depth was less than 300 m, and the detection depth increased to 1700 m after using new technology. Deeply hidden ore bodies were delineated in multiple mining areas. After drilling, deep gold, silver and other metal mines were discovered, which brought significant economic benefits and achieved a breakthrough in prospecting.

Geothermal Detection: The Yangbajing geothermal field in Tibet is a world-famous high-temperature geothermal field. After more than 40 years of development and utilization, shallow geothermal resources are gradually shrinking, and it is urgent to develop deep high-temperature geothermal resources to maintain geothermal power generation or increase installed capacity. In the past, 2D detection was used to form a conceptual model to support the south area re-injection. Six electrical source short-offset transient electromagnetic detection profiles were implemented in the Yangbajing geothermal field, and the electrical structure of the strata within a depth range of 2 km was obtained through inversion. A high-resistance rock wall was found under the China-Nepal Highway, which hindered the heat exchange between the north and south areas, resulting in insufficient deep heat and scarce water resources in the north area, and the opposite in the south area. It is suggested to enhance geothermal energy in the north area.

Engineering Detection: The book also introduces the results of the SOTEM survey conducted in Shunping, Hebei Province, with a detection depth of up to 2000 m. By using the short-time sliding window adaptive singular value decomposition algorithm based on wavelet threshold for denoising, and using the independently developed SOTEMsoft software for inversion, a buried active fault was successfully discovered.

In addition to the national projects undertaken by the author and the team, they also undertake local cooperation and exploration service projects: (1) Jointly funded project of Baiyun Ebo Rare Earth Resources Research and Comprehensive Utilization National Key Laboratory; (2) Shaanxi Coalfield Geology Group Co., Ltd. Major scientific and technological research and development project: Coal mine electrical source short-offset transient electromagnetic method equipment technology; (3) Henan Yu Di Technology Group key scientific research project: Short-offset transient electromagnetic equipment research and development and deep prospecting application; (4) Guangxi Zhuang Autonomous Region key research and development plan project: Short-offset ground-air transient electromagnetic detection technology research and application demonstration; (5) Central government guides local science and technology development fund project: Short-offset transient electromagnetic method fine detection key technology and demonstration; (6) Qinghai talent project; (7) Shaanxi talent project; (8) Open course topic of the Key Laboratory of Coal

Resources Exploration and Comprehensive Utilization of the Ministry of Natural Resources; (9) Research and development project of scientific research instruments and equipment of the Chinese Academy of Sciences.

It has been applied in 80 units at home and abroad. In the process of promoting the method and technology, the author has received guidance from many domestic and foreign peers and experts. They have provided valuable help in method and technology consulting, testing sites, geological data, etc., which has benefited the author a lot. For this, the author would like to express special thanks to the Chinese Geophysical Society, Northwest Nonferrous Geological Mining Group Co., Ltd., Henan Nonferrous Metals Geological Mining Bureau, Qinghai Geological Mining Exploration and Development Bureau, and other units for their support and help in technical procedures. At the same time, thanks to Xi'an Northwest Nonferrous Geo-exploration General Brigade Co., Ltd., Shanxi Coal Geological Geo-detection and Mapping Institute, Jiangsu Geological Exploration Technology Institute, Shaanxi Coalfield Geology Group Co., Ltd., Anhui Exploration Technology Institute, Chang'an University, Qinghai Third Geological Exploration Institute, Beijing Jiedeng Geophysical Exploration Co., Ltd., Hengda Xinchuang (Beijing) Geophysical Technology Co., Ltd., Hunan Jishan High-tech Co., Ltd. and Beijing Zhongke Diyuan Technology Co., Ltd. and other units for their contributions to the engineering of method and technology.

This monograph introduces the data inversion, additional field information extraction technology, and the research progress of deep learning in data processing of short-offset electromagnetic method, which helps the application and promotion of near-source electromagnetic method technology, for scholars to refer to, question or discuss, to promote the progress of near-source electromagnetic method and even the whole geoelectromagnetism. However, given the limited knowledge accumulation and business level of the author, some method and technology content still needs further in-depth research, and there are inevitably errors and omissions, please experts and readers criticize and correct.

Beijing, China
November 2024

Guoqiang Xue
Xin Wu
Weiying Chen
Nannan Zhou

Acknowledgments This work was supported by the National Science Foundation of China (grant no. 41474095, 41874162, 42030106, 42074121, 92262303, and 42274192), the projects from the Ministry of Science and the Technology National Key R&D Program of China (grant nos. 2022YFC2903505), and the Baogang (Group) Company Bayan Obo Mineral Resources Comprehensive Utilization Academician Workstation Research Project (GZ-2023-1-LH-003).

Contents

About the Authors

Guoqiang Xue is a researcher at the Institute of Geology and Geophysics, Chinese Academy of Sciences, a professor at the University of Chinese Academy of Sciences, and the director of the Earth Detection and Information Technology Research Office at the College of Earth Sciences. His primary research focus is on the theory and application of transient electromagnetic detection. He has established near-field detection theory for transient electromagnetic methods, proposed the Short Offset Transient Electromagnetic Method (SOTEM), developed a fine detection equipment system for electrical source transient electromagnetic methods, and created technologies for efficient noise removal based on deep learning, imaging of subsurface interfaces, and constrained imaging of electrical structures. He serves as the associate editor for several journals. He has published 162 SCI papers (47 as first author and 52 as corresponding author), with approximately 3225 citations, and holds 75 authorized patents, including 7 US patents and 19 Chinese patents (as the first inventor). He has received numerous awards, including the National Science and Technology Progress Award (Second Class), the Chinese Academy of Sciences Outstanding Achievement Award.

Xin Wu obtained his Doctorate degree (2018) in Geophysics from the University of Chinese Academy of Sciences. Currently, he is a Associate professor in the Institute of Geology and Geophysics, Chinese Academy of Sciences (subordinate to the Innovation Academy for Earth Science of the Chinese Academy of Sciences), with an emphasis in theory, technology and application of the transient electromagnetic method.

Weiying Chen received the master's degree from Guilin University of Technology in 2009, and the Ph.D. degree from the University of Chinese Academy of Sciences in 2015. He is an Associate Professor with the Institute of Geology and Geophysics, Chinese Academy of Sciences (subordinate to the Innovation Academy for Earth Science of the Chinese Academy of Sciences). His research interests include the forward modeling and data processing of TEM data, and the case history of land, borehole, airborne and marine TEM method.

Nannan Zhou obtained his Doctorate degree (2014) in Geophysics from the University of Chinese Academy of Sciences. Currently, he is a Associate professor in the Institute of Geology and Geophysics, Chinese Academy of Sciences (subordinate to the Innovation Academy for Earth Science of the Chinese Academy of Sciences), with an emphasis in theoretical calculation and applications of the transient electromagnetic method.

Chapter 1
Introduction

1.1 Method Development

For the frequency domain electromagnetic method of grounding source form, our country's electromagnetic workers have actively promoted the electromagnetic frequency sounding method of the former Soviet Union, using single-component apparent resistivity definition and full-field observation, pushing the observation area of the frequency domain artificial source electromagnetic method from the far zone to the middle zone. At the same time, Professor He (2010) of Central South University proposed the wide-area electromagnetic method, defined the wide-area apparent resistivity, used the sequence pseudo-random signal technology, expanded the observation range of the artificial source electromagnetic method, improved the observation speed, accuracy and field efficiency, and formed a new electrical exploration method with unique characteristics (He, 2010). For deep target detection, the grounded wire source transient electromagnetic method can obtain a larger detection depth in near-field observation, compared with CSAMT and LOTEM, it has higher detection accuracy, this research is of great significance.

In fact, when choosing the appropriate transmission waveform, the grounded wire source can also achieve near-source observation, similar to the same point observation method of the loop source. Since the 1950s, former Soviet scholars have established a near-field construction method, which has a larger detection depth and better detection accuracy compared to the far-field construction method. In the 1970s, there were reports of experimental oil exploration results. However, because the theoretical formula of the electric dipole cannot fully adapt to the data interpretation of the near-field TEM, the short-offset transient electromagnetic method of the electrical source has not been widely used for a long time. In recent years, this device and method of the time-domain transient field have once again attracted attention and show good application prospects. The equipment and software development of the electrical source transient electromagnetic method has received more and more investment.

G. Xue et al., *Exploration with Short-Offset Grounded-Wire Transient Electromagnetic Method*, https://doi.org/10.1007/978-981-96-8846-3_1

Xue Guoqiang's team defined this electrical source near-field transient electromagnetic detection as short-offset transient electromagnetic method (short-offset TEM, abbreviated as SOTEM) in 2013. Different from the definition of LOTEM, SOTEM refers to the observation device where the distance from the field point to the source point is roughly equal to or slightly smaller than the detection depth. In addition, in rugged mountainous areas, the layout of long wires is more flexible and convenient than large loops. Conducting breakthrough research under this device is of great significance for the fine detection of deep ore bodies.

As a special working device of the transient electromagnetic method, the development process of the electrical source transient electromagnetic method is tortuous and long. The earliest method used to study the layered earth structure was a method called "Eltran", which originated from a patent applied by American scientist L.W. Blan in 1933. The specific operation is to supply direct current I to the grounded wire at both ends, and then suddenly interrupt the current at $t = 0$, thus generating a decaying transient electromagnetic field underground, which is then received by another wire.

To achieve a higher signal-to-noise ratio and receiver sensitivity, in addition to using high-power transmitters, low-noise low-drift preamplifiers, and signal processing technology, it is also necessary to use longer transmission lines. According to the observation duration, the length of the transmitting electrode is generally 1000–2000 m. To meet the dipole conditions, a long offset device (Long Offset Transient Electromagnetic Method, LOTEM) is usually used, and its transmission and reception distance is 3–6 times the detection depth. However, when LOTEM is used in deep detection, the signal strength drops sharply due to the large transmission and reception distance, the signal-to-noise ratio decreases, which greatly weakens the deep detection capability of LOTEM and reduces the detection accuracy.

In actual work, the grounded wire source transient electromagnetic method (LOTEM) has a greater detection depth compared to the central return line device, and its working mode is similar to CSAMT, and its transmission and reception distance is several times the detection depth. As we all know, as the transmission and reception distance increases, the field strength will decrease. Therefore, based on the detection principle of the time-domain transient electromagnetic method, choosing an appropriate transmission waveform, such as a rectangular pulse, and separating the secondary field in the observation mode after the excitation source is turned off, can enhance the near-field detection capability. For this reason, Xue et al. (2013a; b) proposed the short offset transient electromagnetic method, and named it SOTEM, which extends the measurement area of the grounded wire source transient electromagnetic method from the far area to the middle and near area, and the observation point can be arranged in the range of 0.3–2 times the detection depth from the field source, thereby improving the deep detection capability and resolution of the grounded wire source transient electromagnetic method. Among them, using a 1000-m offset to detect a 2000-m deep coal mine, the position of the water-rich zone has been verified by drilling, indicating that near-source detection has good deep detection capability and higher detection accuracy.

The Electrical-source short-offset transient electromagnetic method, abbreviated as SOTEM, is a newly emerging branch of the transient electromagnetic method (TEM) in recent years. This method uses a long wire grounded at both ends as a transmission source, and observes the pure secondary field within an offset range less than 2 times the detection depth. Compared with the currently widely used loop source TEM, SOTEM has three significant advantages: **Increased detection depth**: The detection depth has significantly increased from about 500 m to about 2000 m; **Enhanced sensitivity**: SOTEM can excite induced currents in the horizontal and vertical directions underground, and can obtain higher detection sensitivity for both low-resistance and high-resistance targets; **Construction convenience**: In mountainous areas with large terrain undulations, SOTEM construction is more convenient and work efficiency is higher. In addition, SOTEM also has three obvious advantages compared to the traditional long offset transient electromagnetic method (LOTEM): **Enhanced signal strength**: Near-source observation enhances the signal strength of the secondary field, which is conducive to ensuring a higher signal-to-noise ratio; **Improved signal resolution**: Small offsets to some extent alleviate the volume effect and improve signal resolution; **Reduced cost**: Small offset observation reduces the requirements for transmitter power, thereby making the detection cost lower.

Although the detection depth of SOTEM is lower than that of LOTEM, in the detection within 2000 m depth, considering factors such as signal strength, detection accuracy, construction efficiency, and cost, SOTEM shows greater advantages. At present, SOTEM has shown great potential in the exploration fields of deep metal mines, coal fields, and hydrogeology.

1.2 Key Issues

For the electromagnetic method, the earth can usually be assumed to be a linear time-invariant system, and the form of its system function is closely related to the resistivity structure of the earth. Under the same earth conditions, the factors that affect the fineness of the electromagnetic detection method mainly include near-field electromagnetic response, multi-source transmission mechanism, and high-resolution signal processing. This project takes the SOTEM method as the research object. This method observes the transient response of the earth after the excitation pulse is turned off, so its source signal is a wideband continuous spectrum signal, with rich spectral components, high spectral resolution, and can achieve high-resolution observation. This makes SOTEM have the basic conditions for deep fine detection. However, in actual work, the detection effect of SOTEM is subject to some limitations. These limitations mainly stem from the variability of geological structures, the specificity of the observation environment, and the complexity of the middle and near-source electromagnetic fields. In fact, through lateral comparison with other methods, it can be found that the factors currently affecting the fineness of SOTEM detection also exist in other artificial source electromagnetic methods. Therefore, the study of these issues will also have a high degree of universality and is worth further exploration.

(1) Field source effect

The study of the field source effect originated from the related research of CSAMT. Because the field source of CSAMT needs to satisfy the plane wave assumption, and the distance between the field source and the observation point is always finite, for some frequency points, the finite offset distance may not guarantee that the source field satisfies the plane wave assumption, thus generating the field source effect problem. In CSAMT, the field source effect is usually classified into three types: non-plane wave effect, shadow effect, and photocopy effect.

For the near-field observation mode, since there is no need to satisfy the plane wave assumption, the non-plane wave effect can be ignored. However, the shadow effect and the photocopy effect will significantly affect the detection effect. The generation of these two effects essentially boils down to the impact of stratum waves on near-field observations. The electromagnetic field emitted from the source has three propagation paths: sky wave, ground wave, and stratum wave. Because the speed of the electromagnetic field propagating underground is slower than in the air, the phase difference between the ground wave and the stratum wave will become larger and larger after leaving the source, causing their wave fronts to gradually approach the horizontal plane, forming a wave S* that is almost vertical downward. In the usual large offset distance electromagnetic method, the stratum wave significantly attenuates, making S* the only excitation source, and its propagation direction is almost vertical downward. When the wave field encounters a low-resistance anomaly during downward propagation, the so-called "shielding effect" will occur. The eddy current generated inside the anomaly will exist for a long time, causing the data to think that the longitudinal extension length of the low-resistance body is greater than the actual situation, thus forming a measurement point shadow area.

When the measurement point is close to the source, in addition to S*, the stratum wave has not completely decayed. Since the stratum wave propagates inside the earth with a nearly hemispherical wave front, when it encounters a low-resistance anomaly, the anomaly will produce a shielding effect on its rear, thus forming a shadow effect. In other words, the prerequisite for the shadow effect is that the stratum wave has not completely decayed where the low-resistance anomaly exists. When the stratum wave propagates underground, its attenuation degree can be estimated by multiples of the wavelength. For the same frequency, the higher the earth's resistivity, the longer the wavelength, and the larger the area where the stratum wave can effectively act. From the perspective of CSAMT, when the earth's resistivity below the source is generally high, it will cause the lower limit of the frequency that satisfies the plane wave assumption (in the far zone) to rise, and the influence range of the stratum wave will therefore expand, thus producing a photocopy effect.

In response to the shadow effect, Newman et al. and Misuhata et al. conducted numerical calculations and discussed strategies to avoid the shadow effect. Boschetto and Hohmann conducted numerical simulations for horizontal electric dipoles, studying the photocopy effect and shadow effect. Yan Shu and Chen Mingsheng studied the shadow effect and photocopy effect in CSAMT detection, and reinterpreted the previous observation data. Yan and Fu proposed a closed solution for the underground electromagnetic field generated by a grounded horizontal electric dipole on a uniform half-space, and used this as a benchmark to assess the degree of stratum wave impact, and then proposed a division standard for CSAMT near, middle, and far field areas. Zhou et al. discussed the impact of the shadow effect and photocopy effect on inversion results in the long grounded wire source electromagnetic method. It can be seen that most of these studies aim to avoid the impact of the photocopy effect and shadow effect in the CSAMT method. However, when conducting time-domain electromagnetic method observations near the source, the response of the stratum wave is strong and always exists. By studying the impact mechanism of the stratum wave, analyzing its impact degree on different field components (relative size and range), and the coupling situation of different targets (body conductivity and burial depth) with the stratum wave, it will lay a good theoretical foundation for solving the field source effect. This is also the main research content of conducting near-source fine electromagnetic detection.

(2) Multi-source emission

In the study of the source field effect, it can be understood that when the stratum wave exists, the shadow effect usually appears at the far end of the line connecting the source and the anomaly, that is, the back end of the anomaly. With the position of the anomaly as the center, the shadow effect is mainly concentrated on the far end of the line connecting the source and the anomaly, while the near end is not affected by the shadow effect. For this reason, a multi-source system can be established, using transmitters at different positions to excite the earth, and verifying each other with data from different sources. Although the response of the stratum wave still exists, this method helps to overcome the shadow effect to a certain extent, thereby more accurately determining the true position of the anomaly.

The study of multi-source detection methods has a long history. In traditional long-offset electromagnetic methods, direct current methods, and other studies, multi-source systems can stimulate the earth in different polarization directions, thereby providing richer underground electrical structure information, and thus a large amount of research results have been obtained. In China, Xi Zhenzhu, Wang Xianxiang, and others have conducted in-depth research on the tensor CSAMT method based on orthogonal sources. Radar, Zhu Weiguo, and others have carried out application demonstration work of tensor CSAMT, all of which have achieved good results. Although multiple transmitters were used in the above research, the main goal is to enhance the detection signal, and usually only consider the overall effect of multi-source excitation. In frequency domain methods, a multi-source transmission system is established, allowing each source to work simultaneously and output different frequency point sequences, thereby obtaining observation data with a higher signal-to-noise ratio in a short time; in time domain methods, the observation signal obtained by superimposing the responses of different sources is regarded as a whole, and the ability of the multi-source system to improve the signal-to-noise ratio is theoretically simulated, and a time-domain full-domain apparent resistivity based on multi-source multi-component data is defined.

When conducting observations in the middle and near source areas, it is necessary to ensure that the positions of different sources are separated from each other and "irradiate" the same observation area at different angles. By comparing the observation results of different sources, the field source effect can be effectively overcome. This need to compare the responses of different sources is essentially a requirement to distinguish the excitation effects of different sources. In existing long-offset methods, it is usually unnecessary to distinguish the responses of different sources in the observation signal, but to process according to the sorted data. However, in short-offset methods, it is necessary to extract the responses of different sources in the observation signal one by one for subsequent processing.

Therefore, this application requirement for a multi-source transmission system is completely new. In practical applications, how to design a transmission system, the specific requirements for the layout of transmission antennas and the waveform of transmission, how to conduct observations, and how to separate responses are all major research contents of near-source fine electromagnetic detection.

(3) Noise processing

After completing the short-offset detection based on multiple transmitters, the first step is to process the observation data to remove most of the noise components in the data. This step is to ensure that the subsequent separation processing of different source responses can be carried out under high signal-to-noise ratio conditions, thereby ensuring the quality of response separation.

Usually, the noise entering the electromagnetic method data can be roughly divided into internal noise and external noise according to the source. Internal noise mainly comes from the instrument hardware system, especially the amplification circuit of the inductive magnetic field sensor, etc.; while external noise comes from outside the system, mainly including natural noise and anthropogenic noise. Natural noise generally refers to atmospheric noise, mainly caused by lightning. As the lightning propagates in the ionosphere-earth waveguide, its time–frequency domain characteristics will change, so atmospheric noise is usually further divided into far-source atmospheric noise and middle and near-source atmospheric noise. Far-source atmospheric noise and instrument internal noise together constitute background noise, which basically appears as white noise within the working frequency range of short-offset transient electromagnetic methods. Middle and near-source atmospheric noise has sporadic short-pulse characteristics. Anthropogenic noise generally refers to power frequency noise, communication system noise, etc. When the observation is in the mining area, the interference of anthropogenic noise is large, the sources are diverse and the characteristics are complex, many noise components are band noise for useful signals, and they usually cannot be removed by notch methods at specific frequencies.

In order to effectively suppress the above noise, researchers have proposed many excellent denoising methods. The most basic noise suppression technology is the "synchronous sampling" technology, which can effectively suppress non-coherent noise with stable randomness. On this basis, Macnae and others proposed to enhance the system's noise suppression performance by optimizing the system sampling function. Nyboe and Sørensen further optimized the gate function and stacking weighting function in the sampling function with Gaussian functions, thereby enhancing the system's performance in suppressing broadband and narrowband noise.

In order to further process short-term and non-generalized stationary random noise, Buselli and Cameron proposed a robust statistical method to remove middle and near-source atmospheric noise. On this basis, Buselli and others further proposed to suppress atmospheric noise, communication noise, and power frequency noise entering the data through far-reference observations. Bouchedda and others proposed two multi-resolution analysis techniques based on stationary wavelet transform to eliminate middle and near-source atmospheric noise in observation data. Reninger and others used singular value decomposition methods to suppress middle and near-source atmospheric noise and conventional anthropogenic noise, while Asten used principal component analysis methods to deal with various noises in airborne time-domain electromagnetic methods. In response to the complex anthropogenic noise in the mining area, Tang Jing Tian and others conducted research on the influence rules of strong noise in the mining area, and proposed mathematical morphology filtering methods, compressed sensing reconstruction methods, and dictionary learning methods for MT, CSAMT and other frequency domain methods. These methods have different focuses and aim to effectively separate strong interference.

In summary, traditional processing methods usually design processing schemes for noise from different sources separately, and then combine these schemes into a processing flow to achieve the processing of multi-source noise. However, this process-oriented processing method often requires a large amount of computational resources, has many process links, loose structure, and low processing efficiency. In addition, although many methods have certain parameter self-adaptive adjustment capabilities, key parameters still depend on manual settings, which require a high level of personal experience from the operator. When the amount of data is large, it is often impossible to ensure consistent processing standards for the entire dataset. Therefore, improving the work efficiency of noise processing in detection data and reducing the influence of the operator's subjective judgment on the data processing results is an important research topic for near-source fine electromagnetic detection.

(4) Fine inversion

There is a mapping relationship between electromagnetic observation data and underground electrical parameter models. When this mapping relationship can be explicitly expressed through data, intermediate parameters such as apparent resistivity and apparent depth can be directly obtained, which is commonly used for rapid imaging processing of exploration site data. However, when the above mapping relationship is implicitly expressed, it is necessary to fit the observed signal through the optimization process, which is called inversion.

At present, the one-dimensional inversion technology of the electromagnetic method is relatively mature. When the underground structure has strong continuity in the lateral direction, quasi-two-dimensional or quasi-three-dimensional inversion can be achieved using boundary value constraints or spatial constraints. Compared with one-dimensional inversion, high-dimensional inversion theoretically can provide more detailed geological structure information extraction. For the inversion of data from each observation point of the short-offset transient electromagnetic method, it is first necessary to perform multiple groups of multi-dimensional time-domain forward modeling according to the partitioning situation of the transmitter, which requires a large amount of computation and high computational resources. This situation is very similar to the inversion of airborne electromagnetic methods. Although many researchers have proposed various acceleration schemes, such as moving footprint method and matrix decomposition method, they still cannot fully meet the needs of actual production work.

For near-source time-domain electromagnetic detection data, current imaging and inversion methods are difficult to achieve fine extraction of underground electrical conditions. If after completing the separation of different source response signals, these response signals are used for comparison, the area affected by the field source effect can be greatly compressed, thereby roughly estimating the position of the abnormal body causing the field source effect. However, in this area, there may still be non-true data anomalies affected by the field source effect, and the existing imaging and inversion methods are still difficult to overcome the influence of the field source effect. This difficulty stems from the working mechanism of existing inversion methods: they belong to model-driven inversion. On the one hand, they rely on the initial model to give the parameter structure of the target model; on the other hand, they rely on stable functionals to constrain the changes in the target model parameters. Therefore, if a more accurate initial model can be obtained, the possibility of successful inversion will be higher. However, in actual work, it is usually very difficult to obtain an accurate initial model, so it depends more on the choice of stable functionals.

The constraint of stable functionals is actually for the high-order features of model parameters, not directly for the model parameters themselves, which results in relatively low convergence efficiency. In addition, the choice of stable functionals still depends on the experience and intervention of the operator. Therefore, relying solely on existing model inversion methods based on stable functionals, it is difficult to effectively estimate the real model. How to effectively overcome the inversion accuracy of existing model-driven inversion methods under the condition of data affected by field source effects, and further improve computational efficiency, overcome subjective regularization problems, and ultimately achieve fine extraction of underground electrical structure information, is the main research content of near-source fine electromagnetic detection.

1.3 Application Situation

Because SOTEM observes in the near-source small offset range, compared with LOTEM, the signal-to-noise ratio of SOTEM signals has significantly improved. However, for the increasing electromagnetic interference, this improvement is still insufficient. The signal-to-noise ratio of data collected in the field with strong interference is still low, so it is particularly important to seek improvements in data processing.

At present, transient electromagnetic instruments at home and abroad, such as the V8 from Phoenix Company in Canada, the GDP32 from Zonge Company in the United States, the Terra TEM from Alpha Geoscience Company in Australia, and the MSD series from Baiyun Instrument Factory in China, use bipolar step current emission gaps, sample thousands of data points in each emission cycle (for example, V8 can collect 3058–3076 data points), enhance useful signals through reverse stacking, cancel DC components and low-frequency interference, and then stack and average more than 3000 sample point data laterally. According to the characteristic that the signal-to-noise ratio of the time channel gradually decreases from early to late, the number of sample points participating in the average gradually increases according to a certain rule, and finally usually obtains 20–40 time channel data. However, this noise suppression method based on the zero-mean interference assumption fails in the environment of colored noise and fixed interference noise.

Although colored noise has attracted widespread attention in the field of electromagnetic exploration, the standard software of domestic and foreign instruments still directly processes and interprets dozens of data after lateral stacking and averaging. Taking the V8 instrument of Phoenix Company as an example, more than 3000 original data are extracted from the output data for spectral analysis, and appropriate noise removal techniques are selected according to the characteristics of interference noise. In addition, different noises cause different interferences to different field components. The multi-channel, multi-component observation method proposed in this article can best avoid or remove interference by selecting appropriate observation fields and combinations of different observation components.

As an important part of the complete electromagnetic exploration system, data interpretation occupies a key position, and the interpretation using apparent resistivity parameters is one of the main means. While developing two-dimensional (2D) and three-dimensional (3D) inversion interpretation to solve complex geological structure problems, research on apparent resistivity algorithms based on the formula of horizontal stratification geology theory is also advancing. Unlike the direct current resistivity method, the relationship between the time-varying electromagnetic field and the underground resistivity is relatively complex. Therefore, when calculating the apparent resistivity, it is necessary not only to degrade the stratified theory formula into a uniform earth formula, but also to further adopt the progressive solution method of the near or far zone. According to the definition of apparent resistivity and the above algorithm, when the earth is a non-uniform half-space or does not meet the field area conditions, there will be a large gap between the stratum section reflected by the apparent resistivity and the actual situation. Therefore, various improvements to apparent resistivity are mainly to achieve the approximation of the apparent resistivity curve to the real earth resistivity section.

For this reason, for the full-area (full-period) formula of the loop source, research is conducted through iterative calculation of apparent resistivity, using algorithms such as direct solution of time shift scaling; for long offset LOTEM, the full-period apparent resistivity of the binary search algorithm is used. Although the full-area (full-period) algorithm has made progress compared to the far and near zone progressive methods, for geological models with more than three layers, the full-period apparent resistivity still appears rough and cannot meet the needs of fine detection. Therefore, the author proposes a method to fit the inversion of the real resistivity and thickness of the stratum from the formula of stratified earth theory.

Unlike the frequency domain CSAMT signal, the time domain TEM signal observed after the step current is cut off is in the form of exponential decay, lacking feature points related to the stratum, which leads to a more serious problem of multiple solutions for fitting inversion. In addition, the number of layers considered in the past fitting inversion is relatively small, and the generated resistivity section does not have enough intuitive expression of the stratum, so it is not common in practical applications. Therefore, it is urgent to improve the fitting inversion algorithm to obtain more accurate geoelectric interpretation sections.

SOTEM is a deep detection method worth further development. By adopting suitable observation devices and data processing methods, it is possible to observe and process all six electromagnetic field components, fully utilizing the advantages of different field components in different geological tasks. Choose the appropriate observation field or its combination according to the purpose of the exploration task, geological conditions, and environmental conditions, in order to obtain more geological information and solve more geological problems. Recently, in the exploration of the Xiaoshan metamorphic core complex mineralization system, we discovered multiple veins in the exploration of the Xiaoshan Kuanping gold and silver mines and other mining areas with a total area of 120 km^2 and a detection depth of 1.7 km; in the detection of the Fengtai mining area, successfully detected the deep structure and hidden ore body of 1.5 km.

The author and his team proposed the SOTEM method and extended the depth of conventional near-field source transient electromagnetic detection from 500 to 1500 m. However, due to the complexity of the deep electromagnetic field itself, the special nature of the observation environment, and the variability of geological structures, there is still a need to further break through the technical bottleneck of transient electromagnetics in fine detection in the full field area. This includes solving a series of practical problems in practical applications, such as the accurate solution of transient electromagnetic fields, the effective removal of electromagnetic noise, and the longitudinal resolution of deep targets, in order to form a series of SOTEM deep fine detection technologies and realize its standardized promotion and application.

In summary, SOTEM inherits the small volume effect of the loop source transient electromagnetic method and overcomes its disadvantage of small exploration depth; at the same time, it inherits the strong field source signal of the grounded source long offset transient electromagnetic method and overcomes its low interpretation accuracy and low work efficiency. This forms a new grounded source full-field detection method, which is technically original, and the project team has achieved certain stage research results.

1.4 Content of This Book

This book introduces the forward and inverse, information extraction technology and its application of the electrical source short offset transient electromagnetic method (SOTEM), including the typical application effects of this method in the fields of metal mines, coal mines, and engineering. The SOTEM technology has achieved good detection results in multiple mining areas. This book aims to provide references for teachers and students of colleges and universities in the geophysical electromagnetic method industry, researchers in scientific research units, and staff in related departments.

This book is divided into 7 chapters. This chapter is the introduction, mainly introducing the development of methods, market demand, and the content of this book. Chapter 2 introduces the SOTEM theory. Chapter 3 introduces the simulation results of typical underground geological bodies, mainly including simulation methods, lithium ore simulation, Baiyun Ebo rare earth ore simulation, coal mine surface detection simulation, and coal mine underground detection simulation. Chapter 4 mainly introduces inversion methods and seismic processing techniques. Chapter 5 mainly introduces the excitation polarization effect and information extraction technology in SOTEM. Chapter 6 mainly introduces the research progress of artificial intelligence in SOTEM data processing. Chapter 7 introduces the SOTEM detection equipment system. Chapter 8 mainly includes the practical problems in the field of resource environment such as metal ore and coal mine prevention and control solved by SOTEM, and gives 9 examples, including: Henan Yexian SOTEM test and salt dissolution detection, Shandong Huafeng coal mine deep water-rich investigation, Xiaoshan mining area gold multi-metal ore detection, Anhui Taoyuan coal mine water body detection, Hebei Shunping city active fault detection, Shaanxi Hancheng deep coal seam water content, Qinghai area wild horse spring iron lead zinc multi-metal ore, Henan coal under bauxite, Anhui Wuwei lead zinc mine, etc.

References

He Jishan. Wide Area Electromagnetic Method and Pseudo Random Signal Electric Method[M]. Beijing: Higher Education Press, 2010.

Xue Guoqiang, Yan Shu, Chen Weiyin. Electric source transient electromagnetic short offset detection method[J]. Journal of Nonferrous Metals, 2013a, 23(9): 2365–2370.

Xue Guoqiang, Chen Weiyin, Zhou Nannan, et al. Ground source transient electromagnetic short offset deep detection technology[J]. Chinese Journal of Geophysics, 2013b, 56(1): 255-261.

Chapter 2
Mechanism of Transient Electromagnetic Field Propagation from Electrical Sources

As the foundation of transient electromagnetic field theory, this chapter mainly introduces the basic principles, observation methods and techniques, and inversion methods. The physical basis of electromagnetic detection is the propagation characteristics of the electromagnetic field in the medium, and its mathematical description is the Maxwell's equations. Based on the Maxwell's equations, this chapter will sort out the current detection transient electromagnetic methods based on different forms of source devices. This chapter also introduces the propagation characteristics of transient electromagnetic fields from electrical sources.

2.1 Basic Equations of Electromagnetic Field

2.1.1 Maxwell's Equations and Physical Quantities

Maxwell's equations are a set of equations that describe the relationship between electric fields, magnetic fields, and charge density and current density. They are based on the principles of Coulomb's law, Biot-Savart law, and Faraday's law. The commonly used Maxwell's equations are simplified and organized by Hertz and Heaviside using mathematical analysis methods. Maxwell's equations have differential and integral forms, and different equation forms are used for different problem requirements.

In Maxwell's equations, the main physical quantities include electric field strength E (V/m), electric displacement vector D (C/m^2), magnetic field strength H (A/m), and magnetic induction strength B (Wb/m^2). At the macroscopic scale, the discrete properties of electric and magnetic fields can be ignored, and they can be considered continuous in space and time. That is, the above four physical quantities are continuous functions of position vector r and time t and have continuous derivatives. In addition, whether as a source or generated by electromagnetic field excitation, charge density ρ (C/m^3) and current density J (A/m^2) can also be assumed to be continuous

G. Xue et al., *Exploration with Short-Offset Grounded-Wire Transient Electromagnetic Method*, https://doi.org/10.1007/978-981-96-8846-3_2

functions of position vector r and time t. Thus, we can have the differential form of Maxwell's equations:

$$\nabla \times \mathbf{E}(\mathbf{r}, t) = -\frac{\partial}{\partial t}\mathbf{B}(\mathbf{r}, t) \tag{2.1a}$$

$$\nabla \times \mathbf{H}(\mathbf{r}, t) = \frac{\partial}{\partial t}\mathbf{D}(\mathbf{r}, t) + \mathbf{J}(\mathbf{r}, t) \tag{2.1b}$$

$$\nabla \cdot \mathbf{D}(\mathbf{r}, t) = \rho(\mathbf{r}, t) \tag{2.1c}$$

$$\nabla \cdot \mathbf{B}(\mathbf{r}, t) = 0 \tag{2.1d}$$

where ρ and J should also satisfy the continuity equation:

$$\nabla \cdot \mathbf{J}(\mathbf{r}, t) + \frac{\partial}{\partial t}\rho(\mathbf{r}, t) = 0 \tag{2.2}$$

Maxwell's equations establish the relationship between electric and magnetic fields through Eqs. (2.1a) and (2.1b): a changing magnetic field generates an electric field (curl equation of electric field); current and changing electric field generate a magnetic field (curl equation of magnetic field). Equations (2.1c) and (2.1d) describe the properties of magnetic and electric fields respectively: charge generates an electric field, free charge density ρ is the divergence source of the electric field (divergence equation of electric field); magnetic flux continuity, i.e., there is no free magnetic charge (divergence equation of magnetic field). Equations (2.1) and (2.2) are five equations, only three of which (can be any three) are independent, and the other two can be derived from the three independent equations. For time-domain or transient electromagnetic fields, it is generally believed that the two curl equations and the continuity equation are independent and can be called basic equations; the two divergence equations are called auxiliary equations.

The differential form of Maxwell's equations is usually considered to reflect the local properties of the electromagnetic field and is only applicable to regions where the medium is continuous. Compared with differential operations, since the integral operation has lower requirements for the integrand, the integral form of Maxwell's equations has a wider range of use. Suppose there is a finite volume V in space, its surface is S, and there is a part A on the surface S with a boundary C, based on Gauss's theorem and Stokes's theorem, the integral form of Maxwell's equations is as follows:

$$\oint_C \mathbf{E}(\mathbf{r}, t) \cdot d\mathbf{l} = -\frac{\partial}{\partial t}\int_A \mathbf{B}(\mathbf{r}, t) \cdot d\mathbf{S} \tag{2.3a}$$

$$\oint_C \mathbf{H}(\mathbf{r}, t) \cdot d\mathbf{l} = \frac{\partial}{\partial t} \int_A \mathbf{D}(\mathbf{r}, t) \cdot d\mathbf{S} + \int_A \mathbf{J}(\mathbf{r}, t) \cdot d\mathbf{S} \tag{2.3b}$$

$$\oint_S \mathbf{D}(\mathbf{r}, t) \cdot d\mathbf{S} = \int_V \rho(\mathbf{r}, t) dV \tag{2.3c}$$

$$\oint_S \mathbf{B}(\mathbf{r}, t) \cdot d\mathbf{S} = 0 \tag{2.3d}$$

The integral form of Maxwell's equations can be applied to discontinuous media regions, deriving the relationships that each field quantity should satisfy in the discontinuous region, that is, the boundary conditions of the electromagnetic field. Assuming that there is a discontinuous region in the medium, the media on both sides of the boundary are denoted as medium 1 and medium 2, and the unit normal vector pointing to the side of medium 1 on the boundary is defined as n, then the conditions that each field quantity should satisfy on the discontinuous interface can be derived from Eq. (2.3) as follows:

$$\mathbf{n} \times (\mathbf{E}_1 - \mathbf{E}_2) = 0 \tag{2.4a}$$

$$\mathbf{n} \times (\mathbf{H}_1 - \mathbf{H}_2) = \mathbf{J}_S \tag{2.4b}$$

$$\mathbf{n} \cdot (\mathbf{D}_1 - \mathbf{D}_2) = \rho_S \tag{2.4c}$$

$$\mathbf{n} \cdot (\mathbf{B}_1 - \mathbf{B}_2) = 0 \tag{2.4d}$$

where J_S and ρ_S are the surface current density and surface charge density on the discontinuous interface, respectively.

Observing Eqs. (2.1), (2.2) and (2.3), they involve a large number of vector functions, but there are only three independent equations, indicating that the electromagnetic field problem cannot be solved using only independent equations. It is also necessary to introduce auxiliary equations describing the relationship between the electromagnetic field and the medium (constitutive relations):

$$\mathbf{D}(\mathbf{r}, t) = \varepsilon_0 \mathbf{E}(\mathbf{r}, t) + \mathbf{P}(\mathbf{r}, t) \tag{2.5a}$$

$$\mathbf{B}(\mathbf{r}, t) = \mu_0 [\mathbf{H}(\mathbf{r}, t) + \mathbf{M}(\mathbf{r}, t)] \tag{2.5b}$$

$$\mathbf{J}(\mathbf{r}, t) = \sigma(\mathbf{r}) \mathbf{E}(\mathbf{r}, t) \tag{2.5c}$$

where: ε_0, μ_0 and σ are the vacuum permittivity ($[1/(36\pi)] \times 10^{-9}$ F/m), vacuum permeability ($4\pi \times 10^{-7}$ H/m) and electrical conductivity of conductive media (S/

m), respectively; P and M are the polarization intensity and magnetization intensity, respectively. For linear media, we have:

$$\mathbf{P}(\mathbf{r},t)=\varepsilon_0\chi_e(\mathbf{r})\mathbf{E}(\mathbf{r},t) \tag{2.6a}$$

$$\mathbf{M}(\mathbf{r},t)=\chi_m(\mathbf{r})\mathbf{H}(\mathbf{r},t) \tag{2.6b}$$

where, χ_e and χ_m are the polarization rate and magnetization rate, respectively. Combining Eqs. (2.5) and (2.6), we can define:

$$\varepsilon(\mathbf{r})=[1+\chi_e(\mathbf{r})]\varepsilon_0=\varepsilon_r(\mathbf{r})\varepsilon_0 \tag{2.7a}$$

$$\mu(\mathbf{r})=[1+\chi_m(\mathbf{r})]\mu_0=\mu_r(\mathbf{r})\mu_0 \tag{2.7b}$$

where, ε is the permittivity, ε_r is the relative permittivity, μ is the permeability, and μ_r is the relative permeability.

2.1.2 Wave Equation

By substituting the constitutive relations into the two basic equations of Maxwell's equations, we can get:

$$\nabla\times\mathbf{E}(\mathbf{r},t)=-\mu(\mathbf{r})\frac{\partial}{\partial t}\mathbf{H}(\mathbf{r},t) \tag{2.8a}$$

$$\nabla\times\mathbf{H}(\mathbf{r},t)=-\varepsilon(\mathbf{r})\frac{\partial}{\partial t}\mathbf{E}(\mathbf{r},t)+\sigma(\mathbf{r})\mathbf{E}(\mathbf{r},t) \tag{2.8b}$$

In this way, the number of unknowns (E and H) is equal to the number of equations, and two equations containing only E and H can be obtained by simplification. Take the curl of the two equations in Eq. (2.8) respectively:

$$\begin{aligned}\nabla\times\nabla\times\mathbf{E}(\mathbf{r},t)&=-\mu(\mathbf{r})\frac{\partial}{\partial t}\nabla\times\mathbf{H}(\mathbf{r},t)\\&=-\mu(\mathbf{r})\frac{\partial}{\partial t}\left[\varepsilon(\mathbf{r})\frac{\partial}{\partial t}\mathbf{E}(\mathbf{r},t)+\sigma(\mathbf{r})\mathbf{E}(\mathbf{r},t)\right]\\&=-\mu(\mathbf{r})\varepsilon(\mathbf{r})\frac{\partial^2}{\partial t^2}\mathbf{E}(\mathbf{r},t)-\mu(\mathbf{r})\sigma(\mathbf{r})\frac{\partial}{\partial t}\mathbf{E}(\mathbf{r},t)\end{aligned} \tag{2.9a}$$

$$\nabla\times\nabla\times\mathbf{H}(\mathbf{r},t)=\varepsilon(\mathbf{r})\frac{\partial}{\partial t}\nabla\times\mathbf{E}(\mathbf{r},t)+\sigma(\mathbf{r})\nabla\times\mathbf{E}(\mathbf{r},t)$$

$$= \varepsilon(\mathbf{r})\frac{\partial}{\partial t}\left[-\mu(\mathbf{r})\frac{\partial}{\partial t}\mathbf{H}(\mathbf{r},t)\right] + \sigma(\mathbf{r})\left[-\mu(\mathbf{r})\frac{\partial}{\partial t}\mathbf{H}(\mathbf{r},t)\right]$$

$$= -\mu(\mathbf{r})\varepsilon(\mathbf{r})\frac{\partial^2}{\partial t^2}\mathbf{H}(\mathbf{r},t) - \mu(\mathbf{r})\sigma(\mathbf{r})\frac{\partial}{\partial t}\mathbf{H}(\mathbf{r},t) \tag{2.9b}$$

Using the vector identity $\nabla \times \nabla \times \mathbf{F} = \nabla\nabla \cdot \mathbf{F} - \nabla^2\mathbf{F}$, and for uniform media, we have $\nabla \cdot \mathbf{E} = 0$, $\nabla \cdot \mathbf{H} = 0$, substituting into Eq. (2.9), we get:

$$\nabla^2\mathbf{E}(\mathbf{r},t) - \mu(\mathbf{r})\varepsilon(\mathbf{r})\frac{\partial^2}{\partial t^2}\mathbf{E}(\mathbf{r},t) - \mu(\mathbf{r})\sigma(\mathbf{r})\frac{\partial}{\partial t}\mathbf{E}(\mathbf{r},t) = 0 \tag{2.10a}$$

$$\nabla^2\mathbf{H}(\mathbf{r},t) - \mu(\mathbf{r})\varepsilon(\mathbf{r})\frac{\partial^2}{\partial t^2}\mathbf{H}(\mathbf{r},t) - \mu(\mathbf{r})\sigma(\mathbf{r})\frac{\partial}{\partial t}\mathbf{H}(\mathbf{r},t) = 0 \tag{2.10b}$$

Equation (2.10) is the fluctuation equation of the electric field and the magnetic field. In Eq. (2.10), the field quantities are functions of time. According to Fourier transform theory, the above field quantities can be regarded as the superposition of time-harmonic fields. Therefore, Eq. (2.10) can be transformed into the frequency domain. In addition, in the frequency domain, the differentiation process of the field quantity $\mathbf{F}' = i\omega\mathbf{F}$. So we have:

$$\nabla^2\mathbf{E}(\mathbf{r}) + \left[\mu(\mathbf{r})\varepsilon(\mathbf{r})\omega^2 - i\mu(\mathbf{r})\sigma(\mathbf{r})\omega\right]\mathbf{E}(\mathbf{r}) = 0 \tag{2.11a}$$

$$\nabla^2\mathbf{H}(\mathbf{r}) + \left[\mu(\mathbf{r})\varepsilon(\mathbf{r})\omega^2 - i\mu(\mathbf{r})\sigma(\mathbf{r})\omega\right]\mathbf{H}(\mathbf{r}) = 0 \tag{2.11b}$$

Equation (2.11) is the frequency domain fluctuation equation of the electric field and the magnetic field. Compared with the time domain fluctuation equation, the frequency domain fluctuation equation eliminates the time variable and becomes a Helmholtz-type equation, which is easier to solve. In addition, the part in the square brackets can be extracted from the equation:

$$k^2 = \mu(\mathbf{r})\varepsilon(\mathbf{r})\omega^2 - i\mu(\mathbf{r})\sigma(\mathbf{r})\omega \tag{2.12}$$

The variable defined by Eq. (2.12) k is the wave number. For geophysical electromagnetics, because the frequencies involved are usually <100 kHz, the first term on the right side of Eq. (2.12) will be much smaller than the second term. Therefore, for electromagnetics, the wave number can usually be simplified as follows:

$$k^2 = -i\mu(\mathbf{r})\sigma(\mathbf{r})\omega \tag{2.13}$$

2.1.3 Basic Theorems of Time Domain Electromagnetic Field

Maxwell's equations are a summary of the basic laws followed by the electromagnetic field. With the deepening of the study of Maxwell's equations, some conclusions with broader significance have been obtained. In this section, we will mainly introduce the Poynting theorem, uniqueness theorem, and causality law, which are of great significance in electromagnetic field theory. To further highlight the key points, this section will simplify the derivation process and focus on the physical meaning of these principles.

(1) **Poynting's Theorem**

Assume that there is a region in the medium (ε, μ), with a volume of V, and its surface is S, then:

$$\int_V \mathbf{E}\cdot\mathbf{J}dV = -\oint_S (\mathbf{E}\times\mathbf{H})\cdot\mathbf{n}dS - \frac{\partial}{\partial t}\int_V \left(\frac{\varepsilon}{2}\mathbf{E}\cdot\mathbf{E} + \frac{\mu}{2}\mathbf{H}\cdot\mathbf{H}\right)dV \tag{2.14}$$

Equation (2.14) is the integral form of Poynting's theorem, where: the integral on the left side of the equation represents the total power loss in volume V, the first term on the right side represents the electromagnetic power flowing into V from the surface *S*, the two terms inside the second term on the right side represent the energy density of the electric field and the magnetic field, so the second term on the right side represents the decrease rate of electromagnetic energy in *V* over time.

Poynting's theorem actually represents the conservation of electromagnetic energy in volume *V*. Energy will not appear out of nowhere, nor will it disappear out of nowhere; if the energy flowing into *V* from the outside is equal to the energy consumed inside *V*, then the total energy inside *V* remains constant.

(2) **Uniqueness Theorem**

The uniqueness theorem mainly solves the uniqueness problem of the solution, that is: when solving the electromagnetic field in a given area, under what conditions is the solution unique? The content of the uniqueness theorem is: for a bounded region *V* (the boundary is *S*), (1) at the moment $t = 0$, the initial value of the electromagnetic field inside is given everywhere; (2) at the moment $t \geq 0$, the tangential component of the electric field on the entire *S* is given, or the tangential component of the magnetic field on the entire *S* is given, or the tangential component of the electric field in a part of *S* and the tangential component of the magnetic field in other areas outside this area are both given, then at $t > 0$, the electromagnetic field inside *V* can be uniquely determined by Maxwell's equations.

Breaking down the definition of the uniqueness theorem, we can see that in order to uniquely determine the electromagnetic field situation in a certain bounded area, we need:

1. Clearly define a moment $t = 0$;
2. At the moment $t = 0$, the existence of the field in the region V must be determined;
3. At $t = 0$, *after this moment, the tangential electric field and tangential magnetic field conditions on the interface of the region V should be fully specified, which can be in three equivalent situations:*

Situation 1: The tangential component of the electric field on the entire boundary surface is fully specified.

Situation 2: The tangential component of the electric field on the entire boundary surface is fully specified.

Situation 3: On the interface, although only a part of the region's electric field tangential component can be specified, the tangential component of the magnetic field for the remaining part can be specified.

After satisfying the above conditions, the electromagnetic field in the region can be uniquely solved.

From this, it can also be seen that the uniqueness of the solution actually has two meanings: first, as long as the above conditions are met, the solution is unique; second, as long as the appropriate solution conditions are given, the solutions obtained by different methods are equivalent. A direct application of this understanding is the "equivalence principle": by setting equivalent sources, changing some characteristics of the solution region while ensuring that the solution results of the rest of the space remain unchanged. The so-called "equivalent source" is key to equivalence, that is, although the form of the source is different, the field generated in the specified region is the same. Using the equivalence principle will help simplify complex problem solving.

(3) **Causality**

For electromagnetics, causality simply means: before the electromagnetic field propagates to a certain area, this area cannot show the effects caused by the excitation of the electromagnetic field. Whether a linear electromagnetic system can be realized is constrained by the causality, requiring that the real part and imaginary part of its response function must satisfy a specific relationship. Suppose a linear time-invariant system, its impulse response is $f(t)$. According to the causality, the impulse response of the system is zero before the moment $t = 0$, that is:

$$f(t) = f(t)u(t) \tag{2.15}$$

where u (t) is the unit step function. Thus, in the frequency domain, we have:

$$F(\omega) = \int_{-\infty}^{\infty} f(t)e^{i\omega t} dt = R(\omega) + iX(\omega) \tag{2.16}$$

where R (ω) and X (ω) are the real and imaginary parts of the system response function $F(\omega)$, and:

$$R(\omega) = -\frac{1}{\pi}\int_{-\infty}^{\infty}\frac{X(\Omega)}{\omega - \Omega}d\Omega \tag{2.17a}$$

$$X(\omega) = \frac{1}{\pi}\int_{-\infty}^{\infty}\frac{R(\Omega)}{\omega - \Omega}d\Omega \tag{2.17b}$$

That is, the real part and the imaginary part of the system response function $F(\omega)$ must satisfy the Hilbert transform relationship. In other words, for a realizable linear time-invariant system, the real part and the imaginary part of its system response function are not independent of each other, but must satisfy a certain relationship, otherwise it will not satisfy the causality. In the classic work "Electromagnetic methods in applied geophysics" (Nabighian, 1991), Nabighian pointed out: within the scope of electromagnetic method research, the earth can be regarded as a linear time-invariant system, and the so-called measurement is to obtain the system response function of the earth. Therefore, as a linear time-invariant system, the real part and the imaginary part of the system response function of the earth must also satisfy the Hilbert transform relationship.

2.2 Basic Principles of Transient Electromagnetic Method

Among many geophysical detection methods, the electromagnetic method can achieve a good balance between structural detection and physical property detection. The electromagnetic method is based on the law of electromagnetic induction, and realizes the detection of the spatial distribution and physical characteristics of underground structures by observing the response of the earth under the excitation of natural or artificial electromagnetic field sources.

The entire process of electromagnetic detection (as shown in Fig. 2.1) consists of five aspects: "source-transmission-reception-noise-solution". Among them: "source" refers to the process of generating a mechanism effect on the earth, the generation of the excitation signal can be natural, or it can be generated by artificial devices (as shown in Fig. 2.1); "transmission" refers to the propagation of the transmitted field, realizing the process of exciting the earth, which is the physical basis for the detection to be established; "reception" refers to the process of sensors and receivers collecting the earth's response signal; "noise" refers to the formation and removal process of noise impact in observation; "solution" refers to the process of solving the electrical structure model of the earth based on the propagation mechanism of the field, on the basis of denoised data. From this, it can be

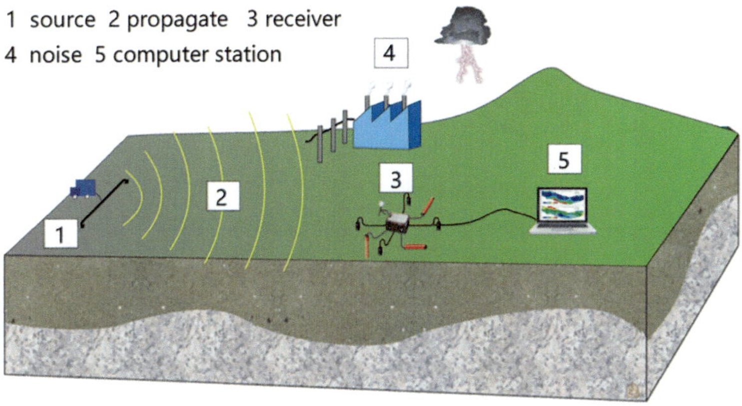

Fig. 2.1 Schematic diagram of the electromagnetic observation process

seen that "transmission" represents the theoretical basis for the realization of detection; "source" and "reception" together lay the data foundation for detection, and its implementation effect is determined by detection equipment; "noise" and "solution" extract the information of underground electrical structure contained in the data, and the upper limit of the information that can be extracted is limited by the data's ability to contain information.

The detection effect that can be achieved by the electromagnetic method is jointly determined by the five aspects of "source-transmission-reception-noise-solution". As shown in Fig. 2.2: "Transmission" as the detection mechanism, determines the maximum detection depth and accuracy capability that the method can theoretically possess; "Source" and "Reception" as equipment, limit the actual detection capability that the method can achieve at the data level; and "Noise" and "Solution" ultimately determine the actual detection capability of the method. Therefore, if we trace back from the downstream end, the research on "Noise" and "Solution" aims to maximize the potential of data mining and make full use of the detection potential contained in the data; "Source" and "Reception" as equipment, their research goal is to be as close as possible to the method requirements, so that the data they produce can be as close as possible to the theoretical limit of the method; and the research on "Transmission" will be at the top, and its research results will fundamentally determine the height of the theoretical limit of the method's detection capability. In fact, the essential differences between the many different methods included under the concept of the electromagnetic method are also constructed on different deconstructions of "Transmission".

For the classification of the electromagnetic method, it is divided into natural source methods and artificial source methods according to the field source, and can be divided into frequency domain methods and time domain methods according to the response properties.

Natural source methods, represented by Magnetotelluric (MT) (Tikhonov, 1950; Cagniard, 1953), use the natural electromagnetic field containing various different

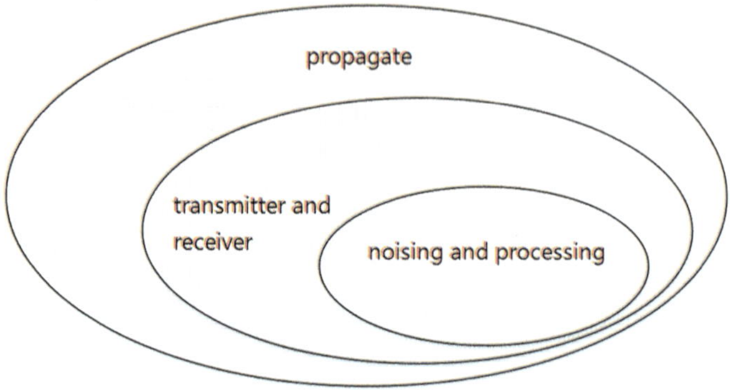

Fig. 2.2 Schematic diagram of the relationship of electromagnetic method capabilities

sources as the excitation source, utilize the magnetic coupling between the alternating currents flowing underground, and extract typical underground structure information by observing the electromagnetic impedance of the earth under steady-state conditions (Kaufman and Keller, 1981). It has considerable advantages in the detection of crustal and upper mantle electrical structures and is currently the only electromagnetic method that can reach the aforementioned great depths besides the seismic method (Di Qingyun et al., 2008). However, because the natural field signal is very weak in the frequency range of about 10^{-3} Hz to 1 kHz, and this frequency range to a considerable extent corresponds to the exploitable depth of human mineral resources at present, MT is limited in meeting resource exploration needs.

Using artificial transmission sources to generate artificial source signals that can cover the above frequency range can effectively overcome the excitation source defects of the above natural source methods, and thus establish the artificial source electromagnetic method. In the early research of artificial source methods, whether in the Western camp or the Soviet camp, the research focus was concentrated on harmonic source excitation (Kaufman and Keller, 1983). The method based on harmonic sources, that is, the frequency domain method we usually talk about, is mainly applicable to discrete single frequency or narrowband electromagnetic signals. Its research analysis method is to hide the time variable, which greatly simplifies various time derivatives in Maxwell's equations, but the applicability of the solution is also limited to steady-state conditions (Changqing and Xili, 2011). Although Soviet scholars had conducted research on time-domain electromagnetic methods before the 1950s, its rise was mainly due to the study of electromagnetic pulses generated by nuclear explosions in the 1960s. Driven by this historic event, the research on time-domain methods in the field of geophysical exploration, that is, the study of transient responses generated by the earth under short pulse source excitation, also flourished. Comparing harmonic sources with pulse sources, when the power and radiation efficiency of the transmission device are the same, the harmonic source will concentrate the transmitted energy on a limited number of discrete frequency points (or narrowband signals), and for each frequency point, a higher signal-to-noise ratio can be obtained. However, because the energy is discretely distributed on a limited number of frequency points, both the spectral resolution and the overall signal bandwidth are relatively limited. Since these parameters are related to the detection accuracy that the method can achieve, therefore, to improve the detection accuracy, either spend more time to excite more frequency points, or use more devices to achieve parallel transmission. Compared with harmonic sources, pulse sources belong to wideband signals, and their spectral resolution is naturally higher than that of harmonic sources, which is conducive to achieving fine detection. The shortcomings of pulse sources are also obvious: if the limited transmission energy is distributed over a wider frequency range, the energy that can be allocated to a single frequency point is relatively less, and under the condition that the environmental noise level remains unchanged, the signal-to-noise ratio (especially the late signal-to-noise ratio) is easy to be limited, which further affects the detection depth. The above analysis shows that: in the usual sense, the frequency domain method using harmonic sources can achieve a greater detection depth, but the resolution is limited; the time domain method using pulse sources can achieve a higher resolution, but the maximum detection depth is limited. Therefore, for the electromagnetic method, it is actually difficult to balance detection depth and resolution.

For a long time, whether it can maximize the realization of the balance between detection depth and resolution has become a focus issue in the international electromagnetic method research. As introduced earlier, under the "source-transmission-reception-noise-solution" framework, if we want to fundamentally solve this problem, we can only work on "transmission". The reason why the existing electromagnetic method is difficult to achieve the balance between detection depth and resolution is mainly because there are too many energy and bandwidth losses in the observation, and ultimately, these losses are the price paid to reduce the difficulty of the "solution" link. For example: the frequency domain method hides the time variable to simplify the solution, so it is only applicable to steady-state conditions, and loses effective bandwidth and spectral resolution in practice; such as CSAMT (Goldstein and Strangway, 1975), LOTEM (Strack, 1990; Yan et al., 2001) and other currently mainstream deep electromagnetic methods, in order to make the excitation field can be regarded as a plane wave vertically incident on the ground, it is required that the observation point is far away from the transmitter, and far-source area observation is carried out, which leads to a significant attenuation of the source signal amplitude and bandwidth during propagation. It can be seen that: in order to improve the detection depth and resolution of the method, on the one hand, it is necessary to use a pulse source (wideband high spectral resolution excitation signal) to carry out time domain observation; on the other hand, it is necessary to observe in areas close to the transmitter (near-source observation) to avoid propagation loss of the source signal.

The basic feature of the SOTEM method is: using a long grounding wire as the transmitting device, using a bipolar pulse with a duty cycle of 1:1 as the main transmitting waveform, and conducting observations in the near-source area. In the following subsections, we will break down the SOTEM method according to its source basis, and introduce the traditional magnetic source transient electromagnetic field, electric source transient electromagnetic field, and the inversion strategy of traditional time domain methods.

2.3 Feasibility Analysis of SOTEM

2.3.1 Feasibility Analysis of Near-Source Detection of Electric Source

In the Controlled Source Audio-frequency Magnetotellurics (CSAMT), which is similar to the type of electric source transient electromagnetic device, the basic requirement for achieving depth measurement is that the receiving point must meet the wave zone condition. Although the wide-area electromagnetic method extends the observable area to the transition zone, it still does not have the ability to measure depth in the near-source area. The widely used loop source transient electromagnetic device, such as the center loop, overlapping loop, etc., is a typical near-source detection (the distance between transmission and reception is zero). This suggests that time-domain electromagnetic methods can achieve near-source detection that frequency-domain methods cannot. To distinguish between CSAMT and electric source transient electromagnetic, and to study the feasibility of near-source detection of electric source transient electromagnetic, we must first understand the establishment and propagation of electromagnetic waves.

According to antenna and electromagnetic wave propagation theory, there are two ways to excite the electromagnetic field: one is directly excited by charge or current, such as the electric field and magnetic field generated by steady current; the other is excited by the mutual induction between the electric field and the magnetic field, that is, the changing magnetic field excites the vortex electric field, and the changing electric field excites the vortex magnetic field again, and the electromagnetic wave is generated and propagated through this mutual induction. In the time-varying electromagnetic field, both of these excitation methods exist. The field of stable charge and current is dominant near the field source, and the field disappears when the source disappears. The change of the field is closely related to the geometric size of the source and the position of the receiving point. This field is called the inherent field of the source, and the direct current depth measurement uses this field. At the same time, due to the mutual induction between electromagnetic fields, a part of the field leaves the field source and radiates outward. This part of the field is called the radiation field, and the variable frequency depth measurement ability of electromagnetic exploration comes from the radiation field.

In the work of CSAMT, the transmitter is supplied with continuous harmonic current, at this time, both the inherent field and the radiation field exist. In order to make the radiation field with variable frequency depth measurement ability dominant, it is necessary to observe in the far field area. Usually, considering the signal-to-noise ratio and the sensitivity of the receiver and other practical situations, the observation point is often arranged at a place 4 to 6 times the detection depth away from the field source. For the time-domain transient electromagnetic field, if continuous waveforms are transmitted (such as LOTEM), the inherent field also exists during the observation period, and the distance between transmission and reception should also be 4–6 times the detection depth like frequency depth measurement, so that the received signal is mainly the radiation field. However, if an intermittent transmission waveform is used to observe after the pulse is turned off, the inherent field disappears after the source is turned off and only the radiation field exists. Therefore, the two fields are separated in time, so that even in places very close to the source, the radiation field with depth measurement ability can be obtained.

The above text analyzes the feasibility of near-source detection of transient fields from the establishment of electromagnetic fields, and the following analyzes from the propagation of electromagnetic waves. As we all know, the propagation paths of electromagnetic waves generated by electric dipoles can be divided into sky waves, ground waves, and ground layer waves. Since long waves and ultra-long waves are used in geophysical exploration, sky waves are generally not considered, only ground waves (represented by S_0) and ground layer waves (represented by S_1 are studied. At a certain moment, due to the path difference, a nearly horizontal wavefront will form near the ground, causing a nearly vertically downward propagating horizontally polarized plane wave. S^*. S_0, S_1 S^* During propagation, they all interact electromagnetically with the underground geological body and reflect the results of the interaction to the ground observation point. In frequency domain CSAMT exploration, the propagation path of electromagnetic waves is from the field source to the receiving point, and it penetrates downwards everywhere, its influence range is the skin depth; at the same time, it reflects the results of interaction with the geological body to the observation point, the data measured at the observation point includes all the geological information within the effective detection range between the field source and the observation point. Among all these geological information, the stratum wave is transmitted directly from the field source to the receiving point, mainly carrying the information between the field source and the receiving point; the ground wave propagates almost vertically downward from the ground, mainly carrying the information directly below the receiving point. Therefore, in CSAMT exploration, we hope to use ground waves more than stratum waves. When the effect of the stratum wave can be ignored, the far-field condition is established, and it is believed that the stratum wave is the main cause of the shadow effect and the field source replication effect in CSAMT measurement. That is to say, in CSAMT exploration, the stratum wave exists as a form of interference wave (Fig. 2.3).

For time-domain electromagnetics, when the primary field is disconnected, a rapidly changing electromagnetic field, known as the secondary field, is generated near the field source and forms eddy currents underground. The secondary field also propagates and diffuses in the two ways mentioned above. The secondary field propagated by the ground wave is transmitted directly from the air to various points on the ground at the speed of light c, and part of the energy is transmitted underground; the secondary field propagated by the stratum wave is transmitted directly from the

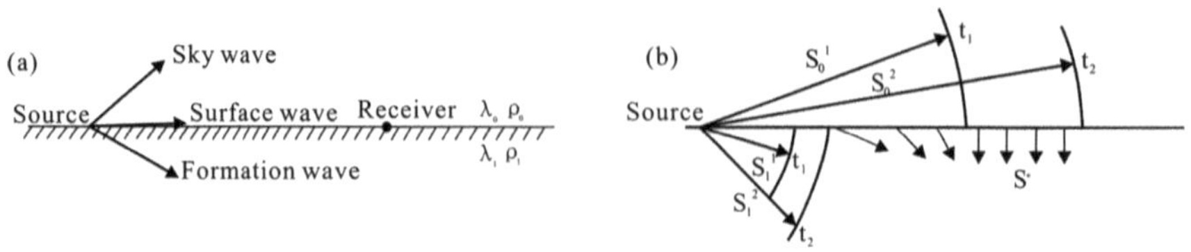

Fig. 2.3 Schematic diagram of electromagnetic wave propagation. **a** Sky wave, ground wave and stratum wave, **b** path difference and wavefront

field source to the underground, and the induced current it excites in the underground space is like "smoke ring" gradually diffusing to the depths of the earth over time. In the early stage of field propagation, the propagation of the ground wave is instantaneously established, while the stratum wave is subject to the impedance of the earth, and the establishment time is relatively late, so these two propagation methods are separated in time, and then these two fields are superimposed on each other; later, the field propagated by the ground wave decays to negligible, at this time the secondary field in the ground mainly comes from the stratum wave. It can be seen that at different times, the proportions of the two propagation methods in the signals received by different position measurement points are not the same. This introduces the concept of early (far-field), middle (transition zone), and late (near-field) periods.

To study in detail the characteristics of the time-domain electromagnetic field in different field areas, it is necessary to combine the offset distance of the field with the decay time of the transient process, and introduce concepts such as "early stage of the far-field", "late stage of the far-field", etc. In the early stage of the transient field far-field, the field has wave zone properties, and the first type of excitation plays a leading role. At this time, for shallow parts, the field has a strong stratification ability. However, in the late stage of the transient field far-field, for the receiving distance r, the total thickness of the layered medium is relatively small compared to the range of eddy currents, making the layer spacing small, and the induction effect between layered media is strong, so the eddy current effects between each layer are averaged, that is, the entire layered cross-section can be equivalent to a layer with a total longitudinal conductivity S. It can be seen that in the late stage of the far-field, only the total longitudinal conductivity and total thickness of each layer can be determined, and it does not have the ability to stratify. Due to this characteristic of the field, the far-field method is generally rarely used. In addition, because the far-field method has a volume effect, it also affects the stratification ability.

In the early stage of the transient field near-field, the early signal amplitude is large, the change is dramatic, it is seriously affected by the detection instrument, and the technical difficulty of accurately detecting the early signal is high. In the late stage of the near-field, the measurement results give a good stratification information of the geoelectric cross-section. The physical process is: in the upper conductive layer, the late stage just appears, that is, the decay process of eddy currents begins to appear, and it is characterized by its longitudinal conductivity S1 when this layer exists, in the deeper conductive layer, due to the "smoke ring" effect, the eddy current is still in the early stage of generation and enhancement. However, due to the strong shielding effect of the decaying eddy current in the first layer, it is difficult or weak to appear the influence of the second layer in the ground observation, with the passage of time, the transient results of the combined influence of the first and second layers can be observed on the ground, and it is characterized by S1 + S2 to characterize its comprehensive influence. At a later time, the comprehensive influence of S1 + S2 + S3 appears, and so on. In this way, as time goes by, all the information that can be measured on the entire cross-section can be obtained.

From the above analysis, it can be seen that in the frequency domain electromagnetic method, in order to utilize the radiation field with frequency measurement capability and the vertically incident ground wave, the distance from the receiving point to the transmitting source must be >4–6 times the detection depth. In the time domain electromagnetic method, if a step waveform current excitation is adopted, the self-field and radiation field can be separated in time, and the measurement of the radiation field is not affected by the self-field, so the radiation field can be observed within a very close range from the transmitting source to achieve the purpose of depth measurement. In addition, the time domain electromagnetic method mainly uses the formation wave components, so observing within a relatively close range from the transmitting source not only has strong stratification ability, but also can reduce the volume effect and better reflect the electrical changes of the formation below the receiving point.

2.3.2 *Analysis of SOTEM Superiority*

The previous section analyzed the feasibility of near-source exploration of transient electromagnetic fields from the perspective of electromagnetic field establishment and propagation. This section studies the superiority of electric source transient electromagnetic near-source detection from the aspects of signal strength, detection sensitivity, and detection depth based on the transient field characteristics of the electric source described in 7.1.

(1) **Signal Strength**

For the horizontal electric field E_x component, as the offset distance increases, the early signal strength drops significantly, while the late signal strength changes little. Therefore, the closer the observation point is to the transmitting source, the stronger the overall signal strength of E_x. The vertical magnetic field H_z decreases in signal strength in the early stage and increases in the late stage as the offset distance increases. Long-offset transient electromagnetics utilizes the characteristics of H_z, receiving at a large offset distance to enhance the late signal strength and achieve deep detection. However, the analysis in the previous section pointed out that the late-stage signal in the far zone does not have stratification ability, and the volume effect is serious, which is not conducive to fine detection. In fact, for exploration within a depth of 2000 m, the actual observation duration generally does not exceed

1s, and for the current sensitivity of the instrument, even if the signal in this time range is observed near the source, the late part can be recognized. Therefore, considering the detection effect and signal strength, observing the electromagnetic field in the area close to the source has a great advantage.

(2) **Detection Sensitivity**

The previous analysis has pointed out that the transient electromagnetic field in different field areas has different interaction modes with the underground, thus having different detection capabilities. In the theory of exploration electromagnetic fields, the electromagnetic field excited by artificial sources is usually divided into three parts: near-field, transition field, and far-field; for time-domain electromagnetic fields, there are also early field, transition period field, and late field divided by the concept of time. We usually define the early condition as $\frac{2\pi r}{\tau} \gg 1$, which is also the far-field condition (where r is the transmit-receive distance, $\tau = 2\pi\sqrt{2\rho t/\mu_0}$ is the transient field parameterization). The so-called early stage refers to the condition that when the transmit-receive distance r is fixed, the time t is very small and can satisfy $\frac{2\pi r}{\tau} \gg 1$, and the far-field refers to the condition that when the time t is fixed, the transmit-receive distance r is very large, thus satisfying $\frac{2\pi r}{\tau} \gg 1$, or both t and r can change, satisfying $\frac{2\pi r}{\tau} \gg 1$. Similarly, $\frac{2\pi r}{\tau} \ll 1$ is referred to as the "near-field" or "late-stage" condition. For near-field methods such as the large loop method for the same point device, there is no distinction between near and far, only early and late.

It can be seen that the field area division of the time-domain transient field is determined by two parameters: offset distance and time. Therefore, even if it is very close to the transmitting source, the field may not necessarily be in the near field; conversely, even if it is far from the transmitting source, the field may not necessarily be in the far field. However, in most cases, the near-source area can still characterize most of the characteristics of the near field, and the far-source area can also characterize most of the characteristics of the far field. Based on this assumption, the following analyzes the sensitivity of the electromagnetic field to the formation under different offset distances from the theoretical formulas of different field areas.

When the near-field and far-field conditions are met, the horizontal electric field E_x has the following approximate expressions:

$$E_x^L(t) = \frac{\mu_0^{3/2} I dx}{12\pi^{3/2}\rho_1^{1/2} t^{3/2}} \quad \text{Near-field (late-time)} \tag{2.18}$$

$$E_x^E(t) = \frac{\rho_1 I dx}{2\pi r^3}\left(3\cos^2\varphi - 2\right) \quad \text{Far-field(early-time)} \tag{2.19}$$

where I is the transmitting current, dx is the dipole source length, ρ_1 is the resistivity of the half-space, φ is the angle between the receiving point and the transmitting source. From the above two formulas, it can be seen that when other parameters are fixed, the field value of the near-field E_x is proportional to $\rho_1^{-1/2}$, while the field value of the far-field E_x is proportional to ρ_1, this indicates that the value of the distant zone

E_x is more dependent on the resistivity of the medium, that is, it is more sensitive to changes in the resistivity of the medium.

The near and far field forms of the vertical magnetic field H_z are as follows:

$$H_z^L(t) = \frac{r\mu_0^{3/2} Idx}{60(\pi t \rho_1)^{3/2}} \sin\varphi \text{ Near-field (late-time)} \tag{2.20}$$

$$H_z^E(t) = \frac{3tIdx}{2\pi\mu_0 r^4}\rho_1 \sin\varphi \text{ Far-field (early-time)} \tag{2.21}$$

It can be seen that when other parameters are fixed, the field value of the near zone H_z is proportional to $\rho_1^{-3/2}$, while the field value of the far zone H_z is proportional to ρ_1. This indicates that the value of the near zone H_z is more dependent on the resistivity of the medium, that is, it is more sensitive to changes in the resistivity of the medium.

(3) **Detection Depth**

Another aspect of evaluating near-source detection capability is detection depth. Spies conducted in-depth and detailed analysis and research on the detection depth of electrical sources. First, similar to the skin depth in the frequency domain, the concept of diffusion depth of transient electromagnetics was proposed, which is defined as the maximum depth that the TEM eddy current field extremum can reach within a given time, and the expression is:

$$d = \sqrt{2t/\sigma\mu_0} \tag{2.22}$$

It can be seen that the diffusion depth of transient electromagnetics mainly depends on time and geoelectric structure, and has nothing to do with offset distance parameters. In theoretical research, the diffusion depth of the electromagnetic field at a certain moment under a specific geoelectric structure can be calculated.

In actual work, the actual detection depth should be considered, that is, the actual detection depth of TEM is defined as the diffusion depth when the signal decays to the noise level under the premise of considering factors such as signal strength, sensitivity of the receiving instrument, accuracy, noise intensity, geoelectric parameters, etc. The actual detection depth of the electrical source is usually estimated according to the following formula:

When in the near zone

$$d = 0.48\left(\frac{Ir\rho AB}{\eta}\right)^{\frac{1}{5}} \tag{2.23}$$

When in the far zone

$$d = 0.28\left(\frac{I\rho AB}{\eta}\right)^{\frac{1}{4}} \tag{2.24}$$

In the formula, η is the minimum distinguishable voltage of the instrument, ρ is the resistivity of the formation, I is the transmission current, AB is the length of the transmission line.

According to formulas (2.23) and (2.24), the actual detection depths of the near and far zones under different parameters were calculated, as shown in Table 2.1. Comparing the results, it was found that the actual detection depth of the near zone is greater than that of the far zone in most cases.

The above analysis results show that the near-field observation has a relatively large actual detection depth, but because the assumptions in the above analysis are too general, it is still impossible to accurately determine the offset range in which the transient electromagnetic of the electric source has the maximum detection depth. Therefore, the concept of effective detection depth is defined. Like other physical exploration methods, the effective detection depth refers to the depth range in which the abnormal field generated by the target layer exceeds the background field level several times, and the existence of the target layer can be distinguished from the observation results.

Suppose the response value (background field) of a uniform half-space (resistivity is $100\Omega \cdot m$) is V_0, the response value (abnormal field) caused by a good conductive thin layer (buried depth is 1000 m, resistivity is $10\Omega \cdot m$) is V_a and suppose $V_a/V_0 = \delta$. Calculate the responses under different offset distances for the two models, and calculate δ. When $V_a/V_0 = \delta \geq 50\%$, it is considered that the existence of the thin conductive layer can be distinguished under this condition, and the detection depth

Table 2.1 Detection depth table for far and near zones

I (A)	Resistivity (Ωm)	AB (m)	Offset distance (m)	Minimum distinguishable voltage (nV)	Far zone depth (m)	Near zone depth (m)
5	50	300	100	60	526.4	501.9
			200			576.5 626.2
			300			626.2
10	100	500	200	60	845.9	842.5
			300			913.7
			500			1012.0
20	300	1000	400	60	1574.5	1590.9 1779.3
			700			
			1000			1910.9
50	1000	4000	1000	60	3783.3	3853.1
			3000			4800
			4000			5084.2

Table. 2.2 Calculation table of effective detection depth

H = 1000 m AB = 1000 m I = 10 A							
Offset distance r (m)	250	500	700	1000	2000	3000	4000
Detection depth d (km)	3.6	4.1	4.2	4.2	4.1	3.9	3.7

under various offset distances is calculated according to formula (2.22), as shown in Table 2.2.

It can be seen that the maximum effective detection depth is obtained within the offset distance of 500–2000 m, that is, when $r/H = 0.5 \sim 2$, the detection depth is the largest.

2.4 Response of Transient Electromagnetic Field of Electric Source

2.4.1 Uniform Half-Space

As shown in Fig. 2.6, there is a horizontal dipole on the ground. The origin of both the Cartesian coordinate system and the cylindrical coordinate system are chosen at the center of the dipole, so that the x axis is consistent with the direction of the dipole, and the z axis is downward. To facilitate the solution, the magnetic vector position A is introduced. The equation satisfied by the vector position A is:

$$\nabla^2 \mathbf{A} - k_1^2 \mathbf{A} = 0 \tag{2.25}$$

where, k_1 is the wave number and $k_1 = \sqrt{-i\omega\mu\sigma}$ (Fig. 2.4).

Since the direction of the dipole is along the x axis, A has an x directional component $\mathbf{A}_x$; due to the presence of a horizontal non-uniform interface (ground), there are

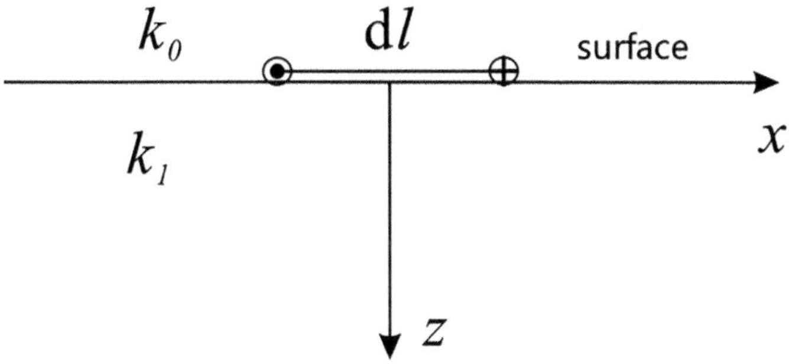

Fig. 2.4 Horizontal electric dipole on a uniform earth surface

accumulated charges at the interface, which makes A have a z directional component $\mathbf{A}_z$.

The Helmholtz equation can be written in cylindrical coordinates as:

$$\begin{aligned}\frac{\partial^2 A_x}{\partial r^2}+\frac{1}{r}\frac{\partial A_x}{\partial r}+\frac{1}{r^2}\frac{\partial A_x}{\partial \varphi^2}+\frac{\partial^2 A_x}{\partial z^2}-k^2A_x=0\\ \frac{\partial^2 A_z}{\partial r^2}+\frac{1}{r}\frac{\partial A_z}{\partial r}+\frac{1}{r^2}\frac{\partial A_z}{\partial \varphi^2}+\frac{\partial^2 A_z}{\partial z^2}-k^2A_z=0\end{aligned} \tag{2.26}$$

Solve by separation of variables, let:

$$A=\sum_{n=0}^{\infty}\cos n\varphi\int_0^{\infty}\left(C_1e^{uz}+C_2e^{-uz}\right)J_n(\lambda r)d\lambda \tag{2.27}$$

where, λ is the wavelength, $r=\sqrt{x^2+y^2}$, $u=\sqrt{\lambda^2+k_1^2}$, $J_n(\lambda r)$ is the n order bulk Bessel function. Use the boundary conditions to find the coefficients C_1, C_2, substitute back into Eq. (2.27) to get the vector position of the lower half space:

$$A_x=\frac{2P_E}{4\pi r^3k_1^2}[1-(1+k_1r)e^{-k_1r}] \tag{2.28}$$

$$A_z=\frac{2P_E}{4\pi r^2}xI_1K_1 \tag{2.29}$$

$$\nabla\cdot\mathbf{A}=-\frac{2P_E}{4\pi r^3}x \tag{2.30}$$

where, $P_E=$ Idx represents the transmitting magnetic moment, I_1 and K_1 are the first order first and second kind modified Bessel functions, and their independent variable is kr.

From $\mathbf{H}=\nabla\times\mathbf{A}$ get $H_x=\frac{\partial A_x}{\partial y}-\frac{\partial A_y}{\partial z}=\frac{\partial A_z}{\partial y}=\frac{\partial A_z}{\partial r}\frac{\partial r}{\partial y}=\frac{y}{r}\frac{\partial A_z}{\partial r}$. Differentiate Eq. (2.29) with respect to r, and use the derivative formula of the imaginary bulk Bessel function to get the components of the magnetic field:

$$H_x=-\frac{P_E}{4\pi r^2}\sin\varphi\cos\varphi[8I_1K_1-k_1r(I_0K_1-I_1K_0)] \tag{2.31}$$

where, I_0 and K_0 are the zero order first and second kind modified Bessel functions. Similarly, from $H_y=\frac{\partial Ax}{\partial z}-\frac{\partial A_z}{\partial x}$ get:

$$H_y=\frac{P_E}{2\pi r^2}[(1-4\sin^2\varphi)I_1K_1+\frac{k_1r}{2}\sin^2\varphi(I_0K_1-I_1K_0)] \tag{2.32}$$

Similarly, from $H_z=\frac{\partial A_y}{\partial x}-\frac{\partial A_x}{\partial y}==\frac{-\partial A_x}{\partial r}\frac{\partial r}{\partial y}=-\frac{y}{r}\frac{\partial A_x}{\partial r}$ get:

$$H_z = \frac{2P_E}{4\pi r^4 k_1^2}\sin\varphi[3-(3+3k_1r+k_1^2r^2)e^{-k_1r}] \tag{2.33}$$

The components of the electric field can be obtained:

$$\begin{aligned} E_x &= i\omega\mu\left(A_x - \frac{1}{k_1^2}\frac{\partial}{\partial x}\nabla\cdot\mathbf{A}\right) \\ &= i\omega\mu\left\{\frac{2P_E}{4\pi r^3k_1^2}\left[1-(1+k_1r)e^{-k_1r}\right]+\frac{1}{k_1^2}\frac{2P_E}{4\pi r^3}\right\} \\ &= \frac{P_E\rho_1}{2\pi r^3}\left[3\cos^2\varphi - 2 + (1+k_1r)e^{-k_1r}\right] \end{aligned} \tag{2.34}$$

From $E_y = -i\omega\mu\frac{1}{k_1^2}\frac{\partial}{\partial y}\nabla\mathbf{A}$ get:

$$E_y = \frac{P_E\rho_1}{2\pi r^3}3\cos\varphi\sin\varphi \tag{2.35}$$

$$E_z = \frac{i\omega\mu Il}{2\pi r}\cos\varphi I_1K_1 \tag{2.36}$$

Use Fourier transform to convert the calculation results in the frequency domain to the time domain:

$$H_z(t) = \frac{P_E}{4\pi r^2}\sin\varphi\left[\frac{3}{\sqrt{\pi}}\frac{e^{-u^2}}{u}+\left(1-\frac{3}{u^2}\right)\Phi(u)\right] \tag{2.37}$$

$$H_x(t) = \frac{P_E}{4\pi r^2}\sin 2\varphi\left\{e^{-u^2/2}\left[I_0\left(\frac{u^2}{2}\right)+2I_1\left(\frac{u^2}{2}\right)\right]-1\right\} \tag{2.38}$$

$$\begin{aligned} H_y(t) = -\frac{P_E}{4\pi r^2}&\left\{\left\{\left[I_0\left(\frac{u^2}{2}\right)+2I_1\left(\frac{u^2}{2}\right)\right]\cos 2\varphi\right.\right. \\ &\left.\left.-I_1\left(\frac{u^2}{2}\right)\right\}e^{-u^2/2}-\cos 2\varphi\right\} \end{aligned} \tag{2.39}$$

$$E_x(t) = \frac{\rho_1 P_E}{2\pi r^3}\left[\Phi(u)-\frac{2}{\sqrt{\pi}}ue^{-u^2}\right] \tag{2.40}$$

$$E_y(t) = \frac{3\rho_1 xzP_E}{2\pi r^3}\left[\Phi(u)-\frac{2}{\sqrt{\pi}}ue^{-u^2}\left(1+\frac{2}{3}u^2\right)\right] \tag{2.41}$$

$$E_z(t) = \frac{3\rho_1 xzP_E}{2\pi R^3}\left[\Phi(u_z)-\frac{2}{\sqrt{\pi}}u_ze^{-u_z^2}\left(1+\frac{2}{3}u_z^2\right)\right] \tag{2.42}$$

where, $r = \sqrt{x^2 + y^2}$, $R = \sqrt{r^2 + z^2}$, $u = \frac{2\pi r}{\tau}$, $\tau = 2\pi\sqrt{2\rho t/\mu_0} = \sqrt{2\pi\rho t \times 10^7}$, $u_z = \frac{R}{2}\sqrt{\frac{\mu_0}{t\rho_1}}$, $\phi(u) = \frac{2}{\sqrt{\pi}}\int_0^{u(t)} e^{-t^2}\mathrm{d}t$ is a probability function.

2.4.2 Layered Earth

It is known that the two-dimensional Fourier transform of the special solution of the electric dipole vector potential in the x direction is

$$\mathbf{A} = \frac{P_E}{2u_0}e^{-u_0(z+h)}\vec{u}_x \tag{2.43}$$

As can be seen from the previous analysis, the electric dipole generates both a vertical electric field:

$$E_z^P = \frac{1}{\hat{y}_0}\frac{\partial^2 A_x}{\partial x \partial z} = -\frac{P_E}{2\hat{y}_0}ik_x e^{-u_0(z+h)} \tag{2.44}$$

and a vertical magnetic field

$$H_z^P = -\frac{\partial \overline{A}_x}{\partial y} = -\frac{P_E}{2}\frac{ik_y}{u_0}e^{-u_0(z+h)} \tag{2.45}$$

Therefore, the electromagnetic field of the electric dipole source has both TE and TM components. By decomposing the primary field into TE and TM components, we can deal with the TE and TM fields separately.

Because only the TM polarization mode has a vertical electric field, and:

$$E_z = \frac{1}{\hat{y}}\left(\frac{\partial^2}{\partial z^2} + k^2\right)A_z \tag{2.46}$$

Therefore, by setting Eq. (2.44) equal to Eq. (2.46) of E_z, the TM coefficient A_p can be obtained, that is:

$$A_P = -\frac{I\mathrm{ds}}{2}\frac{ik_x}{k_x^2 + k_y^2} \tag{2.47}$$

Because only the TE polarization mode has a vertical magnetic field, and

$$H_z = \frac{1}{\hat{z}}\left(\frac{\partial^2}{\partial z^2} + k^2\right)F_z \tag{2.48}$$

Therefore, by setting Eq. (2.45) of H_z^p equal to Eq. (2.48) of H_z, the TE coefficient F_p can be obtained, that is

$$F_p = -\frac{\hat{z}_0 P_E}{2u_0}\frac{ik_y}{k_x^2 + k_y^2} \tag{2.49}$$

By substituting the above expressions of A_p and F_p into Eq. (2.50), the expressions of TM and TE potentials between the earth and the electric dipole can be obtained.

$$\mathbf{A} = \frac{1}{4\pi^2}\int_{-\infty}^{\infty}\int_{-\infty}^{\infty} A_p e^{-u_0 h}\left(e^{-u_0 z} + r_{TM}e^{u_0 z}\right)e^{i(k_x x + k_y y)}\mathrm{d}k_x \mathrm{d}k_y$$

$$\mathbf{F} = \frac{1}{4\pi^2}\int_{-\infty}^{\infty}\int_{-\infty}^{\infty} F_p e^{-u_0 h}\left(e^{-u_0 z} + r_{TE}e^{u_0 z}\right)e^{i(k_x x + k_y y)}\mathrm{d}k_x \mathrm{d}k_y \tag{2.50}$$

In the above equations, (x, y, z) are the coordinates of the receiving point; r is the distance from the receiving point to the dipole; r_{TM} and r_{TE} are the reflection coefficients under TM and TE modes, including $\hat{y}_0$, $\hat{z}_0$ and u_0, the specific algorithm is given by Eqs. (2.51)–(2.56); $J_1(\lambda r)$ and $J_0(\lambda r)$ are the first-order and zero-order Bessel functions respectively.

r_{TE} and r_{TM} are given by the following two equations:

$$r_{TE} = \frac{Y_0 - \hat{Y}_1}{Y_0 + \hat{Y}_1} \tag{2.51}$$

$$r_{TM} = \frac{Z_0 - \hat{Z}_1}{Z_0 + \hat{Z}_1} \tag{2.52}$$

In the equation, $Y_0 = \frac{u_0}{\hat{z}_0}$, $Z_0 = \frac{u_0}{\hat{y}_0}$, and $\hat{z}_0 = i\omega\mu_0$, $\hat{y}_0 = i\omega\varepsilon_0$; for an N-layer earth, there is a recursive formula

$$\hat{Y}_n = Y_n\frac{\hat{Y}_{n+1} + Y_n \tanh(u_n h_n)}{Y_n + \hat{Y}_{n+1}\tanh(u_n h_n)} \tag{2.53}$$

$$\hat{Y}_n = Y_n \tag{2.54}$$

$$\hat{Z}_n = Z_n\frac{\hat{Z}_{n+1} + \hat{Z}_n \tanh(u_n h_n)}{Z_n + \hat{Z}_{n+1}\tanh(u_n h_n)} \tag{2.55}$$

$$\hat{Z}_n = Z_n \tag{2.56}$$

In the equation, $Y_n = \frac{u_n}{\hat{z}_n}$, $Z_n = \frac{u_n}{\hat{y}_n}$; $u_n = \left(k_x^2 + k_y^2 - k_n^2\right)^{1/2}$, $k_n^2 = -\hat{z}_n\hat{y}_n = \omega^2\mu_n\varepsilon_n - i\omega\mu_n\sigma_n$, $k_x^2 + k_y^2 = \lambda^2$. Therefore, starting from the bottom layer and gradually moving upwards, we can obtain $\hat{Y}_1$ and $\hat{Z}_1$.

The expressions of TM and TE potentials between the earth and the electric dipole obtained from Eq. (2.39) are:

$$A(x, y, z,) = -\frac{P_E}{8\pi^2}\int_{-\infty}^{+\infty}\int_{-\infty}^{+\infty}\left[e^{-u_0(z+h)} + r_{TM}e^{u_0(z-h)}\right]\frac{ik_x}{k_x^2 + k_y^2}e^{i(k_x x + k_y y)}dk_x dk_y \tag{2.57}$$

$$F(x, y, z,) = -\frac{\hat{z}_0 P_E}{8\pi^2}\int_{-\infty}^{+\infty}\int_{-\infty}^{+\infty}\left[e^{-u_0(z+h)} + r_{TE}e^{-u_0(z-h)}\right]\frac{ik_y}{u_0\left(k_x^2 + k_y^2\right)}e^{i(k_x x + k_y y)}dk_x dk_y \tag{2.58}$$

Substitute the above two formulas into the equation

$$\begin{aligned}
TM_z : E_x &= \frac{1}{\hat{y}}\frac{\partial^2 A_z}{\partial x \partial z}\\
E_y &= \frac{1}{\hat{y}}\frac{\partial^2 A_z}{\partial y \partial z}\\
E_z &= \frac{1}{\hat{y}}\left(\frac{\partial^2}{\partial z^2} + k^2\right)A_z TE_z : E_x = -\frac{\partial F_z}{\partial y}\\
E_y &= \frac{\partial F_z}{\partial x}\\
E_z &= 0
\end{aligned} \tag{2.59}$$

Thus, the components of the electric field can be obtained. When the electric dipole is laid on the ground ($h = 0$), on the ground (z = 0) or above the ground, the x component of the electric field is:

$$\begin{aligned}
E_x = &-\frac{P_E}{8\pi^2\hat{y}_0}\int_{-\infty}^{+\infty}\int_{-\infty}^{+\infty}[1 - r_{TM}]e^{u_0 z}\frac{u_0 k_x^2}{k_x^2 + k_y^2}e^{i(k_x x + k_y y)}dk_x dk_y\\
&-\frac{\hat{z}_0 P_E}{8\pi^2}\int_{-\infty}^{+\infty}\int_{-\infty}^{+\infty}[1 + r_{TE}]e^{u_0 z}\frac{k_y^2}{u_0\left(k_x^2 + k_y^2\right)}e^{i(k_x x + k_y y)}dk_x dk_y
\end{aligned} \tag{2.60}$$

In order to simplify the following discussion about the long grounded wire source, it is necessary to decompose the electromagnetic response into items related to the current source and items related to the charge source. To achieve this decomposition, the following substitution is needed

$$\frac{k_y^2}{k_x^2+k_y^2}=1-\frac{k_x^2}{k_x^2+k_y^2}$$

Hence, we have

$$E_x=-\frac{P_E}{8\pi^2}\int_{-\infty}^{+\infty}\int_{-\infty}^{+\infty}\left[\frac{u_0}{\hat{y}_0}-\frac{\hat{z}_0}{u_0}-\frac{\hat{z}_0}{u_0}r_{TE}-\frac{u_0}{\hat{y}_0}r_{TM}\right]e^{u_0z}\frac{k_x^2}{k_x^2+k_y^2}e^{i(k_xx+k_yy)}dk_xdk_y$$
$$-\frac{\hat{z}_0P_E}{8\pi^2}\int_{-\infty}^{+\infty}\int_{-\infty}^{+\infty}[1+r_{TE}]e^{u_0z}\frac{1}{u_0}e^{i(k_xx+k_yy)}dk_xdk_y \tag{2.61}$$

Using the equation

$$\int_{-\infty}^{\infty}\int_{-\infty}^{\infty}F\left(k_x^2+\mathrm{k}_y^2\right)e^{i(k_x\mathrm{x}+\mathrm{kyy})}dk_xdk_y=2\pi\int_0^{\infty}F(\lambda)\lambda\mathrm{J}_0(\lambda\rho)d\lambda \tag{2.62}$$

Convert the Fourier transform into Hankel transform, and take z as zero, to get the electric field on the ground

$$E_\mathrm{x}=-\frac{P_E}{4\pi}\frac{\partial}{\partial x}\frac{x}{r}\int_0^{+\infty}\left[(1-r_{TM})\frac{u_0}{\hat{y}_0}-(1+r_{TE})\frac{\hat{z}_0}{u_0}\right]J_1(\lambda r)\mathrm{d}\lambda$$
$$-\frac{\hat{z}_0P_E}{4\pi}\int_0^{+\infty}(1+r_{TE})\frac{\lambda}{u_0}J_0(\lambda\rho)\mathrm{d}\lambda \tag{2.63}$$

The reason for choosing the above form of solution is that for a very important excitation source like the grounded wire, the differentiation operation for x in the first term no longer exists when integrating along the long wire, and only the remaining part of the first term at the two grounding points (taking opposite signs) needs to be calculated. After similar derivation, we have:

$$E_y=-\frac{P_E}{4\pi}\frac{\partial}{\partial x}\frac{y}{\rho}\int_0^{+\infty}\left[(1-r_{TM})\frac{u_0}{\hat{y}_0}-(1+r_{TE})\frac{\hat{z}_0}{u_0}\right]J_1(\lambda\rho)d\lambda \tag{2.64}$$

Similarly, the expressions for each component of the magnetic field can be obtained. The x component of the magnetic field is

$$H_x = \frac{P_E}{4\pi}\frac{\partial}{\partial x}\frac{y}{\rho}\int\limits_0^{+\infty}(r_{TM} + r_{TE})e^{u_0 z}J_1(\lambda\rho)d\lambda \tag{2.65}$$

The y component of the magnetic field is

$$\begin{aligned} \mathrm{H_y} = & -\frac{P_E}{4\pi}\frac{\partial}{\partial x}\frac{x}{\rho}\int\limits_0^{+\infty}(r_{TM} + r_{TE})e^{u_0 z}J_1(\lambda\rho)\mathrm{d}\lambda \\ & - P_E\int\limits_o^{+\infty}(1 - r_{TE})e^{u_0 z}\lambda J_0(\lambda\rho)\mathrm{d}\lambda \end{aligned} \tag{2.66}$$

The z component of the magnetic field is

$$H_z = \frac{P_E}{4\pi}\frac{y}{\rho}\int\limits_0^{+\infty}(1 + r_{\mathrm{TE}})e^{u_0 z}\frac{\lambda^2}{u_0}J_1(\lambda\rho)d\lambda \tag{2.67}$$

The field of the long wire source can be divided into two parts, one part is excited by the long wire, and the other part is excited by the two power supply point electrodes. For E_x integrated along the long wire (x direction), the $\partial/\partial x$ operation in the first term no longer exists. This term is the field generated by the grounding electrode. When calculating, only the positive and negative grounding items need to be calculated; the second term is the field excited by the long wire part. When performing numerical calculations, the integral along the long wire is changed to the accumulation of the fields generated by the sub-units with a small length of Δx located at $(x_n, 0, 0)$. The calculation expressions are obtained:

$$\begin{aligned} E_x = & -\frac{I}{4\pi}\sum_{m=1}^{2}(-1)^m\frac{x - x_m^s}{r_m}\int\limits_0^{\infty}\left[(1 - r_{TM})\frac{u_0}{\hat{y}_0} - (1 + r_{TE})\frac{\hat{z}_0}{u_0}\right]J_1(\lambda r_m)d\lambda \\ & - \frac{I}{4\pi}\sum_{n=1}^{N}\Delta x\int\limits_0^{\infty}(1 + r_{TE})\frac{\hat{z}_0}{u_0}\lambda J_0(\lambda r_n)d\lambda \end{aligned} \tag{2.68}$$

$$E_y = -\frac{I}{4\pi}\sum_{m=1}^{2}(-1)^m\frac{y}{r_m}\int\limits_0^{\infty}\left[(1 - r_{TM})\frac{u_0}{\hat{y}_0} - (1 + r_{TE})\frac{\hat{z}_0}{u_0}\right]J_1(\lambda r_m)d\lambda \tag{2.69}$$

$$E_z = \frac{I}{4\pi}\sum_{m=1}^{2}(-1)^m\int\limits_0^{\infty}[(1 + r_{TM})]\frac{\lambda}{\hat{y}_0}e^{u_0 z}J_0(\lambda r_m)d\lambda \tag{2.70}$$

$$H_x = \frac{I}{4\pi}\sum_{m=1}^{2}(-1)^m \frac{y}{r_m}\int_0^\infty (r_{TM} + r_{TE})e^{u_0 z}J_1(\lambda r_m)d\lambda \tag{2.71}$$

$$H_y = -\frac{I}{4\pi}\sum_{m=1}^{2}(-1)^m \frac{x}{r_m}\int_0^\infty (r_{TM} + r_{TE})e^{u_0 z}J_1(\lambda r_m)d\lambda$$
$$-\frac{I}{4\pi}\sum_{n=1}^{N}\Delta x\int_0^\infty (1 - r_{TE})e^{u_0 z}\lambda J_0(\lambda r_n)d\lambda \tag{2.72}$$

$$H_z = \frac{Iy}{4\pi}\sum_{n=1}^{N}\frac{\Delta x}{r_n}\int_0^\infty (1 + r_{TE})e^{u_0 z}\frac{\lambda^2}{u_0}J_1(\lambda r_n)d\lambda \tag{2.73}$$

In the formulas, x_m^s is the x-coordinate of the grounding electrode; r_m is the distance from the receiving point to the grounding electrode.

The frequency domain response of the above components can be obtained through Hankel transform and digital filtering. After obtaining the frequency domain response, the time domain response can be obtained using Fourier transform.

2.5 Characteristics of Transient Electromagnetic Field Propagation of Electric Sources

In the transient electromagnetic method of electric dipole sources, due to the presence of grounding electrodes, the earth becomes part of the source, forming a circuit together with the long wire. Compared to the current flowing along the source in the ungrounded loop source in the horizontal or vertical direction, the current direction in the long wire-earth system is much more complex, with not only horizontal induced current but also vertical induced current. This leads to the complexity of the corresponding electromagnetic field diffusion and distribution. Studying the diffusion and distribution characteristics of the transient field of electric dipole sources is of great importance for a deep understanding of the principles of the transient electromagnetic method and exploring the effectiveness and applicability of the transient electromagnetic detection method of electric dipole sources.

2.5.1 Underground Induced Current Diffusion Characteristics

The response of the transient electromagnetic field observed on the ground is generated by the induced eddy currents in the earth. According to the research of Lewis and Lee (1978), Nabighian (1979), Reid and Macnae (1998), Wang (2002), etc., we know that the induced current in the earth can be approximately equivalent to a circular current loop, that is, the equivalent current can be vividly represented as a "smoke ring" blown out by the source. The diffusion process of the "smoke ring" over time can be described by the equipotential lines of the underground electric field. Ruan and Xu (1996), Yan et al. (2002) used the finite difference direct time domain numerical calculation method to obtain the transient secondary induced electric field generated by the loop source underground at different times. Using the method given by Kerry (2009), the induced current generated by the electric dipole source underground can also be calculated. Figures 2.5 and 2.6 are the diffusion patterns of horizontal induced current and vertical induced current underground at different times (1–3 ms) of the electric dipole source.

Let's analyze the characteristics of these two induced currents. The horizontal induced current of the electric dipole source consists of two parts: the upper induced horizontal current and the lower induced horizontal current. The upper horizontal current spreads in the x and y directions in the shape of the source AB, and its

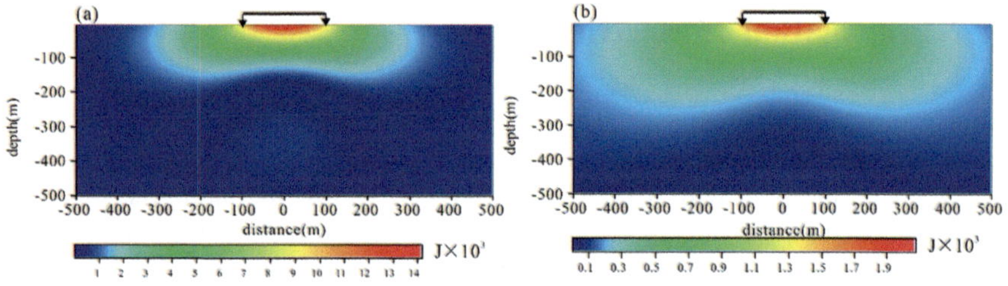

Fig. 2.5 Diffusion diagram of horizontal induced current of electric dipole source over time. **a** time = 1 ms, **b** time = 3 ms

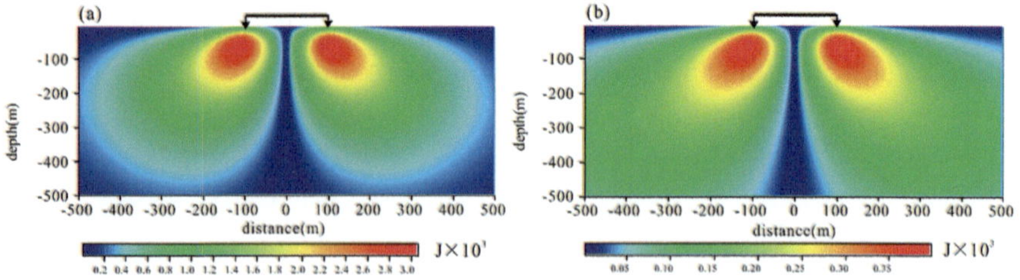

Fig. 2.6 Diffusion diagram of vertical induced current of electric dipole source over time. **a** time = 1 ms, **b** time = 3 ms

maximum current value is concentrated near the field source within the full period detection range. The lower induced horizontal current, also known as the "return current", spreads downward faster but has a smaller amplitude than the upper current. Because the maximum value of the horizontal current is always close to the field source, when there is an inhomogeneous body near the field source, the electromagnetic response of the electric dipole source will produce a static offset effect (Newman, 1989). The vertical induced current has shown many important characteristics. First, the maximum value of the vertical induced current is concentrated under the two grounding electrodes and gradually spreads outward and downward. There is no vertical induced current directly under the midpoint of the source; the intensity of the vertical induced current decays faster with time than the horizontal induced current. The intensity of the vertical current at 3 ms is nearly 10 times weaker than at 1 ms, while the horizontal induced current has only decayed about 7 times. By comparing the amplitude strengths of the two induced currents, it is found that at a given time, the amplitude of the maximum value of the vertical induced current is only about 20% of the horizontal induced current, but at a given depth, the amplitude of the maximum value of the vertical induced current is always greater than the horizontal component. Therefore, it can be inferred that the response generated by this rapidly spreading and highly concentrated vertical induced current can distinguish targets within a certain depth earlier than other components. Secondly, the direction of the maximum value of the vertical induced current spreading downward is ~45° to the ground, and we know that the maximum value of the eddy current field of the loop source is transmitted downward at a 30° angle to the ground, that is, the vertical induced current of the electric dipole source is more concentrated below the field source than the horizontal induced current generated by the transient electromagnetic of the loop source.

These characteristics of the horizontal and vertical induced currents of the electric dipole source determine its detection capabilities for different target bodies: the horizontal induced current is more sensitive to low-resistivity target bodies, while the vertical induced current is more sensitive to high-resistivity target bodies. It is precisely because of the superiority of the electric dipole source in the detection of high-resistivity bodies that this method plays an irreplaceable role in the exploration of oil and gas fields, mined-out areas, and other high-resistivity target bodies.

2.5.2 *Ground Electromagnetic Field Propagation and Distribution Characteristics*

The diffusion characteristics of underground induced currents determine the diffusion and distribution of the ground electromagnetic field. Studying the diffusion and distribution rules of the transient electromagnetic field of the electric dipole source on the ground over time is of great significance for analyzing the characteristics of the electromagnetic field and selecting suitable electromagnetic components for observation. As mentioned earlier, the electric dipole source placed on the ground can generate all six electromagnetic field components in a Cartesian coordinate system, they are E_x, E_y, E_z and H_x, H_y, H_z Although these six fields all contain the electrical structure information of the underground medium and have the ability to detect strata, in actual field work, the E_z component is not easy to observe. Therefore, in the following calculations and analyses, the E_z component is temporarily not considered. Using the method described in the previous section, we calculated the responses of the five components E_x, E_y, H_x, H_y, H_z at different times and locations in a uniform half-space (since the response of E_y in a uniform half-space is zero, the model used for calculating E_y is a uniform half-space with a low-resistivity thin layer), and plotted the single-point decay curves and distribution plane diagrams at two moments (Figs. 2.7, 2.8, 2.9, 2.10, 2.11, 2.12, 2.13, 2.14, 2.15, and 2.16).

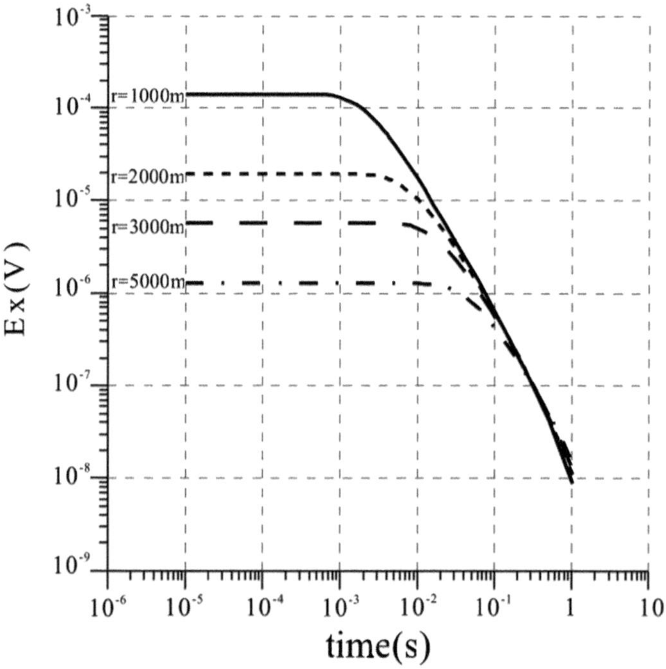

Fig. 2.7 Decay curve of E_x at different offsets

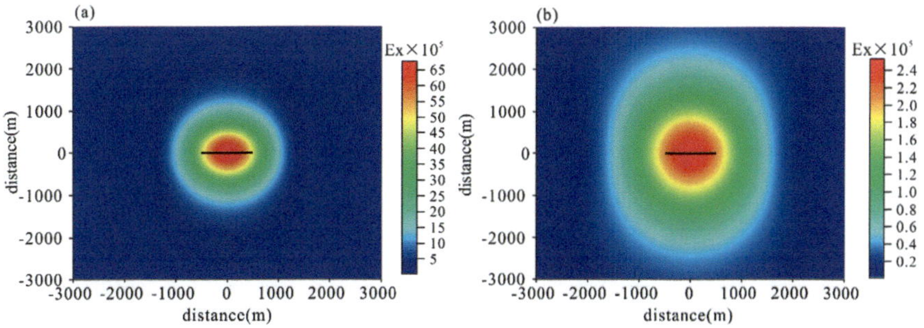

Fig. 2.8 Diffusion plane diagram of E_x at different times. **a** Time = 1 ms, **b** time = 10 ms

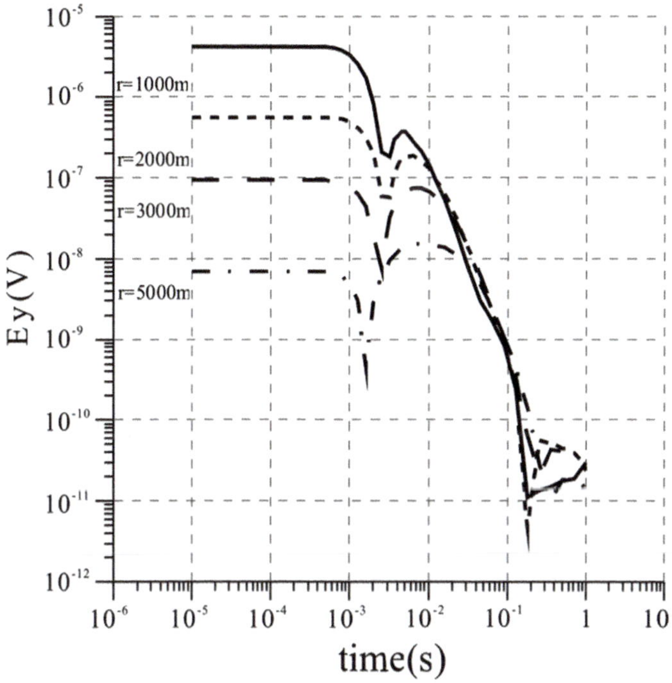

Fig. 2.9 Decay curve of E_y at different offsets

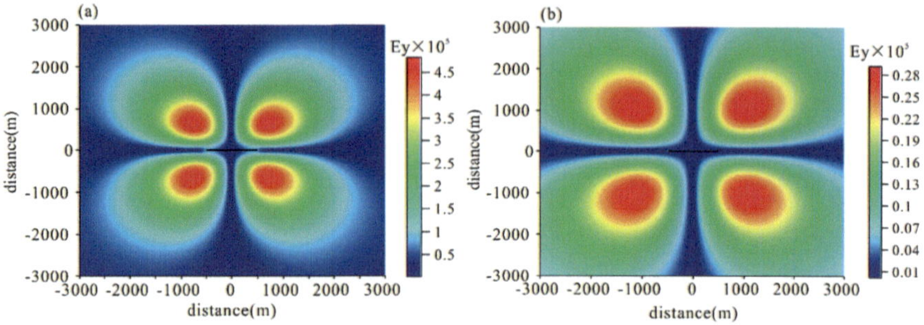

Fig. 2.10 Diffusion plane diagram of E_y at different times. **a** Time = 1 ms, **b** time = 10 ms

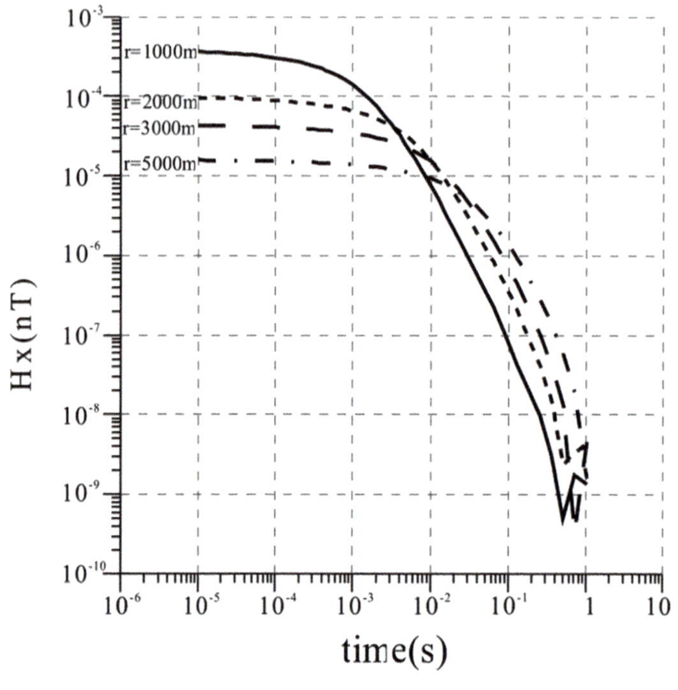

Fig. 2.11 Decay curve of H_x at different offsets

From Fig. 2.7, it can be seen that as the offset increases, the signal strength of the horizontal electric field E_x decreases sharply in the early time period, while the response strength at different offsets tends to be consistent in the late time period. The plane distribution diagram (Fig. 2.8) more clearly shows the diffusion and distribution characteristics of E_x: it is evenly distributed in all directions on the plane, the maximum electric field within the entire detection range is concentrated near the transmitter, and the farther away from the transmitter, the lower the signal strength,

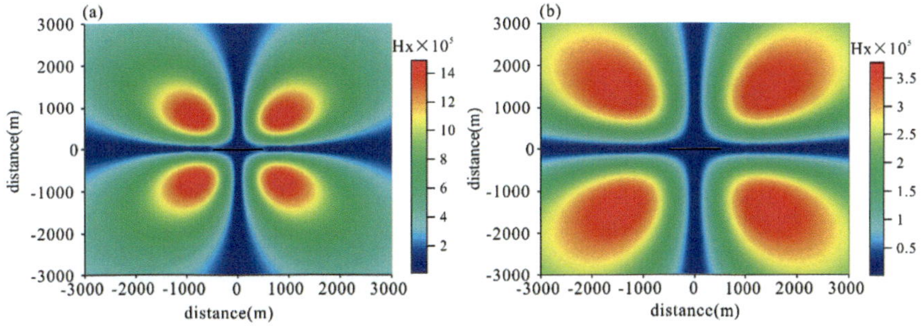

Fig. 2.12 Diffusion plane diagram of H_x at different times. **a** Time = 1 ms, **b** time = 10 ms

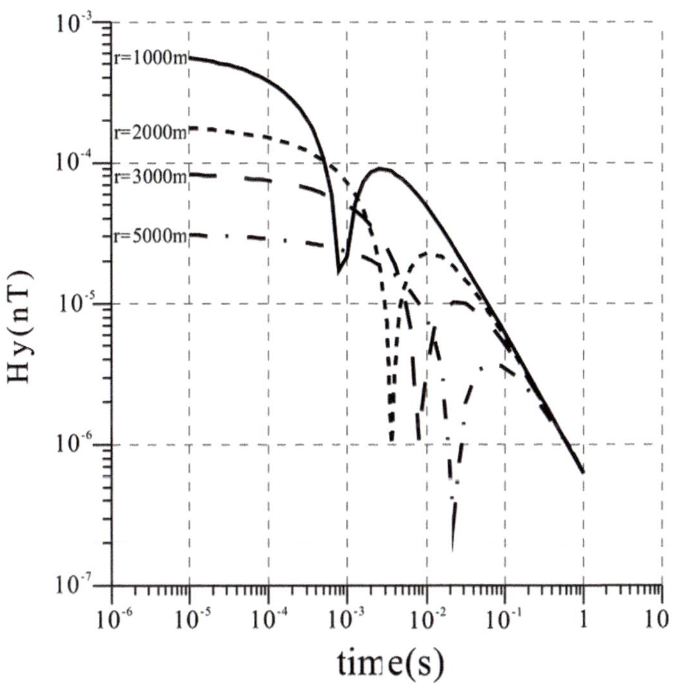

Fig. 2.13 Decay curve of H_y at different offset distances

which is consistent with the previous analysis that the upper horizontal induced current is concentrated near the transmitter.

Due to the lack of stability of the algorithm, the response curve of the late stage of E_y oscillates, as shown in Fig. 2.9. However, it can still be seen that as the offset increases, the strength of the early response of E_y significantly decreases, while the signal strength of the late stage is not much different, indicating that the extreme values of the E_y field are concentrated near the transmitter throughout the period and

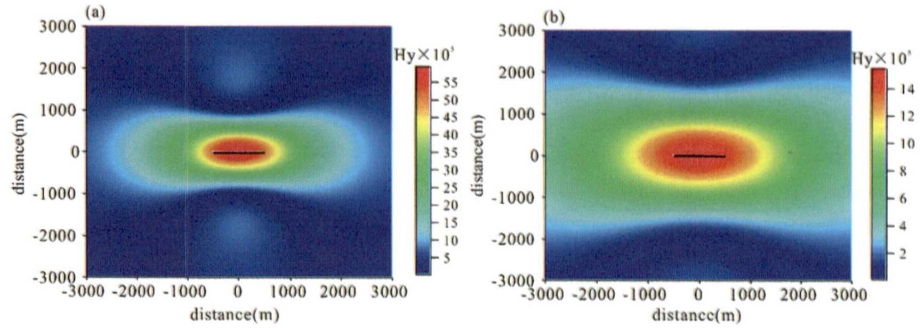

Fig. 2.14 Diffusion plane diagram of H_y at different times. **a** Time = 1 ms, **b** time = 10 ms

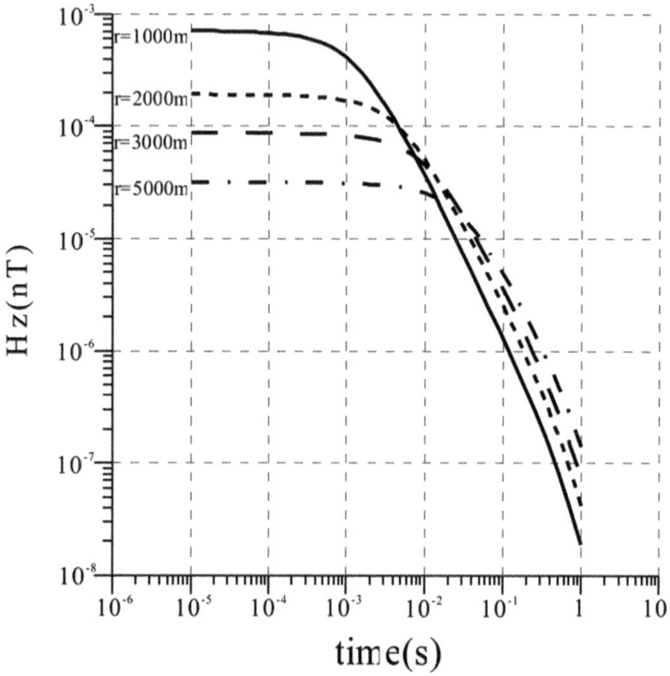

Fig. 2.15 Decay curve of H_z at different offset distances

do not diffuse outward over time, which is similar to the properties exhibited by E_x. The most obvious feature of the curves in Fig. 2.9 is that each decay curve has a sudden drop to a minimum point between 1 and 10 ms, and this phenomenon does not occur in other geological models (half-space with high-resistivity thin layer), so it can be inferred that the low-resistivity thin layer in the model is the cause of the rapid signal decay. The sensitive and dramatic response of E_y to the low-resistivity thin layer can be utilized. The plane distribution diagram shown in Fig. 2.10 shows that the

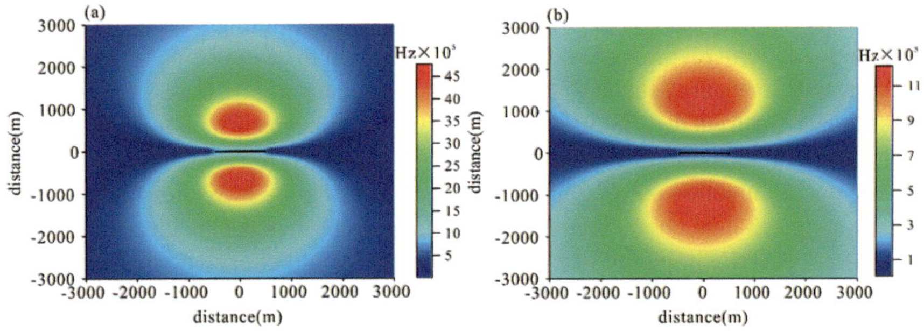

Fig. 2.16 Diffusion plane diagram of H_z at different times. **a** Time = 1 ms, **b** time = 10 ms

field value of E_y is extremely unevenly distributed, with extreme values concentrated in the four quadrants. The horizontal electric field E_y is only related to the source term, so according to the principle of vector superposition of electromagnetic fields, the field in the equatorial direction towards $\phi = 90°$ and the axial direction $\phi = 0°$ is zero due to the two electrodes of opposite polarity, and the field value is maximum at $\phi = 45°$.

The horizontal magnetic field H_x is also only related to the source term, so the field distribution is the same as E_y. Very similar. However, as can be seen from Fig. 2.6, as the offset distance increases, H_x the strength of the early signal decreases while the strength of the late signal increases, indicating that H_x the extreme values are gradually spreading out over time, which is different from E_y.

The calculation results of the horizontal magnetic field H_y show that within the calculation time range, H_y has positive and negative field quantities in both directions. The early path field value is negative, and the late path field value is positive (Fig. 2.13). From the plane distribution Fig. 2.14, it can also be seen that the positive horizontal magnetic field is concentrated near the transmitter, which is because H_y is excited by the positive induced current in the transmitter that is consistent with the current direction; while the negative horizontal magnetic field is concentrated at a certain distance from the transmitter on both sides, this is because the H_y here is induced by the negative "return current". H_y The field spreads over time, but the extreme values are always concentrated at the line source, which is because the extreme values of the vertical induced current that generates the horizontal magnetic field are spreading vertically downward along the plane where the transmitter is located.

The signal strength of the vertical magnetic field H_z decreases with the increase of offset distance in the early stage, and increases with the increase of offset distance in the late stage (Fig. 2.15). Figure 2.16 shows that the extreme values of H_z are distributed on both sides of the transmitter and move gradually away from the transmitter with time along the vertical line of the transmitter.

By analyzing and comparing the above five components, it can be found that E_x and H_z these two components are relatively evenly and simply distributed in the plane, especially in the direction along the transmitter wire and its vertical line, which is very important for ensuring that the response amplitudes measured at each measurement point on the same measurement line tend to be uniform. The responses of E_y and H_x are only related to the power source, causing the field distribution to show a clear quadrant distribution, which is extremely uneven in the direction parallel to the transmitter. H_y is due to the effect of the return current, there are two fields with opposite directions throughout the period, causing the measurement response to have a sign change phenomenon. Therefore, in order to make the electromagnetic response observed simple and easy to handle, we should choose components with uniform and simple field distribution as much as possible.

2.5.3 *Transient Response of Electrical Sources in Typical Geoelectric Sections*

In various studies of electromagnetic sounding in one-dimensional layered earth, according to the size relationship between the resistivities of each layer, there are six typical resistivity distribution structures for two-layer and three-layer geological structures (Cheng, 2007). Among them, the two-layer geological structure is divided into two types according to the relationship between the resistivities of the upper and lower strata: low-resistivity basement type (D type, $\rho_1 > \rho_2$) and high-resistivity basement type (G type, $\rho_1 < \rho_2$); the three-layer geological structure is divided into four types, namely low-resistivity interlayer type (H type, $\rho_1 > \rho_2 < \rho_3$), high-resistivity interlayer type (K type, $\rho_1 \langle \rho_2 \rangle \rho_3$), increasing resistance stratum type (A type, $\rho_1 < \rho_2 < \rho_3$), and decreasing resistance stratum type (Q type, $\rho_1 > \rho_2 > \rho_3$), their resistivity distribution is shown in Fig. 2.19. These six geoelectric structures plus the uniform half-space model can represent many simplified typical actual stratum structures. According to the previous analysis, the horizontal electric field E_x and the vertical magnetic field H_z In practical measurement work, the advantages are significant. This article only calculates the transient responses generated by these two components on the surfaces of the above six types of geoelectric structures, and studies the laws of the influence of changes in parameters such as formation resistivity and thickness on the transient field (Fig. 2.17).

In all the following calculation results, the source parameters and the position of the receiving point are the same, that is, the source length is 1 km, the emission current is 10 A, and the receiving point is (0, 1000). During the calculation, when the thickness of the middle layer changes, the ratio of the resistivity of the cover layer to the resistivity of the middle layer (i.e., ρ_1/ρ_2) is fixed at 10 or 0.1; when the resistivity of the middle layer changes, the thickness of the cover layer and the middle layer are fixed at 500 and 100 m respectively.

Figure 2.18 is the calculation result of D-type formation E_x and H_z, where a and c are the response curves when the resistivity ratio of the base to the cover layer changes continuously, and b and d are the response curves when the thickness of the cover layer changes continuously. It can be seen that with the change of formation

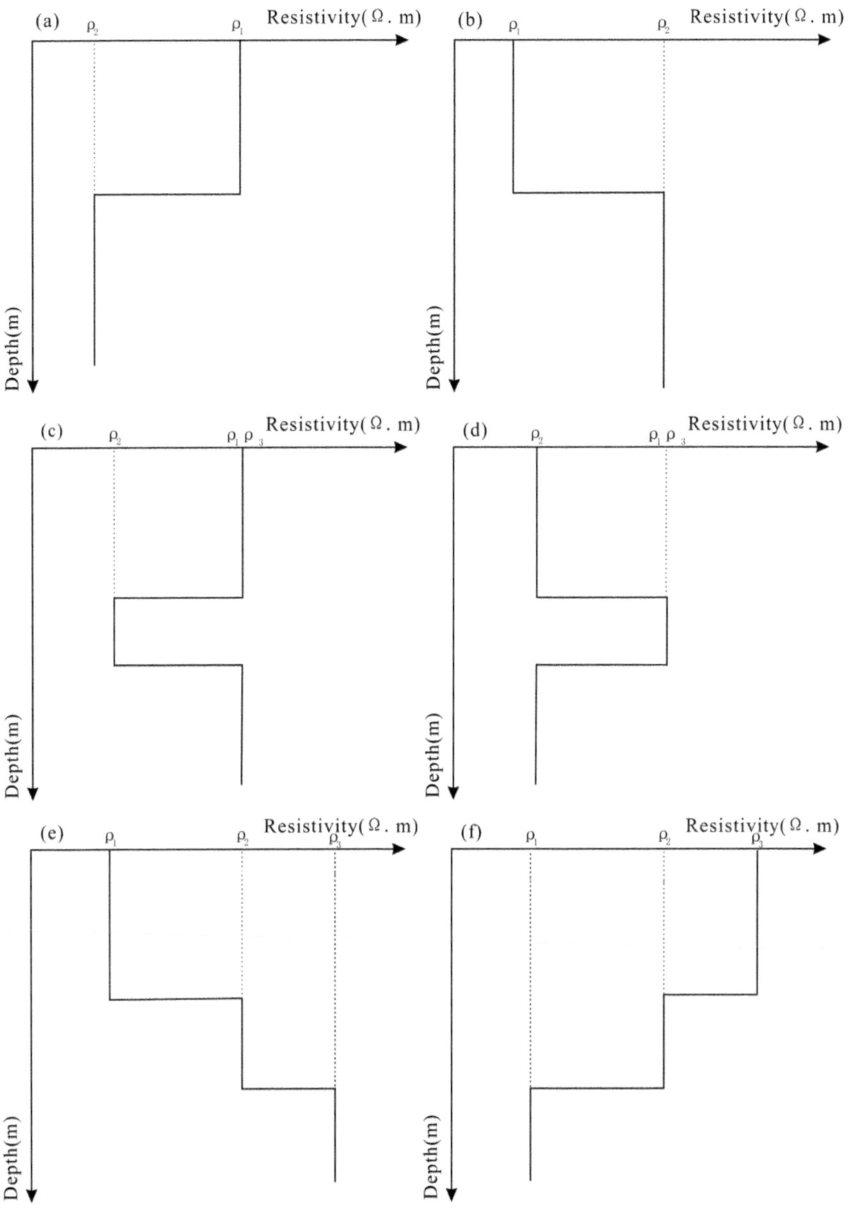

Fig. 2.17 Schematic diagram of resistivity distribution of typical geoelectric structures. **a** D-type formation, **b** G-type formation, **c** H-type formation, **d** K-type formation, **e** A-type formation, **f** Q-type formation

parameters, H_z the curve shape changes to a greater extent than E_x; in addition, H_z the change in the curve shape only occurs after a period of time, and the curves before this time overlap each other, while E_x the change in the curve shape occurs throughout the time period. H_z The change characteristics of the curve are easy to understand. The electromagnetic field must pass through a certain period of time before it can penetrate the cover layer and detect the underlying formation. Therefore, the early stage of the curve carries the information of the constant resistivity cover layer. And E_x the component contains information carried by the return current from the deep part at an early stage, so even within a very early time range, the E_x response generated by models with the same surface resistivity is not the same.

In the three-layer model, with the change of the parameters of the middle layer, the H-type and Q-type formations with a relatively low-resistivity middle layer have a

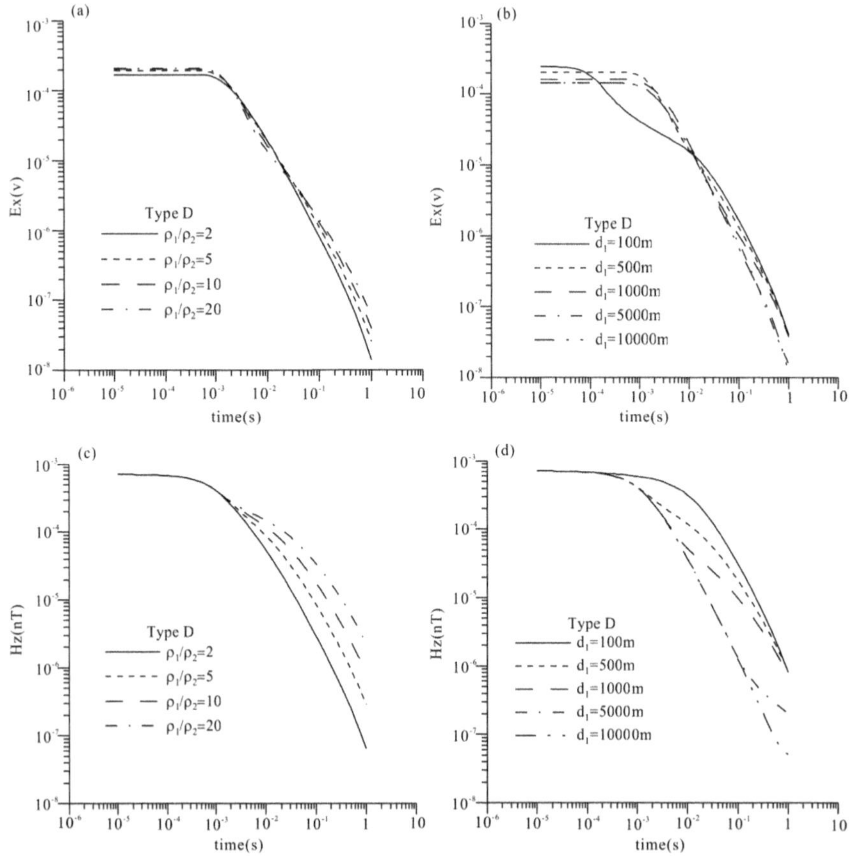

Fig. 2.18 D-type formation E_x, H_z response curve. **a** D-type E_x response curve when the base resistivity is different, **b** D-type E_x response curve when the cover layer thickness is different, **c** D-type H_z response curve when the base resistivity is different, **d** D-type H_z response curve when the cover layer thickness is different

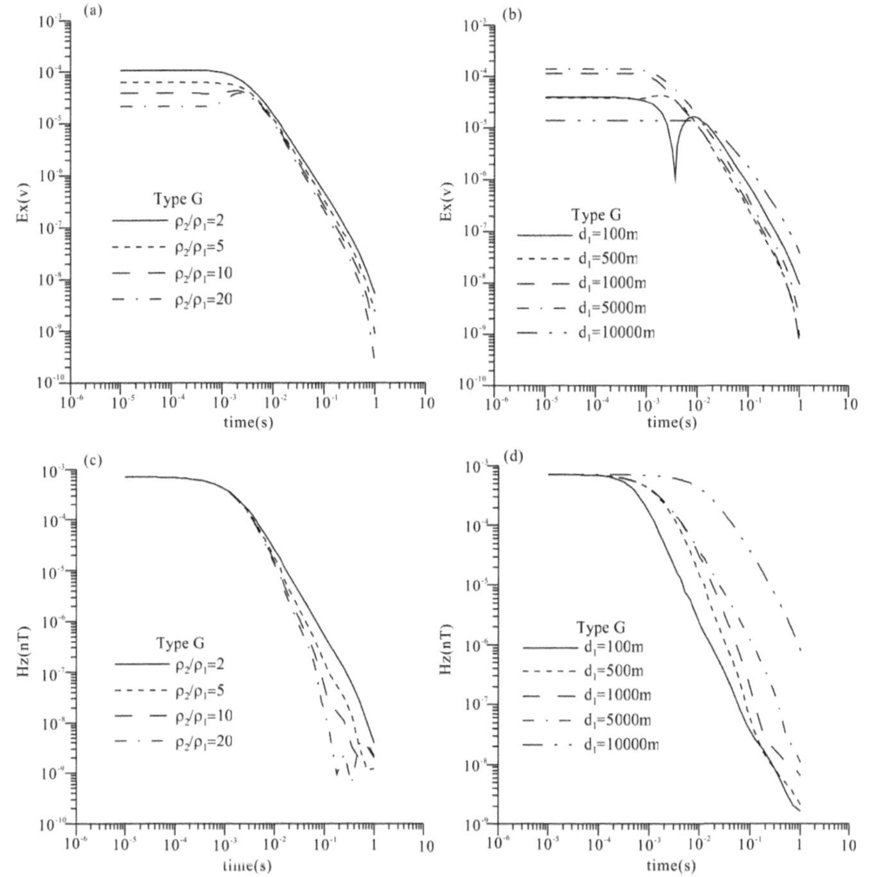

Fig. 2.19 G-type formation E_x, H_z response curve. **a** G-type E_x response curve when the base resistivity is different, **b** G-type E_x response curve when the cover layer thickness is different, **c** G-type H_z response curve when the base resistivity is different, **d** G-type H_z response curve when the cover layer thickness is different

larger change in the H_z response curve, while the K-type and A-type formations with a relatively high-resistivity middle layer have a larger change in the E_x response curve. This confirms the previous analysis that the electric field component is more sensitive to high-resistivity bodies, and the magnetic field component is more sensitive to low-resistivity bodies. The calculation results show that for different types of formations, it is necessary to select the appropriate component for detection (Figs. 2.19, 2.20, 2.21, 2.22, 2.23).

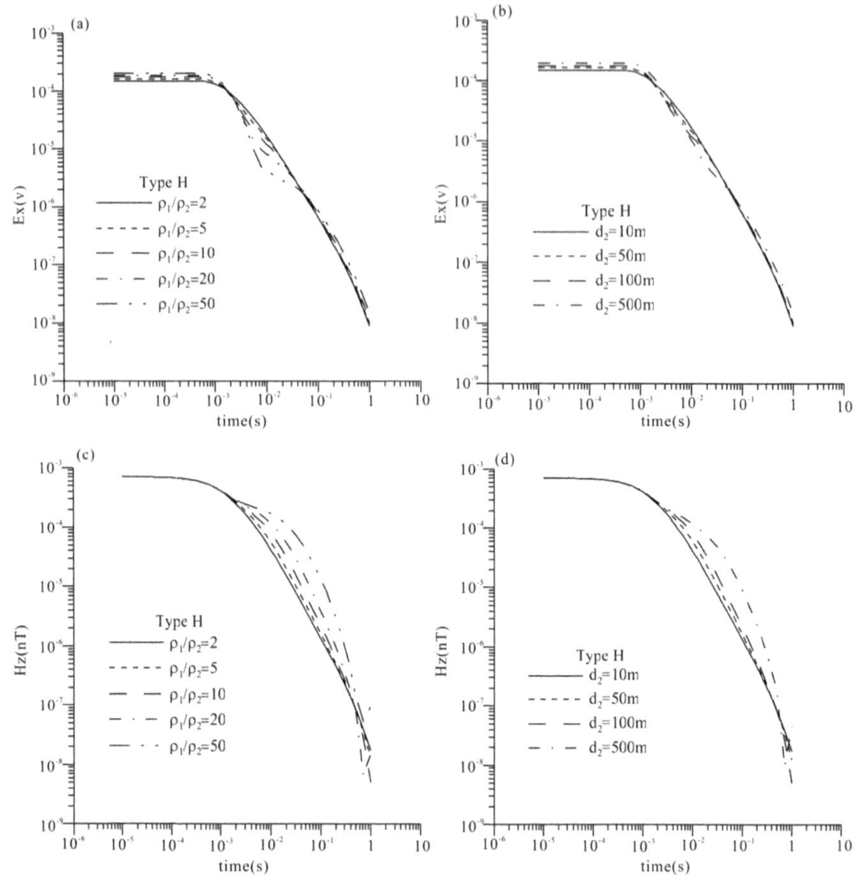

Fig. 2.20 H-type stratum E_x, H_z response curve. **a** H-type response curve when the resistivity of the middle layer varies, **b** H-type response curve when the thickness of the middle layer varies E_x, **c** H-type response curve when the resistivity of the middle layer varies, **d** H-type response curve when the thickness of the middle layer varies H_z

Simulations were carried out on the horizontal electric field and vertical magnetic field derivative components in the uniform half-space medium. The model parameters are: transmitter length 1000 m, current 10 A, observation area is 4000 m × 5000 m. The changes in the dH_z/dt, E_x component response [taking the response logarithm log10(dH_z/dt), log10(E_x)] over time are extracted, as shown in Fig. 2.24.

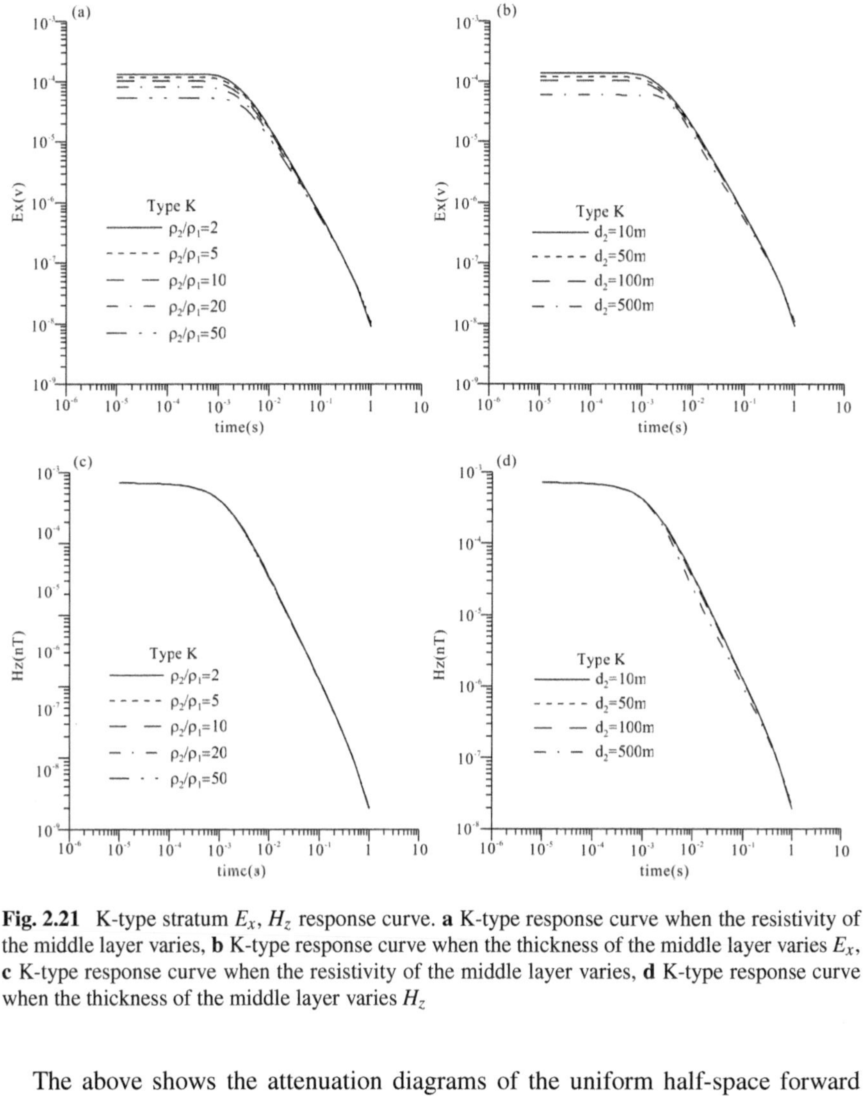

Fig. 2.21 K-type stratum E_x, H_z response curve. **a** K-type response curve when the resistivity of the middle layer varies, **b** K-type response curve when the thickness of the middle layer varies E_x, **c** K-type response curve when the resistivity of the middle layer varies, **d** K-type response curve when the thickness of the middle layer varies H_z

The above shows the attenuation diagrams of the uniform half-space forward response signals for each component (Figs. 2.24 and 2.25). It can be seen that the signal received on the ground gradually attenuates over time: on the time scale, as time passes, the signal strength received on the ground gradually attenuates, and the maximum amplitude of the dH_z/dt component at 0.01 ms and 1 s are $10^{-0.8}$ and $10^{-4.8}$ respectively, a difference of 4 orders of magnitude, while the maximum and minimum amplitudes of the electric field E_x component within these two times are $10^{-2.9}$ and 10^{-5}, a difference of 2 orders of magnitude, indicating that the amplitude of the dH_z/dt component signal attenuates faster; on the spatial scale, it can also be found that the propagation speed of the dH_z/dt is faster than that of the E_x component.

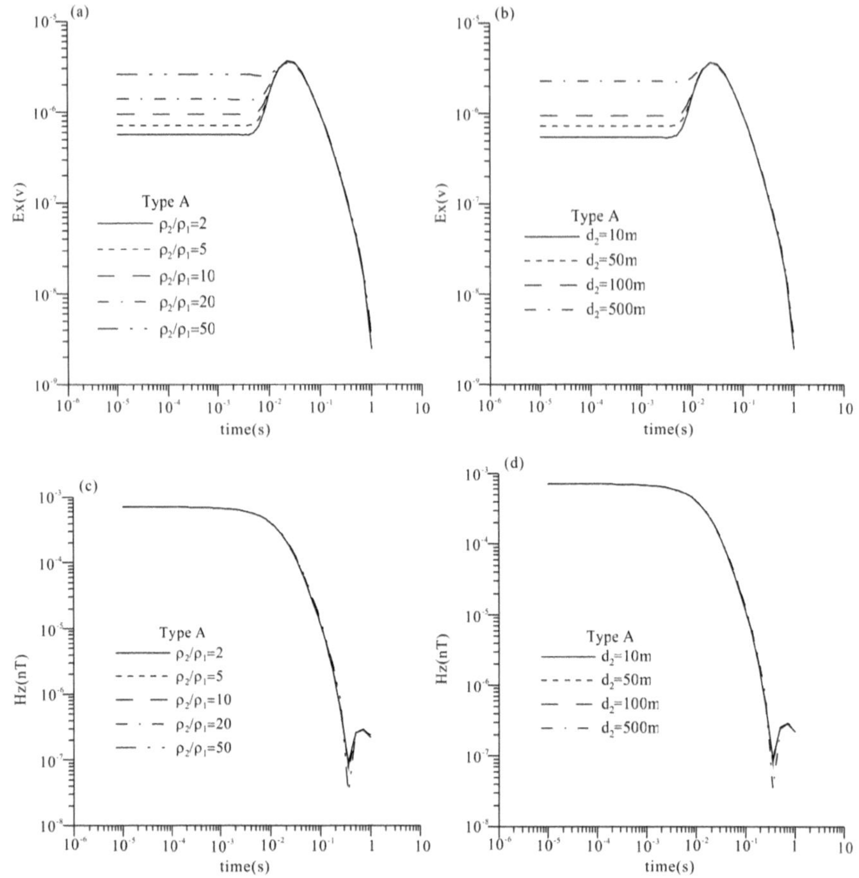

Fig. 2.22 A-type stratum E_x, H_z response curve. **a** A-type response curve when the resistivity of the middle layer varies, **b** A-type response curve when the thickness of the middle layer varies E_x, **c** A-type response curve when the resistivity of the middle layer varies, **d** A-type response curve when the thickness of the middle layer varies H_z

To further discuss the time variation rules of the high and low resistance anomaly responses of SOTEM for one-dimensional layered media, the time–space distribution of anomaly responses, and to deepen the understanding of the resolution of commonly used components of SOTEM for high and low resistance strata, the best observation areas of each component, this paper simulates the rules of response changes with offset distance and anomaly depth for uniform half-space, H-type and K-type models, as follows.

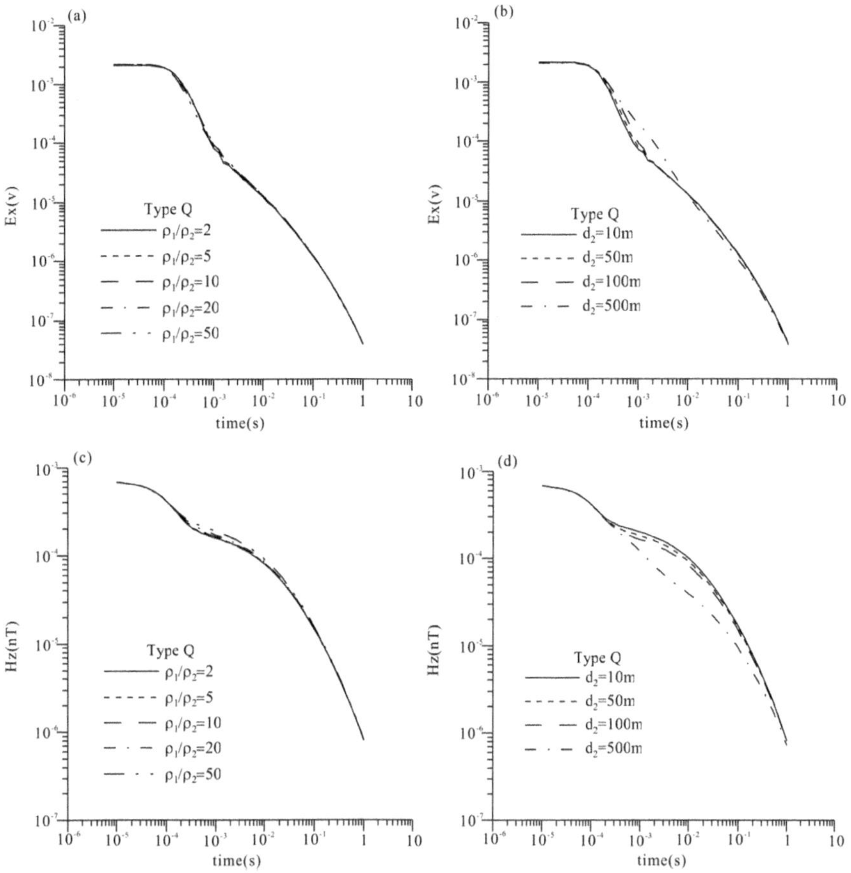

Fig. 2.23 Q-type stratum E_x, H_z response curve. **a** Q-type response curve when the resistivity of the middle layer varies, **b** Q-type response curve when the thickness of the middle layer varies E_x, **c** Q-type response curve when the resistivity of the middle layer varies, **d** Q-type response curve when the thickness of the middle layer varies H_z

As shown in Fig. 2.26, for the uniform half-space model, the signal amplitudes of both the electric field E_x and the magnetic field dH_z/dt decrease with the increase of offset distance. Within the observation range (100–2500 m), E_x attenuates by more than 2 orders of magnitude, while dH_z/dt attenuates by nearly 4 orders of magnitude, indicating that the magnetic field is more affected by the offset distance, and the signal space attenuates faster, which indirectly shows that the dH_z/dt component has a larger volume effect. The dH_z/dt component signal changes throughout the sampling time range (0.00001–1 s), while the E_x component only affects the early signal range within the entire observation range, and the signal overlaps and no longer changes around 0.1 s, indicating that the offset distance has a broader impact on the magnetic field signal and a more concentrated impact on the electric field signal.

Fig. 2.24 Planar distribution of ground E_x intensity at different times after SOTEM source shutdown. **a** t = 0.01 ms, **b** t = 0.3162 ms, **c** t = 31.62 ms, **d** t = 1000 ms

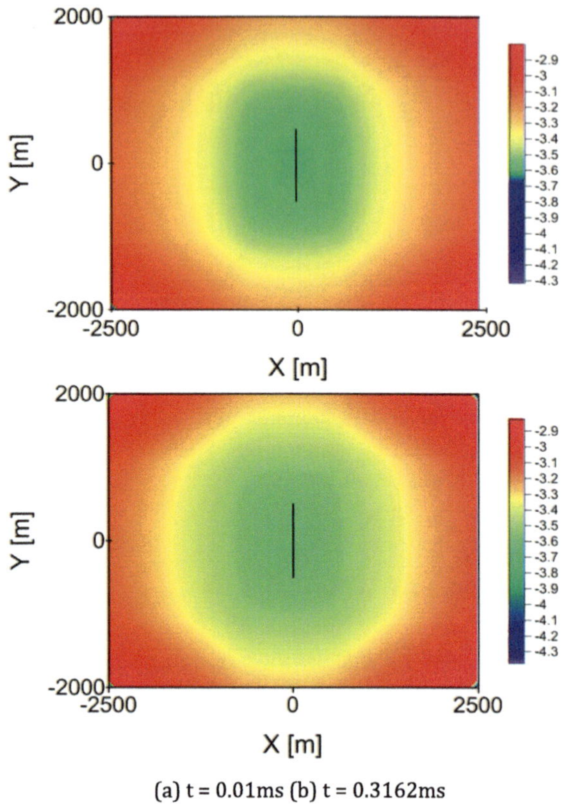

As shown in Fig. 2.27, for low-resistivity anomalies (H model), both the signal amplitudes of E_x and dH_z/dt decrease with the increase of offset. From the relative anomaly curve, it can be seen that the two component anomaly signals have different time channel distributions. The magnetic field mainly appears between 0.001 and 0.1 s, while the electric field is between 0.00001 and 0.1 s, with a wider anomaly signal distribution. The maximum anomaly amplitude of the magnetic field is greater than the E_x component, that is, both the magnetic field and the electric field are sensitive to low-resistivity bodies, but the magnetic field is slightly superior. For the K-type model (Fig. 2.28), the anomaly signal changes are more complex, the E_x component attenuation curve shows a process of first rising and then falling, which will occur in actual exploration. Similarly, the dH_z/dt anomaly signal only appears between 0.001 and 0.05 s, while the electric field changes before 0.01 s, and the amplitude is greater than the magnetic field signal, especially for large offsets, indicating that for high-resistivity bodies, the E_x component has more advantages.

Fig. 2.24 (continued)

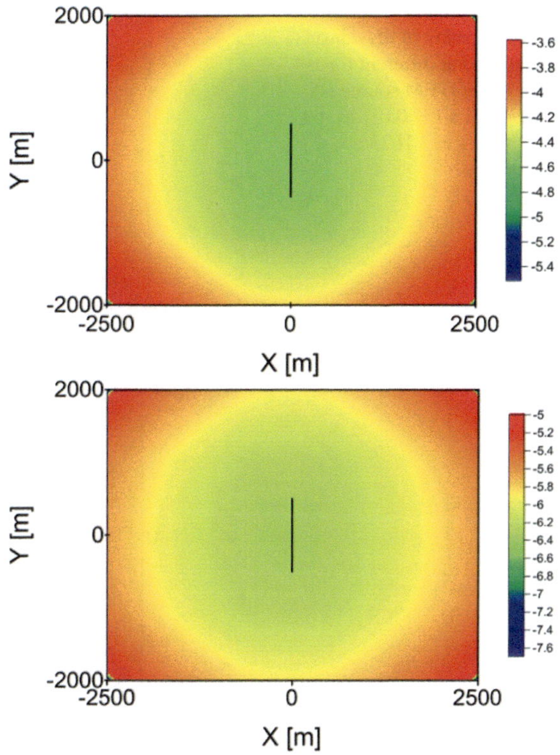

(c) t = 31.62ms (d) t = 1000ms

The following shows the pattern of observed signals changing with the depth of the anomaly layer.

As can be seen from Figs. 2.29 and 2.30, as the depth of the anomaly layer increases, the amplitude of the anomaly signal gradually decreases. For the H-type model, as the depth of the low-resistivity anomaly layer increases, the signals of E_x and dH_z/dt change accordingly. The response to shallow anomalies is very obvious, but as the depth increases, the response amplitude decays rapidly; for the K-type model, dH_z/dt Only when the burial depth is shallow, there is a significant signal change, but as the depth of the high-resistivity anomaly layer increases, the anomaly signal decreases sharply and is not clearly distinguishable. However, for the E_x component, the signal amplitude changes with the depth, indicating that it has a strong ability to distinguish high-resistivity bodies. In Fig. 2.30b, the shallow high-resistivity signal negative anomaly causes relative anomaly distortion.

Fig. 2.25 Planar distribution of the intensity of dH_z/dt on the ground at different times after the SOTEM source is turned off. **a** t = 0.01 ms, **b** t = 1 ms, **c** t = 3.1623 ms, **d** t = 1000 ms

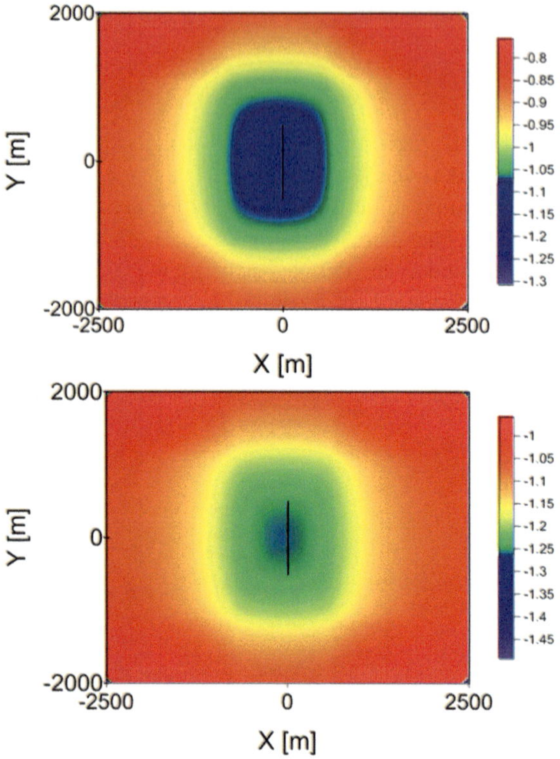

a.t = 0.01ms (b) t = 1ms

The following continues to give the influence of the SOTEM two-component response to the change in resistivity of the abnormal high and low resistance strata. The abnormal layer burial depth is fixed, and the abnormal layer resistivity is gradually changed. The response curve is shown in the figure below.

Fig. 2.25 (continued)

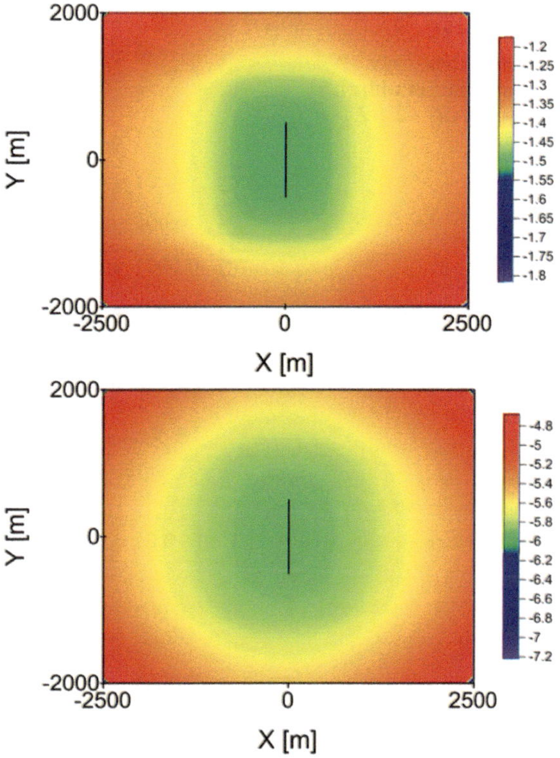

(c) t = 3.1623ms (d) t = 1000ms

From Figs. 2.31 and 2.32, for the H model, the dH_z/dt component is more sensitive to the resistivity response change of the low-resistivity layer, while the E_x component is quite obvious for the response change of the low and high resistance layers. For the K model, the magnetic field anomaly response signal does not show much difference, that is, the dH_z/dt component is not sensitive to the resistivity response change of the high-resistivity layer. The above once again proves the detection advantage of E_x for the high-resistivity layer.

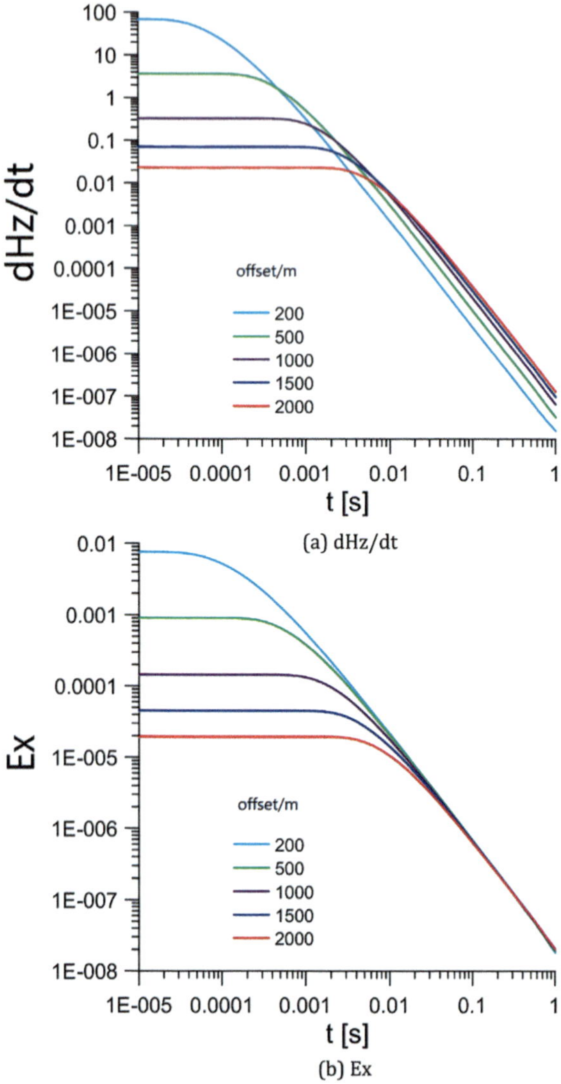

Fig. 2.26 The response of dH_z/dt and E_x components in uniform half-space medium changes with offset distance. **a** dH_z/dt, **b** E_x

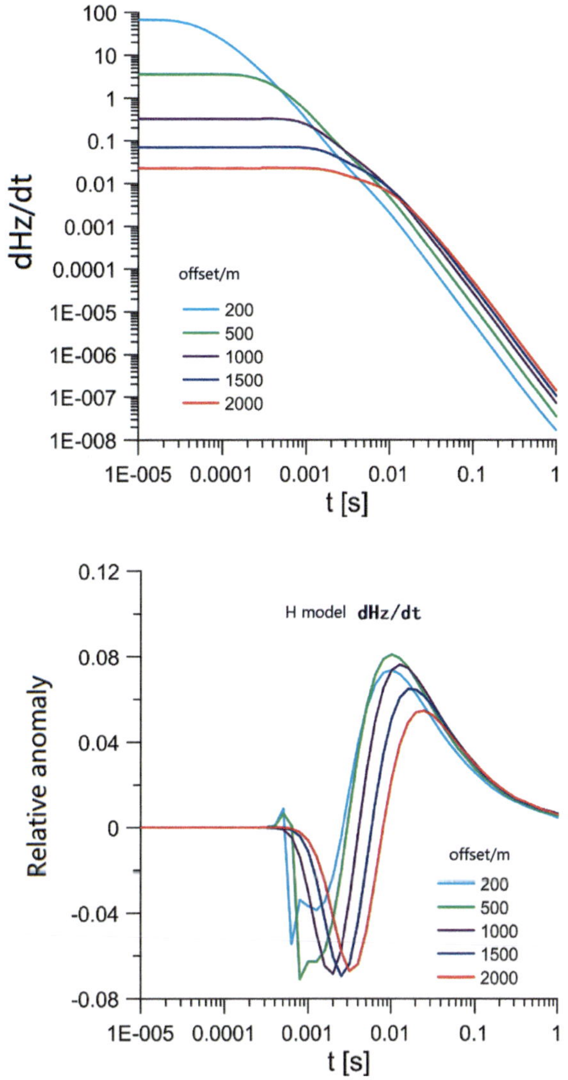

(a) $\mathrm{d}H_z/\mathrm{d}t$

Fig. 2.27 H-type model $\mathrm{d}H_z/\mathrm{d}t$ and E_x component response and relative anomaly variation curve with offset. **a** $\mathrm{d}H_z/\mathrm{d}t$, **b** E_x

Fig. 2.27 (continued)

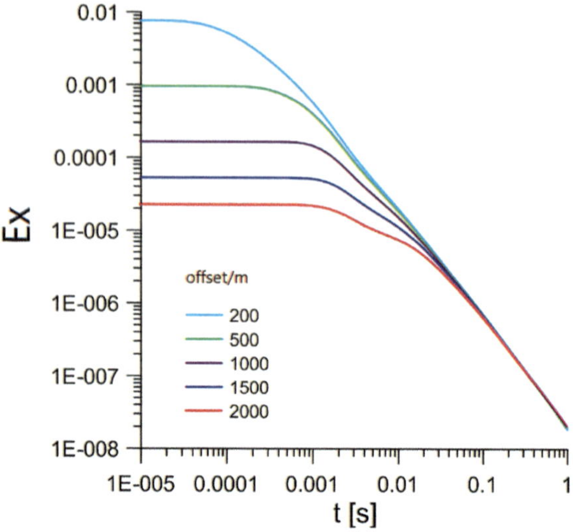

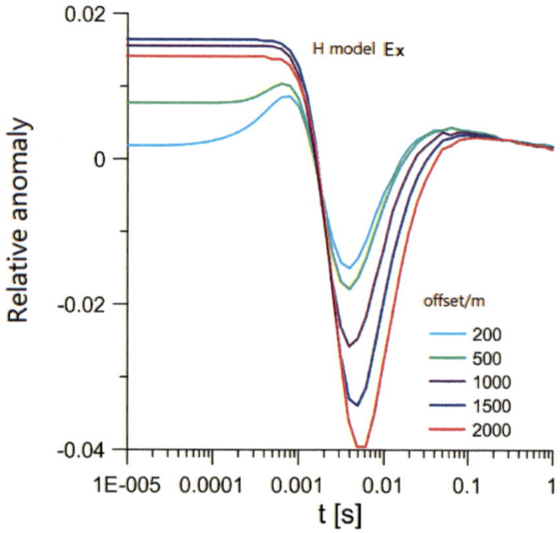

(b) E_x

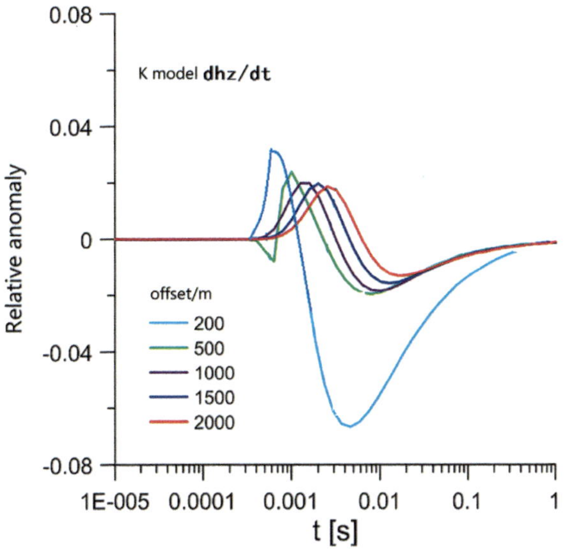

(a) dH_z/dt

Fig. 2.28 K-type model dH_z/dt and E_x component response and relative anomaly variation curve with offset. **a** dH_z/dt, **b** E_x

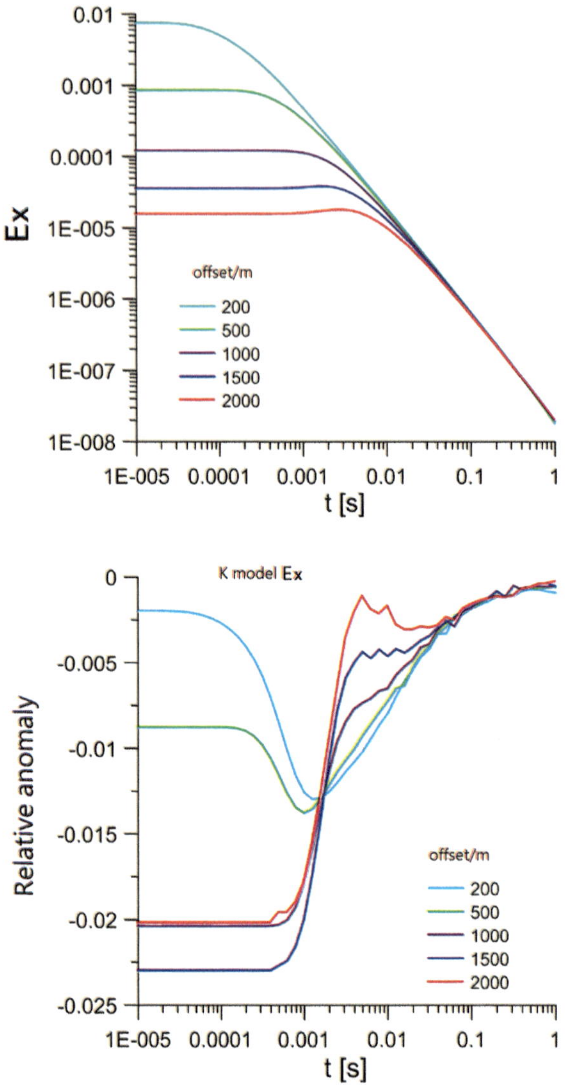

(b) E_x

Fig. 2.28 (continued)

Fig. 2.29 H-type model dH_z/dt and E_x component response and relative anomaly variation curve with anomaly layer depth. **a** dH_z/dt, **b** E_x

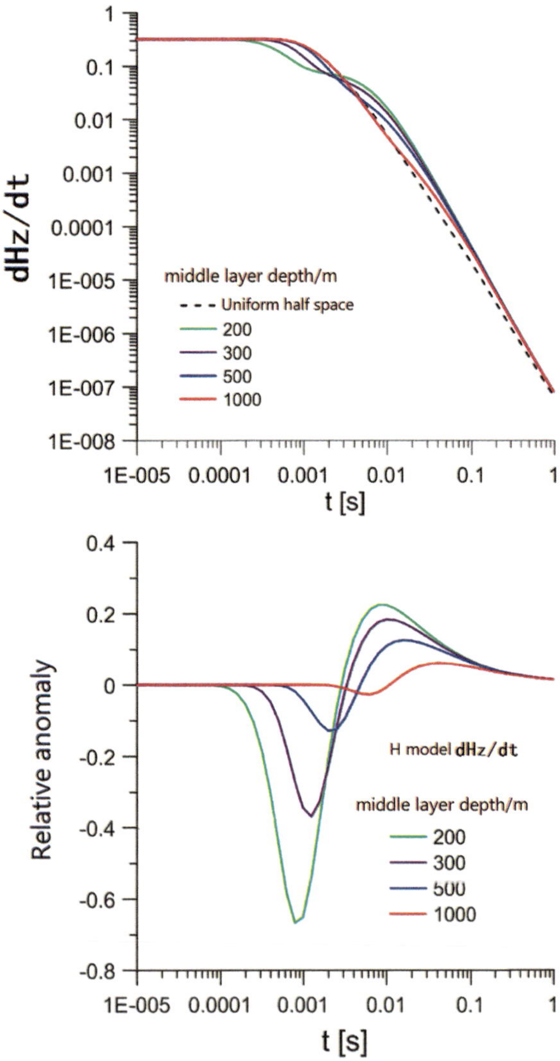

(a) dH_z/dt

Fig. 2.29 (continued)

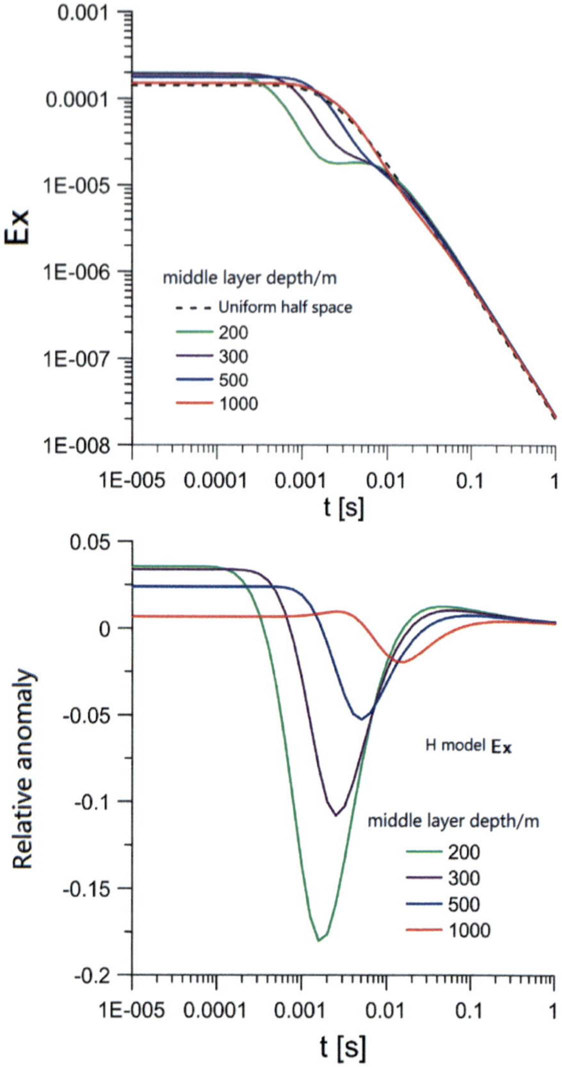

(b) E_x

Fig. 2.30 K-type model dH_z/dt and E_x component response and relative anomaly variation curve with anomaly layer depth. **a** dH_z/dt, **b** E_x

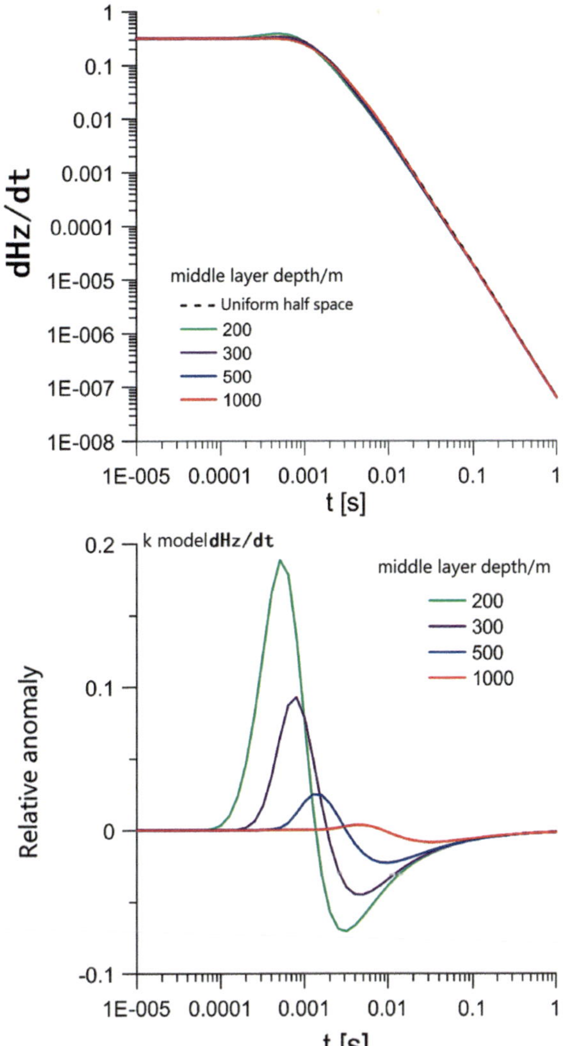

a.d H_z /d t

Fig. 2.30 (continued)

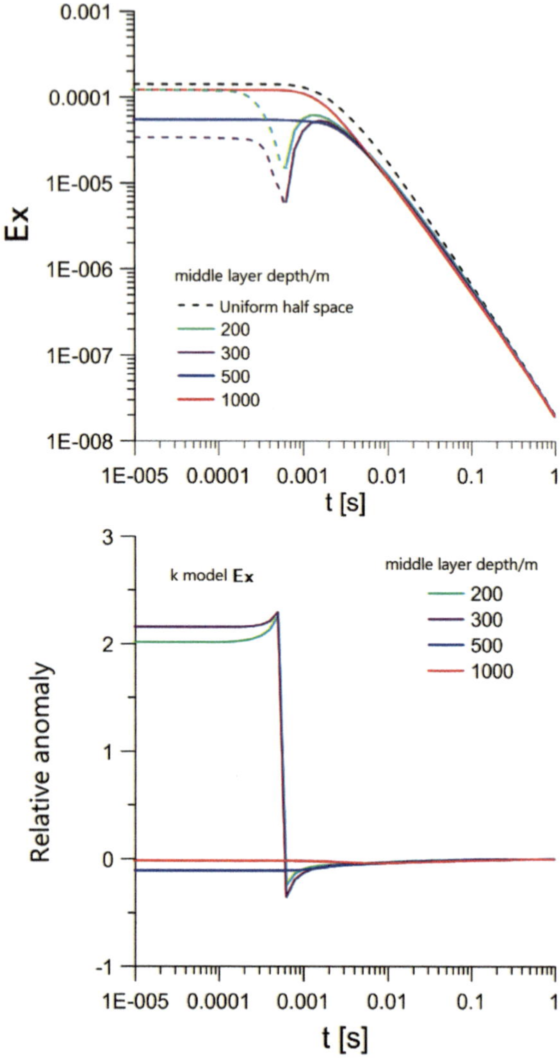

(b) E_x

Fig. 2.31 H model E_x and dH_z/dt component response and relative anomaly with resistivity change curve. **a** dH_z/dt, **b** E_x

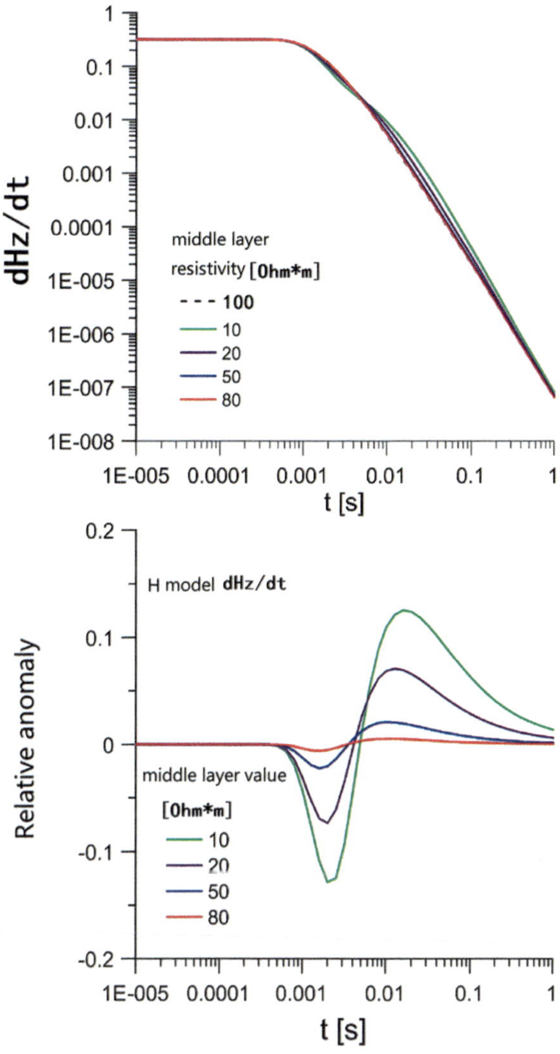

(a) dH_z/dt

Fig. 2.31 (continued)

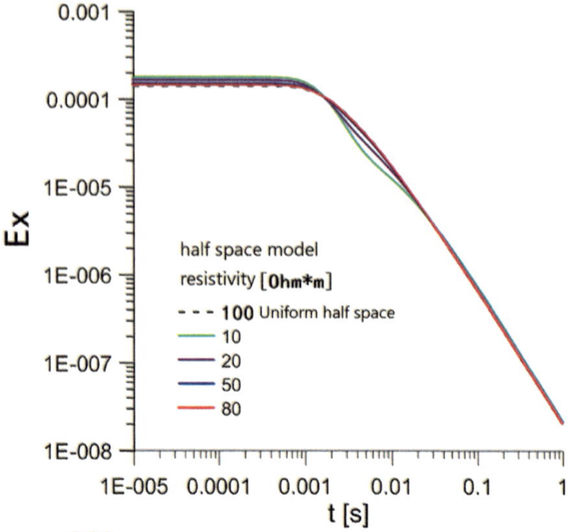

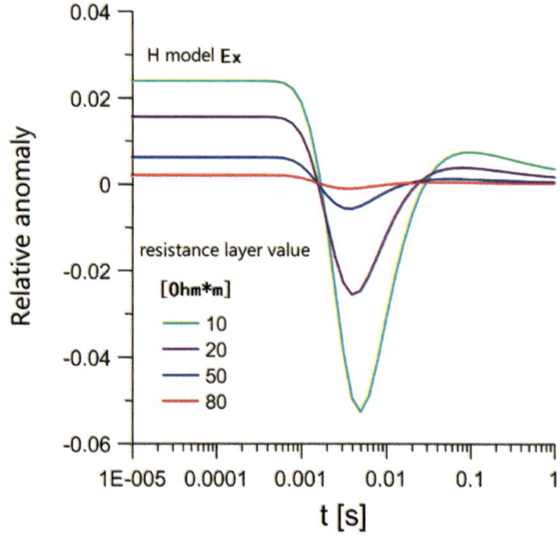

(b) E_x

Fig. 2.32 K model E_x and dH_z/dt component response and relative anomaly with resistivity change curve. **a** dH_z/dt, **b** E_x

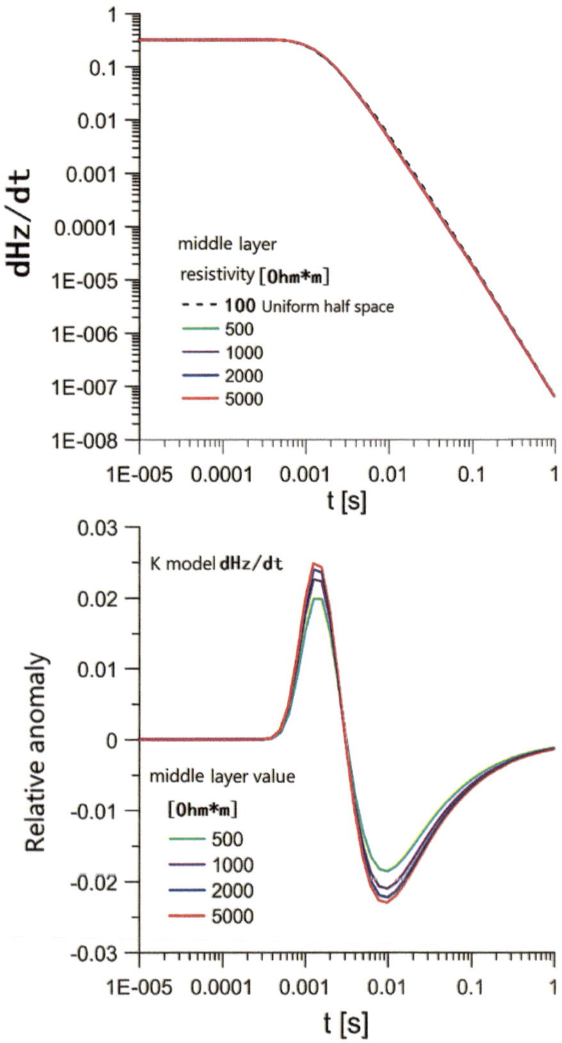

a.dH_z/dt

Fig. 2.32 (continued)

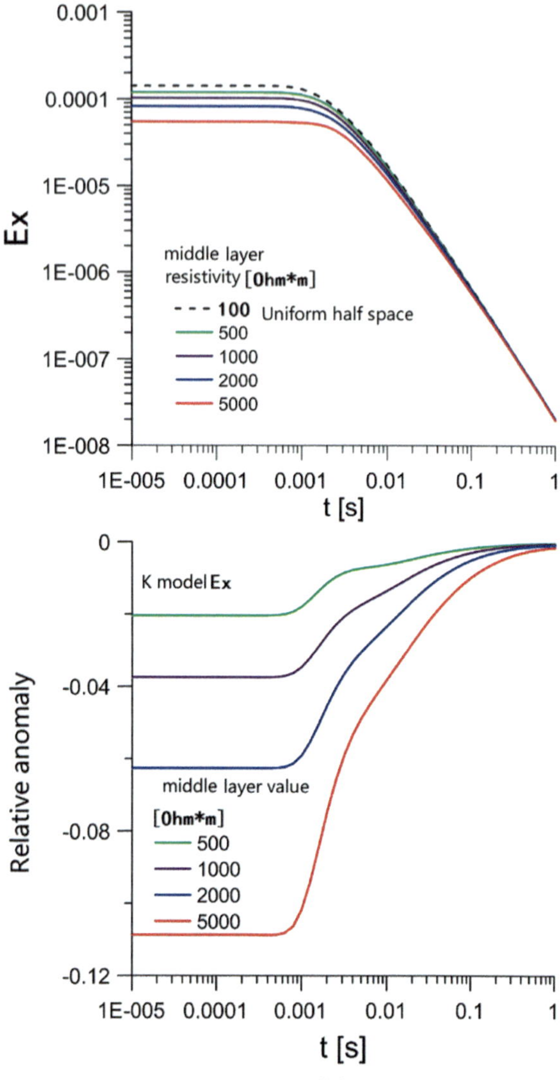

(b) E_x

References

Cagniard L. 1953. Basic theory of the magnetotelluric method of geophysical prospecting. Geophysics, 18(1): 605–635.

Cheng Zhiping. Electrical Prospecting Course[M]. Beijing: Metallurgical Industry Press, 2007.

Di Q, Wang R, et al. 2008. Forward modeling, inversion and application of Controlled Source Audio Magnetotelluric data (in Chinese). Science Press

Goldstein M A, Strangway D W. 1975. Audio-frequency magnetotellurics with a grounded electric dipole source. Geophysics, 40(1): 669–683.

Kaufman A A, Keller G V. 1981. The magnetotelluric sounding method. Amsterdam- Oxford-New York: Elsevier scientific publishing company.

Kaufman A A, Keller G V. 1983. Frequency and transient soundings. Amsterdam- Oxford-New York: Elsevier scientific publishing company.

Nabighian M N. 1979. Quasi-static transient response of a conducting half-space -an approximate representation. Geophysics, 44: 1700–1705

Nabighian M N. 1991. Electromagnetic Methods in Applied Geophysics: Volume 2, Application, Parts A and B. Society of Exploration Geophysicists

Newman G A, Anderson W L, Hohmann G W. 1989. Effect of conductive host on borehole transient electromagnetic responses. Geophysics, 54(5): 598–608.

Ruan Baiyao, Xu Shizhe. Finite difference solution of line source two-dimensional time domain electromagnetic response. Journal of Geology in Higher Education, 1996, 2(4): 437-447.

Strack K M, Lüschen E, Köta A W. Long-offset transient electromagnetic (LOTEM) depth soundings applied to crustal studies in the Black Forest and Swabian Alps, Federal Republic of Germany. Geophysics , 1990, 55(7): 834–842

Tikhonov A. 1950. On Determining Electrical Characteristics of the Deep Layers of the Earth's Crust. Doklady, 73: 295–297

Yan Liangjun, Hu Wenbao, Chen Qingli, et al. Application effect of long offset transient electromagnetic depth measurement method in confirming local structure in carbonate rock covered area[J]. Seismology and Geology, 2001, 23(2): 271-286.

Yan Shu, Chen Mingsheng, Fu Junmei. Direct time domain numerical analysis of transient electromagnetic field[J]. Chinese Journal of Geophysics, 2002, 45(2): 275-284.

Chapter 3
Three-Dimensional Forward Numerical Simulation of Transient Electromagnetic Response from Electrical Sources

One-dimensional forward calculations can yield exact solutions for electromagnetic response; however, the range of electromagnetic field problems that can be effectively characterized by analytical solutions is quite limited. This limitation does not suffice for the numerical simulation of complex geological scenarios. Additionally, layered models are only appropriate for simple stratigraphic structures. To analyze local anomalies and complex target bodies, it is essential to employ three-dimensional numerical simulation methods. These methods allow for the establishment of models tailored to various geological conditions, thereby providing electromagnetic responses that better reflect actual conditions.

Many geophysicists are dedicated to the three-dimensional forward calculation of transient electromagnetics. The primary methods utilized include the integral equation method, time-domain finite difference method (Wang and Hohmann 1993; Feng 2013), finite element method (Kuo and Cho 1980; Um et al. 2010; Hui 2011; Hua 2015; Jun 2016), and the SLDM algorithm (Druskin et al. 1988; Zaslavsky et al. 2011), among others. This text will primarily introduce the principles and solution processes of the time-domain three-dimensional vector finite element method and the three-dimensional time-domain finite difference method, laying the groundwork for high-dimensional simulation methods in subsequent chapters.

G. Xue et al., *Exploration with Short-Offset Grounded-Wire Transient Electromagnetic Method*, https://doi.org/10.1007/978-981-96-8846-3_3

3.1 Time-Domain Three-Dimensional Vector Finite Element Calculation

The finite element method (FEM) is a numerical technique that begins with the governing equations and boundary conditions. It divides the computational domain into multiple elements and discretizes the boundary value problem of partial differential equations into a set of finite element equations using interpolation within each element, ultimately leading to a solution (Ming 1998). This method was initially proposed by Turner and Clough to address problems in elasticity mechanics (Turner et al. 1956). Coggon (1971) derived the electromagnetic variational equation using the principle of minimum energy, marking the introduction of FEM into geophysics. Kuo and Cho (1980) were the first to apply the three-dimensional time-domain vector finite element method to simulate the response of low-resistance ore bodies beneath low-resistance shielding layers. Since then, significant research and analysis on the three-dimensional time-domain finite element method have been conducted by geophysicists both domestically and internationally, yielding fruitful results (Biao et al. 2014).

Generally, the finite element method is categorized into two types: the scalar finite element method and the vector finite element method. The scalar finite element method, also known as the node finite element method, assigns degrees of freedom to each node, resulting in a single value on the common edge of adjacent elements. However, when medium discontinuities occur in the computational area, this method fails to satisfy the continuity conditions of the electromagnetic field, leading to the phenomenon of "pseudo-solutions." Conversely, the vector finite element method, or edge finite element method, assigns degrees of freedom to each edge, automatically ensuring the continuity of the tangential components of the electric and magnetic fields while accommodating discontinuities in the normal direction. This effectively avoids the "pseudo-solution" phenomenon. As illustrated in the element partitioning diagram of the two methods (3.1), the vector finite element method has 12 unknowns per element, whereas the scalar finite element method has 24. Thus, the vector finite element method has fewer unknowns, significantly reducing the computational storage requirements and enhancing efficiency.

The calculation methods for three-dimensional transient electromagnetic vector finite element forward modeling can be divided into two categories: the indirect method and the direct method. The indirect method first obtains the response in the frequency domain and then converts it to the time domain using time–frequency conversion techniques. However, this approach is limited by the accuracy of the time–frequency conversion method, which can lead to inaccuracies in the time-domain electromagnetic response and may not adequately simulate the emission waveform. In contrast, the direct method discretizes the time-dependent partial derivative terms using difference methods and directly solves the electromagnetic field in the time domain. This method allows for consideration of the excitation source's emission waveform, aligning closely with actual conditions and providing a more effective means of simulating and solving practical problems. The method employed in this

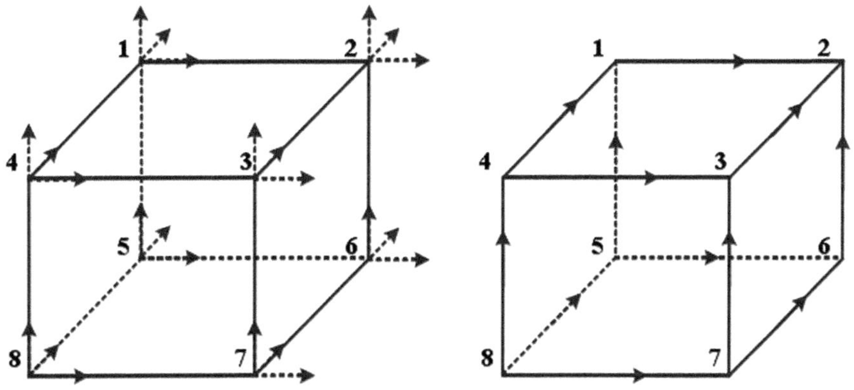

Fig. 3.1 Comparison of element partitioning between scalar finite element method and vector finite element method

paper is the direct time-domain solution of the three-dimensional vector finite element (Fig. 3.1).

3.1.1 Variational Solution

There are two primary approaches to deriving the finite element equation: the variational (or Ritz) method and the weighted residual (or Galerkin) method. In this section, we will first utilize the weighted residual method to define the residual r corresponding to the electric field control equation:

$$r = \nabla \times \nabla \times \mathbf{E} + \sigma \mu_0 \frac{\partial \mathbf{E}}{\partial t} + \mu_0 \frac{\partial \mathbf{J}_s}{\partial t} \tag{3.1}$$

let r satisfy in the calculation area Ω:

$$\int_{\Omega} f \cdot r d\Omega = 0 \tag{3.2}$$

where f represents the vector basis function. Substituting (3.1) into (3.2) gives:

$$\int_{\Omega} f \cdot \left(\nabla \times \nabla \times \mathbf{E} + \sigma \mu_0 \frac{\partial \mathbf{E}}{\partial t} + \mu_0 \frac{\partial \mathbf{J}_s}{\partial t} \right) d\Omega = 0 \tag{3.3}$$

According to the vector analysis identity:

$$\mathbf{B} \cdot (\nabla \times \mathbf{A}) = \mathbf{A} \cdot (\nabla \times \mathbf{B}) + \nabla \cdot (\mathbf{A} \times \mathbf{B}) \tag{3.4}$$

where, $\boldsymbol{A}$ and $\boldsymbol{B}$ are vectors. According to (3.4), we expand the first term of the integral in Eq. (3.3) into two terms as follows:

$$\int\limits_{\Omega} f \cdot (\nabla \times \nabla \times \mathbf{E}) d\Omega = \int\limits_{\Omega} (\nabla \times \mathbf{E}) \cdot (\nabla \times f) d\Omega + \int\limits_{\Omega} \nabla \cdot ((\nabla \times \mathbf{E}) \times f) d\Omega \tag{3.5}$$

According to Gauss's theorem:

$$\int\limits_{e} \nabla \cdot \mathbf{A} de = \oint_{\alpha} \mathbf{n}_{\alpha} \cdot \mathbf{A} d\alpha \tag{3.6}$$

where, $\boldsymbol{A}$ is a three-dimensional vector, α is the outer boundary of the calculation area, and n $_{\alpha}$ is the unit normal vector of the outer boundary α. Using (3.6), the volume integral of the second term on the right side of (3.5) is converted into a surface integral:

$$\int\limits_{\Omega} \nabla \cdot ((\nabla \times \mathbf{E}) \times f) d\Omega = \int\limits_{a} \mathbf{n}_{\alpha} \cdot ((\nabla \times \mathbf{E}) \times f) da \tag{3.7}$$

Then according to the formula:

$$\mathbf{A} \cdot (\mathbf{B} \times \mathbf{C}) = (\mathbf{A} \times \mathbf{B}) \cdot \mathbf{C} \tag{3.8}$$

And adding the boundary conditions at infinity, the final finite element equation is as follows:

$$\int\limits_{\Omega} \left[(\nabla \times f) \cdot (\nabla \times \mathbf{E}) + \sigma \mu_0 f \cdot \frac{\partial \mathbf{E}}{\partial t} \right] d\Omega + \int\limits_{a} \mathbf{n}_{\alpha} \cdot ((\nabla \times \mathbf{E}) \times f) da = - \int\limits_{\Omega} \mu_0 f \cdot \frac{\partial \mathbf{J}_s}{\partial t} d\Omega \tag{3.9}$$

When the calculation area Ω is sufficiently large, the electric field value at the boundary α can be considered negligible. Specifically, this means that the tangential component of the electric field at the boundary is zero.

3.1.2 Excitation Source and Boundary Conditions

(1) Excitation source

In transient electromagnetics with electrical source, the excitation of the electromagnetic field is achieved through an externally applied electric dipole or a finite-length wire. When an electric dipole is utilized, the current density can be expressed as:

$$\mathbf{J}_e(\mathbf{r}, t) = \frac{d\mathbf{p}(t)}{dt}\delta\left(\mathbf{r} - \mathbf{r}^*\right) \tag{3.10}$$

where, $\mathbf{p}$ is the electric dipole moment, $\mathbf{r}^*$ is the spatial position of the excitation source. The function δ(x) is known as the Dirac Delta function, representing the waveform function of the emission current. In practice, to enhance the electromagnetic signal and improve the signal-to-noise ratio, a long grounded wire carrying a large current is typically used as the excitation source. Additionally, an appropriate waveform function is chosen based on the specific conditions.

(2) Boundary conditions

When the medium is discontinuous, the boundary conditions that must be satisfied at the discontinuous surface are specified in Eq. (2.4), which will not be reiterated here.

3.1.3 Spatial Division and Time Discretization

(1) Spatial division

One advantage of the finite element method is its ability to construct unstructured grids, allowing for effective simulation of irregular geometric shapes and underground resistivity anomalies. This capability helps to mitigate errors associated with stepwise time-domain finite difference modeling. In this article, the calculation area is divided into a finite number of tetrahedral elements *e*. Within each tetrahedron e, the electric field vector at the point (x, y, z) can be represented using the electric field and vector basis functions associated with each edge (Debiao et al. 2014):

$$E_e(x, y, z, t) = \sum_{j=1}^{J} f_j(x, y, z)E_j^e(t) \tag{3.11}$$

J is the number of edges of the element *e*, the number of edges of the tetrahedral element is 6, j value ranges from 1 to 6. The vector basis function f_j can be written as (Jianming 1998):

$$f_j = l_j\left(f_{j1}\nabla f_{j2} - f_{j2}\nabla f_{j1}\right) \tag{3.12}$$

Fig. 3.2 Schematic diagram of tetrahedral element node and edge numbering

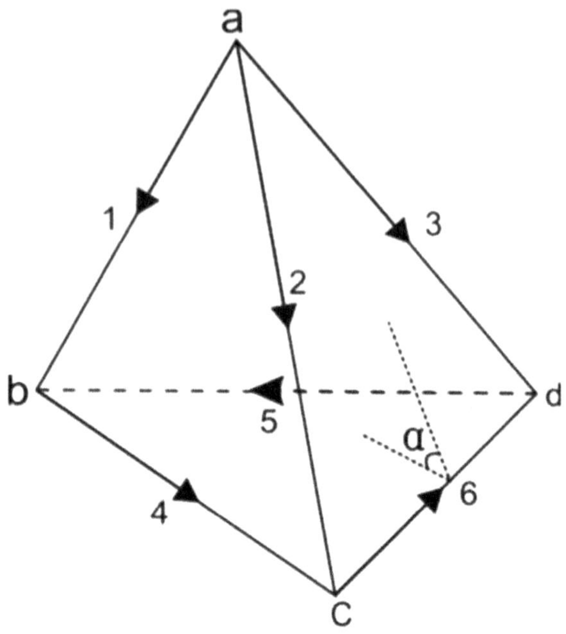

where, j_1 and j_2 are the vertex numbers of the tetrahedron, their values range from 1 to 4. l_j is the length of an edge, f_{j1} and f_{j2} are the scalar basis functions of the two nodes corresponding to the edge, and the edge number and node number are shown in Fig. 3.2.

Substituting (3.11) into (3.9), we get:

$$\sum_{j=1}^{J}\int_{\Omega}\left[(\nabla \times f)\cdot\left(\nabla \times f_j\mathbf{E}_j\right)+\sigma\mu_0 ff_j\frac{\partial \mathbf{E}_j}{\partial t}\right]d\Omega=-\int_{\Omega}\mu_0 f\cdot\frac{\partial \mathbf{J}_s}{\partial t}d\Omega \tag{3.13}$$

In the formula, the curl of the vector basis function can be expressed as follows (Jianming 1998):

$$\nabla \times f_j = 2l_j\left(\nabla f_{j1}\times \nabla f_{j2}\right) \tag{3.14}$$

The term on the right side of the equation is present only in the field source distribution unit.

(2) Time discretization

This paper employs the method of backward difference to discretize the formula (3.9):

$$\frac{\partial \mathbf{E}_n}{\partial t_n}=\frac{\mathbf{E}_n-\mathbf{E}_{n-1}}{t_n-t_{n-1}} \tag{3.15}$$

Substituting (3.15) into (3.9), we get:

$$\int_{\Omega} (\nabla \times f_i) \cdot (\nabla \times f_j)\mathbf{E_n} d\Omega + \sigma\mu_0 \int_{\Omega} f_i \frac{E_n - E_{n-1}}{t_n - t_{n-1}} d\Omega = -\mu_0 \int_{\Omega} f_i \cdot \frac{\mathbf{J_n} - \mathbf{J_{n-1}}}{t_n - t_{n-1}} d\Omega \quad (3.16)$$

After simplification, we get:

$$\int_{\Omega} (\nabla \times f_i) \cdot (\nabla \times f_j) d\Omega \mathbf{E_n} + \frac{\sigma\mu_0}{t_n - t_{n-1}} \int_{\Omega} f_i \cdot f_j d\Omega \mathbf{E_n} = \sigma\mu_0 \int_{\Omega} f_i \frac{\mathbf{E_{n-1}}}{t_n - t_{n-1}} d\Omega - \mu_0 \int_{\Omega} f_i \cdot \frac{\mathbf{J_n} - \mathbf{J_{n-1}}}{t_n - t_{n-1}} d\Omega \quad (3.17)$$

Write the above formula in matrix form:

$$\mathbf{A_e E_e^n} = \mathbf{b_e} \quad (3.18)$$

where,

$$\mathbf{A_e} = \int_{\Omega} \left[(\nabla \times f_i) \cdot (\nabla \times f_j) + \frac{\sigma\mu_0}{t_n - t_{n-1}} f_i \cdot f_j \right] d\Omega \quad (3.19)$$

$$\mathbf{b_e} = \frac{\sigma\mu_0}{t_n - t_{n-1}} \int_{\Omega} \mathbf{E_{n-1}} \cdot f_i d\Omega - \frac{\mu_0}{t_n \quad t_{n-1}} \int_{\Omega} f_i \cdot (\mathbf{J_n} - \mathbf{J_{n-1}}) d\Omega \quad (3.20)$$

In the formula, $\mathbf{E_e^n}$ is the column vector formed by the projection of the electric field at the nth moment on each edge of the tetrahedron.

$\mathbf{J}_n$ is the current density provided by the outside world at the nth moment. For matrix $\mathbf{A}_e$ it can be divided into two parts $\mathbf{A}_{1e}$ and $\mathbf{A}_{2e}$, where,

$$\mathbf{A_{1e}} = \iint_{\Omega} [(\nabla \times f_e^i) \cdot (\nabla \times f_e^j)] d\Omega \quad (3.21)$$

$$\mathbf{A_{2e}} = \frac{\sigma\mu_0}{t_n - t_{n-1}} \iint_{\Omega} f_i \cdot f_j d\Omega \quad (3.22)$$

Similarly, matrix $\mathbf{b_e}$ can also be divided into two parts, $\mathbf{b_{e1}}$ and $\mathbf{b_{e2}}$, where:

$$\mathbf{b_{e1}} = \frac{\sigma\mu_0}{t_n - t_{n-1}} \int_{\Omega} \mathbf{E_{n-1}} \cdot f_i d\Omega \quad (3.23)$$

$$\mathbf{b}_{e2} = \frac{\mu_0}{t_n - t_{n-1}} \int_{\Omega} (\mathbf{J}_\mathbf{n} - \mathbf{J}_{\mathbf{n}-1}) \cdot f_i d\Omega \tag{3.24}$$

To achieve a good simulation effect, the selection of the time step must consider the following convergence conditions (Jun 2016):

$$\Delta t \leq \alpha CFL \frac{\Delta_{\min}}{v_{\max}} \tag{3.25}$$

In the formula, α is an empirical constant; $\Delta_{\min}$ is the minimum grid size; $v_{\max}$ is the maximum propagation speed of the electromagnetic field in the medium. When using p order basis function, the CFL condition is usually taken as $1/(2p + 1)$.

3.1.4 Solution of the Equation System

Solving large-scale linear sparse matrix equations is the final step of finite element analysis. The resulting equation system can be represented in one of the following two forms (Jianming 1998):

$$[K]\{\Phi\} = \{b\} \tag{3.26}$$

or

$$[A]\{\Phi\} = \lambda[B]\{\Phi\} \tag{3.27}$$

Equation (3.26) is deterministic, derived from non-homogeneous differential equations or non-homogeneous boundary condition problems, and is typically associated with scattering, radiation, and other deterministic issues involving elements or excitations. On the other hand, Eq. (3.27) is of an eigenvalue type, originating from homogeneous control differential equations and homogeneous boundary conditions. Once the $\{\Phi\}$ equation set is solved, the required parameters can be calculated, allowing for further analysis and application in relevant fields.

The methods for solving large-scale linear sparse matrix equations mainly include direct methods and iterative methods. Direct methods mainly include Gaussian elimination, pivot selection, and triangular decomposition, etc. This method can theoretically obtain the exact solution of the equation group, but it requires a large amount of computer memory. Iterative methods mainly include Jacobi iteration, Gauss–Seidel iteration, over-relaxation iteration, and conjugate gradient method, etc. This method starts from an initial preset value, reduces the residual error through successive iterations, gradually makes the solution meet the specified accuracy, and obtains an approximate solution. Iterative methods require less computer memory, but often require longer computation time. Compared with the iterative method to solve the equation group, the direct method to solve the large-scale equation group has the advantages of high accuracy and good stability, but its memory demand is large and the solution efficiency is low. In recent years, with the rapid development of computer hardware and parallel computing, the efficiency of the direct method has been significantly improved and has been widely used in the three-dimensional numerical simulation of the electromagnetic method. In the actual simulation process, the best solution plan needs to be selected according to the number of degrees of freedom of the problem solution and the configuration of computer memory.

3.2 Controlled Source Transient Electromagnetic Field Three-Dimensional Time Domain Finite Difference Method

Numerical simulation of forward modeling is an important means of studying the transient electromagnetic response. The commonly used three-dimensional forward methods include the integral equation method (SanFilipo and Hohmann et al. 1985; Changchun et al. 1994; Xingong et al. 2000; Zhdanov et al. 2006), finite element method (Sugeng 1998; Um et al. 2010; Yin et al. 2016; Li et al. 2017, 2018), and finite difference method (Wang and Hohmann 1993; Commer and Newman 2004; Huifeng et al. 2013).The advantage of the integral equation method is that it only requires discretizing the anomalous area, without needing to compute the field over the entire region. However, it encounters complex mathematical challenges, such as solving the Green's function and the matrix equation when dealing with complex models, making it generally suitable for simulating the electromagnetic response of simpler models. There are two approaches to solving the transient electromagnetic field using the finite element method. One approach is to obtain the transient electromagnetic response directly in the time domain, while the other involves first solving the large-scale linear equation group in the frequency or Laplace domain and then transforming the results to the time domain. The finite difference method typically solves directly in the time domain, applying finite difference approximations to the time and spatial partial differential operators for the two Maxwell curl equations. The boundary conditions for the ground and air are generally implemented using the upward extension of the potential field. Additionally, many scholars employ methods such as solving in the virtual wave field domain or using the Krylov subspace algorithm for three-dimensional simulations of the transient electromagnetic field (Changkai et al. 2020).

Unlike other forward simulation methods, the finite difference method does not require solving large-scale equation groups, resulting in faster solution speeds and better suitability for parallel computing. It is gradually becoming a commonly used method for electromagnetic field forward simulation. Wang and Hohmann (1993) established a three-dimensional time-domain electromagnetic field forward algorithm using the improved Dufort-Frankel finite difference method. This algorithm uses the analytical solution of the uniform half-space transient electromagnetic field as the initial condition and employs the upward extension of the potential field to handle the ground-air boundary, avoiding direct subdivision calculations of the air layer. In cases where surface resistivity is uneven or the analytical solution is unavailable, Huifeng et al. (2013) introduced the loop source current density into the iterative equation to replace the initial condition excitation source, enabling the calculation of the transient electromagnetic field for any waveform emission current. Building on Wang and Hohmann's algorithm, Commer and Newman (2004) applied the three-dimensional finite difference method to the controlled source transient electromagnetic field. They directly solved the Poisson equation of the stable current field to obtain the initial electric field value and applied the divergence-free condition to the time derivative of the magnetic field and the induced current density in later stages to enhance the accuracy of the late field.

3.2.1 Control Equation and Difference Discretization

The time-domain finite difference method is the primary approach for three-dimensional numerical simulation of the transient electromagnetic field. The grid unit it employs is illustrated in Fig. 3.3 (Yee 1966). This grid configuration naturally satisfies the continuity condition of the field components on the three-dimensional surface. Under quasi-static conditions, the Maxwell equations are given by (Wang and Hohmann 1993)

$$\nabla \times \mathbf{E} = -\frac{\partial \mathbf{B}}{\partial t} \tag{3.28}$$

$$\nabla \times \mathbf{H} = \sigma \mathbf{E} + \tilde{\varepsilon}\frac{\partial \mathrm{E}}{\partial t} \tag{3.29}$$

$$\nabla \mathbf{D} = 0 \tag{3.30}$$

$$\nabla \mathbf{B} = 0 \tag{3.31}$$

where, E is the electric field strength, B is the magnetic induction strength, H is the magnetic field strength, σ is the medium conductivity, $\tilde{\varepsilon}$ is the artificially added dielectric constant, which can be significantly increased (by several orders of magnitude compared to the actual dielectric constant of the medium) to allow for larger time steps; it is also known as the virtual dielectric constant.

The central difference method can be applied to discretize Eq. (3.29) to obtain the difference format for the electric field

$$E^{n+1} = \frac{2\tilde{\varepsilon} - \sigma \Delta t_n}{2\tilde{\varepsilon} + \sigma \Delta t_n} E^n + \frac{2\Delta t_n}{2\tilde{\varepsilon} + \sigma \Delta t_n} \nabla \times H^{n+\frac{1}{2}} \tag{3.32}$$

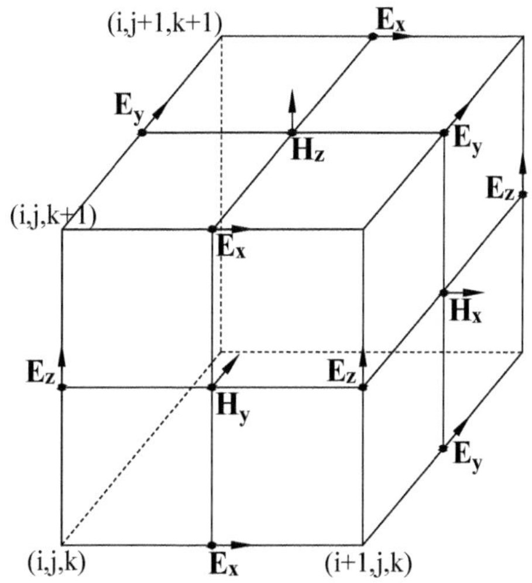

Fig. 3.3 Schematic diagram of the positions of various field components in the Yee unit discrete space (Yee 1966)

Discretizing Eq. (3.28) can yield

$$B^{n+\frac{1}{2}} = B^{n-\frac{1}{2}} - \frac{\Delta t_{n-1} + \Delta t_n}{2} \nabla \times E^n \tag{3.33}$$

Equation (3.31) indicates that only two of the three directional components of the magnetic induction intensity are independent; one component can be derived from the other two. Wang and Hohmann (1993) employed Eq. (6) to calculate the magnetic induction intensities B_x and B_y while simulating transient electromagnetic fields, subsequently using Eq. (3.31) to determine the magnetic field B_z.

The finite difference method in the time domain for the electromagnetic field utilizes an explicit difference format. The sizes of the spatial and time steps must satisfy certain conditions to ensure iteration stability. The numerical stability of the finite difference iteration in the time domain is governed by the Courant condition (Yee 1966):

$$\Delta t \leq \frac{\Delta \sqrt{\mu \tilde{\varepsilon}}_{\min}}{\sqrt{3}} \tag{3.34}$$

where $\Delta_{\min}$ is the minimum value of the spatial grid spacing, μ is the magnetic conductivity of the medium.

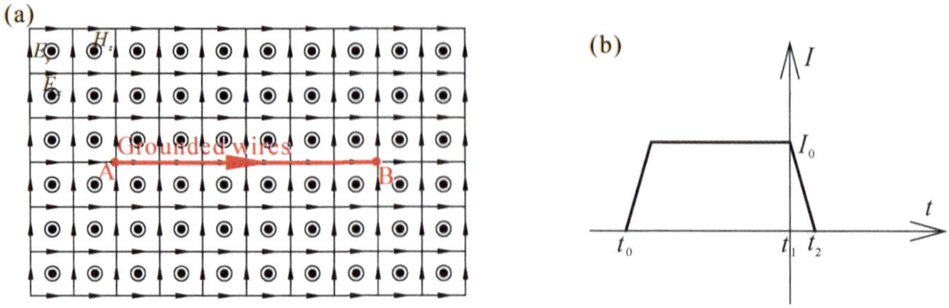

Fig. 3.4 Schematic diagram of the electrical source

3.2.2 Loading of the Transmitter

Wang and Hohmann (1993) employed the analytical solution of the uniform half-space at the initial moment as the initial field when calculating the loop source. Commer and Newman (2004) derived the initial value of the electric field by directly solving the Poisson equation for the stable current field when calculating the electrical source. These methods circumvent the need to calculate the active area. Haifeng et al. (2013) introduced the current density of the loop source into the iterative equation to replace the initial condition excitation source, enabling the calculation of the transient electromagnetic field while considering the shutdown time condition. According to the Yee unit grid, when applying the grounded wire source, the wire source can coincide with the spatial position of the electric field component, as illustrated in Fig. 3.4a. In the active area, Eq. (3.29) can be modified as follows (Haifeng et al. 2013):

$$\nabla \times H = \sigma E + \tilde{\varepsilon}\frac{\partial E}{\partial t} + J_s \tag{3.35}$$

where, J_s is the source current density. The difference equation for the electric field under the quasi-static approximation in the active area can be derived by discretizing the Eq. (3.35).

The transmitter waveform can be defined as a trapezoidal wave (see Fig. 3.3b) or any other waveform. This method directly computes the electric field in the grid where the transmitting loop is located using the difference equation for the active area, thereby avoiding the need to calculate the initial electromagnetic field.

3.2.3 Boundary Conditions

Boundary conditions are a significant factor influencing the accuracy of the late field. Effective absorption boundaries can enhance the accuracy of the solution while also reducing the number of grids. This article employs the convolution perfectly matched layer (CPML) absorption boundary condition (Kuzuoglu and Mittra 1996; Roden and Gedney 2000) at the truncation boundary. The CPML absorption boundary not only exhibits the properties of the perfectly matched layer (PML) absorption boundary but also effectively absorbs low-frequency fields.

In a lossy medium, the frequency domain equation in the CPML region is given by (Chew and Weedon 1994)

$$\nabla_s \times H = \sigma E + j\omega\tilde{\varepsilon}E \tag{3.36}$$

$$\nabla_s \times E = -j\omega B \tag{3.37}$$

$$\nabla_s \cdot B = 0 \tag{3.38}$$

where, σ is the conductivity in the truncation area, ω is the angular frequency of the electromagnetic field, ∇_s operator is

$$\nabla_s = \vec{x}_0 \frac{1}{s_x}\frac{\partial}{\partial x} + \vec{y}_0 \frac{1}{s_y}\frac{\partial}{\partial y} + \vec{z}_0 \frac{1}{s_z}\frac{\partial}{\partial z} \tag{3.39}$$

where s_x, s_y, s_z are the coordinate scaling factors. When the coordinate scaling factor is a complex number, the electromagnetic wave entering the CPML area decays exponentially. In CPML, its value is (Kuzuoglu and Mittra 1996)

$$s_w = \kappa_w + \frac{\sigma_w}{\alpha_w + j\omega\varepsilon_0} \quad (w = x, y, z) \tag{3.40}$$

Equations (3.36) and (3.37) can be transformed into time domain forms to derive the difference expressions in the CPML area, where the complex terms can be addressed using the convolution recursive method (Roden and Gedney 2000). To enhance the accuracy of the late field, Eq. (3.38) must be incorporated into the iteration process. Specifically, first use Eq. (3.37) to calculate the magnetic induction intensities B_x and B_y, and then proceed to solve for the magnetic field strength B_z.

3.2.4 Algorithm Verification

(1) Uniform Half-Space Model

The uniform half-space model is employed to verify the accuracy of the algorithm. The resistivity of the underground medium is set at 100 Ω·m, with a grounding wire source of 200 m in length deployed on the surface. The current waveform is a trapezoidal wave, featuring a width of 100 ms and rising and falling edge durations of 10 us. The receiving point is positioned at the equatorial location of the transmitter, with reception occurring at offset distances of 200, 500 and 1000 m. Non-uniform grid partitioning is utilized, comprising 120 × 120 × 120 grids, with the outer 10 layers designated as CPML boundary layers. The minimum grid size is 20 m, and the distance between the CPML boundary interface and the transmitter is 2500 m. The verification results are illustrated in Fig. 3.5, where the time-domain finite difference solution is compared with the analytical solution. The maximum relative error between the two solutions is 2%. After 1 ms, the relative error drops to less than 0.3%. Due to the absorption effect of the CPML boundary, no significant reflections occur at the boundary of the late grid, indicating that the finite difference algorithm employed in this study demonstrates high accuracy.

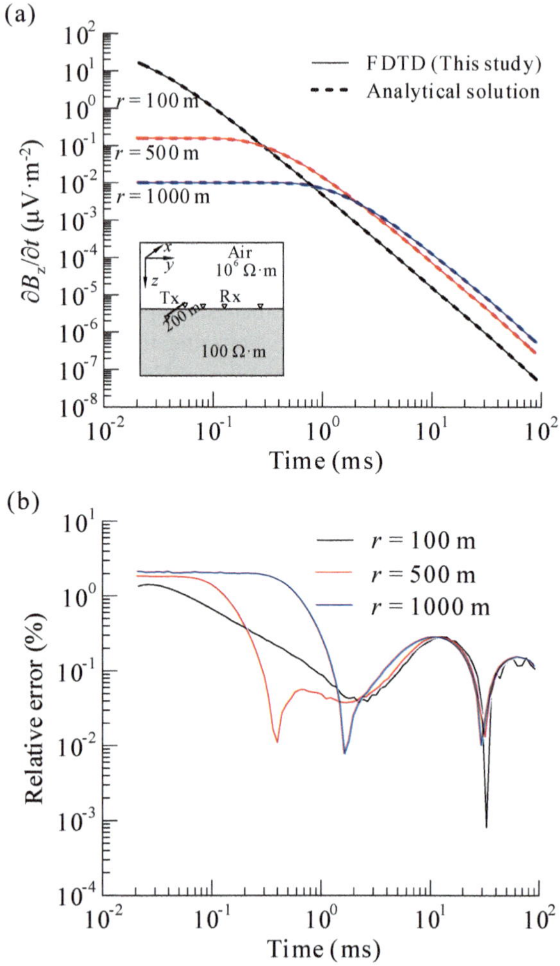

Fig. 3.5 Comparison of time-domain finite difference solution and analytical solution

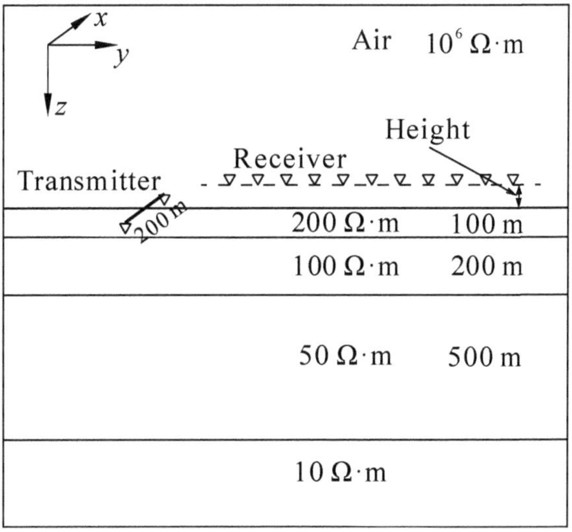

Fig. 3.6 Schematic diagram of layered half-space model

(2) Layered Half-Space Model

This paper also employs a layered half-space model to verify the reliability of the algorithm. The model used is illustrated in Fig. 3.6. A grounding wire source, 200 m in length, is deployed on the surface, with the current waveform designed as a trapezoidal wave. The width of the trapezoidal wave is 1 s, and the rising and falling edges each have a duration of 10μs. The resistivity of the air is set at 10^{6}Ω·m. Figure 3.7 presents the comparison results of the finite difference solution and the one-dimensional solution for offset distances ranging from 200 to 2500 m from the transmitter equator. As demonstrated in Fig. 3.7, the finite difference numerical solution aligns closely with the one-dimensional solution at all times. Due to the implementation of the CPML absorbing boundary, the field values at the boundary maintain high accuracy for times exceeding 0.96 s.

3.3 Implementation of Time-Domain Three-Dimensional Calculation

Based on the preceding theory, this paper implements a three-dimensional time-domain vector finite element forward simulation of the electrical source transient electromagnetic field using the COMSOL Multiphysics platform. COMSOL Multiphysics is a commercial finite element software that originated from the MATLAB toolbox. It simulates real physical phenomena by solving partial differential equations or systems of such equations. Over many years of development, the software has integrated a basic module along with eight specialized modules, enabling the modeling and simulation of single or multiple physical fields in areas such as electromagnetic analysis, fluid flow, heat conduction, and structural mechanics.

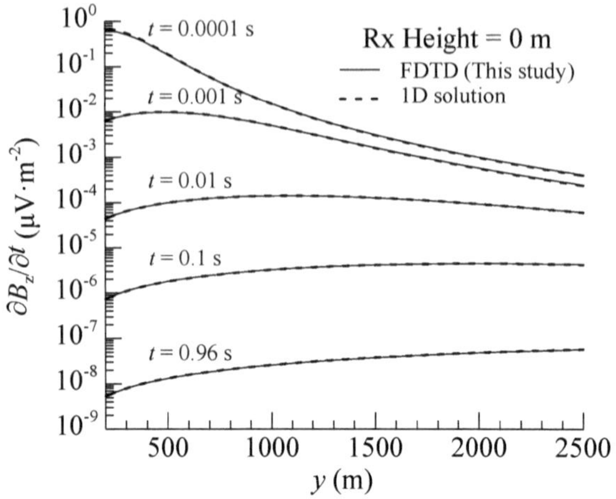

Fig. 3.7 Comparison of time-domain finite difference solution and one-dimensional solution

The COMSOL Multiphysics interface is user-friendly and straightforward, allowing for complete visual modeling throughout the process. It integrates well with CAD and other graphic processing software, facilitating the establishment of complex models. Users can define material properties, source terms, and various boundary conditions by setting related parameters and physical quantities, thus avoiding the need to modify underlying code. This capability enables rapid and flexible adjustments to the physical model, significantly enhancing modeling efficiency.

COMSOL Multiphysics not only provides comprehensive fast modeling and post-processing analysis functions but also integrates seamlessly with MATLAB, offering a user-friendly environment for secondary development using various MATLAB functions and tools. The software encompasses visual modeling, grid partitioning, equation and boundary setting, time step selection, solver determination, and post-processing, as illustrated in Fig. 3.8.

For simulating electromagnetic fields, COMSOL Multiphysics offers two relevant modules: the AC/DC Module and the RF Module. This paper primarily addresses low-frequency electromagnetic field problems with frequencies below 1000 Hz. Since the wavelengths in the solution area are often much smaller than the radiation wavelengths of electromagnetic waves in the air, the AC/DC Module is selected. This module solves Maxwell's equations using the quasi-static approximation. Within this module, users can choose different interfaces to simulate the dynamic characteristics of static, quasi-static, transient, and time-harmonic electromagnetic fields. The typical interfaces and their functional features are summarized in (Table 3.1).

In this study, the focus is on the characteristics of electric and magnetic field responses in the x, y, and z directions under time-domain conditions. Therefore, the mf interface was ultimately chosen, and a three-dimensional model was implemented.

The geometric model used in this study is implemented as a cube, as shown in Fig. 3.9.

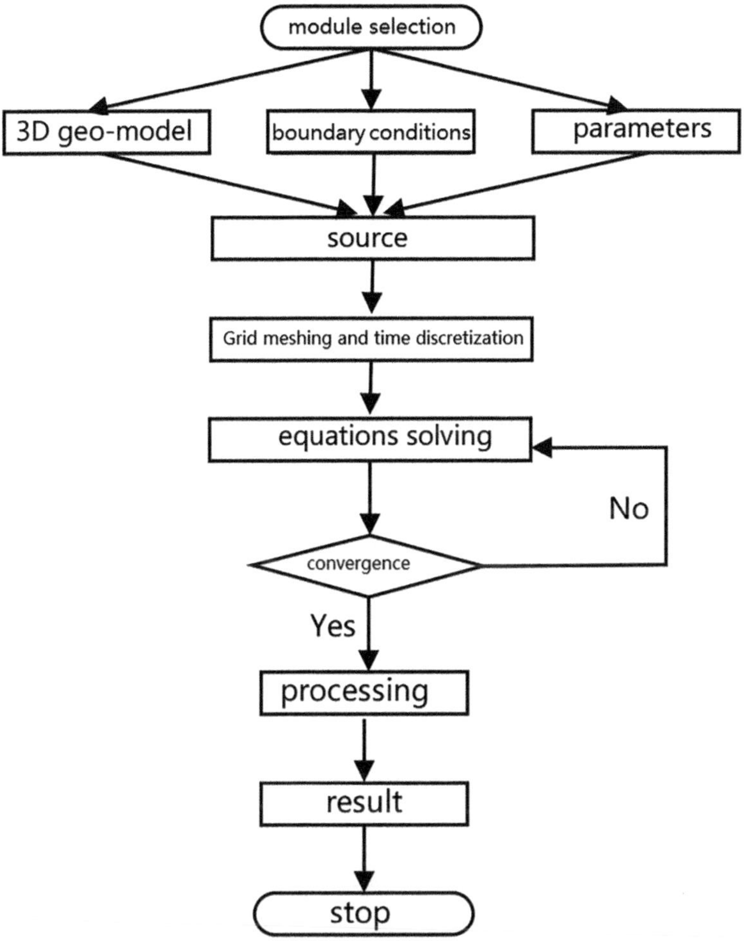

Fig. 3.8 COMSOL Multiphysics forward process diagram

The model is primarily divided into two components: the upper layer represents the air layer, with a volume of 10 × 10 × 1 km (length × width × height), while the lower layer represents the earth, with a volume of 10 × 10 × 4 km (length × width × height). The conductivity of the air layer is set to $\sigma = 10^{-8} S/m$, and the uniform earth conductivity is $\sigma = 10^{-2} S/m$. The relative magnetic permeability and relative dielectric constant are both set to 1. To enhance solution accuracy while ensuring computational efficiency, the earth is divided into multiple blocks, with grid division becoming progressively denser from the outside toward the center. The central area is designated as the key calculation domain, with a volume of 2.5 × 2.5 × 0.6 km (length × width × height).

After completing the geometric model settings, excitation sources can be applied through various geometric structures. This study primarily utilizes point electric

Table 3.1 Electromagnetic interfaces and their functional characteristics in AC/DC

Interface name	Identifier	Based on variables	Field components				Preset studies		
			Magnetic field	Electric field	Magnetic potential	Current density	Static	Time domain	Frequency domain
Static Electric Field	es	V		x, y, z			√	√	√
Current	ec	V		x, y, z		x, y, z	√	√	√
Current, shell	ecs	V		x, y, z		x, y, z	√	√	√
Magnetic field	mf	A	x, y, z	x, y, z	x, y, z	x, y, z	√	√	√
Magnetic field formula	mfh	H	x, y, z				√	√	√
Magnetic field and electric field	mef	VA	x, y, z	x, y, z	x, y, z	x, y, z	√		√
Magnetic Field, No Current	mfnc	V_m	x, y, z				√	√	
Rotating machinery, magnetic field	rmm	V_m	x, y	z	z	z	√	√	
Circuit	cir	n/a					√	√	√

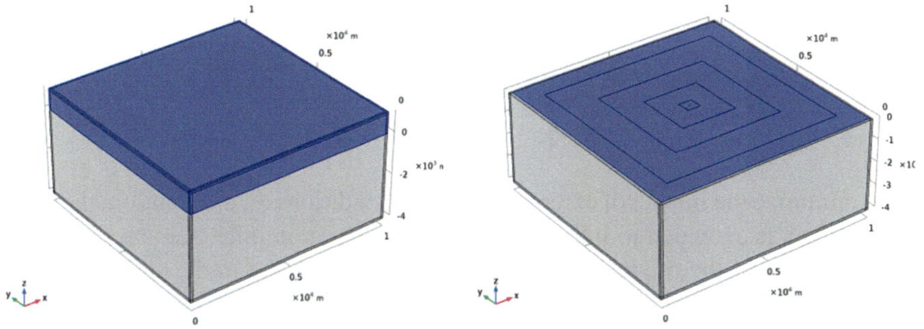

Fig. 3.9 Geometry of the three-dimensional forward model

dipole sources and long wire sources. When loading the electric dipole source, both the electric dipole moment and its direction must be defined. In contrast, the long wire source only requires the specification of the current magnitude. The excitation source used in this study is a trapezoidal wave, which accounts for the rising and falling edges of the excitation current. Figure 3.10 presents a schematic diagram of the waveform of the emission current utilized in this study.

The waveform function of the trapezoidal wave is as follows:

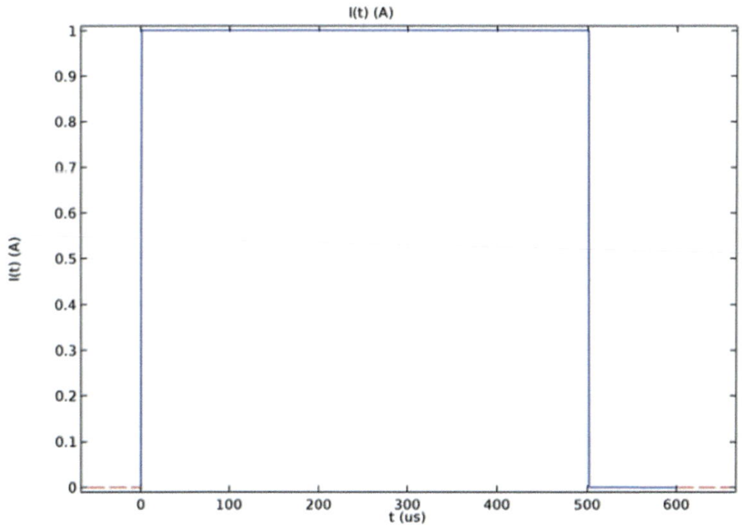

Fig. 3.10 Schematic diagram of the trapezoidal emission current waveform

$$I(t) = \begin{cases} \frac{t}{t_1} \cdots\cdots 0 \le t < t_1 \\ 1 \cdots\cdots t_1 \le t \le t_2 \\ \frac{t-t_3}{t_2-t_3} \cdots\cdots t_2 \le t \le t_3 \\ 0 \cdots\cdots t \ge t_3 \end{cases} \tag{3.41}$$

To minimize the impact of external boundary conditions on the solution, the finite solution area is extended to infinity by incorporating an infinite element domain in the outer layer of the model. The specific setting parameters in the software are as follows:

$$\Delta s = 1000 \times dGeomChar \tag{3.42}$$

$$\Delta p = dGeomChar \tag{3.43}$$

where, *dGeomChar* is the actual size of the model. By adjusting Δs, We can adjust the scaling factor of the model. In this simulation, the existing geometric model was enlarged by a factor of 1000 using the infinite element domain.

When utilizing the direct time domain vector finite element method for three-dimensional simulations, both grid division and time discretization are crucial for solution accuracy. Finer grid division enhances accuracy but also increases the order of the large-scale linear sparse equation system, raising the degrees of freedom and significantly reducing calculation speed. Conversely, if the grid is too coarse, it may not meet the accuracy requirements. The same considerations apply to the choice of time step size. In this study, a tetrahedral unstructured grid was employed for division, with a denser grid in the key research area. Additionally, to ensure calculation stability, we controlled the ratio of the sizes of all adjacent grids within the calculation area to remain within 2. The results of the model division are illustrated in Fig. 3.11.

If a non-uniform grid is used for grid division, the choice of time step size can be based on $\Delta t = \frac{\lambda}{15c}$, where λ is the shortest wavelength corresponding to the highest

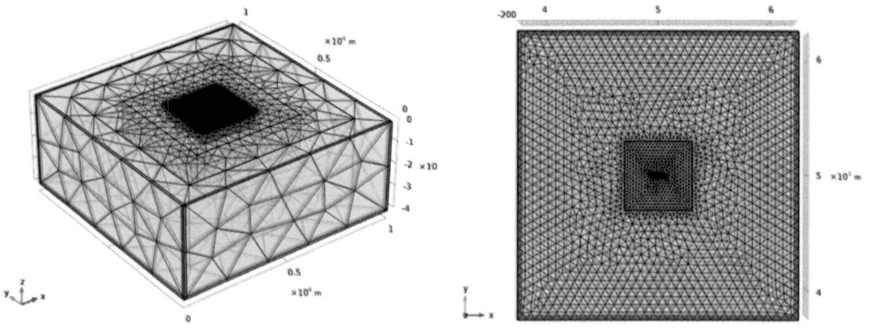

Fig. 3.11 Schematic diagram of grid division

frequency. For the short-offset transient electromagnetic method, low-frequency electromagnetic waves are primarily utilized, with frequencies below 1000 Hz. For a 1000 Hz electromagnetic wave, the wave field in a vacuum is approximately 3×10^5 m, at this time the step size used for three-dimensional simulation should not be greater than 7×10^{-5} s.

After completing the grid partitioning, an appropriate solver can be selected to address the system of equations. Compared to iterative methods, direct methods for solving large systems of equations offer advantages in terms of high accuracy and good convergence. This simulation employs a direct method known as PARDISO, which stands for Parallel Direct Sparse Solver. PARDISO utilizes parallel solving technology, maximizing the capabilities of multi-core, multi-thread workstations, and providing both efficiency and stability. It is particularly suitable for solving multiple right-hand side linear systems of the form Ax = b. The solver supports various sparse matrix storage schemes and can automatically select or manually specify the use of complete or incomplete LU, LDL, or LLT parallel decomposition algorithms based on the characteristics of the sparse matrix.

This calculation was implemented using COMSOL Multiphysics5.3 version, on a 64-bit system computer, with an Inter[R] Xeon[R] CPU E5-2620 v4 2.1 GHz processor, and 64 G of memory. To improve calculation speed and accelerate convergence, the relative tolerance was set to 0.01, and the time range for solving was 10^{-6} ~ 10^{-2} s. The degrees of freedom solved in the model were 1,124,148, occupying 16.79 G of physical memory and 19.33 G of virtual memory, with a solving time of 102905 s. Figure 3.12 shows the distribution of ground E_x, H_x, H_y, and H_z. Figure 3.13 shows the 3D slice of the H_z component under uniform half-space conditions.

To verify the correctness and calculation accuracy of the 3D vector finite element forward results, this section compares the analytical solution under uniform half-space conditions with the 3D simulation results, as illustrated in Figs. 3.14, 3.15 and 3.16. Since there is no analytical solution available for layered media, we use the comparison between one-dimensional linear filtering forward results and three-dimensional forward results to validate the accuracy of the three-dimensional outcomes. The one-dimensional filtering solution derives the reflection coefficients between layers using a layered recursive formula. It then solves the integral of the Bessel function through the Hankel transform and transforms the results into the time domain via the time–frequency transform method (Guptasarma 1982; Kong 2007).

First, we utilize a uniform half-space for verification. The uniform half-space model uses a conductivity of $\sigma = 10^{-2} S/m$. On the uniform ground, a dipole is chosen as the emission source, with an emission current of 50 A. The center coordinates of the source are (5000, 4000, 0), while the center coordinates of the observation point are (5000, 4200, 0), resulting in an offset distance of 200 m. The calculation results are compared with the analytical solution of the dipole axial electric field (Weir 1980).

For further verification, two typical three-layer models, designated as H and K, are selected. The solution from the three-dimensional transient electromagnetic time domain finite element method is compared with the linear digital filter solution of the step pulse. The length of the emission source is 1000 m, arranged along the equator,

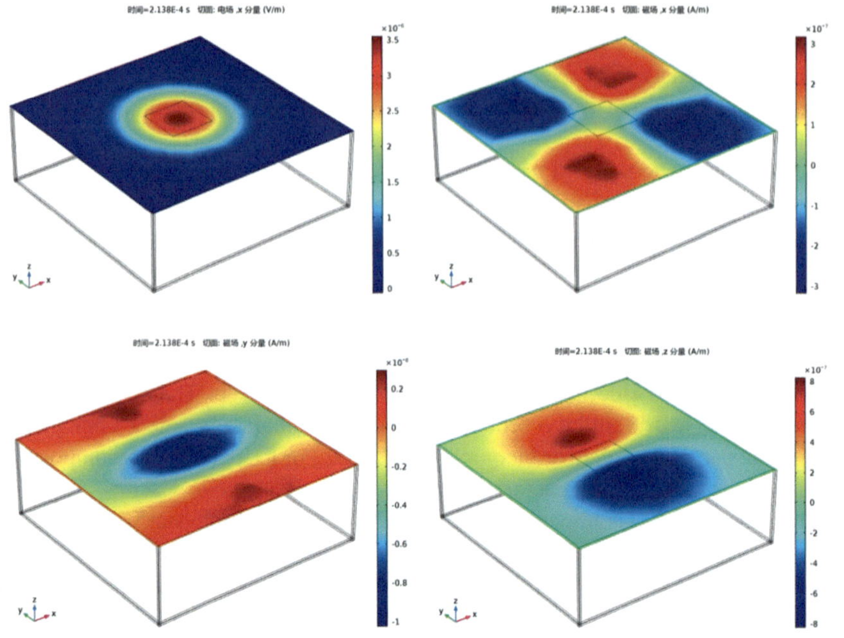

Fig. 3.12 Distribution of ground response components in uniform half-space

with an excitation current of 50 A. The offset distance between the center of the emission source and the observation point is 200 m. Both the rising and falling edges of the pulse are 1 μs, with a total duration of 5 ms. The geoelectric parameters of the two models are presented in Table 3.2.

Through the above comparison, it is observed that the relative error between the analytical solution of the uniform half-space and the results of the three-dimensional simulation is less than 4%, indicating a good fit. The results of the three-dimensional vector finite element forward simulation align closely with the one-dimensional linear filter results, exhibiting a relative error of less than 2%. This alignment is consistent for both the electric field and magnetic field components. It is worth noting that, since the three-dimensional forward simulation accounts for the shutdown time, the error during the early fitting may be relatively large. However, overall, the results of the three-dimensional forward simulation are considered valid.

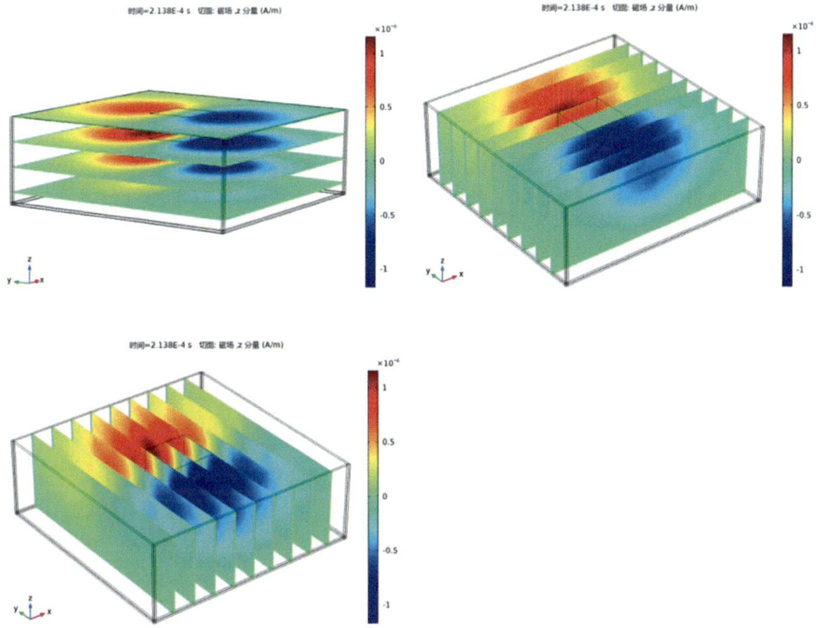

Fig. 3.13 3D response slice of H_z component in uniform half-space

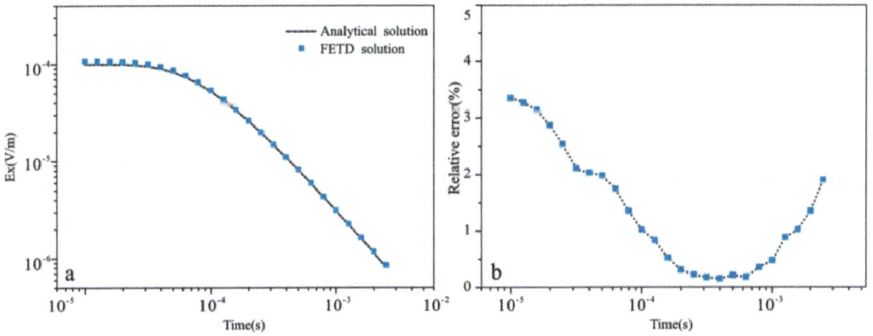

Fig. 3.14 Comparison of E_x component analytical solution and 3D response in uniform half-space (**a** Comparison of E_x component attenuation curves; **b** Error analysis)

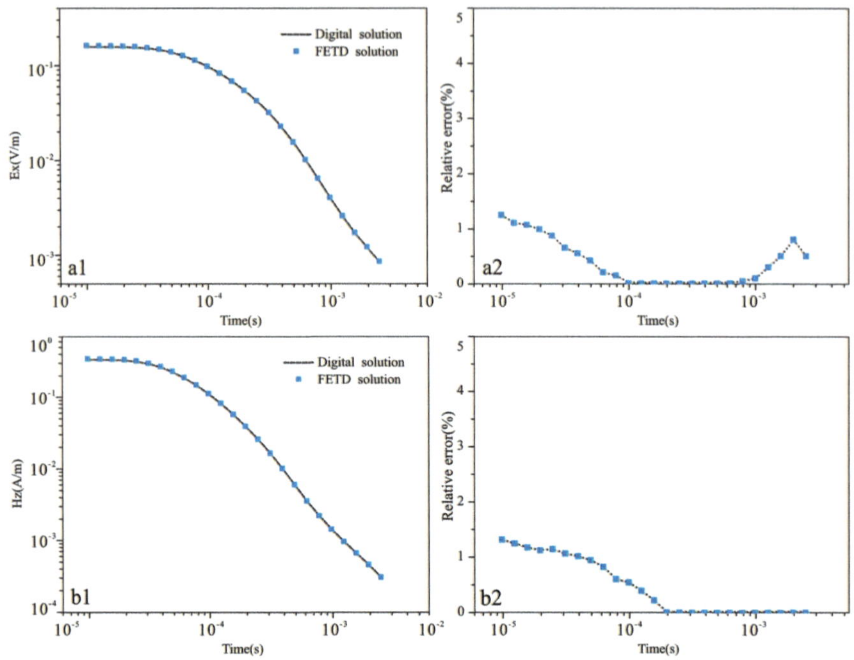

Fig. 3.15 H model E $_x$ and H $_z$ component 1D numerical solution and 3D response comparison analysis (a1. Comparison of E_x component attenuation curves; a2. Error analysis; b1. Comparison of E_x component attenuation curves; b2. Error analysis)

3.4 Electromagnetic Field Characteristics of Coal Mine Water-Rich Body SOTEM

3.4.1 Water-Filled Mining Area

Figure 3.17 presents a geoelectric model of a water-filled mining area, constructed based on actual coalfield geological data. The thickness of the coal seam cover is 300 m, while the thickness of the coal seam itself is 20 m. The dimensions of the mining area are 400 × 400 m, with a horizontal distance of 600 m between the emission source and the center of the mining area. The resistivity parameters are: $\rho_1 = 100\ \Omega\cdot\text{m}$, $\rho_2 = 400\ \Omega\cdot\text{m}$, $\rho_3 = 150\ \Omega\cdot\text{m}$, $\rho_4 = 1\ \Omega\cdot\text{m}$. The length of the emission source is 1000 m, oriented along the x-axis, with an emission current of 1 A.

Using the three-dimensional time-domain finite difference algorithm for forward simulation calculations, Fig. 3.18 illustrates the distribution of the electric field E_x anomaly field (the difference between the total field and the background field) on the ground at different times in the water-filled mining area. As shown in the figure, at 1 ms, the E_x anomaly field exhibits a negative maximum value directly above the anomaly body. Over time, the E_x anomaly field not only maintains a negative

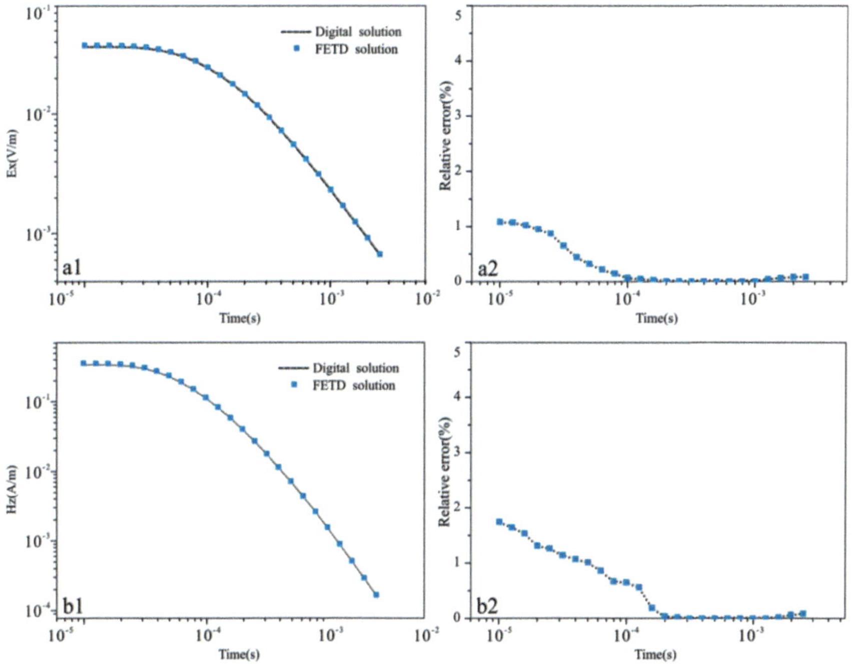

Fig. 3.16 K model E_X and H_Z component 1D numerical solution and 3D response comparison analysis (a1. Comparison of E_X component attenuation curves; a2. Error analysis; b1. Comparison of E_X component attenuation curves; b2. Error analysis)

Table 3.2 Geoelectric parameters of two typical models

	Layer 1		Layer 2		Layer 3	
	Thickness [M]	Resistivity ($\Omega \cdot m$)	Thickness [M]	Resistivity ($\Omega \cdot m$)	Thickness [M]	Resistivity ($\Omega \cdot m$)
H	300	100	50	10	/	100
K	300	100	50	1000	/	100

maximum value above the anomaly body but also develops a positive maximum value on both sides of it. The contour lines are elongated in the y-direction, influenced by the accumulated charge generated by the induced current on the surface of the anomaly body. Therefore, the analysis of the transient electromagnetic field of the electrical source must consider not only the distribution of the induced current but also the impact of the accumulated charge on the surface of the anomaly body.

Figure 3.19 shows the distribution of the ∂Bz/∂t anomaly field generated by the water-filled goaf at different times on the ground surface. As observed in the figure, at 1 ms, the ∂Bz/∂t anomaly field exhibits both negative and positive maximum values on either side of the anomaly body (near and far from the transmitter, respectively).

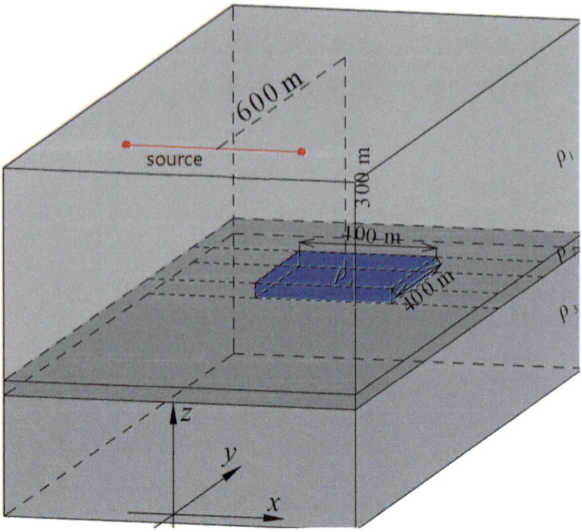

Fig. 3.17 Model of water-filled mining area

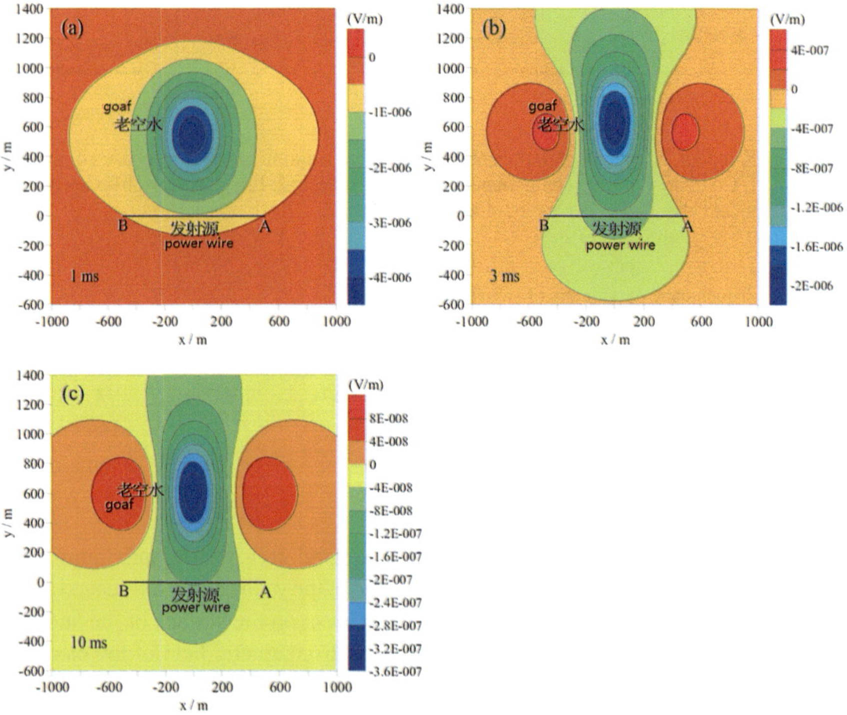

Fig. 3.18 Plane distribution map of electric field Ex anomaly field generated by the goaf at different times

As time progresses, a positive anomaly gradually emerges above the anomaly body and spreads outwards. Ultimately, a positive maximum value develops on the side of the anomaly body that is farthest from the transmitter, while a negative maximum value appears on the side closest to the transmitter. Consequently, compared to the loop source, the diffusion and distribution of the SOTEM magnetic field are more complex (Fig. 3.19).

Three survey lines are arranged near the goaf, as shown in Fig. 3.20. Figure 3.21 is the multi-measurement curve of the electric field Ex on the three survey lines. For comparison, the response curve of the layered background model is also drawn in the figure.

As can be seen from the figure, the responses of the x-component of the electric field on the three survey lines are basically similar. The multi-measurement curve of the electric field Ex is lower in amplitude at the position corresponding to the anomaly body, and the Ex response of the three survey lines can reflect the lateral changes of the anomaly body.

Figure 3.22 is the multi-measurement curve of the magnetic field ∂Bz/∂t on the three survey lines. For comparison, the response curve of the layered background model is also included in the figure. Figure 3.22a is the ∂Bz/∂t response of survey line 1. In the early stage, the response amplitude at the position of the anomaly body is

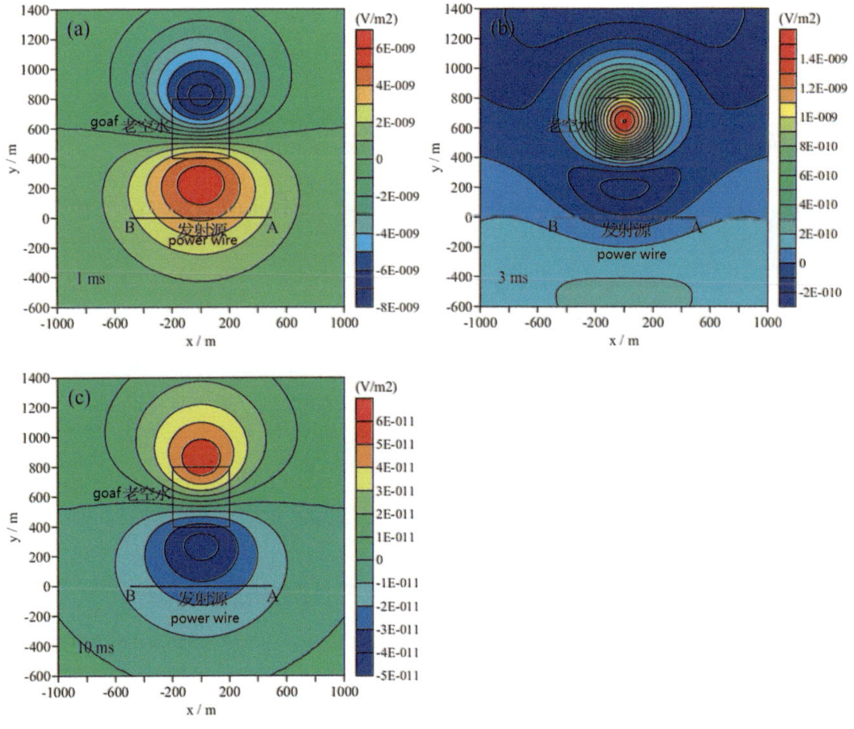

Fig. 3.19 Plane distribution map of ∂Bz/∂t anomaly field generated by the goaf at different times

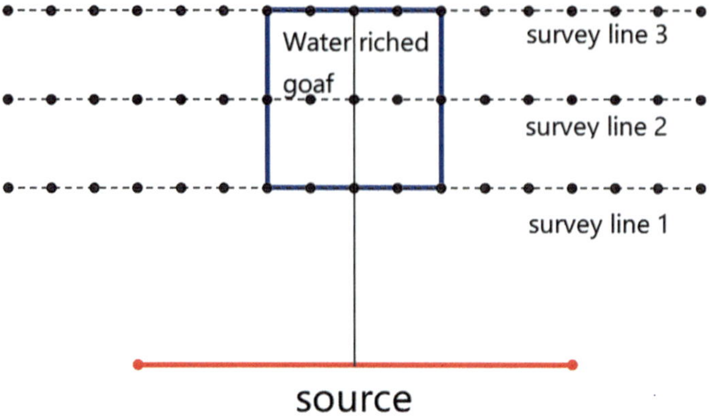

Fig. 3.20 Layout of survey lines in the water-filled goaf

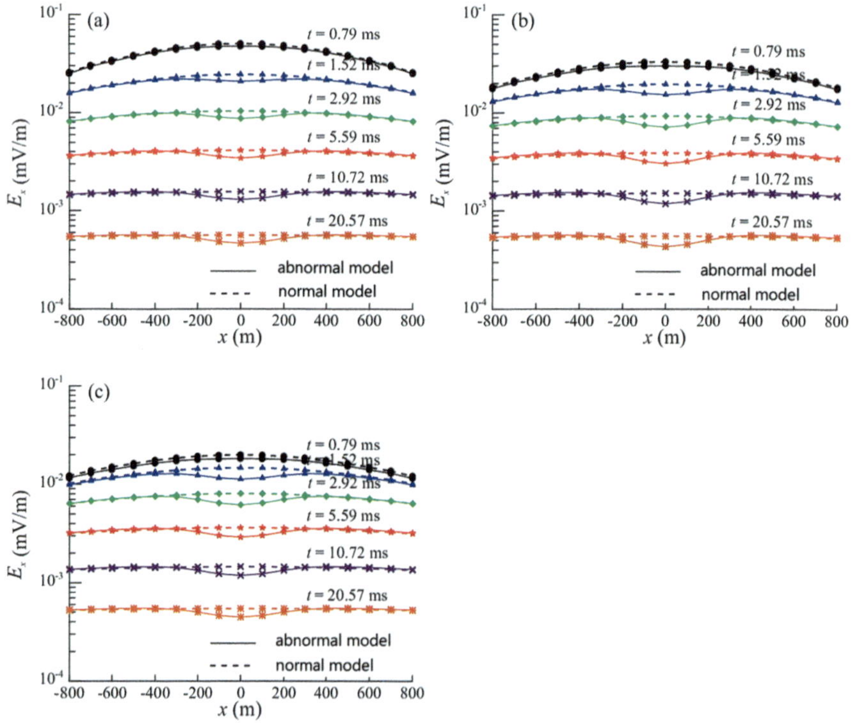

Fig. 3.21 Multi-measurement response of electric field Ex in the goaf

higher than the background value. However, in the later stage, the response amplitude at the position of the anomaly body is significantly lower than the background value. Figure 3.22b is the ∂Bz/∂t response of survey line 2. In the early stage, the response amplitude at the position of the anomaly body is lower than the background value.

As time progresses, the response amplitude at the position of the anomaly body gradually exceeds the background value. Ultimately, the response of the old void water aligns with the response curve of the background model. Figure 3.22c is the ∂Bz/∂t response of survey line 3. In the early stage, the response amplitude at the position of the anomaly body is lower than the background value. However, in the later stage, the response amplitude at the position of the anomaly body is significantly higher than the background value. Therefore, the ∂Bz/∂t responses observed on the three survey lines exhibit significant differences. This analysis provides a theoretical basis for interpreting actual detection data.

Figure 3.23 illustrates the decay curve of the SOTEM transient electromagnetic response influenced by the goaf, with the observation point positioned directly above the center of the goaf. For comparison, the response curve of the layered background model is also included. As shown in Fig. 3.23a, the decay curve of the electric field E_x response influenced by the old void water is higher than the background response in the early stage but lower in the later stage. Figure 3.23b presents the decay curve of the magnetic field ∂Bz/∂t response, Its decay pattern closely resembles that of the loop source response: initially, the response curve of the old void water model aligns with that of the background model, but as time progresses, the response curve of the old void water model diverges from the background model. Overall, the ∂Bz/∂t

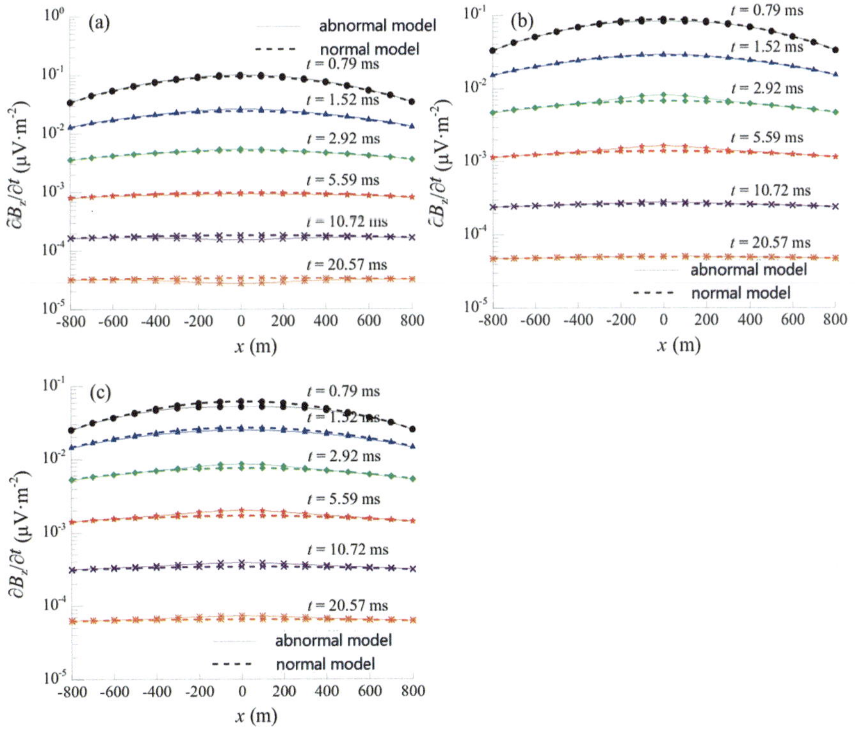

Fig. 3.22 Multi-measurement response of magnetic field ∂Bz/∂t in the goaf

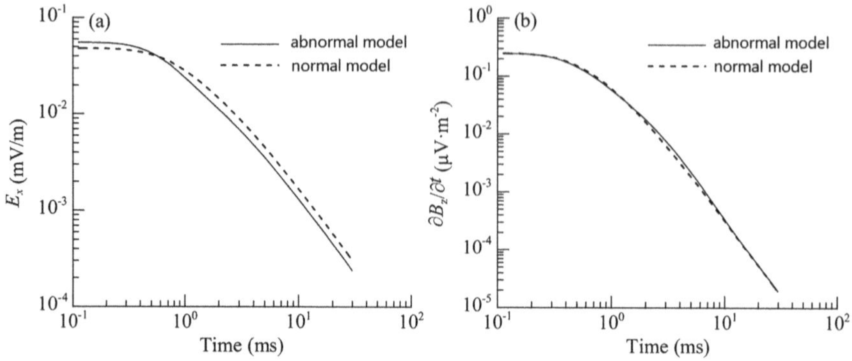

Fig. 3.23 Decay curve of SOTEM transient electromagnetic response under the influence of the goaf

curve under the influence of the old void water is lower than the background model. Therefore, although the ∂Bz/∂t responses on different survey lines have significant differences, the decay law of ∂Bz/∂t directly above the center of the goaf is basically similar to the loop source.

3.4.2 *Water-Filled Subsidence Column*

Figure 3.24 depicts a geoelectric model of a water-filled subsidence column, constructed using actual coalfield geological data. The thickness of the coal seam cover layer is 300 m, while the thickness of the coal seam itself is 20 m. The subsidence column has a depth of 200 m and a radius of 100 m, with a horizontal distance of 600 m between its column axis and the transmitter. The resistivity parameters are: $\rho_1 = 100\ \Omega\cdot\text{m}$, $\rho_2 = 400\ \Omega\cdot\text{m}$, $\rho_3 = 150\ \Omega\cdot\text{m}$, $\rho_4 = 1\ \Omega\cdot\text{m}$. The length of the emission source is 1000 m, it is arranged along the x-axis, and the emission current is 1 A.

Figure 3.25 illustrates the distribution of the electric field anomaly field (the difference between the total field and the background field) generated by the sinkhole column at various times on the ground. As observed in the figure, at 1 ms, the E_x anomaly field exhibits a negative maximum value directly above the anomaly body. As time progresses, not only does the E_x anomaly field maintain a negative maximum value directly above the anomaly body, but it also develops a positive maximum value on either side of the anomaly body (as shown in Fig. 3.25b and c. This phenomenon is attributed to the induced current generating accumulated charges on the surface of the anomaly body.

Figure 3.26 shows the distribution of the ∂Bz/∂t anomaly field generated by the sinkhole column at different times on the ground. As can be seen from the figure, at 1 ms, the ∂Bz/∂t anomaly field generates both negative and positive maximum values

Fig. 3.24 Water-containing sinkhole column model

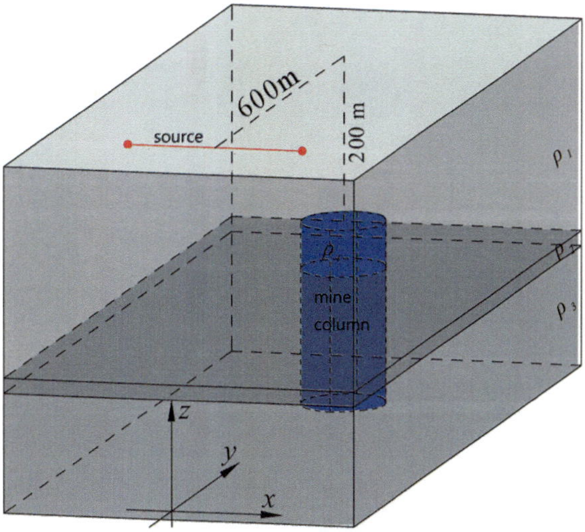

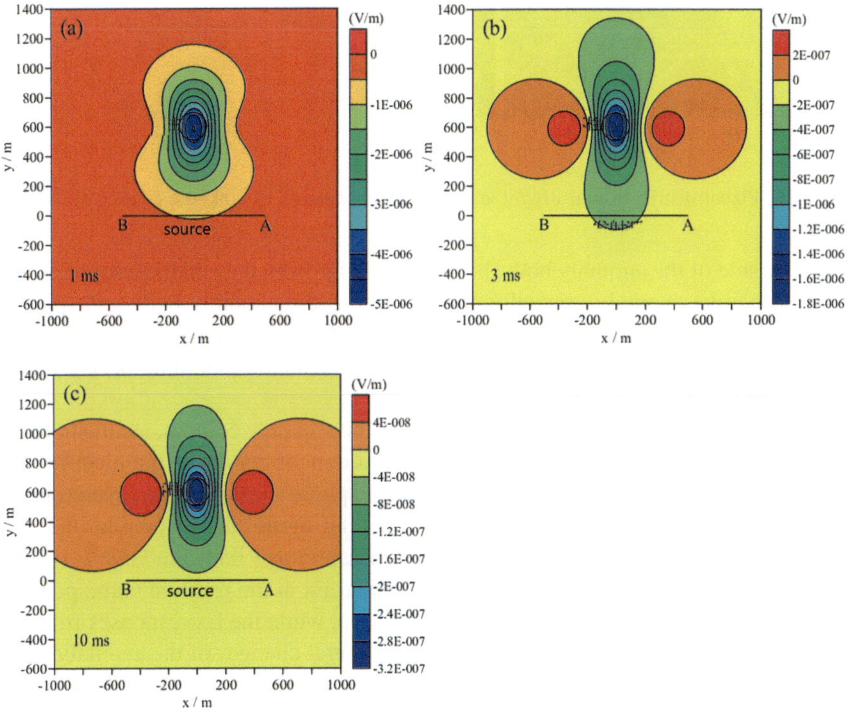

Fig. 3.25 Planar distribution of electric field Ex anomaly field generated by sinkhole column at different times

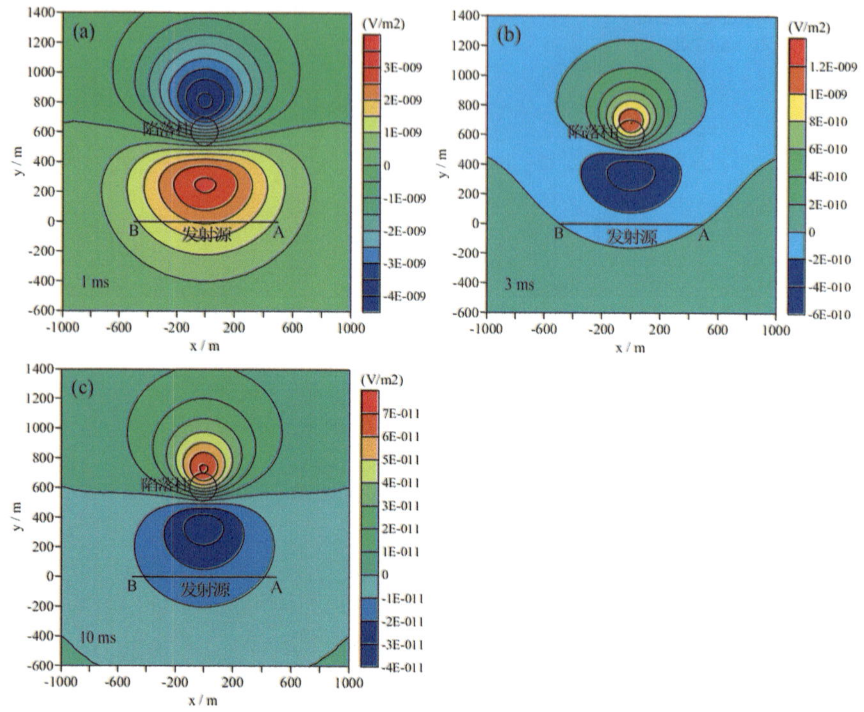

Fig. 3.26 Planar distribution of ∂Bz/∂t anomaly field generated by sinkhole column at different times

on either side of the anomaly body, both near and far from the emission source. Over time, a positive anomaly gradually emerges above the anomaly body and spreads. Ultimately, a positive maximum value is generated on the side of the anomaly body farthest from the emission source, while a negative maximum value occurs on the side nearest to the emission source.

Three measurement lines are arranged near the sinkhole column, as illustrated in Fig. 3.27. Figure 3.28 presents the multi-measurement curves of the electric field Ex on these three lines. For comparison, the response curves of the layered background model are also included. As shown in the figure, the responses of the x-component of the electric field on the three measurement lines are largely similar. The multi-measurement curve of the electric field Ex at the position corresponding to the sinkhole column exhibits a lower amplitude, while the Ex responses from the three measurement lines effectively reflect the lateral changes of the anomaly body.

Figure 3.29 shows the multi-measurement curves of the magnetic field ∂Bz/∂t on the three measurement lines. For comparison, the response curves of the layered background model are also included in the figure. Figure 3.29a shows the ∂Bz/∂t response of measurement line 1. The response curves of the sinkhole column at all times closely align with the background model's response, indicating that the abnormal response cannot be detected on this measurement line. Figure 3.29b shows the ∂Bz/∂t response of measurement line 2. In the early stages, the response amplitude at the sinkhole column position is lower than the background value. Over

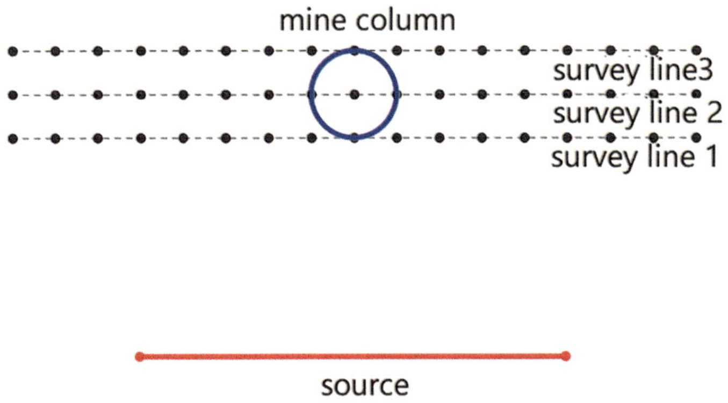

Fig. 3.27 Layout of measurement lines for water-containing sinkhole column

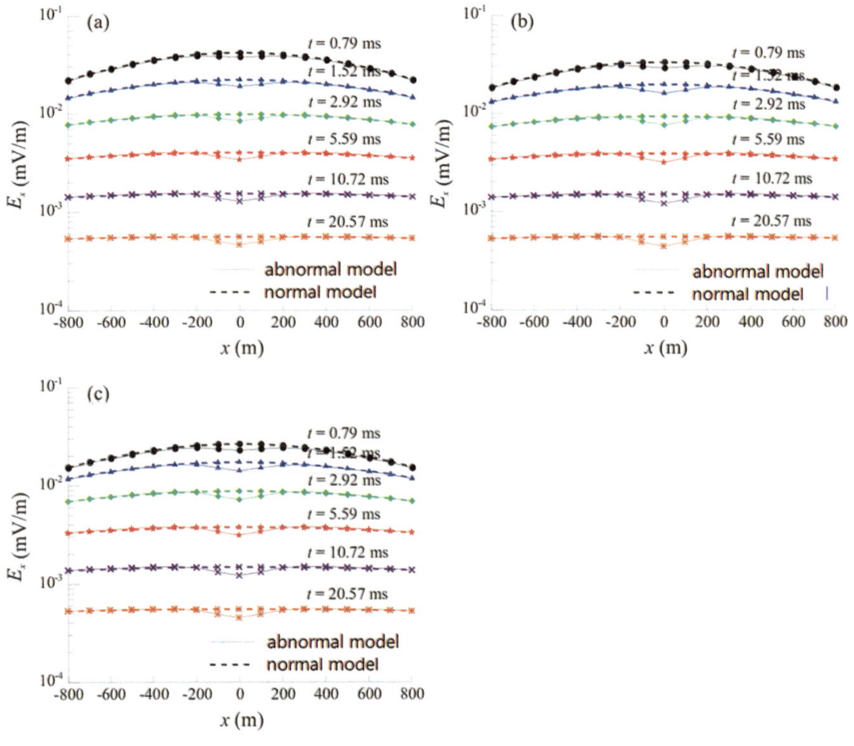

Fig. 3.28 Multi-measurement response of electric field Ex in water-containing sinkhole column

time, the response amplitude at this position gradually exceeds the background value. Ultimately, the response of the sinkhole column closely aligns with the response curve of the background model. Figure 3.29c shows the ∂Bz/∂t response of measurement line 3. In the early stage, the response amplitude at the sinkhole column position is

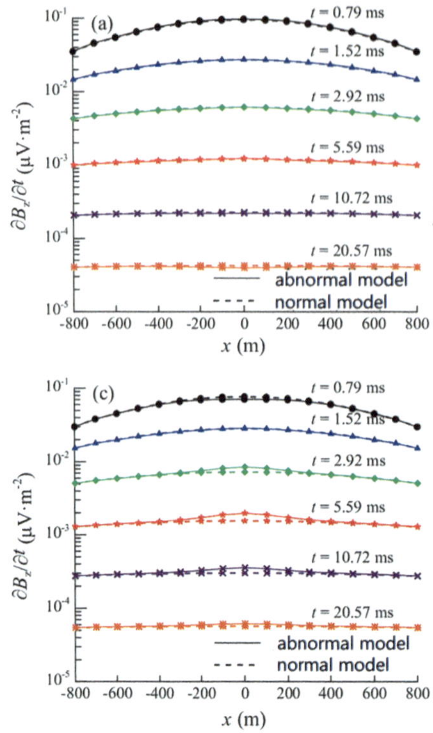

Fig. 3.29 Multi-measurement response of magnetic field ∂Bz/∂t in water-containing sinkhole column

lower than the background value; however, in the later stage, the response amplitude becomes significantly higher than the background value. Therefore, the observed ∂Bz/∂t responses on the three measurement lines have significant differences, and the above analysis provides a theoretical basis for the interpretation of actual detection data.

Figure 3.30 illustrates the decay curve of the SOTEM transient electromagnetic response influenced by a sinkhole column, with the observation point situated directly above the center of the sinkhole. For comparison, the response curve of the layered background model is also included. As seen in Fig. 3.30a, the decay curve of the electric field Ex response under the influence of the sinkhole column is higher than the background response in the early stage, but lower in the late stage. Figure 3.30b depicts the decay curve of the magnetic field ∂Bz/∂t response, which exhibits a similar trend to the loop source response. Initially, the response curve of the sinkhole column model coincides with that of the background model, but as time progresses, the response curve of the sinkhole column model begins to diverge from the background model's curve. Overall, the ∂Bz/∂t curve under the influence of the sinkhole column is lower than the background model. Therefore, although there are large differences in the ∂Bz/∂t responses on different measurement lines, the decay pattern of ∂Bz/∂t directly above the center of the sinkhole column is similar to that of the loop source.

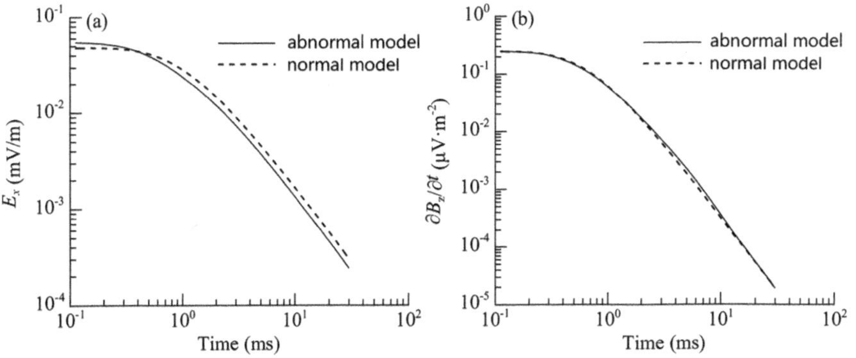

Fig. 3.30 Decay curve of SOTEM transient electromagnetic response under the influence of a water-filled sinkhole column

3.5 Characteristics and Curves of SOTEM Response in Geothermal Model

According to previous geological surveys, a low-resistance layer (3.28 Ω·m) about 200 m thick exists within 450 m depth in the Yangbajing geothermal field, confirmed by drilling as a shallow thermal reservoir (Wenhua et al. 1985). Additionally, the near-surface media in the northern and southern areas consist mainly of altered granite and Quaternary sediments, covering the shallow thermal reservoir (Duoji et al. 1997). A high-resistivity, high-density Paleozoic metamorphic rock beneath the China-Nepal Highway extends below 2000 m, separating the north and south areas of the Yangbajing geothermal field and blocking the lateral migration of geothermal fluid from northwest to southeast (Liao Zhijie et al. 1999). Combining the current SOTEM inversion results, a simplified model of the Yangbajing geothermal field was established for further research, as shown in Fig. 3.16. In the figure, the brown shallow layer represents the near-surface altered granite and Quaternary sediments, with a resistivity of 200 Ω·m. The red area below the cover layer indicates the shallow and deep thermal reservoirs, with the shallow reservoir extending 3000 m (y-axis) and two deep reservoirs each extending 1000 m, both having a resistivity of 5 Ω·m. The purple area in the middle represents the Paleozoic metamorphic rock under the China-Nepal Highway, with a resistivity of 1000 Ω·m. The light blue sections represent the geothermal fluid on both sides, with a resistivity of 20 Ω·m, extending throughout the entire calculation domain. The dark blue area represents the surrounding medium, with a resistivity of 500 Ω·m. The source is arranged along the x-axis, 2000 m long, with its center point at the origin. The center point of the observation area is also the origin, measuring 3000 × 2300 m, with both line spacing and point spacing set to 100 m. The emission current is 20 A (Fig. 3.31)

Figure 3.32 shows the horizontal electric field *Ex* and vertical induced electromotive force *dBzdt* response curves of the Yangbajing geothermal field model within the range of [0.05, 100] ms (referencing the actual field exploration time window of [1,

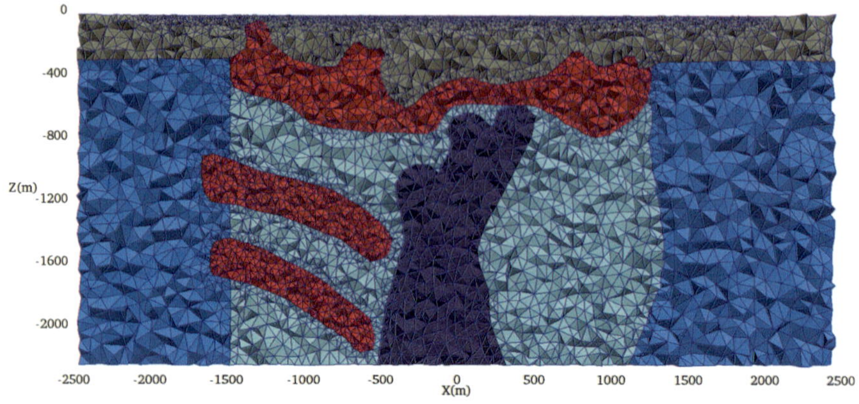

Fig. 3.31 Cross-sectional schematic of the Yangbajing geothermal field model

100] ms). The observation point coordinates are (−1000, −450, 0), (−500, −450, 0), (−1000, −250, 0), (−500, -250, 0), (−1000, −50, 0), and (−500, −50, 0). The solid and dashed lines in the figure represent the response curves when the deep thermal reservoir is present or absent. As shown in Fig. 3.17, the deep thermal r Ex response and dBz/dt response is not significant. This indicates that the deep thermal reservoir may be distinguishable through the SOTEM horizontal electric field component, but it cannot be identified through the vertical induced electromotive force response.

The following analysis focuses on the influence of near-surface altered granite and Quaternary sediments (shallow cover) and shallow thermal reservoirs on the detection of deep thermal reservoirs. Figures 3.33 and 3.34a–c show the xoy plane distribution of the horizontal electric field Ex and vertical induced electromotive force dBz/dt responses of the Yangbajing geothermal field model, which simultaneously

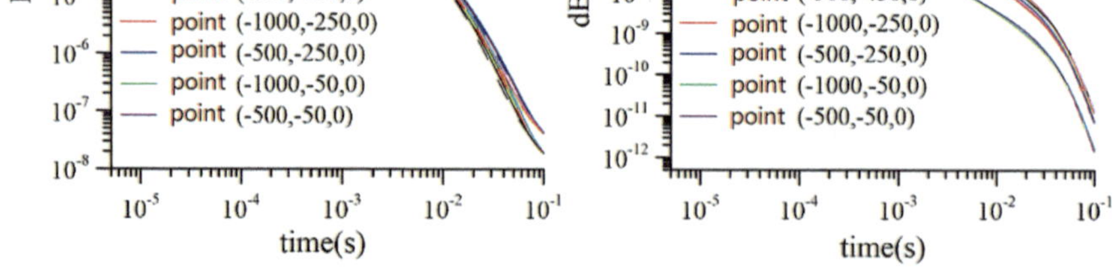

Fig. 3.32 Response curves of the Yangbajing geothermal field model when the deep thermal reservoir exists or does not exist **a** Ex **b** dBz/dt

contains both shallow cover and shallow thermal reservoir, only shallow cover, or only a shallow thermal reservoir at t = 1, 10, and 50 ms. Comparing figures (a) and (b), we find that the shallow thermal reservoir affects both the distribution pattern and amplitude of the Ex and dBz/dt responses. Notably, at t = 50 ms, the Ex response of the model without a shallow thermal reservoir shows a relatively low value near x = – 1000, which is inferred to be caused by deep heat storage. Comparing figures (a) and (c), we observe that the impact of shallow coverage on Ex and dBz/dt responses is concentrated in the early stage (t = 1 ms), with almost no effect on the late-stage (t = 10 ms and 50 ms) responses. The main effect is on the response amplitude, with little impact on the distribution shape. Overall, the shallow heat storage with lower resistivity has a greater impact on Ex and dBz/dt responses. Regardless of the presence of shallow coverage and shallow heat storage, dBz/dt responses cannot distinguish deep heat storage. In other words, the multi-layer heat storage structure of the Yangbajing geothermal field affects the detection depth of SOTEM. It is suggested to reduce the base frequency emission and increase the emission current in future explorations.

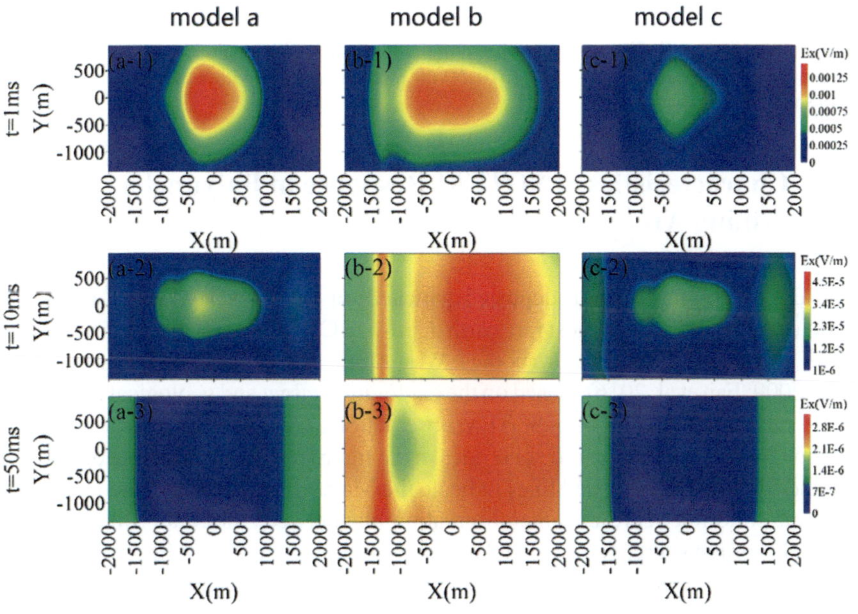

Fig. 3.33 Distribution of Ex responses for different Yangbajing geothermal field models (a-1) ~ (a-3) t = 1, 10, 50 ms model responses with shallow coverage and shallow heat storage (b-1) ~ (b-3) t = 1, 10, 50 ms model responses with shallow coverage but without shallow heat storage (c-1) ~ (c-3) t = 1, 10, 50 ms model responses without shallow coverage but with shallow heat storage

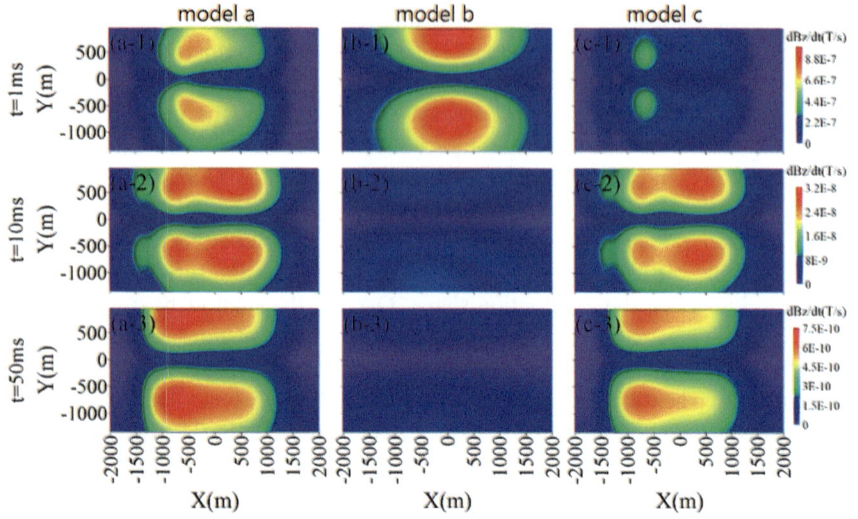

Fig. 3.34 Distribution of dBz/dt responses for different Yangbajing geothermal field models at t = 1, 10 and 50 ms (a-1) ~ (a-3) t = 1, 10, 50 ms model responses with shallow coverage and shallow heat storage (b-1) ~ (b-3) t = 1, 10, 50 ms model responses with shallow coverage but without shallow heat storage (c-1) ~ (c-3) t = 1, 10, 50 ms model responses without shallow coverage but with shallow heat storage

3.6 Characteristics of SOTEM Response in Baiyunebo Mining Area

The Baiyunebo multi-metal deposit is situated on the northern edge of the North China Craton and is the largest rare earth deposit in China, as well as a significant iron mine. It includes three main ore bodies: West Mine, Main Mine, and East Mine (Jian et al. 2009; Fan et al. 2016). Based on the exploration target and geological conditions in the survey area, two types of SOTEM observation devices were primarily designed, as shown in Fig. 3.35. The red solid lines in the figure represent the emission sources S1 and S2. The black dashed lines indicate the Main Mine and East Mine pits, while the blue solid line represents the observation line between the Main Mine and East Mine. Emission source S1 is located on the west side of the Main Mine, oriented in a north–south direction, and emission source S2 is positioned on the north side of the mine pit, oriented in an east–west direction. Both sources have a length of 2000 m, and the observation line is also 2000 m long.

(1) Influence of Main East Mine Pit

The Main Mine and East Mine pits significantly impact the observed signal. First, we analyze their effects on the horizontal electric field (Ey or Ex) and vertical induced electromotive force (dBz/dt) using a pit model of uniform half-space with slate as the background, set to a resistivity of 500 Ω·m. Figures 3.36 and 3.37 illustrate

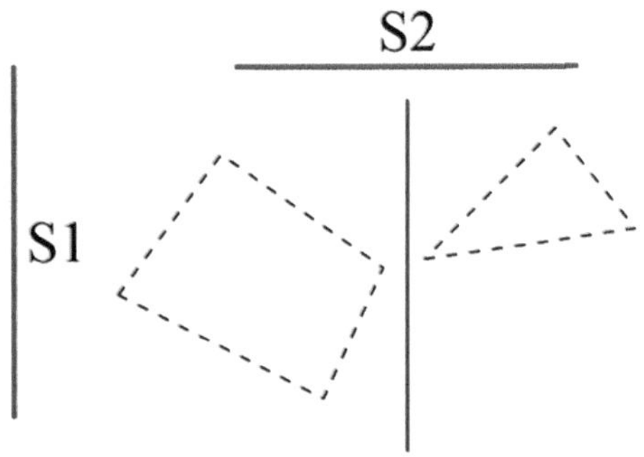

Fig. 3.35 Schematic diagram of SOTEM observation device

the distribution of the ground horizontal electric field (Ey/Ex) and vertical induced electromotive force (dBz/dt) responses for emission sources S1 and S2, with the black dashed lines indicating the positions of the Main Mine and East Mine pits. As shown in the figures, similar to a valley terrain, when the maximum values of Ey or Ex and dBz/dt do not yet diffuse into the pit area (t = 0.1 ms), the pit accelerates field diffusion. Once the maximum values enter the Main Mine or East Mine pits (t = 1, 5 and 10 ms), the pits hinder further diffusion, resulting in local maxima at their positions. Clearly, the Main Mine and East Mine pits exert a more pronounced effect on the horizontal electric field, with the pit closer to the emission source having a greater impact. Conversely, the farther pit influences the vertical induced electromotive force more significantly. Specifically, when the emission source is S1, the Main Mine pit has a stronger effect on the horizontal electric field Ey, while the East Mine pit affects the vertical induced electromotive force dBz/dt more. When the emission source is S2, the East Mine pit has a greater impact on the horizontal electric field Ex, whereas the Main Mine pit influences the vertical induced electromotive force dBz/dt more.

Additionally, the Main Mine and East Mine significantly affect the horizontal electric field and vertical induced electromotive force responses along the measurement line between the two mines. We examine observation points A (166,500, 4,639,500, 0), B (166,500, 4,640,000, 0), C (166,500, 4,640,500, 0), and D (166,500, 4,641,000, 0) as examples. As illustrated in Fig. 3.23, observation points B and C are located between the Main Mine and East Mine. From Fig. 3.38a–b, when the source is S1, both mines increase the amplitude of the early Ey and dBz/dt responses along the observation line while slightly reducing the amplitude of the late responses. Overall, the mines' impact on the response is primarily concentrated in the early stage, with more pronounced effects observed at points B and C. In contrast, when the source is S2, the mines exert a weaker influence on the response compared to when the source is S1, with minimal differences between the early and late responses. Here, the early Ex and dBz/dt responses at observation points A and B are enhanced, while the late responses are weakened. Simultaneously, the Ex responses at points C and D, as well as the dBz/dt response at point C, generally weaken, and the late dBz/dt response at point D shows a slight enhancement.

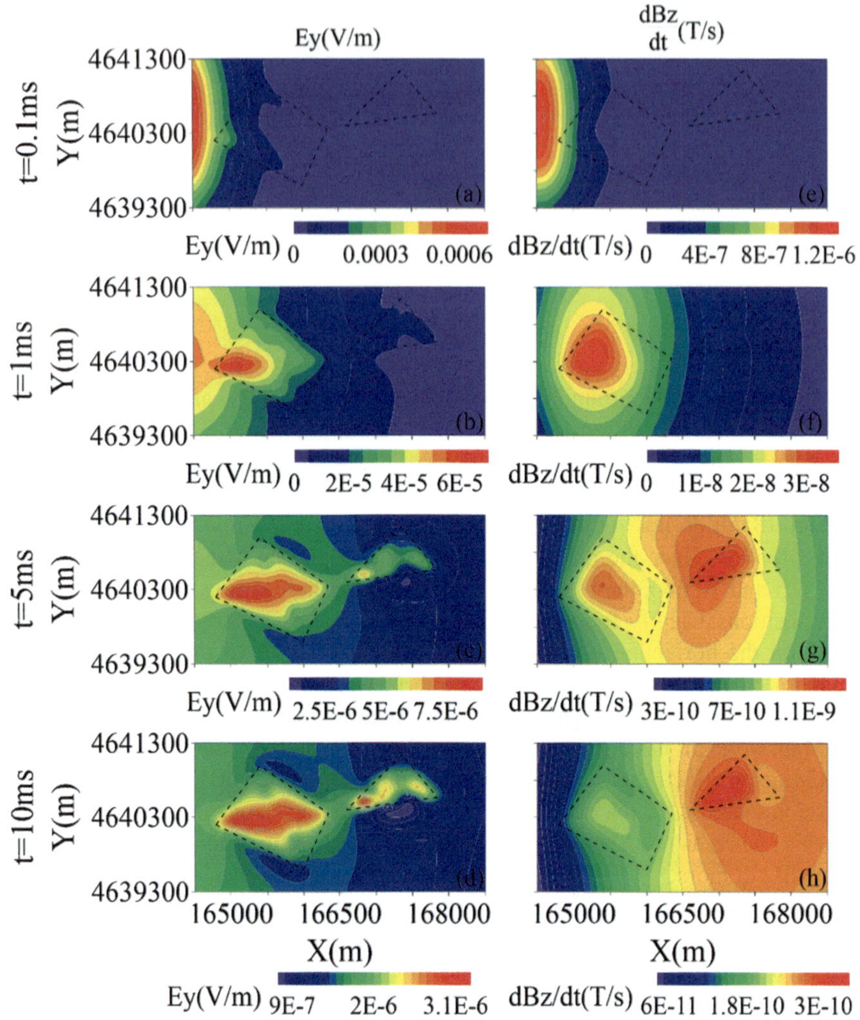

Fig. 3.36 Distribution of pit model responses when the emission source is S1 **a**, **b**, **c**, **d** $t = 0.1$, 1, Ey response at 5, 10 ms **e**, **f**, **g**, **h** $t = 0.1$, 1, 5, 10 ms dBz/dt response

(2) Construction suggestions

Based on the geological conditions of the Baiyun Ebo mining area, a polymetallic deposit model has been established, as shown in Fig. 3.39. In Fig. 3.39a, the ground boundary line represents the boundary between the iron ore body and the surrounding dolomite (blue) and slate (light blue). The encrypted sections in the 3D cross-section of Fig. 3.39b from left to right depict the East Mine iron ore body, the iron ore body between the East Mine and Main Mine, labeled as Fe3, Fe2, and Fe1, respectively, with Fe2 being the target for subsequent exploration. According to available

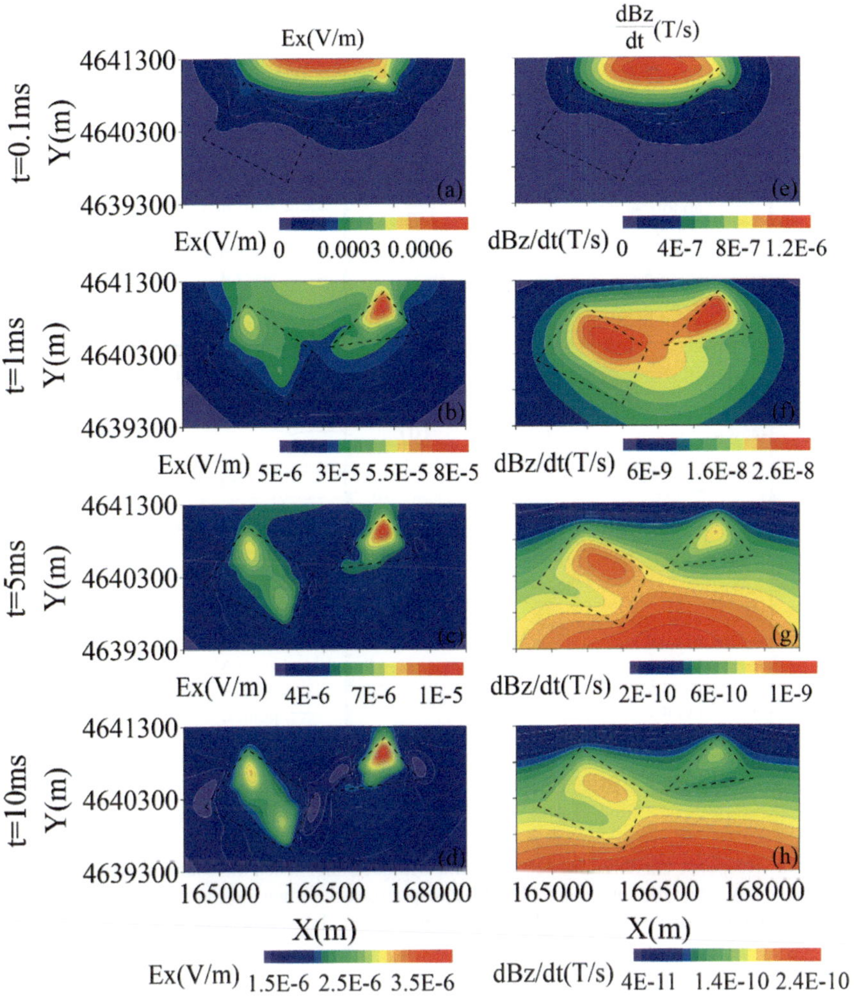

Fig. 3.37 Response distribution of the mine model when the source is S2 **a**, **b**, **c**, **d** $t = 0.1, 1, 5, 10$ ms Ex response **e**, **f**, **g**, **h** $t = 0.1, 1, 5, 10$ ms dBz/dt response

geological and drilling data, the Fe2 ore body is assumed to be 300 m long in the east–west direction (from 166,250 to 166,550), 200 m long in the north–south direction (top surface at 4,640,800–4,641,000, bottom surface at 4,639,850–4,640,050), and extends 995 m vertically (from 55 to 1050). The dip angles for Fe1, Fe2, and Fe3 are 55°, 45°, and 60°, respectively, with the strike taken as the approximate true value. The orientations of the dolomite and slate align with that of the iron ore body. Additionally, the resistivities of dolomite, slate, and the iron ore body are approximately 2000 Ω·m, 500 Ω·m, and 50 Ω·m, respectively. The following analysis will

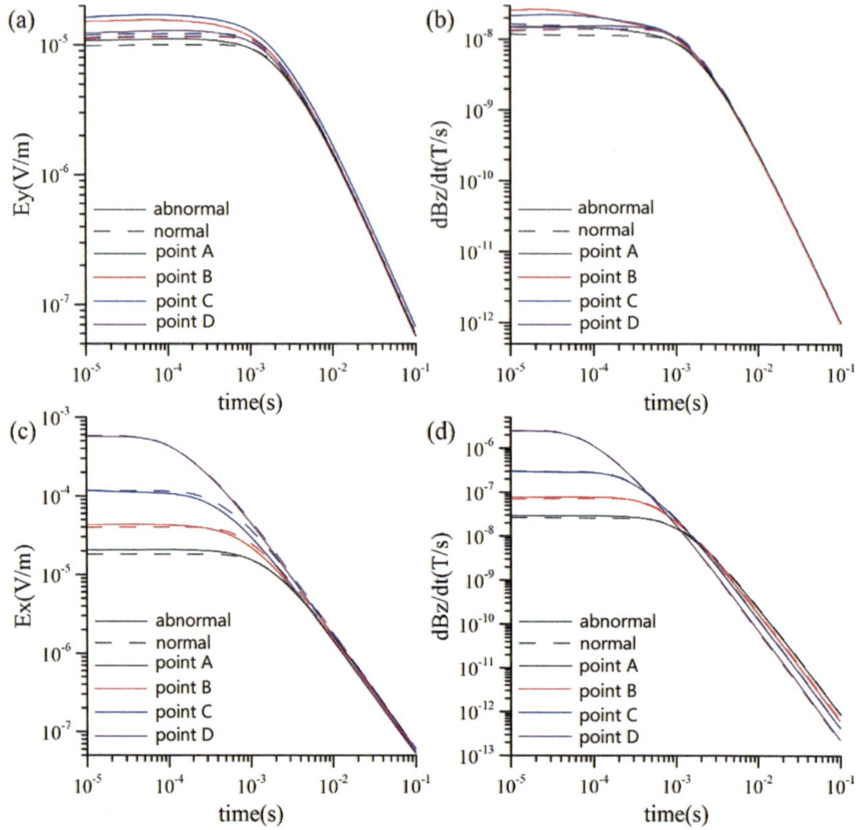

Fig. 3.38 Response curves of models with and without mines when the source is S1 and S2 **a** Ey response curve when the source is S1 **b** dBz/dt response curve when the source is S1 **c** Ex response curve when the source is S2 **d** dBz/dt response curve when the source is S2

focus on the response curves and relative anomaly curves of the Baiyun Ebo polymetallic deposit model at observation points E (166,500, 4,639,300, 0), F (166,500, 4,639,780, 0), G (166,500, 4,640,300, 0), H (166,500, 4,640,780, 0), and I (166,500, 4,641,300, 0), as well as the model without Fe2.

Figures 3.40 and 3.41 illustrate the response curves and relative error curves for the Baiyun Ebo polymetallic deposit model at observation points E, F, G, H, and I, corresponding to sources S1 and S2. The analysis reveals the following key observations:

(1) Whether the source is S1 or S2, the relative anomaly of the horizontal electric field (Ey or Ex) and the induced electromotive force (dBz/dt) at observation points E and F is close to 0. This implies that the presence of the Fe2 ore body cannot be determined at these locations;

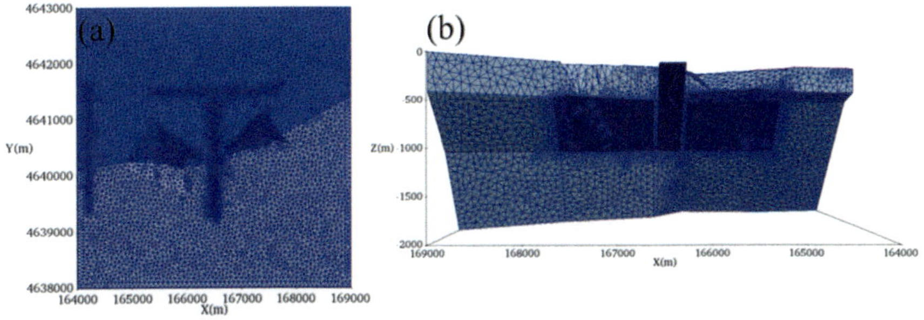

Fig. 3.39 Schematic diagram of Baiyun Ebo polymetallic deposit model **a** Plan view **b** 3D cross-section view

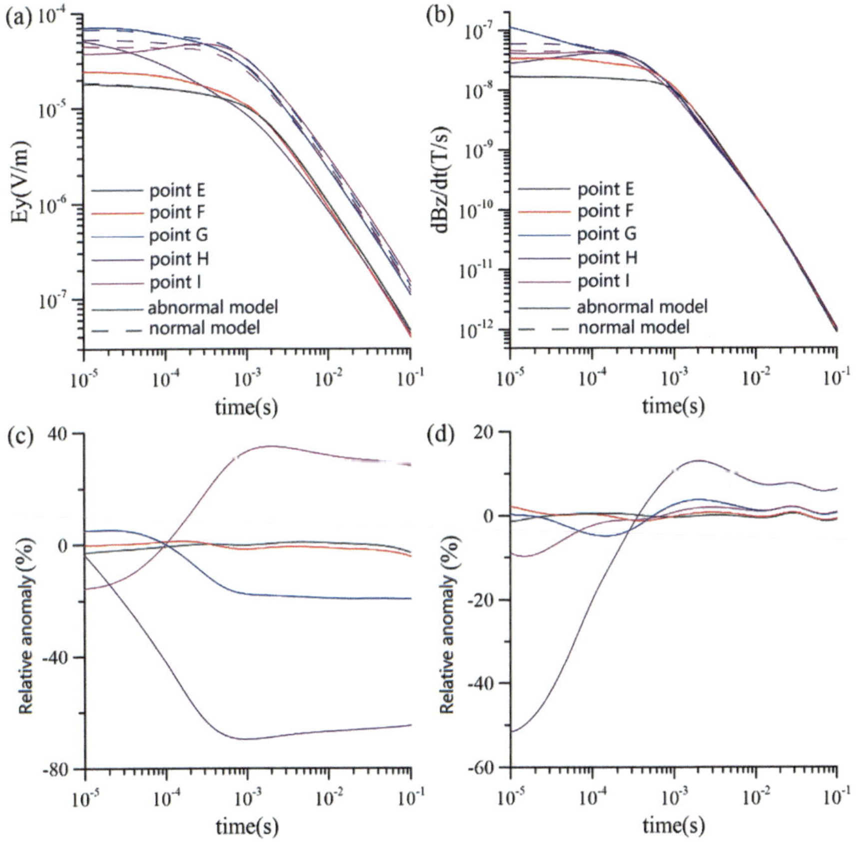

Fig. 3.40 Response curves and relative anomaly curves of Ey and dBz/dt for the Baiyun Ebo multi-metal ore deposit model when the emission source is S1 **a** Ey response curve **b** dBz/dt response curve **c** Ey response relative anomaly curve **d** dBz/dt response relative error curve

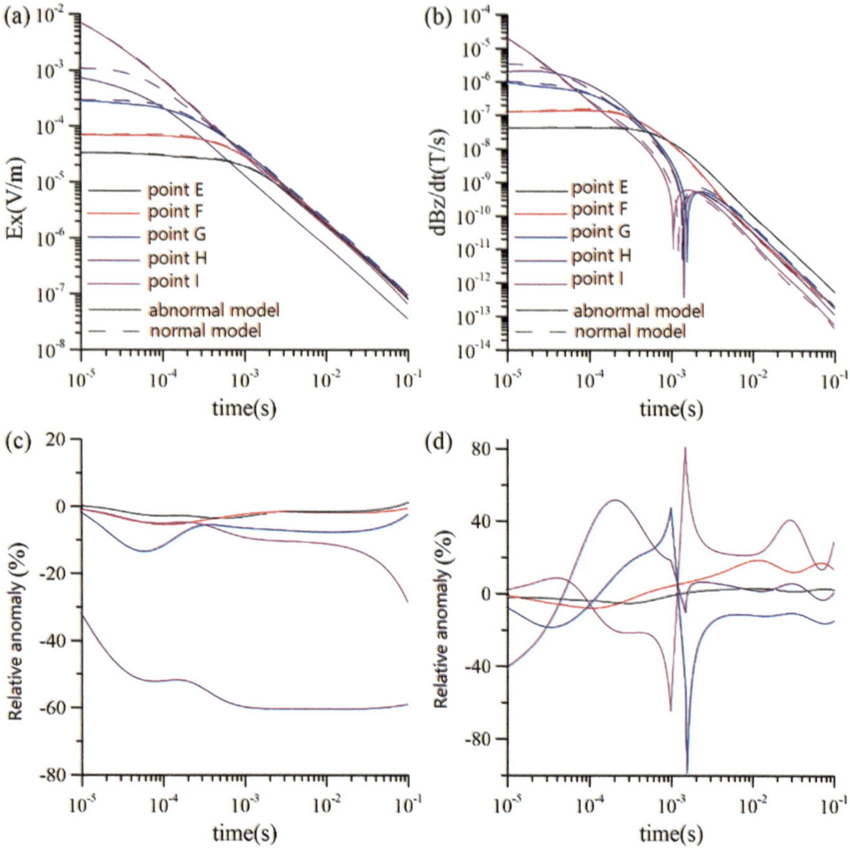

Fig. 3.41 Response curves and relative error curves of Ex and dBz/dt for the Baiyun Ebo multi-metal ore deposit model when the emission source is S2 **a** Ey response curve **b** dBz/dt response curve **c** Ey response relative anomaly curve **d** dBz/dt response relative error curve

(2) Whether the source is S1 or S2, the existence of Fe2 ore can be determined by the decrease in the amplitude of the horizontal electric field (Ey or Ex), resulting in a negative relative anomaly, and the increase in the amplitude of the late induced electromotive force (dBz/dt), resulting in a positive relative anomaly at observation points G and H. However, for source S2, the response curves at observation points G and H show a significant "negative value" phenomenon, which is unfavorable for subsequent data interpretation.

(3) When the emission source is S1, the Ey and dBz/dt at observation point I first weaken and then strengthen. In contrast, when the emission source is S2, the Ex response at observation point I weakens while the late dBz/dt response strengthens. This pattern can also be used to indicate the existence of the ore body; however, the relative anomaly at this point is much smaller than that at observation points G and H.

Overall, both types of observation devices can effectively distinguish between the main ore and the Fe2 ore in the eastern region, with the horizontal electric field exhibiting stronger resolution. Considering the influence of the mine pit and the geological conditions in the survey area, it is recommended to use the construction device with the emission source S1 and to observe the horizontal electric field (Ex) and the vertical induced electromotive force (dBz/dt). Additionally, it is advisable to account for the effects of the mine pit and the shape of the emission source during the later data processing.

3.7 Characteristics of SOTEM Response to Pegmatite Lithium Ore Model

As a new industrial resource, lithium is often referred to as the "green high-energy metal of the twenty-first century" and "white oil." It is an indispensable raw material for low-carbon development (Lijun et al. 2017; Mingguo et al. 2019; Sovacool et al. 2020). In recent years, the rapid advancement of information technology and electric vehicles has significantly increased the demand for metallic lithium across various countries, sparking a research boom in lithium ore exploration. So far, the discovered lithium ores are primarily of two types: brine type (66%) and pegmatite type (26%), with sedimentary types being less common (only 8%) (Kesler et al. 2012; Hu Xiaojun and Huan 2021; Xiwangwan et al. 2023). Current mining, selection, and smelting technologies mainly focus on pegmatite lithium ore, which has a high degree of development and substantial output.

However, conventional methods such as gravity, magnetic, direct current, and audio frequency magnetotelluric techniques are not ideal for detecting pegmatite lithium ore. This is due to two main reasons: First, pegmatite lithium ores exhibit significant variations in physical properties (density, resistivity, etc.) and morphology depending on their ore-forming environments. Second, pegmatite lithium ore is typically found in deep metamorphic rocks of the Archean granite-greenstone belt, Paleozoic basic intrusive rocks, Mesozoic slate and schist, and granite that is contemporaneous with the pegmatite vein. The ore bodies often take the form of vesicles, lenses, or veins (Wang Denghong et al. 2017; Kezhang et al. 2021). The physical property differences between the parent rock pegmatite and surrounding rock are not pronounced, and the scale of mineralization is generally small. Consequently, conventional geophysical exploration methods are primarily employed for preliminary regional geological surveys or prospecting area forecasts (Xiaofang et al. 2015, 2019, 2021; Huang et al. 2018; Rong et al. 2020; Junbo et al. 2022).

Currently, the application of the time-domain electromagnetic method in pegmatite lithium ore exploration is limited. As a new type of deep, high-precision detection device, SOTEM has proven instrumental in the exploration of various mineral resources. Thus, it is essential to discuss and analyze the detection capabilities of SOTEM for pegmatite lithium ore.

(1) Layered pegmatite lithium ore model

First, we analyze the resolution capability of SOTEM for high-resistance pegmatite lithium ore bodies situated in a relatively high-resistance background using a simple layered model. Referring to the model of the Jiajika rare metal ore field used by Junbo (2019), we designed the pegmatite lithium ore model shown in Fig. 3.42 (notably, this section does not consider the fourth series cover layer). In Fig. 3.42a, the black solid line represents the emission source arranged along the x-axis, which is 2000 m long with its center coordinate at (0, 1500, 0). The red and blue solid line boxes indicate the top surface of the pegmatite lithium ore and the ground, respectively. The pegmatite lithium ore has a 70° inclination angle, runs along the y-axis, measures 1000 m in length (from 2000 to 3000), is 100 m wide along the x-axis, and has a height of 400 m in the vertical direction (z-axis). The projection coordinate of its center point on the ground is (500, 2500, 0), with a resistivity of 8000 Ω·m. As illustrated in Fig. 3.42b, the surrounding rock of the ore body is assumed to be slate, 1000 m thick, with a resistivity of 2000 Ω·m, while the deeper magma rock is 3000 m thick, with a resistivity of 15,000 Ω·m.

Figure 3.43 illustrates the distribution of ground Ex and dBz/dt responses at times t = 0.01, 0.05, 0.2, 0.8 and 1 ms. Figure 3.44 further provides the Ex and dBz/dt response curves at the observation point (0, 2500, 0), comparing scenarios with and without the presence of lithium ore. From Fig. 3.43a–e, it is evident that the high-resistance, inclined plate-shaped lithium ore body accelerates the diffusion of the Ex field in the positive direction of the y-axis. This results in the Ex field contour line protruding in that direction at t = 0.01, 0.05, and 0.2 ms, with a corresponding increase in the response amplitude at the observation point. At t = 0.8–1 ms, a distinct high-value area forms around the anomaly. In terms of the dBz/dt response, the influence of the pegmatite lithium ore body is more complex. When the dBz/dt field diffuses toward the ore body at t = 0.01 ms, the diffusion speed of the dBz/dt field in the positive direction of the y-axis increases, causing the dBz/dt contour line to protrude in that direction and resulting in a higher ground response amplitude. However, as the maximum value of the dBz/dt field approaches the anomaly at t = 0.05–0.2 ms, the lithium ore body hinders further diffusion of the field maximum in the positive direction of the y-axis. This causes the dBz/dt response contour line to bend slightly toward the negative direction of the y-axis, leading to a decrease in the ground response amplitude. At t = 0.8–1 ms, the dBz/dt contour line appears to be "repelled" by the lithium ore body, deforming in either the positive or opposite direction of the y-axis.

Overall, both the horizontal electric field (Ex) and the vertical induced electromotive force (dBz/dt) components of SOTEM demonstrate a degree of resolution capability for high-resistance pegmatite lithium ore situated in relatively high-resistance media. However, the resolution ability of the horizontal electric field is notably stronger.

(2) Dome-shaped pegmatite lithium ore model

For actual pegmatite lithium ore deposits, the underlying magma rock typically appears as a dome, with pegmatite veins distributed around the top of the dome. Additionally, there is often a shallow Quaternary cover in the mining area. This shallow

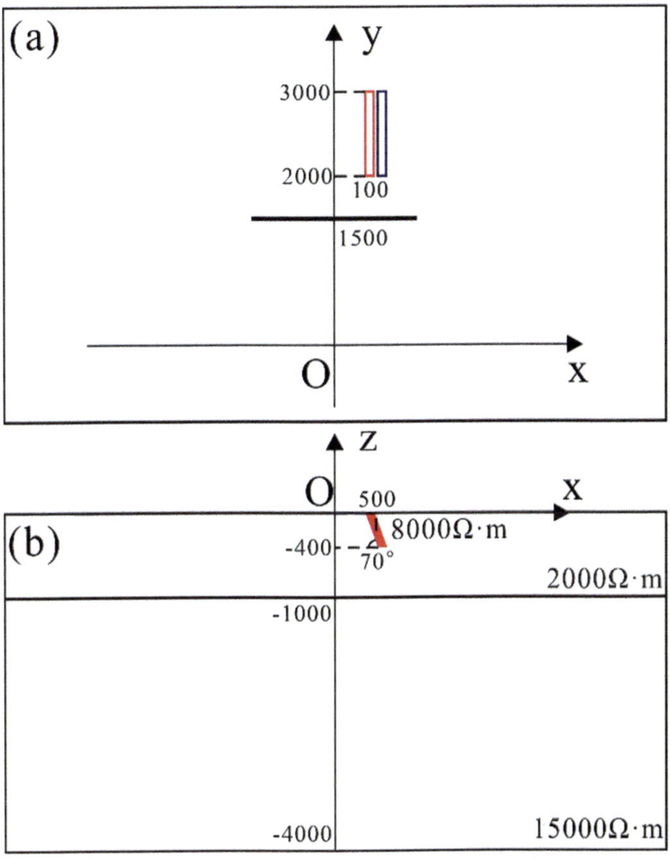

Fig. 3.42 Schematic diagram of pegmatite lithium ore model **a** Ground projection map **b** XOZ cross-section map

low-resistance cover and common geological structures can affect the SOTEM response, thereby interfering with the identification of anomalies. Therefore, the following further analyzes the interference of dome-shaped magma rock and shallow cover on the detection of pegmatite lithium ore.

As shown in Fig. 3.45, the shallow brown area represents the shallow Quaternary cover, which has a thickness of 200 m and a resistivity of 500 Ω·m (refer to Junbo Fan 2019). The light blue area represents the surrounding rock (also slate, with the same resistivity), while the cyan area indicates the dome-shaped magma rock, characterized by a resistivity of 15,000 Ω·m. The portion of the magma rock exposed at the surface measures 800 m long (from $x = -500$ to 300) and 200 m wide (from $y = 200$ to 400). The red inclined plate-shaped body represents the pegmatite lithium ore vein, with the same geometric and physical parameters. To clearly distinguish the influence of the dome-shaped magma rock and shallow cover on the identification of pegmatite lithium ore, this section calculates the responses of the dome-shaped

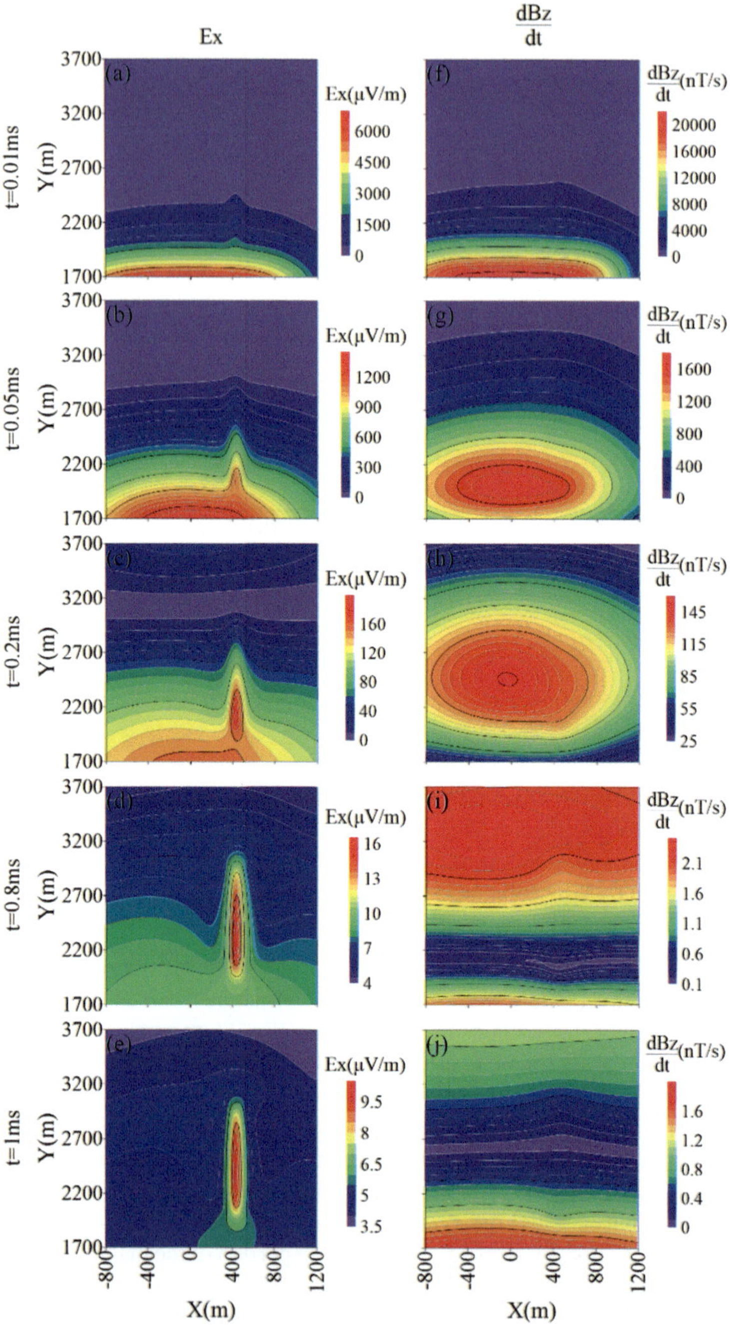

Fig. 3.43 Distribution of responses to the pegmatite lithium ore model at different times **a**–**e** $t = 0.01, 0.05, 0.2, 0.8, 1$ ms Ex response **f**–**j** $t = 0.01, 0.05, 0.2, 0.8, 1$ ms dBz/dt response

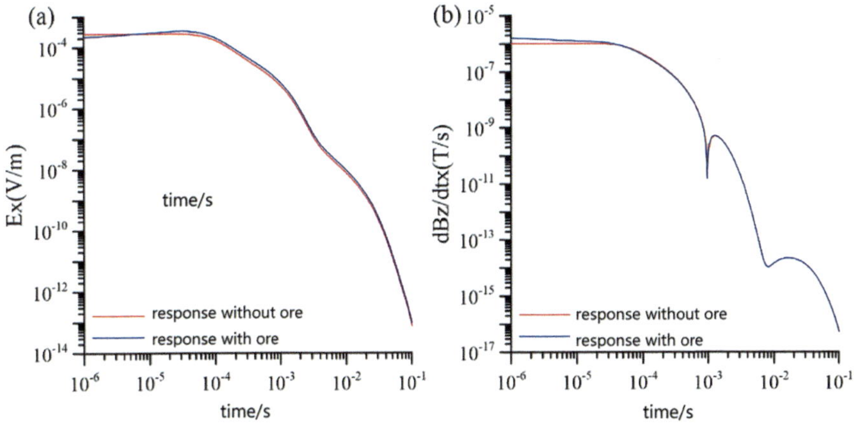

Fig. 3.44 Response curves of the pegmatite lithium ore model at the observation point (0, 2500, 0) **a** Ex **b** dBz/dt

pegmatite lithium ore under three scenarios: (1) including both the lithium ore body and shallow cover, (2) including the lithium ore body but excluding the shallow cover, and (3) excluding both the lithium ore body and shallow cover. Figure 3.46 first presents the Ex response distribution of the three models at different times. From Fig. 3.46a–c, it is evident that the dome-shaped magma rock influences the Ex field by accelerating field diffusion and producing a clear positive anomaly near the dome, particularly noticeable at the interfaces where $x = -500$ and $x = 300$. Comparing Fig. 3.46a–c with (d–f), although the dome accelerates the diffusion of the Ex field, it does not impair the ability to distinguish pegmatite lithium ore. Similarly, comparing Fig. 3.46d–f with (g–i), the shallow Quaternary cover layer hinders the downward diffusion of the Ex field. However, the position of the pegmatite lithium ore can still be inferred from the high-value anomaly in the Ex plane response distribution at $t = 0.2$–1 ms.

Figure 3.47 further illustrates the dBz/dt response of the three models at different times. From Fig. 3.47a, d, at $t = 0.01$ ms, the dBz/dt contour lines converge near the high-resistance dome, and the contour gradient increases, indicating that the high-resistance dome has an attractive effect on the dBz/dt field. Simultaneously, this attractive effect diminishes the influence of the lithium ore body. At $t = 0.2$–1 ms, the dBz/dt field contour lines are clearly convex in the positive direction of the y-axis, suggesting that the high-resistance dome accelerates the diffusion of the surrounding dBz/dt field, producing a repulsive effect on it. Although the influence of the high-resistance dome interferes with the ground dBz/dt response in detecting the lithium ore body to some extent, it does not completely obscure the abnormal response generated by the high-resistance pegmatite lithium ore body. Similarly, in Fig. 3.47g–i, the shallow Quaternary cover slows down the diffusion of the dBz/dt field. However, the dBz/dt response at $t = 0.2$–1 ms still retains the ability to distinguish the lithium ore body.

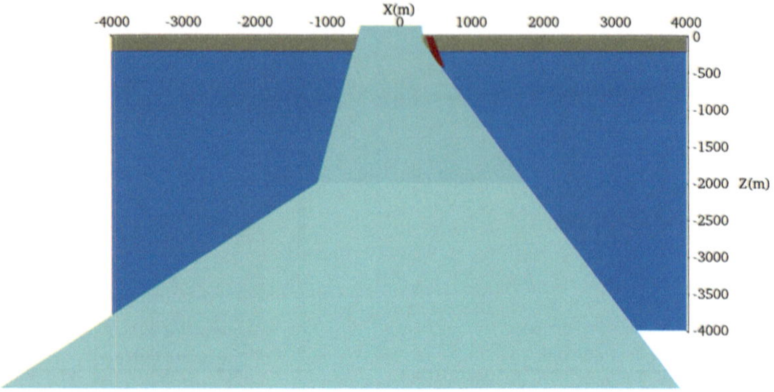

Fig. 3.45 Schematic diagram of the dome-shaped pegmatite lithium ore model

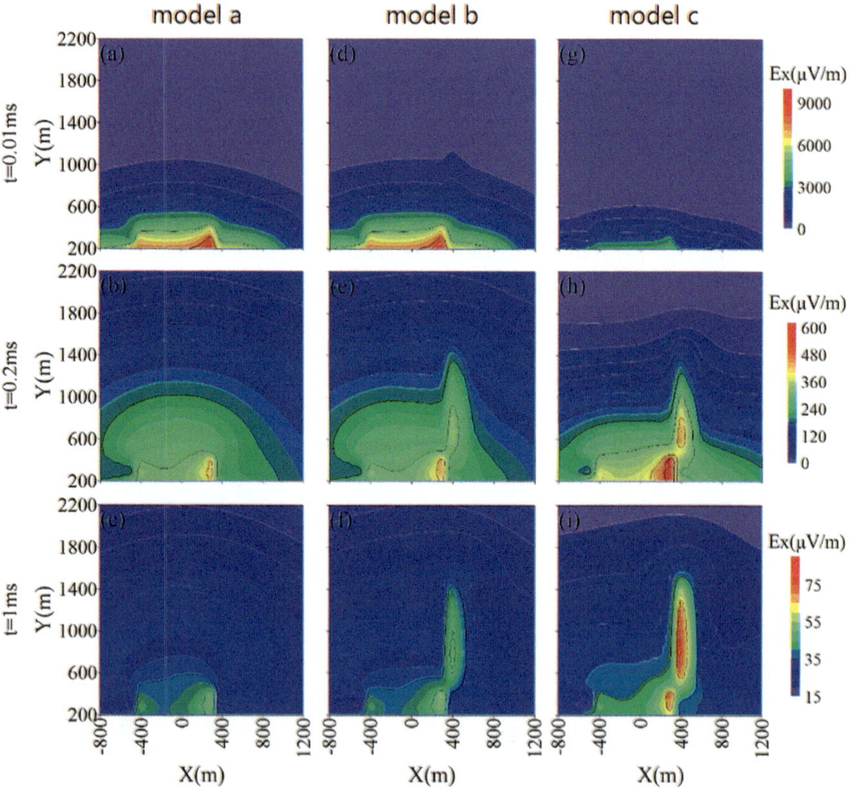

Fig. 3.46 Distribution of Ex response at different times for the dome-shaped pegmatite lithium ore model **a**, **b**, **c** $t = 0.01, 0.2, 1$ ms response of the model without lithium ore body and shallow cover **d**, **e**, **f** $t = 0.01, 0.2, 1$ ms response of the model without shallow cover but with lithium ore body **g**, **h**, **i** $t = 0.01, 0.2, 1$ ms response of the model with both lithium ore body and shallow cover

In summary, SOTEM can effectively be used for the exploration of pegmatite lithium deposits. While the Ex field is susceptible to various influencing factors, it possesses a strong ability to distinguish lithium ore bodies. The presence of dome-shaped high-resistance magmatic rock and shallow Quaternary cover does not compromise the Ex field's distinguishing capability. Therefore, under field conditions, it is advisable to prioritize the observation of the horizontal electric field (Ex) component. However, considering the large differences in the physical properties and shapes of pegmatite lithium ore bodies formed in different ore-forming environments, it is recommended to establish a deposit model based on existing geological data before field exploration, demonstrate the ability of the SOTEM method to distinguish ore bodies through three-dimensional forward simulation, and select the appropriate construction plan and construction parameters by analyzing response characteristics.

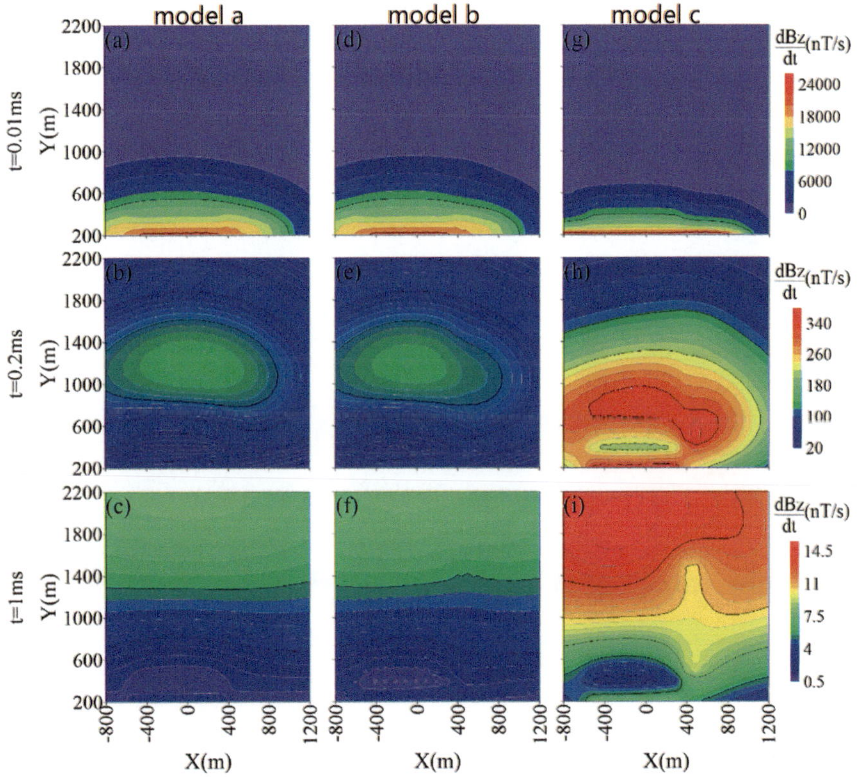

Fig. 3.47 Distribution of dBz/dt response at different times for the dome-shaped pegmatite lithium ore model **a**, **b**, **c** $t = 0.01$, 0.2, 1 ms response of the model without lithium ore body and shallow cover **d**, **e**, **f** $t = 0.01$, 0.2, 1 ms response of the model without shallow cover but with lithium ore body **g**, **h**, **i** $t = 0.01$, 0.2, 1 ms response of the model with both lithium ore body and shallow cover

References

Cox L H, Zhdanov M S. Advanced computational methods of rapid and rigorous 3-D inversion of airborne electromagnetic data[J]. Communications in Computational Physics, 2008, 3(1): 160-179.

Guoqiang, X.; Xiu, L. The Technology of TEM Tunnel Prediction Imaging. Chinese J. Geophys. 2008, 51, 894–900.

Maaø F A. Fast finite-difference time-domain modeling for marine-subsurface electromagnetic problems[J]. Geophysics, 2007, 72(2): A19–A23.

SanFilipo W A, Hohmann G W. Integral equation solution for the transient electromagnetic response of a threedimensional body in a conductive half-space[J]. Geophysics, 1985, 50(5): 798–809.

Xue Guoqiang, Li Xiu, Di Qingyun. Progress in the theory and application of transient electromagnetic method[J], Progress in Geophysics, 2008, 22(4): 1195-1200.

Um E S, Harris J M, Alumbaugh D L. 3D time-domain simulation of electromagnetic diffusion phenomena: A finite-element electric-field approach[J]. Geophysics, 2010, 75(4): F115-F126.

Wang, T. & Hohmann, G.W., 1993. A Finite-difference, Time-domain Solution for Three-dimensional Electromagnetic Modeling. Geophysics, 58: 797–809.

Chapter 4
SOTEM Inversion

The essence of electromagnetic data inversion is to find a reasonable geophysical model that matches the observed results through mathematical procedures, enabling the quantitative acquisition of underground structure distribution from the observed electromagnetic field. Although there are various electromagnetic detection methods, the core of their data inversion imaging algorithms remains the same. Generally speaking, electromagnetic data inversion imaging methods can be divided into three major categories: direct imaging methods, deterministic inversion methods, and stochastic inversion methods (as shown in Fig. 4.1).

Direct imaging methods obtain underground geological structures through direct transformation or equivalent substitution of observed data, including Bostick imaging of frequency domain electromagnetic methods, direct imaging methods of conductivity in time domain electromagnetic methods, S-inversion, Q transformation, offset imaging methods, and others. These methods are characterized by their fast imaging speed. Deterministic inversion methods include the least squares inversion method, Gauss–Newton method, and conjugate gradient method. These methods establish a regularized objective function, search for the optimal model along the gradient direction of the objective function, and aim to minimize the objective function. To find the optimal model that aligns with geological structures, constrained inversion methods and joint inversion methods have been further developed. Stochastic inversion methods include simulated annealing, particle swarm optimization, Bayesian inversion, and artificial intelligence inversion methods. These approaches do not rely on gradient information; instead, they conduct random searches in model space, allowing for the attainment of the global optimal solution of the objective function. Due to the extensive forward calculations involved, these methods are primarily applied in one-dimensional scenarios. With advancements in computer technology, stochastic inversion methods are gaining increasing attention, and some researchers have further applied them to two-dimensional and even three-dimensional models.

G. Xue et al., *Exploration with Short-Offset Grounded-Wire Transient Electromagnetic Method*, https://doi.org/10.1007/978-981-96-8846-3_4

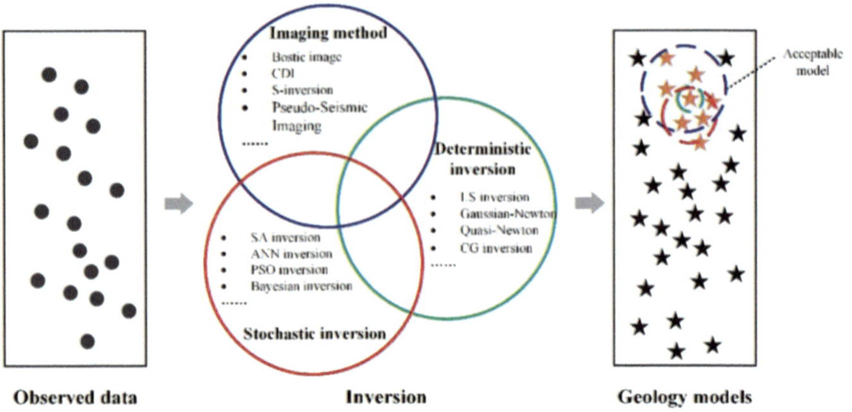

Fig. 4.1 Electromagnetic data inversion methods

4.1 Imaging Methods

4.1.1 Conductivity Depth Imaging Method

Conductivity depth imaging is a time-domain approximate inversion method. Regardless of the shape and electrical properties of the medium, its electromagnetic response can be represented by the sum of decay exponents, allowing for the calculation of apparent conductivity that varies with depth. The methods mainly include: the Maxwell mirror source calculation method, an approximate imaging method based on the depth of the maximum current layer in the half-space model, the step current response method calculated from any emission current waveform off-time data using convolution or deconvolution, the thin plate difference method, and the lookup method based on the pseudo-layer half-space model. Since the effective penetration depth of the electromagnetic field in the earth also depends on frequency or time, a function of apparent resistivity and depth can be derived through algebraic operations. Taking the thin plate method as an example, an equivalent conductive thin plate is introduced in free space. The plate is located at a certain depth from the surface and has a uniform integral conductivity S. For the step response of the vertical magnetic field excited by a vertical magnetic dipole, at sufficiently late times and near the source, the rate of change of the vertical magnetic field with time is analyzed is

$$\frac{\partial H_z^H(t)}{\partial t}\Big|_{z=-d_0} \approx -\frac{3M}{16\pi\mu_0 S}\frac{1}{(d_0+t/\mu_0 S)^4} \tag{4.1}$$

In order to accurately determine the depth d and conductivity S of the conductive thin plate in the formula, Tartaras et al. (2000) and Zhdanov (2002) further proposed

S-transformation and S-inversion based on the thin plate method. S-transformation involves deriving the formula, allowing for the simultaneous equation solution to be obtained directly.

$$S = \frac{16\pi^{1/3}}{\left(3M\sum n\right)^{1/3}\mu_0^{4/3}}\frac{(|V|)^{5/3}}{\left(|V|'\right)^{4/3}}d = \frac{1}{\mu_0 S}\left(-\frac{4|V|}{|V|'} - t\right) \tag{4.2}$$

where $|V| = \frac{3M\Sigma n}{16\pi S}\frac{1}{(d+t/\mu_0 S)^4}$, is the induced voltage received by the ideal receiving coil. S-transformation has the advantage of not requiring iterations or initial models; however, it also introduces potential errors into the interpretation results. Both S-transformation and S-inversion are characterized by their fast calculation speed and relatively accurate interpretation results. Xue and Deng (2007) and Xue et al. (2008) and Qi et al. (2022) further applied this method to transient electromagnetic tunnel ahead detection. Qi et al. (2022) also introduced this method into the data interpretation of multi-source semi-airborne transient electromagnetic detection, achieving positive application results in both cases.

4.1.2 Bostick Inversion

Bostick inversion is a method for direct imaging of frequency domain magnetotelluric data. The principle of this method is that the apparent resistivity at a specific frequency is solely related to the depth corresponding to that frequency and is independent of the electrical properties of the medium outside that depth.

Rodríguez et al. (2010) extended the Bostick imaging method of magnetotellurics to two dimensions, utilizing regularized Hopfield artificial neural networks to manage series and parallel magnetic resistances and their influence functions for analysis. The adaptive weighted average approximation method retains the partial nonlinearity of the original problem. By employing series and parallel impedances, this approach eliminates the need to determine the optimal rotation angle to identify TE and TM modes. Although the inversion results of the Bostick method may lack precision, its calculation speed is rapid, allowing it to intuitively illustrate the changes in underground geological structures with depth. It is primarily used for regional geoelectric structure research.

4.1.3 Offset Imaging

Offset imaging refers to seismic exploration theory, which directly obtains fast and accurate electrical parameters of underground complex media structures through offset methods based on different time or frequency electromagnetic response functions observed on the ground. It is mainly divided into the direct electromagnetic field offset method and the virtual wave field method.

The direct electromagnetic field offset method is an imaging technique that reverses the principles of seismic wave field analysis from the ground to the underground. Electromagnetic offset can be conducted in both the time domain and frequency domain, and the imaging results can reflect the position and shape of the electrical anomaly body. Zhdanov et al. (1996) primarily conducted experimental research on models of isolated and multiple electrical anomaly bodies, resulting in good imaging effects.

The virtual wave field method uses the mathematical relationship between the diffusion equation of the electromagnetic field and the virtual wave equation to transform the diffusion equation into the virtual wave equation. Since the virtual wave field shares characteristics with the seismic wave field, imaging interpretation methods from seismic exploration can be applied to it. According to Lee et al., there is an integral correspondence between the diffusion equation of transient electromagnetic fields in the time domain and the seismic wave equation.

Research on the transformation between the virtual wave field and the electromagnetic field can be divided into forward transformation and reverse transformation. Forward transformation research analyzes the characteristics of the virtual wave field, while reverse transformation research performs wave field imaging analysis.

In the virtual wave field domain, three-dimensional curved surface extension and synthetic aperture imaging methods are commonly used techniques. Synthetic aperture imaging technology borrows from radar synthetic aperture methods, synthesizing smaller aperture data into a larger equivalent transmitter coil, thereby improving resolution.

Taking the center point i as an example, select the measurement points on the left and right sides of point from −N to N to correlate with the center point. The normalized cross-correlation coefficient can be expressed as:

$$\lambda = \frac{\sum_{j=1}^{m} U(r_i, t_j) U(r_{i+k}, t_j - q)}{\left[\sum_{j=1}^{m} \left(U(r_i, t_j)\right)^2 \sum_{j=1}^{m} \left(U(r_{i+k}, t_j)\right)^2\right]^{1/2}} \tag{4.3}$$

where, m is the number of time channels for each measurement point, λ is the correlation degree of two columns of data, and the final synthetic value of the virtual wave field at the center point can be expressed as:

$$U'(r_i, t_j) = \sum_{k=-N}^{N\sum} \lambda k_k^m U\left(k_{j_{k\max}^m}\right) \tag{4.4}$$

Based on the Kirchhoff integral formula, energy from the same diffraction point scattered across various measurement points on the ground can be collected and transferred to the corresponding underground diffraction point, ultimately realizing three-dimensional curved surface extension Kirchhoff offset imaging of the virtual wave field.

$$U(r,t) = \frac{1}{4\pi}\iint_{xy} \frac{1}{r}\frac{\partial}{\partial n}[\mathbf{U}] - \frac{1}{vr}\frac{\partial r}{\partial n}\left[\frac{\partial \mathbf{U}}{\partial t}\right] ds \tag{4.5}$$

where, $[\mathbf{U}]$ is the delay value of the wave field, r is the distance from the calculation point to the integral scanning point, n is the normal vector of the boundary.

In the virtual wave field domain, in addition to conventional speed analysis and tomographic imaging techniques based on travel time analysis, full waveform inversion imaging can also be performed to enhance the detection of complex surrounding electrical structures.

The objective function of the full waveform inversion of the virtual wave field can be expressed as:

$$\Phi = \frac{1}{N}\sum_{1}^{N}(d - \mathbf{F}(m))^2 \tag{4.6}$$

where, N is the number of measurement points, $\boldsymbol{F}$ represents the virtual wave field forward operator, m is the model parameter of a single measurement point, d represents the measured data.

The development of reverse time migration technology, which is based on similar theory, has significantly advanced the theoretical refinement and practical application of full waveform inversion technology. The gradient of the objective function can be obtained using the adjoint state method:

$$\begin{cases} \lambda(T) = 0 \\ \frac{\partial \lambda(T)}{\partial t} = 0 \\ m\frac{\partial^2 \lambda}{\partial t^2} - \nabla^2 \lambda = \sum_{1}^{N} \mathbf{F}(d - \mathbf{F}(m)) \\ \mu_0 = m\frac{\lambda(0)}{\partial t} \\ \mu_1 = m\lambda(0) \end{cases} \tag{4.7}$$

where, λ is the adjoint state variable.

The full waveform inversion method of the transient electromagnetic virtual wave field uses model forward fitting to match the virtual wave field over the entire time period and can simultaneously detect the position and water content of underground anomalies. Compared to the traditional virtual wave field offset imaging method, the inversion results obtained by the full waveform inversion method exhibit better resolution and greater development potential.

4.2 Determined Inversion Methods

4.2.1 Based on Gradient Methods

(A) *Least Squares Method*

The most basic method of electromagnetic inversion is to minimize the square sum of the error between the model response and the observed data. Let vector $\mathbf{d} = (d_1, d_2, d_3, \ldots, d_N)$ represents the observed electromagnetic response, let vector $\mathbf{m}$ represent the geoelectric model parameters, generally the logarithm of resistivity or conductivity, that is, $\mathbf{m} = (\log(\rho_1), \log(\rho_2), \log(\rho_3), \ldots, \log(\rho_\mathrm{M}))^\mathrm{T}$ or $\mathrm{m} = (\log(\sigma_1), \log(\sigma_2), \log(\sigma_3), \ldots, \log(\sigma_\mathrm{m}))^\mathrm{T}$; let $\mathbf{f} = (f_1, f_2, f_3, \ldots, f_\mathrm{n})^\mathrm{T}$ represent the response of the model parameters, that is,

$$\mathbf{f} = \mathbf{A}(\mathbf{m}) \tag{4.8}$$

where, $\mathbf{A}$ is the forward operator. Let $\mathbf{m}^0$ be the initial value of the model parameter $\mathbf{m}$, $\mathbf{f}^0$ is the response of the initial model. The forward is nonlinear, and the Taylor expansion is used for local linearization. The first-order Taylor expansion of the model disturbance response at $\mathbf{m}^0$ can be expressed as:

$$\mathbf{f} = \mathbf{f}^0 + \mathbf{J}\Delta\mathbf{m} \tag{4.9}$$

where, $\Delta\mathbf{m} = \mathbf{m} - \mathbf{m}^0$ is the change in model parameters, $\mathbf{J}$ is the $\mathrm{n} \times \mathrm{m}$ partial derivative Jacobian matrix:

$$J = \begin{bmatrix} \frac{\partial f_1}{\partial m_1} & \frac{\partial f_1}{\partial m_M} \\ \frac{\partial f_N}{\partial m_1} & \frac{\partial f_N}{\partial m_M} \end{bmatrix}. \tag{4.10}$$

The error between the model response and the observed data can be expressed as:

$$\mathbf{e} = \mathbf{d} - \mathbf{f}^0 - \mathbf{J}\delta \tag{4.11}$$

According to the least squares method, let the change in model parameters, find the minimum square error, that is,

$$\phi = \mathbf{e}^T\mathbf{e} = \left(\mathbf{d} - \mathbf{f}^0 - \mathbf{J}\Delta\mathbf{m}\right)^T\left(\mathbf{d} - \mathbf{f}^0 - \mathbf{J}\Delta\mathbf{m}\right) \rightarrow \min \tag{4.12}$$

By deriving the objective function for the model parameters and making it zero, the solution of the model parameter change amount is obtained

$$\Delta\mathbf{m} = \left(\mathbf{J}^T\mathbf{J}\right)^{-1}\mathbf{J}^T\left(\mathbf{d} - \mathbf{f}^0\right) \tag{4.13}$$

Equation (4.13) is the least squares solution of the model parameter update amount.

(B) ***SVD Decomposition Method***

Because the matrix in formula (4.13) $\mathbf{J}^T\mathbf{J}$ is ill-conditioned or singular, causing the model solution to tend to infinity. To solve this problem, the matrix $\mathbf{J}$ is usually decomposed by SVD, that is, $\mathbf{J} = \mathbf{UWV}^T$, at this time if the matrix $\mathbf{J}$ is close to singular, then some singular values are very small after SVD decomposition, and the error of the final solution is very large. At this time, replace the small singular values in $\mathbf{W} = \mathrm{diag}(\mathrm{w}_1, \mathrm{w}_2, \mathrm{w}_3, \ldots, \mathrm{w}_\mathrm{q})$ with zero to get a new matrix, that is, $\mathbf{W}_e^{-1} = diag(1/w_1, 1/w_2, 1/w_3, \ldots, 0, 0, 0)$, at this time the solution of the model update amount is obtained

$$\Delta\mathbf{m} = \mathbf{V}\mathbf{W}_\mathbf{e}^{-1}\mathbf{U}^\mathbf{T}\left(\mathbf{d} - \mathbf{f}^0\right) \tag{4.14}$$

The SVD decomposition method has low requirements for the initial model and demonstrates good convergence stability. Pedersen (2004) applied the SVD decomposition method to one-dimensional MT inversion of data, while Yu et al. (2013) utilized this method for the inversion of TEM detection data with IP effects.

(C) ***Damped Least Squares of SVD Decomposition***

To improve the singularity of the matrix in the least squares solution process, Marquardt (1963) proposed the damped least squares method. By introducing a damping factor into formula (4.13), the singularity of the matrix $\mathbf{J}^T\mathbf{J}$ was improved, resulting in

$$\Delta\mathbf{m} = \left(\mathbf{J}^T\mathbf{J} + \beta\mathbf{I}\right)^{-1}\mathbf{J}^T\left(\mathbf{d} - \mathbf{f}^0\right) \tag{4.15}$$

where, β is the damping factor.

Essentially, the singular value decomposition method improves the singular values by canceling out the equations that lead to small singular values, while the damped least squares method increases the singular values by adding a constant to the diagonal elements of the matrix $\mathbf{J}^T\mathbf{J}$. Combining these two methods, we get the least squares solution of SVD decomposition:

$$\Delta\mathbf{m} = \mathbf{V}diag\left(\frac{\lambda_j}{\lambda_j^2 + \beta}\right)\mathbf{U}^\mathbf{T}\left(\mathbf{d} - \mathbf{f}^0\right) \tag{4.16}$$

where, λ_j is the singular value of the matrix $\mathbf{J}$. The damped least squares method is used to solve ill-posed inversion problems and has been widely applied in electromagnetic detection inversion (Candansayar, 2008; Zhdanov, 2002). Although it offers fast computation speeds, the inversion results are significantly influenced by the choice of the damping factor.

(D) ***Gauss-Newton Method***

The main idea of Tikhonov regularization is to introduce a penalty function term into the objective function to obtain a stable and unique solution. Under this concept, the objective function can be expressed as

$$\varphi = \varphi_d(\mathbf{m}) + \lambda\mathbf{R}(\mathbf{m}) \quad \rightarrow min \ \ \varphi_d = \frac{1}{2}(\mathbf{A}(\mathbf{m}) - \mathbf{d})^T\mathbf{W}_d^T\mathbf{W}_d(\mathbf{A}(\mathbf{m}) - \mathbf{d}) \quad (4.17)$$

where, φ_d is the mismatch function, used to describe the fitting difference between the observed data and the model response. $\mathbf{W}_d$ is the data weight matrix; **R** (**m**) is the regularization term, used to describe the prior information or constraint information of the model; λ is the regularization factor, used to balance the data fitting error and model constraint. Equation (4.17) uses the Taylor expansion formula (4.17) to get

$$\varphi(\mathbf{m} + \Delta\mathbf{m}) = \varphi(\mathbf{m}) + \nabla_{\mathbf{m}}\varphi(\mathbf{m})^T\Delta\mathbf{m} + \frac{1}{2}\Delta\mathbf{m}^T\mathbf{H}(\mathbf{m})\Delta\mathbf{m} + \mathrm{O}\|\Delta\mathbf{m}\|^3 \quad (4.18)$$

where, **H** (**m**) is the Hessian matrix

$$\mathbf{H}(\mathbf{m}) = \nabla^2\varphi(\mathbf{m}) \quad (4.19)$$

Derive formula (4.18) ignoring the high-order terms and set the derivative to zero to get

$$\mathbf{H}(\mathbf{m})\Delta\mathbf{m} = -\nabla_{\mathbf{m}}\varphi(\mathbf{m}) \quad (4.20)$$

From formula (4.20), to obtain the solution for the update step length, the Hessian matrix needs to be computed first. According to the definition in formula (4.19) of the Hessian matrix, and combining it with formula (4.17), we can derive:

$$\mathbf{H}(\mathbf{m}) = \nabla^2_{\mathbf{m}}\varphi_d(\mathbf{m}) + \lambda\mathbf{R}''(\mathbf{m}) \quad (4.21)$$

where,

$$\nabla^2_{\mathbf{m}}\varphi_d(\mathbf{m}) = \mathbf{J}(\mathbf{m})^T\mathbf{W}_d^T\mathbf{W}_d\mathbf{J}(\mathbf{m}) + \frac{\partial\mathbf{J}(\mathbf{m})^T\mathbf{W}_d^T\mathbf{W}_d(\mathbf{A}(\mathbf{m}) - \mathbf{d})}{\partial\mathbf{m}} \quad (4.22)$$

J (**m**) is the sensitivity matrix. Considering that the second term in formula (4.22) contributes less, the second term is discarded, which is the Gauss–Newton approximation method. At this time, the step length iteration update formula (4.20) is

$$\left(\mathbf{J}(\mathbf{m})^T\mathbf{W}_d^T\mathbf{W}_d\mathbf{J}(\mathbf{m}) + \lambda\mathbf{R}''(\mathbf{m})\right)\Delta\mathbf{m} = -\nabla_{\mathbf{m}}\varphi(\mathbf{m}) \quad (4.23)$$

By iteratively solving formula (4.23), the update step length $\mathbf{\Delta m}$ can be obtained.

The Gauss–Newton method directly utilizes second derivative information, resulting in relatively fast convergence. Haber et al. (2007) and Oldenburg et al. (2013) improved convergence by constructing a preconditioning factor in the iterative step length equation using the quasi-Newton method. Man et al. (2023) and Zhang et al. (2019) applied the Gauss–Newton method in transient electromagnetic inversion with excitation effects and marine controlled source inversion, achieving favorable inversion results. Ren et al. (2023) proposed an inversion method that combines stochastic gradient descent with Gauss–Newton, reducing the computational load of airborne transient electromagnetic 3D inversion.

(E) ***Quasi-Newton Method***

From formulas (4.23), (4.20), for electromagnetic detection methods, the calculation amount of directly obtaining the inverse of $\mathbf{H}(\mathbf{m})$ is huge. The quasi-Newton method can use a positive definite symmetric matrix $\tilde{\mathbf{H}}$ to approximate $\mathbf{H}(\mathbf{m})$ inverse, to avoid direct calculation of the Hessian matrix inverse. Using the L-BFGS quasi-Newton method, in the k + 1 iteration step, the matrix $\mathbf{H}(\mathbf{m})$ inverse approximation $\tilde{\mathbf{H}}_{k+1}$ is as follows

$$\begin{aligned}\tilde{\mathbf{H}}_{k+1} &= \left(\mathbf{V}_k^T\mathbf{V}_{k-1}^T\cdots\mathbf{V}_{k-m}^T\right)\tilde{\mathbf{H}}_0(\mathbf{V}_{k-m}\cdots\mathbf{V}_{k-1}\mathbf{V}_k)\\ &\quad+\left(\mathbf{V}_k^T\cdots\mathbf{V}_{k-m+1}^T\right)\rho_{k-m}\mathbf{s}_{k-m}\mathbf{s}_{k-m}^T+(\mathbf{V}_{k-m+1}\cdots\mathbf{V}_k)\\ &\quad+\cdots+\mathbf{V}_k^T\rho_{k-1}\mathbf{s}_{k-1}\mathbf{s}_{k-1}^T\mathbf{V}_k+\rho_k\mathbf{s}_k\mathbf{s}_k^T\end{aligned} \tag{4.24}$$

where, $\mathbf{s}_k = \mathbf{m}_{k+1} - \mathbf{m}_k$, $\mathbf{y}_k = \nabla_{\mathbf{m}}\varphi(\mathbf{m})_{k+1} - \nabla_{\mathbf{m}}\varphi(\mathbf{m})_k$, $\rho_k = \frac{1}{\mathbf{s}_k^T\mathbf{y}_k}$, $\mathbf{V}_k = \left(\mathbf{I} - \rho_k\mathbf{y}_k\mathbf{s}_k^T\right)$, $\tilde{\mathbf{H}}_0$ is the initial positive definite matrix. As can be seen from the formula, the LBFGS method only needs to store $m+1$ vectors $\{\mathbf{s}_i, \mathbf{y}_i\}_{i-k-m}^k$ and 1 $\tilde{\mathbf{H}}_0$ matrix to find $\tilde{\mathbf{H}}_k$.

The quasi-Newton method offers improved inversion calculation speeds and outperforms the nonlinear conjugate gradient method regarding calculation time and model inversion resolution, making it more suitable for large-scale three-dimensional electromagnetic inversion. Haber (2005) utilized the quasi-Newton method for large-scale electromagnetic data inversion, while Avdeev et al. (2009) and Cao et al. (2018) applied it to geoelectromagnetic three-dimensional inversion. Lavoué et al. (2014) even implemented the quasi-Newton method in full waveform inversion for ground penetrating radar. However, they noted that the quasi-Newton method struggles to accurately scale the descent direction for different parameter types. While it avoids the direct computation of the Hessian matrix inverse, it still involves a significant computational load.

(F) ***Conjugate Gradient***

The essence of the conjugate gradient method is to divide the search for the minimum value of the target function into multiple directions, then find the extreme value in each direction and combine them. The conjugate gradient method only uses the first-order derivative information of the function, specifically the Jacobian matrix, which avoids the need to calculate the Hessian matrix in Newton-like methods and results in faster convergence. It is an effective algorithm for solving large-scale nonlinear optimization problems. Wang et al. (1994) applied the conjugate gradient method in transient electromagnetic three-dimensional inversion; Newman and Alumbaugh (2000) performed geoelectromagnetic three-dimensional inversion using the nonlinear conjugate gradient method; Cox et al. (2010) and Kamm et al. (2014) applied the conjugate gradient method in the inversion of airborne electromagnetic data. Kelbert et al. (2008) even utilized the conjugate gradient inversion method in the study of global long-period electromagnetic induction. Open-source geoelectromagnetic inversion programs like ModEM also employ the nonlinear conjugate gradient algorithm.

For the objective function (4.17) of electromagnetic inversion, the algorithm process of the conjugate gradient method is as follows:

$$
\begin{aligned}
\mathbf{r}_n &= \mathbf{A}(\mathbf{m}_n) - \mathbf{d} \\
\mathbf{I}_n^{\alpha_n} &= \mathbf{I}_n^{\alpha_n}(\mathbf{m}_n) = \mathbf{J}_{mn}^T \mathbf{W}_d^T \mathbf{W}_d \mathbf{r}_n + \alpha_n \mathbf{R}'(\mathbf{m}) \beta_n^{\alpha_n} = \frac{\left\| I_n^{\alpha_n} \right\|^2}{\left\| I_{n-1}^{\alpha_{n-1}} \right\|^2}, \\
\mathbf{I}_n^{\alpha_n} &= \mathbf{I}_n^{\alpha_n} + \beta_n^{\alpha_n} \mathbf{I}_{n-1}^{\alpha_{n-1}}, \\
\mathbf{I}_0^{\alpha_0} &= \mathbf{I}_0^{\alpha_0} k_n^{\alpha_n} = \frac{\left(\mathbf{I}_n^{\alpha_n} \mathbf{I}_n^{\alpha_n}\right)}{\left\{ \left\| \mathbf{W}_d \mathbf{J}_{m_n} \widetilde{\mathbf{I}}_n^{\alpha_n} \right\|^2 + \alpha \left\| \mathbf{W}_m \widetilde{\mathbf{I}}_n^{\alpha_n} \right\|^2 \right\}} \mathbf{q} \\
\mathbf{m}_{n+1} &= \mathbf{m}_n - \widetilde{k}_n^{\alpha_n} \widetilde{\mathbf{I}}_n^{\alpha_n}
\end{aligned}
\tag{4.25}
$$

where, $\mathbf{W}_{\mathrm{m}}$ is the model parameter weight.

Liu and Yin (2013) conducted a three-dimensional inversion of the frequency domain airborne electromagnetic method and compared the operational efficiency and effectiveness of the quasi-Newton method and the nonlinear conjugate gradient method. They concluded that the quasi-Newton method has certain advantages over the nonlinear conjugate gradient method in terms of calculation time and model inversion resolution.

4.2.2 *Constrained Inversion*

Electromagnetic inversion has non-uniqueness; therefore, introducing prior information or constraint data into the inversion process can enhance the accuracy of the solution.

(A) ***OCCAM Inversion***

The OCCAM algorithm, also known as the smooth constraint inversion method, seeks the smoothest model that satisfies the target fitting residual through the deviation principle. It has several advantages, including independence from the initial model, good convergence stability, and avoidance of over-interpretation. The model roughness matrix is introduced in the regularization term, meaning that

$$\mathbf{R}(\mathbf{m}) = (\partial\mathbf{m})^T(\partial\mathbf{m}) \tag{4.26}$$

where ∂ represents the roughness matrix, which has various definitions. For the commonly used first-order roughness model, its form is

$$\partial = \begin{bmatrix} 0 & & & & 0 \\ -1 & 1 & & & \\ & -1 & 1 & & \\ & & \ddots & \ddots & \\ 0 & & & -1 & 1 \end{bmatrix} \tag{4.27}$$

Constable et al. (1987) first proposed the concept of OCCAM inversion and applied it to the inversion of geoelectromagnetic data. Subsequently, Degroot-Hedlin and Constable (1990) further conducted OCCAM inversion for two-dimensional geoelectromagnetic data. Liu et al. (1991), Sattel (2005), and Vallée and Smith (2009) applied OCCAM to the inversion of airborne electromagnetic data. Key (2009) and Moghadas et al. (2015) conducted inversion studies on marine controlled-source electromagnetic data using the OCCAM algorithm. Li et al. (2015) performed a one-dimensional inversion study on the axial electric field component of the Multi-channel Transient Electromagnetic Method (MTEM) using the OCCAM algorithm. Chen et al. (2017) studied and analyzed the effects of one-dimensional OCCAM inversion applied to three-dimensional SOTEM model data.

(B) ***Structural Constraints***

Structural constraints refer to the limitations placed on the spatial structure of model parameters during the inversion process. Auken and Christiansen (2004) proposed the Lateral Constraint Inversion (LCI), which ensures that the model parameters remain laterally continuous during inversion. Compared to one-dimensional single-point inversion, LCI can yield more complex layered quasi-two-dimensional inversion results. The essence of LCI is to further introduce a regularization term for lateral constraints in the objective function, meaning that

$$\mathbf{R}(\mathbf{m}) = (\mathbf{Lm})^T(\mathbf{Lm}) \tag{4.28}$$

where, $\mathbf{L}$ is the lateral constraint matrix.

To accommodate different geological structure models, Cai et al. (2014) proposed a weighted LCI inversion method based on conventional LCI. Viezzoli et al. (2008) extended the LCI method to Spatial Constraint Inversion (SCI), which constrains the model to maintain continuity in both lateral and vertical sections. Vignoli et al. (2015, 2017) further introduced sharp boundary constraints into SCI inversion, enabling the acquisition of sharp boundary quasi-three-dimensional inversion results.

(C) *Multi-scale Inversion*

Using weak regularization can improve the fit to the data and possibly increase resolution, but it may reduce the stability of the inversion. To stably recover various underground structures, performing multi-scale inversion of the model is essential. Wavelet transform can convert the spatial domain model into two sets of wavelet domain coefficients: one set contains the approximation coefficients, while the other set includes the detail coefficients, which represent the global and local detail information of the model, respectively. By applying wavelet transform, the resistivity parameters of the inversion are transformed into the wavelet domain, where the model parameters of the inversion become wavelet domain coefficients. The wavelet transform can be expressed as

$$\tilde{\mathbf{m}} = \mathbf{W}_w \mathbf{m} \tag{4.29}$$

where, $\tilde{\mathbf{m}}$ are the wavelet domain coefficients, $\mathbf{W}_w$ is the wavelet transform matrix. Considering the orthogonality of the wavelet transform matrix and according to the chain rule, the wavelet domain sensitivity matrix can be obtained as

$$\mathbf{J} = \mathbf{J}\mathbf{W}_w^{-1} \tag{4.30}$$

where, $\mathbf{W}_w^{-1}$ is the inverse wavelet operator. Since wavelet coefficients have a sparse property, Donoho (2006), Simons et al. (2011) proved that it is easier to obtain sparse solutions in inversion by using L1 norm constraints on sparse domain coefficients. Therefore, the regularization term generally uses the L1 norm, and the data fitting term still retains the L2 norm, then the objective function is transformed into

$$\varphi(\tilde{\mathbf{m}}) = \|\mathbf{A}(\tilde{\mathbf{m}}) - \mathbf{d}\|_2^2 + \|\tilde{\mathbf{m}}\|_1 \quad \rightarrow min \tag{4.31}$$

Nittinger et al. (2016) introduced the multi-scale inversion method using wavelet transform into the inversion of geomagnetic data. In their work, Su et al. (2019) applied this method to the one-dimensional inversion of time-domain airborne electromagnetic data. They pointed out that sparse constraint inversion based on wavelet transform features high efficiency and resolution, allowing for accurate determination of underground electrical boundaries. However, it is affected by the characteristics of the wavelet waveform, which can sometimes produce redundant structures at the boundary. The concept of multi-scale transform inversion can also be applied to curvelet transform and shearlet transform.

(D) *Physical Property Constraints*

Electromagnetic data often acquire prior information about the detection area before inversion. This information can be introduced through the minimum model in the regularization term:

$$R(\mathbf{m}) = \left\|\mathbf{m} - \mathbf{m}_{ref}\right\|_2^2 \tag{4.32}$$

where, $\mathbf{m}_{ref}$ is the prior model, containing known prior information.

Generally, the upper and lower bounds of the resistivity model parameters are constrained. Specifically, model parameter values that exceed this range are directly forced to the boundary values during inversion. Alternatively, model parameters can be transformed using the upper and lower cut-off boundary values

$$\overline{\mathbf{m}} = \ln\left(\frac{\mathbf{m} - m^{-}}{m^{+} - \mathbf{m}}\right) \tag{4.33}$$

where, m^{-} and m^{+} are the upper and lower boundaries of the model parameters, respectively. Through this transformation, using the properties of the logarithmic function, it ensures that the model parameters are always within the given range during the inversion process.

When inverting the distribution of resistivity values, Multinary constraint inversion can be employed. This method was first developed by Zhdanov et al. (2013) based on binary and ternary inversion. Luan et al. (2020, 2023) applied it to the inversion of ground electromagnetic methods and ground-to-air electromagnetic methods. The Multinary transformation is constructed using the error function, which transforms the model parameters into a new model space:

$$\hat{m}_i = f_E(m_i) = \frac{1}{2}\sum_{j=1}^{P}\left[1 + E\left(m_i - m^{(j)}\right)\right] + \varepsilon m_i \tag{4.34}$$

where, $\mathbf{m} = [m_1, m_2, \ldots, m_{N_m}]^T$ are the original model parameters, $\hat{\mathbf{m}} = [\hat{m}_1, \hat{m}_2, \ldots, \hat{m}_{N_m}]^T$ are the transformed new model parameters. $E\left(m_i - m^{(j)}\right)$ is the error function, whose derivative is a Gaussian function. Since this transformation is for model parameters, the sensitivity matrix can be obtained using the chain rule.

$$\hat{\mathbf{J}}(\hat{m}) = \mathbf{J}(m) \cdot \frac{\partial m}{\partial \hat{m}} = \mathbf{J}(m)g(m)^{-1} \tag{4.35}$$

where, $g(m_i) = \frac{\partial \hat{m}_i}{\partial m_i} == \sum_{j=1}^{P} G\left(m_i - m^{(j)}\right) + \varepsilon$, and $G\left(m_i - m^{(j)}\right)$ is a Gaussian function.

4.2.3 Joint Inversion

Joint inversion of multiple methods can reduce the ambiguity in parameter estimation. Combining ground TEM with MT can compensate for the static effect of shallow ground structures on MT. Additionally, merging ground MT data with airborne ZTEM data can alleviate the shortcomings of a single method in resistivity calibration, enabling the recovery of deep structures from MT data and fine near-surface structures from ZTEM data. This joint inversion of different electromagnetic methods for resistivity is accomplished by merging the inversion target functions to construct a new target function.

$$\varphi(\mathbf{m}_1, \mathbf{m}_2) = \alpha\varphi_d(\mathbf{m}_1) + \beta\varphi_d(\mathbf{m}_2) + \lambda\mathbf{R}(\mathbf{m}_1, \mathbf{m}_2) \tag{4.36}$$

However, the method of directly merging target functions for inversion does not incorporate structural coupling into the inversion process. Gallardo and Meju (2003) were the first to introduce cross-gradients to constrain the structural similarity of different model parameters. Since then, the cross-gradient method has been widely applied in joint inversion (Gallardo and Meju, 2007; Meju et al., 2018; Wu et al., 2020). The cross-gradient function can be defined in the inversion target function as follows:

$$\mathbf{t}(x, y, z) = \nabla\mathbf{m}_1(x, y, z) \times \nabla\mathbf{m}_2(x, y, z) \tag{4.37}$$

where, ∇ is the gradient operator. The gradient direction represents the steepest descent direction of the physical parameters at the current point, while the gradient value indicates the rate of change of the model parameters. If the directions of two gradient vectors are the same (parallel), the cross-gradient function of the two vectors is zero; otherwise, it is non-zero. This behavior allows the cross-gradient method to effectively capture and enforce structural similarities between different model parameters during the inversion process. In joint inversion, the cross-gradient term is

$$\mathbf{R}_{cg} = \mathbf{t}^{\mathrm{T}}\mathbf{t} \tag{4.38}$$

The cross-gradient term is introduced into the target function of each respective single-method inversion through a weight factor, forming two joint inversion target functions, namely

$$\begin{aligned} \varphi^{(1)}(\mathbf{m}_1) &= \varphi_d^{(1)}(\mathbf{m}_1) + \lambda_1\mathbf{R}^{(1)}(\mathbf{m}) + \eta_1\mathbf{R}_{cg}\varphi^{(2)}(\mathbf{m}_2) \\ &= \varphi_d^{(2)}(\mathbf{m}_2) + \lambda_2\mathbf{R}^{(2)}(\mathbf{m}) + \eta_2\mathbf{R}_{cg} \end{aligned} \tag{4.39}$$

Or the two joint target functions are merged to construct the overall target function

$$\begin{aligned} \varphi(\mathbf{m}_1, \mathbf{m}_2) = {} & \alpha_1\varphi_d^{(1)}(\mathbf{m}_1) + \alpha_2\varphi_d^{(2)}(\mathbf{m}_2) \\ & + \lambda_1\mathbf{R}^{(1)}(\mathbf{m}_1) + \lambda_2\mathbf{R}^{(2)}(\mathbf{m}_2) + \eta\mathbf{R}_{cg} \end{aligned} \tag{4.40}$$

where, $\varphi_d^{(1)}(\mathbf{m}_1)$ and $\varphi_d^{(1)}(\mathbf{m}_1)$ are the fitting functions of model parameters $\mathbf{m}_1$ and $\mathbf{m}_2$; $\mathbf{R}^{(1)}(\mathbf{m}_1)$ and $\mathbf{R}^{(2)}(\mathbf{m}_2)$ are regularization terms; η is the weight of the cross-gradient term. By minimizing the above target function, a structurally consistent inversion result is obtained.

Furthermore, Zhdanov et al. (2012, 2021) proposed a more general method based on Gramian constraints for the joint inversion of multiple geophysical datasets. This method minimizes the determinant of the Gram matrix formed by different model parameters. The Gramian constraint can be applied to model parameters, and its construction form is as follows:

$$R_G = \begin{vmatrix} \left(\mathbf{m}^{(1)}, \mathbf{m}^{(1)}\right) & \left(\mathbf{m}^{(1)}, \mathbf{m}^{(2)}\right) \\ \left(\mathbf{m}^{(2)}, \mathbf{m}^{(1)}\right) & \left(\mathbf{m}^{(2)}, \mathbf{m}^{(2)}\right) \end{vmatrix} \tag{4.41}$$

It can also be applied to the gradient of model parameters:

$$R_G = \begin{vmatrix} \left(\nabla\mathbf{m}^{(1)}, \nabla\mathbf{m}^{(1)}\right) & \left(\nabla\mathbf{m}^{(1)}, \nabla\mathbf{m}^{(2)}\right) \\ \left(\nabla\mathbf{m}^{(2)}, \nabla\mathbf{m}^{(1)}\right) & \left(\nabla\mathbf{m}^{(2)}, \nabla\mathbf{m}^{(2)}\right) \end{vmatrix} \tag{4.42}$$

The target function of joint inversion then becomes

$$\begin{aligned} \varphi(\mathbf{m}_1, \mathbf{m}_2) = \alpha_1\varphi_d^{(1)}(\mathbf{m}_1) + \alpha_2\varphi_d^{(2)}(\mathbf{m}_2) \\ + \lambda_1\mathbf{R}^{(1)}(\mathbf{m}_1) + \lambda_2\mathbf{R}^{(2)}(\mathbf{m}_2) + \eta\mathbf{R}_G \end{aligned} \tag{4.43}$$

The Gramian constraint method does not require prior knowledge; instead, it determines the nature of these relationships during the inversion process. This allows for the consideration of both linear and nonlinear relationships between different physical parameters of the geological model.

4.2.4 Anisotropic Inversion

Ignoring the anisotropy of the underground medium in electromagnetic observation data can lead to erroneous interpretation results. Research has been conducted on the anisotropic inversion of marine controlled-source electromagnetic data. Wang et al. (2018) integrated the technical advantages of unstructured finite element methods, primary field separation, and direct solution techniques to achieve high-precision, rapid three-dimensional anisotropic inversion of marine controlled-source electromagnetics. Meju et al. (2019) demonstrated that by setting the cross product of the horizontal resistivity gradient and the vertical resistivity gradient to zero at critical geological boundaries, it is possible to achieve objective geological orientation or structural customization inversion in three-dimensional anisotropic strata. Additionally, Cao et al. (2018) studied the anisotropic forward and inverse problems in terrestrial electromagnetic methods, while Su et al. (2021) focused on the three-dimensional anisotropic inversion of airborne electromagnetic data. Collectively, these studies indicate that the deep Earth exhibits anisotropic properties.

4.2.5 Parallel Inversion

In the detection of complex three-dimensional geological structures, a significant amount of electromagnetic detection data is often collected. Despite the availability of various parallel frameworks, MPI and OpenMP are popular choices for processing electromagnetic detection data. MPI builds a distributed data structure and distributes the inversion process of electromagnetic detection data across the entire cluster. Newman and Commer (2008) utilized MPI's parallel method in controlled source electromagnetic inversion, accelerating the process through model and data decomposition. Wang et al. (2021) applied MPI's parallel acceleration in inverting MTEM, resulting in a tenfold increase in inversion calculation speed. OpenMP, on the other hand, does not require rewriting data programs, fully utilizes the multi-core performance of a single computer, and is simpler to implement. Kirkegaard and Auken (2015) parallelized the forward calculation and sensitivity calculation of each one-dimensional model in airborne time-domain electromagnetic inversion using OpenMP. Zhang et al. (2021) and Liu et al. (2016) further employed OpenMP parallel acceleration in three-dimensional time-domain electromagnetic inversion. In frequency domain electromagnetic inversion, OpenMP is generally used to distribute the calculation process of different frequencies across different cores to accelerate the inversion process.

In recent years, direct solution technology has significantly advanced electromagnetic three-dimensional inversion. Mainstream tools include MUMPS, Paradiso, and SuperLU. The advantages of applying direct solution technology to three-dimensional electromagnetic problems are: (1) it can overcome the influence of the large condition number of the equations in the forward process and achieve stable solutions; (2) in multi-source and multi-time channel electromagnetic inversion, the forward left-end matrix remains the same, and changes in sources and time channels only affect the right-end items, allowing for acceleration by saving the matrix decomposition results and replacing the right-end items. Haber and Schwarzbach (2014) used MUMPS to parallel solve large-scale airborne time-domain electromagnetic data, while Cai et al. (2022) completed the three-dimensional inversion of time-domain electromagnetic data using the Paradiso solver. However, the disadvantage of direct solution technology is that it occupies a large amount of memory and is challenging to apply to the solution of super-large scale three-dimensional models. It needs to be combined with model dimension reduction to handle large-scale problem calculations. Iterative methods, on the other hand, require less memory and can achieve fast solutions for large-scale three-dimensional forward and inverse electromagnetic problems with efficient preconditioning technology.

4.3 Stochastic Inversion Method

4.3.1 Simulated Physical Process Algorithms

This type of method is primarily based on optimization techniques derived from physical processes, such as the simulated annealing algorithm and the gravity search algorithm. Wang et al. (2012) applied the simulated annealing algorithm to the inversion of the controlled source audio-frequency magnetotelluric method. By fixing all parameters, except for the resolution difference, to their average values and repeatedly performing the simulated annealing search, improved results were achieved. Stochastic inversion methods do not require the calculation of the sensitivity matrix, allowing the objective function to be set to the L1 norm without any approximations. This approach enables the attainment of a sparse solution for the model parameters.

4.3.2 Swarm Intelligence Algorithms

Swarm intelligence optimization algorithms are inversion techniques based on the group behavior of natural species. Representative algorithms include genetic algorithms, differential evolution algorithms, and particle swarm algorithms. The basic process of these algorithms involves generating a certain number of groups, where each individual represents a set of solutions to the inversion objective function of electromagnetic data. The group is continuously updated and iterated according to specific rules, ultimately selecting the set of solutions that minimizes the inversion objective function as the final inversion result. For example, the Particle Swarm Optimization (PSO) method has been successfully applied to electromagnetic methods in both underground and surface environments. Leveraging the global convergence of the algorithm, particle swarm inversion of electromagnetic data has been developed and utilized for quantitative interpretation of electromagnetics.

Particle Swarm Optimization adjusts its speed and trajectory in the search space based on the best individual position and the best group position information, and each iteration moves closer to the global optimum. Assuming in a j-dimensional search space, M particles form a swarm, the spatial position of the i-th particle can be represented as $X_i = \left(X_{i1}, X_{i2}, \ldots, X_{ij}\right), i = 1, 2, \ldots., M$. The particle swarm equation can be written as:

$$V_{ij}^{n+1} = \omega V_{ij}^{n} + c_1 r_1 \left(Pbest_i^n - X_{ij}^{n}\right) + c_2 r_2 \left(Gbest^n - X_{ij}^{n}\right) \tag{4.44}$$

$$X_{ij}^{n+1} = X_{ij}^{n} + V_{ij}^{n+1} \tag{4.45}$$

where, Pbest and Gbest represent the best individual position and the best group position respectively. ω is the weight of the previous speed; c1 and c2 are positive integers, called acceleration coefficients; r1 and r2 are random numbers varying from 0 to 1; n is the number of iterations.

4.3.3 Bayesian Inversion Algorithms

The Bayesian inversion method characterizes the likelihood of model parameters across different intervals as a posterior probability distribution, incorporating prior information and observed data. Unlike deterministic inversion theory, Bayesian inversion quantitatively evaluates the uncertainty of inversion results and the interrelationship among model parameters by extracting the marginal probability distribution of model parameters, maximum posterior solutions, average solutions, correlation coefficients, and more. It reflects the constraint capabilities of observed data and prior information on model parameters through the posterior probability distribution. In recent years, Bayesian methods have been successfully applied to electromagnetic data inversion. To enhance computational efficiency, the Bayesian inversion of electromagnetic data has been developed using a one-dimensional forward operator, primarily for the quantitative interpretation and reliability assessment of artificial source electromagnetic detection data. Some researchers have combined two-dimensional parameterization with a one-dimensional forward operator to perform cross-dimensional Bayesian inversion, achieving favorable results.

Generally in Bayesian methods, the norm of the model vector (such as smoothness and sparsity) and prior information can be encapsulated in the prior distribution. In this case, the Bayesian equation can be written as

$$p(\mathbf{m}|\mathbf{d}) \propto p(\mathbf{d}|\mathbf{m})p(\mathbf{m}) \tag{4.46}$$

where p(m) is the prior distribution, which defines the model space of all models allowed by prior information. The prior distribution is independent of the observed data. After obtaining electromagnetic detection data, the models in the prior distribution are reweighted according to the likelihood function p(d|m) to obtain the posterior probability of the model. Sampling and weighting are carried out in turn to obtain the posterior distribution of each model.

In Bayesian methods, the posterior distribution offers a set of models constrained by the fitting of residuals and prior information. This approach not only provides model parameter estimates typically offered by deterministic inversion but also quantifies the uncertainty of parameter estimation influenced by multiple solutions to inverse problems, observation errors, and noise. Consequently, it supplies more comprehensive information for data interpretation. The prior distribution in Bayesian inversion facilitates the introduction of multiple model spaces, making it suitable for the joint inversion of various parameters. For instance, it can be applied to parallel sampling of resistivity and polarization parameters in transient electromagnetic (TEM) data, as well as for the joint inversion of artificial source electromagnetic data and magnetotelluric (MT) data.

4.3.4 Neural Network Inversion

In recent years, the continuous rise of artificial intelligence has led to significant advancements in the theory and algorithms of artificial neural networks. Deep learning, which enables machines to learn the inherent rules and features of sample data, allows for analytical learning capabilities similar to those of humans. This technology can recognize problems, images, sounds, and more, making it a prominent focus in the fields of artificial intelligence and machine learning. Consequently, deep learning neural networks have garnered widespread attention once again. As neural networks have evolved into deep neural networks (DNNs) with substantial non-linear mapping capabilities, they are increasingly applied to solve inversion problems. The integration of deep learning neural network algorithms with geophysical exploration has emerged as a hot research area (Fig. 4.2).

Through the weighted summation of each node of the neural network, forward propagation is realized. Backward propagation mainly propagates the error of the output layer back to the hidden layer nodes, and then continuously adjusts the weights w and biases b. The specific process takes the error calculation of hidden layer node f as an example. It can be seen that the error of node f is related to node i and node j. According to the principle of distribution according to work, the error of hidden layer node f can be expressed as:

$$e_{h1} = \frac{w_{11}^3}{w_{11}^3 + w_{21}^3 + w_{31}^3} \cdot e_{o1} + \frac{w_{12}^3}{w_{12}^3 + w_{22}^3 + w_{32}^3} \cdot e_{o2} \tag{4.47}$$

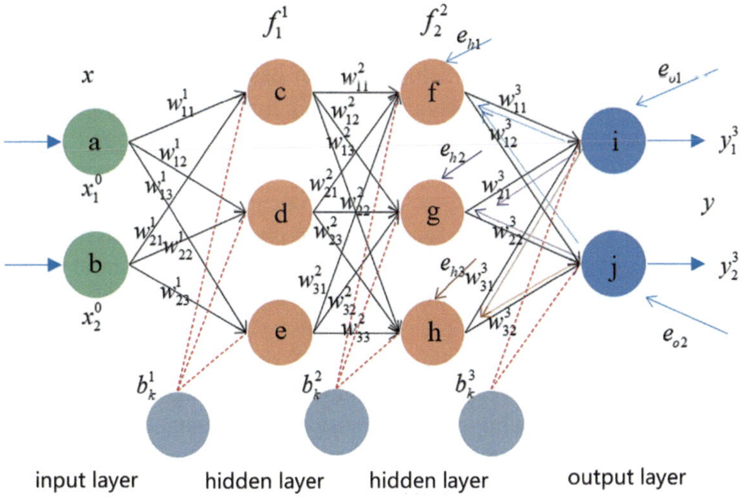

Fig. 4.2 Neural network input and output

Convolutional neural networks (CNNs) can be applied to full waveform inversion, effectively handling geophysical detection data with uncertainty. Additionally, CNNs enable rapid inversion and imaging of the electrical structure of underground geological bodies, as well as the denoising of transient electromagnetic data.

Beyond CNNs, other deep learning-based neural networks are also employed in various geological exploration tasks. For instance, denoising autoencoders (DAEs) are used for electromagnetic data denoising, adaptive CNNs are utilized for resistivity inversion, and long short-term memory (LSTM) networks are applied for real-time inversion of transient electromagnetic data. These advancements highlight the increasing significance of deep learning techniques in enhancing geophysical exploration and data interpretation.

4.3.5 Fusion Algorithm

The fusion algorithm refers to the combination of two or more algorithms to overcome the limitations of a single approach. This can involve fusing different stochastic algorithms or integrating stochastic algorithms with deterministic ones.

For instance, Moghadas et al. (2020) proposed a fusion of neural networks with Bayesian algorithms. In Bayesian inversion of electromagnetic methods, sufficient sampling of the model space is required, which necessitates extensive forward computations, thereby reducing the efficiency of large-scale regional surveys. To address this, Moghadas et al. replaced the forward computation of the electromagnetic method with a neural network. Once trained, the network can quickly produce forward response results for various models. The inversion process still employs Bayesian methods, enabling rapid and accurate determination of underground conductivity distribution. This approach has been successfully applied to electromagnetic induction (EMI) data in the Alken Enge area of Denmark. Additionally, recognizing that neural network methods can be sensitive to initial parameter settings and prone to local minima, researchers have sought to balance convergence speed and inversion quality. This is achieved by integrating the neural network algorithm with particle swarm optimization, genetic algorithms, differential evolution algorithms, and others. The initial weights and thresholds of the neural network can be determined through particle swarm and genetic algorithms, thereby enhancing the global optimization capabilities of the neural network.

In another example of fusion, He et al. (2022) and Xue et al. (2023) developed a model and data-driven inversion algorithm (PSO-GD algorithm) for the three-dimensional inversion of semi-airborne transient electromagnetics. This approach combines particle swarm optimization (PSO) with gradient descent (GD). This algorithm first uses PSO to fit the inversion to ensure global convergence, and then uses the GD algorithm to accelerate the fitting process of the inversion. The cosine probability function is used as the weighting function of the model update equation in the new algorithm to ensure a smooth transition from global convergence to accelerated convergence. Thus, the new particle swarm equation

$$V_i^{k+1} = \left(1 - w^k\right)\left[wV_i^k + \alpha_1^k \gamma_1 \left(P_i - x_i^k\right) + \alpha_2^k \gamma_2 \left(G - x_i^k\right)\right] + w^k \left(\alpha_3^k \gamma_3 \delta m\right) \quad (4.48)$$

where, G is the sensitivity matrix, $\delta m = \left(G^T G\right)^{-1} G^T \delta d$; w is the weight matrix. Guo et al. even applied deep residual convolutional neural network technology to the joint inversion of audio magnetotelluric and seismic data, learning structural similarity and resistivity-velocity relationship based on prior knowledge, with faster convergence speed and lower fitness function values.

Different deterministic algorithms can also be combined to enhance performance. For example, Egbert (2012) proposed integrating the computational and storage efficiency of the conjugate gradient method with the Occam scheme for regularization and step size control. The inversion results from magnetotelluric data demonstrated that this combination of algorithms significantly reduced the number of iterations required for inversion, leading to faster convergence overall.

There are several approaches to reducing inversion challenges and non-uniqueness in artificial source electromagnetic method data. In cases of limited detection data, it is essential to leverage prior information to constrain the inversion process. Techniques such as wavelet transforms and multinary transforms can integrate known information into the inversion, yielding results that align with geological requirements. Anisotropic inversion and the incorporation of excitation polarization effects can extract additional information, facilitating the accurate detection of anomalous structures. Joint inversion methods can combine the strengths of various detection data, leading to precise imaging of underground structures. Additionally, fusing multiple inversion methods can enhance the advantages of different approaches. The integration of stochastic and deterministic inversion methods allows for a comprehensive inversion driven by both data and models, effectively addressing the challenges of excessive forward computations in stochastic inversion and the risk of local minima in deterministic inversion.

Another promising direction for improving the speed and accuracy of electromagnetic inversion is the efficient large-scale model three-dimensional constrained inversion that considers anisotropy and excitation polarization effects. The rapid inversion of electromagnetic data using artificial intelligence technology is expected to play a crucial role in future developments. Currently, AI applications for inversion imaging primarily focus on one-dimensional data. In the future, constructing a large training set and a universal training network model for two-dimensional and three-dimensional data can enhance the rapid imaging capabilities of electromagnetic detection data through artificial intelligence. Furthermore, combining stochastic inversion methods based on artificial intelligence with deterministic gradient-based inversion can quickly and reliably reveal the underground electrical structure distribution.

4.4 SOTEM Data Pseudo-Seismic

The transient electromagnetic method effectively avoids interference from the primary field by directly observing the secondary field generated by underground media. Its advantages—such as a simple setup, a wide frequency range of transmitted signals, and high detection efficiency—make it valuable for detecting water-filled goafs and deep metal ore bodies (Viezzoli et al., 2008; Di et al., 2019; Jiang et al., 2019; Xue et al., 2020a). However, traditional data inversion methods have limited interpretability, leading to poor resolution when identifying deep concealed ore bodies. Given the characteristics of deep metal ore bodies—such as small volume, strong interference, and challenging detection—there is an urgent need to enhance the processing and interpretation methods for transient electromagnetic data.

Seismic exploration offers high spatial resolution, particularly in identifying the distribution range, burial depth, and spatial position of deep geological bodies (Stewart et al., 2003; Wang Yanfei et al., 2018; Xue et al., 2020b). However, the complex spatial occurrence of metal ore bodies, combined with low signal-to-noise ratios, complicates the acquisition of reliable reflection wave data. While multiple coverage observation systems can improve the signal-to-noise ratio, increasing the survey line length and the number of reception channels also significantly raises exploration costs.

By integrating the high spatial resolution of seismic exploration with the high sensitivity of low-resistance bodies in electromagnetic exploration, the signal-to-noise ratio of transient electromagnetic data can be effectively enhanced, reducing exploration costs and optimizing the detection of deep ore bodies. Mathematical processing methods that transform the transient electromagnetic field into a wave field (Lee et al., 1987; Mittet, 2010) can suppress or eliminate dispersion and attenuation features during electromagnetic wave propagation, amplifying propagation characteristics in the electromagnetic responses. This approach combines the sensitivity of transient electromagnetic methods to underground minerals with the high geometric resolution of seismic exploration, improving the sensitivity of the transient electromagnetic field to changes in the position and properties of anomalous bodies (Xue et al., 2011; Jiulong et al., 2016; Li et al., 2018; Kailiang et al., 2023).

Through integral transformation, the transient electromagnetic field satisfying the diffusion equation can be converted into a pseudo-seismic wave field that adheres to the wave equation. This transformation is described by a first-type Fredholm integral equation, which presents a high degree of ill-posedness, posing a core challenge to accurately extracting the virtual wave field (Xue et al., 2006; Huasen et al., 2016; Kerui et al., 2023). While traditional regularization methods can mitigate the singularity of the kernel function, the use of regularization factors can significantly impact the initial arrival time, wave width, and overall morphology of the virtual wave field, compromising stability (Junjie et al., 2018).

Deep learning neural network methods have gained traction across various fields due to their high-precision learning and prediction capabilities. In seismic inversion, convolutional neural networks (CNNs) excel in inverting wave speed and wave impedance (Das et al., 2019; Sun, 2020; Wu et al., 2022; Yaru et al., 2022). CNNs also demonstrate robust performance in electromagnetic data inversion, predicting the location and resistivity characteristics of geological anomalies in both the time domain and frequency domain (Puzyrev, 2019; Moghadas, 2020). The U-Net architecture, a generalized form of CNN, employs an encoding–decoding structure for upsampling and downsampling data, offering advantages in computation speed and stability compared to traditional CNNs (Pan et al., 2024). Combining U-Net with electromagnetic data inversion can significantly enhance both inversion speed and resolution (Li et al., 2021). However, while U-Net is effective in image classification, it may overfit or underfit in regression predictions (Yanqi et al., 2023), complicating complex data prediction tasks. Composite networks can address these issues by enhancing the feature extraction capabilities of training samples through network integration, deepening the correlation between input data and labels (Moghadas, 2020; Chi et al., 2023).

To achieve high-precision extraction of the virtual wave field, this paper combines deep learning neural network methods with wave field transformation techniques. We propose a composite neural network model, C-Unet, which integrates a convolutional network module prior to the input of data into the U-Net. A training set and corresponding labels are constructed by calculating the transient electromagnetic field and pseudo-seismic wave field under the same resistivity model. Finally, the reliability of the proposed method is analyzed and validated through numerical simulations and actual case studies.

4.4.1 Transient Electromagnetic Field and Virtual Wave Field

To obtain reliable transient electromagnetic and pseudo-seismic wave field data, it is essential to utilize a forward method based on the same physical model and variable parameters. This approach serves as the foundation for implementing the neural network effectively.

(1) **Forward Modeling of Transient Electromagnetic Data**

Taking the large loop source transient electromagnetic method as an example, the transient electromagnetic field formed by an isotropic horizontal layered earth is:

$$H_z = \frac{Ia}{2}\int_0^{\infty}\left[e^{-u_0(z+h)} + r_{TE}e^{u_0(z-h)}\right]\frac{\lambda^2}{u_0}J_1(\lambda a)d\lambda \tag{4.49}$$

where, λ is the integral variable, J_1 is the first order Bessel function, r_{TE} is the reflection coefficient, h is the height of the receiving device, I represents current, a represents coil radius.

(2) **Forward Modeling of Virtual Wave Field Data**

The wave equation satisfied by the virtual wave field can be expressed as:

$$\nabla^2\mathbf{U}(q) - \mu\sigma\frac{\partial^2\mathbf{U}(q)}{\partial q^2} = 0 \tag{4.50}$$

where, U represents the virtual wave field, q is the virtual time, μ and σ represent magnetic permeability and electrical conductivity respectively.

The central difference format of the equation can be expressed as:

$$\mathbf{U}_{i,j}^{k+1} = \frac{\left(\frac{\mathbf{U}_{i+1,j}^{k}+\mathbf{U}_{i-1,j}^{k}-2\mathbf{U}_{i,j}^{k}}{\Delta x^2} + \frac{\mathbf{U}_{i,j+1}^{k}+\mathbf{U}_{i,j-1}^{k}-2\mathbf{U}_{i,j}^{k}}{\Delta z^2} - \mathbf{F}(q)\right)}{\mu\sigma}\Delta q^2 + 2\mathbf{U}_{i,j}^{k} - \mathbf{U}_{i,j}^{k-1} \tag{4.51}$$

The Perfectly Matched Layer (PML) is currently one of the more effective artificial boundary conditions, designed to absorb incident waves from various angles. By establishing a PML region around the model, different areas can be assigned distinct attenuation coefficients. When the virtual wave field propagates into the PML area, it is absorbed and attenuated, effectively reducing or eliminating boundary reflections. However, conventional PML boundary conditions can generate noticeable false reflections for grazing and low-frequency waves. To address this, the auxiliary differential equation method is employed to implement complex frequency shift perfectly matched layer boundary conditions for the virtual wave field. This method is relatively straightforward in principle, as it does not require the separation of the second-order equation wave field, and it demonstrates effective absorption characteristics for grazing and low-frequency waves (Yingjie, 2016).

The expression for the auxiliary differential equation complex frequency shift perfectly matched layer boundary condition of the virtual wave field can be represented as:

$$\begin{aligned}
\left(\alpha_x + \frac{\partial}{\partial t} + d_x\right)\mathbf{U}_{x_3} &= d_x^2 d_x' \frac{\partial \mathbf{U}}{\partial x} + d_x^2 \alpha_x' \frac{\partial \mathbf{U}}{\partial x}\left(\alpha_x + \frac{\partial}{\partial t} + d_x\right)\mathbf{U}_{x_2} \\
&= \left(2d_x d_x' \frac{\partial \mathbf{U}}{\partial x} + 2d_x\alpha_x' \frac{\partial \mathbf{U}}{\partial x} + d_x^2\frac{\partial^2 \mathbf{U}}{\partial x^2}\right) \\
&\quad - \mathbf{U}_{x3}\left(\alpha_x + \frac{\partial}{\partial t} + d_x\right)\mathbf{U}_{x_1} \\
&= \left(d_x' \frac{\partial \mathbf{U}}{\partial x} + 2d_x\frac{\partial^2 \mathbf{U}}{\partial x^2}\right) - \mathbf{U}_{x_2} \\
\mu\sigma\frac{\partial^2 \mathbf{U}}{\partial t^2} &= \frac{\partial^2 \mathbf{U}}{\partial x^2} + \frac{\partial^2 \mathbf{U}}{\partial z^2} - \mathbf{U}_{x_1}
\end{aligned} \tag{4.52}$$

where, α and d represent the absorption boundary parameters respectively.

4.4.2 Establishment of Transient Electromagnetic Field and Seismic Wave Field Model

The transient electromagnetic field value Hz in formula (4.49) and the seismic wave field value U in formula (4.51) both have variable resistivity. Based on the same resistivity model, the electromagnetic field value and the corresponding seismic wave field value can be obtained.

Seismic exploration methods have a rich and complex longitudinal wave speed model. The empirical formula can be used to achieve the conversion between seismic longitudinal waves and resistivity (Meju et al., 2003):

$$Log_{10}\rho = mLog_{10}V_p + c \tag{4.53}$$

where, ρ represents resistivity, V_p represents seismic longitudinal wave speed, m and c are empirical parameters, in this paper, $m = 3.88$, $c = -11$.

4.4.3 Deep Learning Convolutional Neural Network

Neural networks increase learning ability through activation functions. Commonly used activation functions include: sigmoid function, tanh and Relu function. Among them, the Relu function is the most commonly used activation function, as shown below:

$$y = \begin{cases} x, \ x > 0 \\ 0, \ x \leq 0 \end{cases} \tag{4.54}$$

The loss function is used to calculate the error between the network output and the label. Backpropagation is used to modify the relevant parameters of the deep learning neural network. This paper uses the mean square error function as the loss function:

$$\phi = \frac{1}{N}\sum_{i=1}^{n}\left(d_{lab}^{i} - d_{pre}^{i}\right) \tag{4.55}$$

where, N represents the number of models, d_{lab} represents the label data, d_{pre} represents the predicted data.

This paper mainly employs U-Net and CNN-U-Net for wave field conversion. Below is a brief introduction to these two network methods and their related parameters:

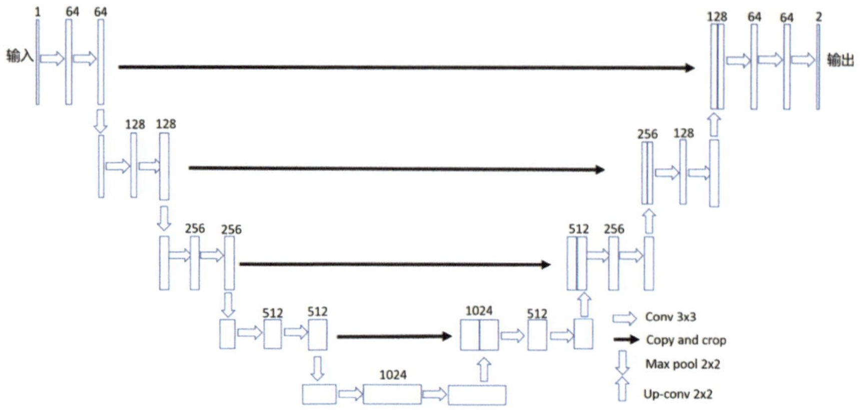

Fig. 4.3 Unet architecture

(1) **Unet Architecture**

The U-Net network is a widely utilized encoder-decoder architecture, characterized by its distinctive U-shaped structure, as illustrated in Fig. 4.3. This design effectively extracts features at multiple levels and seamlessly integrates both global and local information (Yu et al., 2023).

Figure 4.4 illustrates the wave field transformation process using the U-Net network. Initially, the seismic velocity model is converted into an approximate resistivity model through an empirical formula. Subsequently, the corresponding transient electromagnetic and seismic data are generated based on this resistivity model. The U-Net network is then employed to train the transient electromagnetic data alongside the seismic labels. The loss function governs the backpropagation process and determines the conditions for iteration termination.

(2) **C-Unet Architecture**

The U-Net network can effectively implement the conversion between seismic data and velocity models (Yang and Ma, 2019). However, unlike the seismic velocity model data, the transient electromagnetic data from different models show small differences, making it difficult to directly use the U-Net network to capture transient electromagnetic features. To enhance the learning capability of U-Net, a Convolutional Neural Network (CNN) is added to the architecture. The network architecture of the CNN is shown in Fig. 4.5. This architecture includes six convolutional layers and pooling layers, along with one fully connected layer (Yang et al., 2021).

A new neural network is constructed based on U-Net and CNN. The output results from the CNN are fed into the U-Net, and the parameters of the C-U-Net network are updated through the loss function. The inversion process is illustrated in Fig. 4.6.

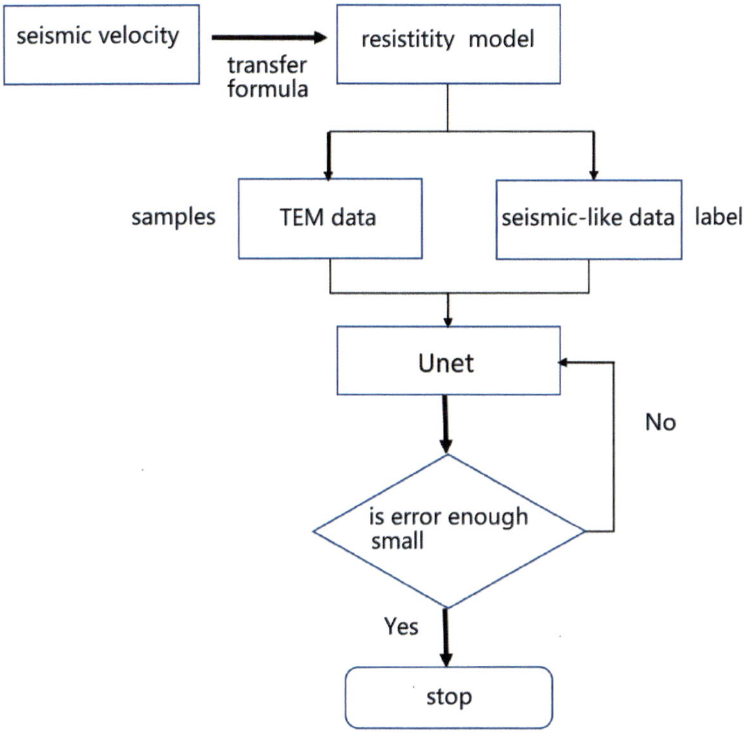

Fig. 4.4 Unet training process

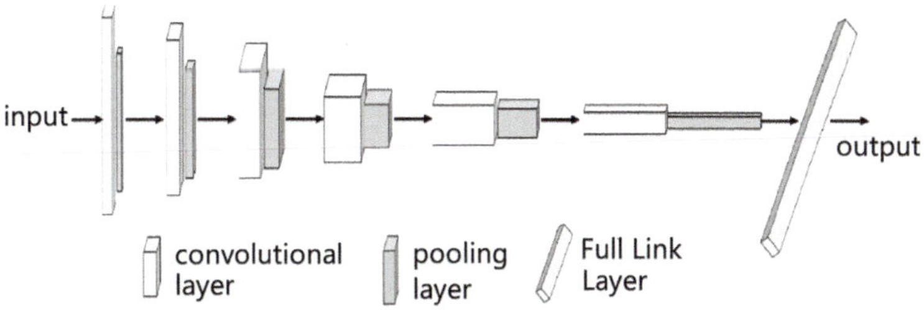

Fig. 4.5 Convolutional neural networks (CNN) module

4.4.4 Numerical Simulation

This section will mainly verify the model simulation of seismic parameter transformation and network inversion results.

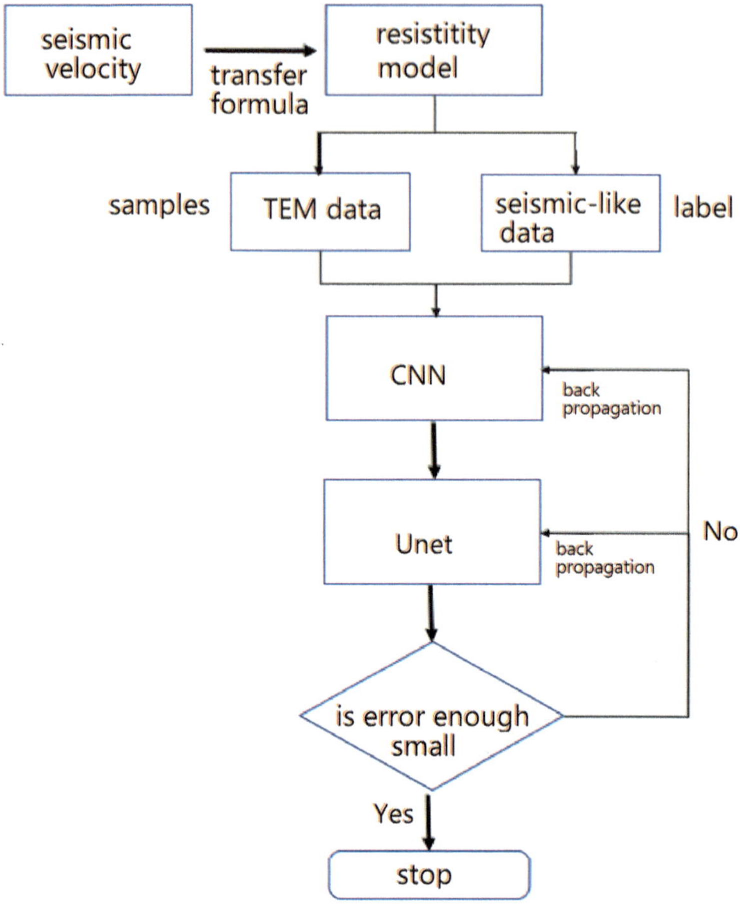

Fig. 4.6 C-Unet training process

(1) **Model Construction**

Figure 4.7 is an example of a deep learning seismic velocity model, in units of m/s. Through the empirical formula (4.49), 1700 seismic models are converted into approximate resistivity models, and then the corresponding transient electromagnetic field and virtual wave field datasets are obtained using formulas (4.50) and (4.53). The model size used is uniformly 301 × 201, with grid sizes dx = 10 m, dz = 20 m. The virtual wave field source is located at model (100, 1).

(2) **Model Training**

Figure 4.8 displays the results of deep learning wave field conversion. In this figure, (a) represents the seismic P-wave velocity model, (b) shows the approximate resistivity model, and (c) and (d) depict the corresponding normalized transient electromagnetic field and virtual wave field data, respectively. Panels (e) and (f) present the

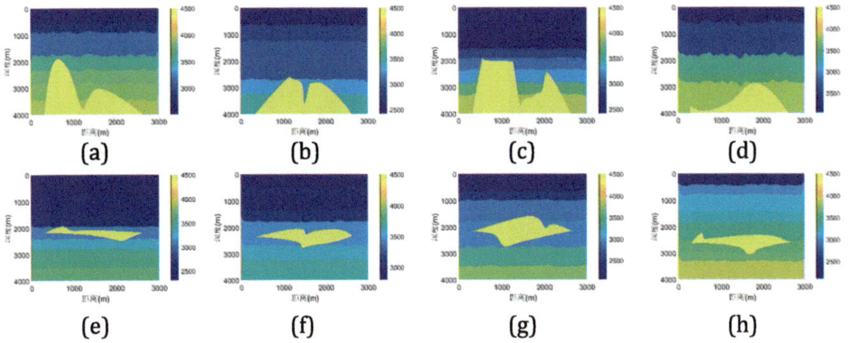

Fig. 4.7 Construction of deep learning seismic velocity model. **a–h** Different seismic velocity models

seismic wave field images obtained from the inversion using the U-Net and C-Unet networks. Comparing (e) and (f), it is evident that the seismic wave field produced by C-Unet aligns more closely with (d).

Figure 4.9 illustrates the iteration errors of the two network methods over 100 iterations. The blue curve represents the root mean square error of the U-Net network, while the red curve corresponds to the root mean square error of the C-Unet. It is evident that the iteration error of C-Unet is smaller and converges more quickly.

To more intuitively demonstrate the effectiveness of the method used in this paper, a simple anomaly model is established. Figure 4.10a presents the resistivity model, while Fig. 4.10b and c show the simulated seismic prediction labels and prediction data, respectively. It is evident that Fig. 4.10c exhibits stronger energy near 0.4 s and a weaker response near 0.5 s, which corresponds well with Fig. 4.10b.

4.4.5 Field Case Study

To verify the effectiveness of the method proposed in this paper, transient electromagnetic data were analyzed for wave field conversion. The field measurement site is located in the eastern part of Xiong'an New Area, Xiong County, Hebei Province. This region is rich in geothermal resources, covering a geothermal field area of 320 km^2. Since the establishment of Xiong'an New Area, the China Geological Survey has actively conducted geothermal energy exploration, resulting in substantial research outcomes (Hui et al., 2023).

The strata in the survey area can be classified from oldest to youngest as follows: (1) Paleozoic: Mainly consisting of Ordovician and Cambrian rocks, primarily limestone and dolomitic limestone. (2) Mesozoic: Predominantly Jurassic and Cretaceous, with the Cretaceous primarily composed of mudstone and conglomerate, while the Jurassic mainly includes mudstone and sandstone, occasionally containing coal seams. (3) Cenozoic: Comprising Paleogene, Neogene, and Quaternary layers.

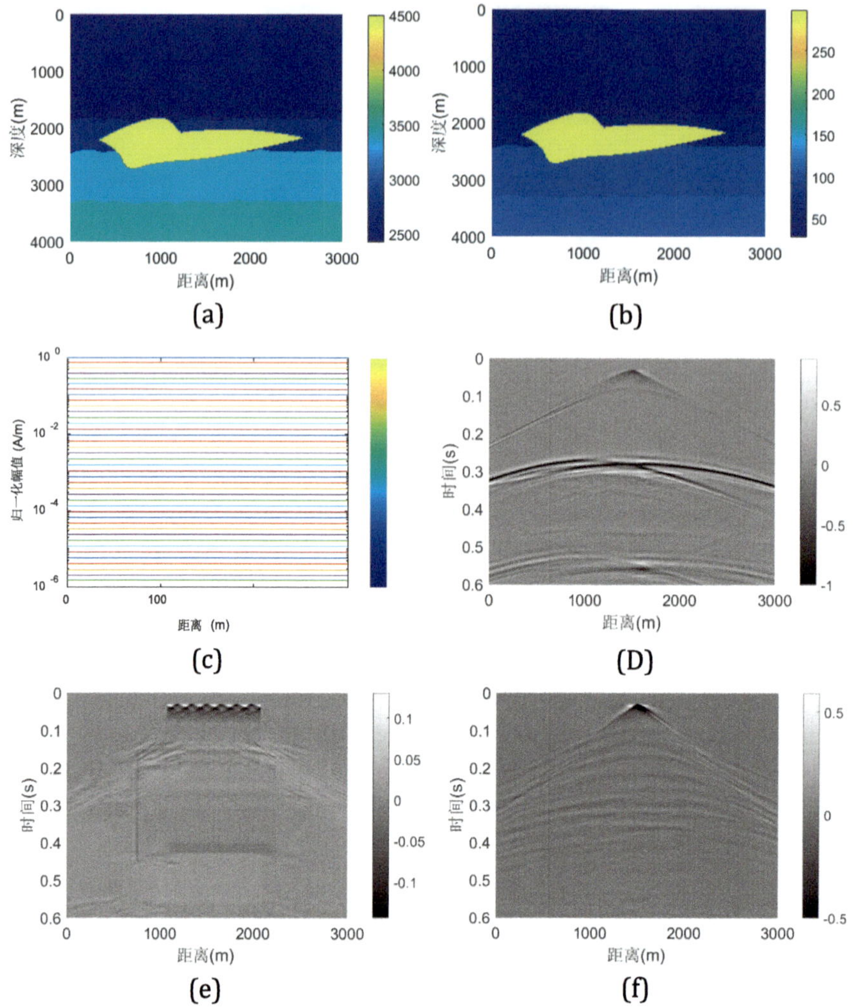

Fig. 4.8 Results of deep learning wave field conversion. **a** Seismic velocity model, **b** approximate resistivity model, **c** normalized transient electromagnetic data, **d** theoretical value of simulated seismic wave field, **e** Unet prediction, **f** C-Unet prediction

The Paleogene mainly consists of sandstone, mudstone, and silty sandstone; the Neogene features sandstone and gravel rock, with a thickness ranging from 200 to 1000 m; and the Quaternary is characterized by sandy soil, sub-clay, and sand layers, with an approximate thickness of 400 m.

The layout of the survey line is illustrated in Fig. 4.11. The survey line extends 13,000 m, with transmitters distributed along the north and south sides over a length of 2 km. The transmitters are activated at various positions in accordance with the survey line.

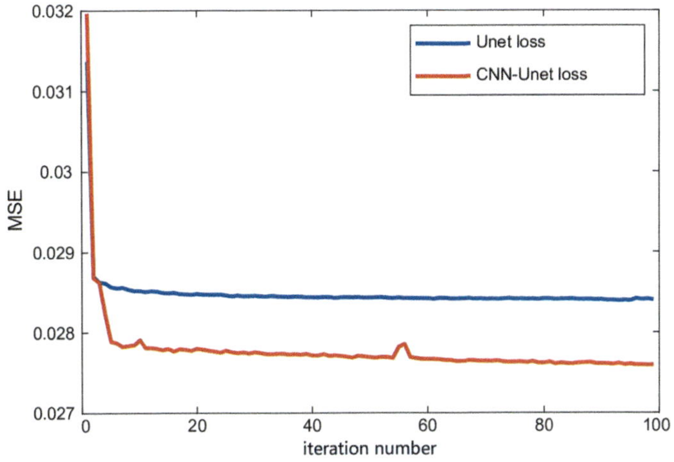

Fig. 4.9 Iteration fitting error

Figure 4.12 presents the seismic time domain stratigraphic division interpretation for the survey line extending from 14,000 to 26,000 m. In this interpretation, the shallow layer corresponds to the Quaternary, while the middle layer interface is primarily associated with the Neogene, and the deeper interface is mainly linked to the Paleogene. Notably, a fault structure is observed near point number 22,000. The red dashed area in the figure indicates the interpretation area of the transient electromagnetic data.

Figure 4.13 is the inversion imaging result of the survey line 20,320–25,720 m, where (a) is the transient electromagnetic resistivity inversion result. The figure shows that the area above 700 m exhibits high resistance, with resistivity decreasing with depth below 700 m. In the horizontal range of 21,500–22,000 m and the depth range of 500–1000 m, the resistivity contour line displays a distinct uplift phenomenon. Compared to Fig. 4.12, the terrain in this area is primarily characterized by stratification. In (a), in the horizontal range of 22,000–22,500 m and the depth range of 1000–1500 m, the resistivity contour line has a clear depression, which is in good agreement with the fault position in this area in Fig. 4.10 (horizontal 22,000–23,000 m, depth 500–1200 m).

Figure 4.13b shows the simulated seismic offset imaging result, utilizing the Born approximation offset imaging processing (Fan et al., 2022). The red dashed line indicates the fault position in the figure. The figure reveals a prominent same-phase axis in the shallow layer at ~200 m depth, which aligns well with the shallow Quaternary stratum depicted in Fig. 4.12. In the horizontal range of 20,320–22,500 m and the depth range of 700–900 m, there is a clear same phase axis, which corresponds well with the stratified feature in the corresponding area in Fig. 4.12, and the same phase axis near the horizontal 22,500 m is discontinuous in this depth range. The fault position in Fig. 4.13b (horizontal 22,500–23,500 m, depth 700–1200 m) corresponds well with the negative value of the wave field, compared with the fault position in

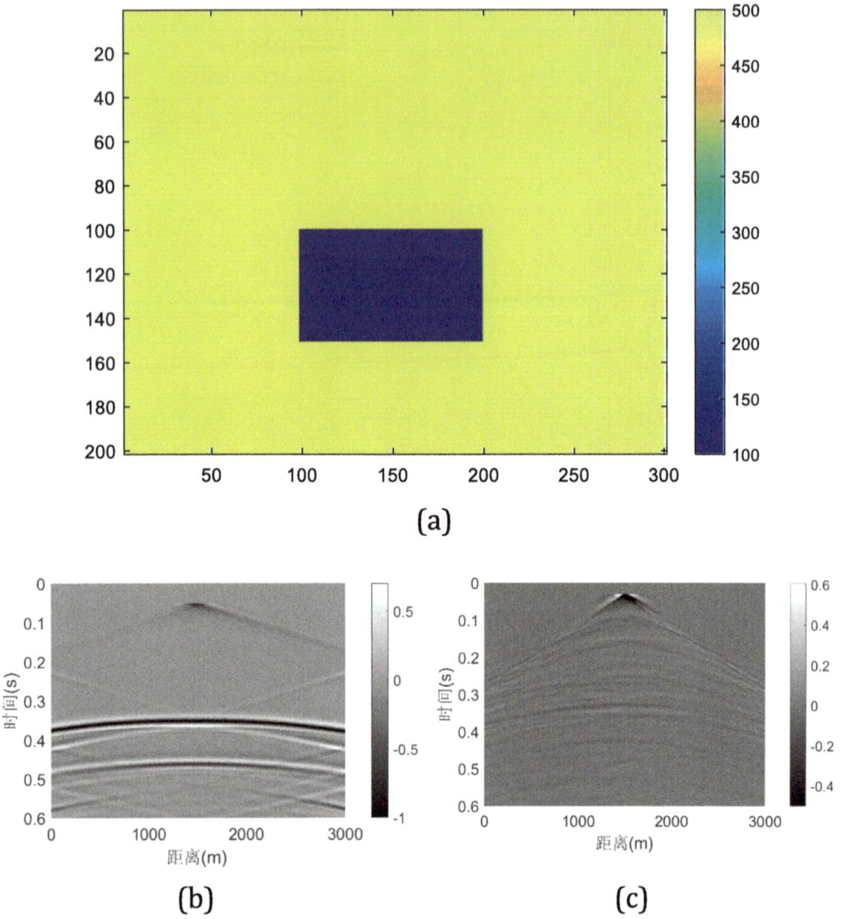

Fig. 4.10 Anomaly body model prediction. **a** Resistivity model, **b** theoretical value of simulated seismic wave field, **c** simulated seismic prediction value

Fig. 4.11 Layout of transient electromagnetic survey line

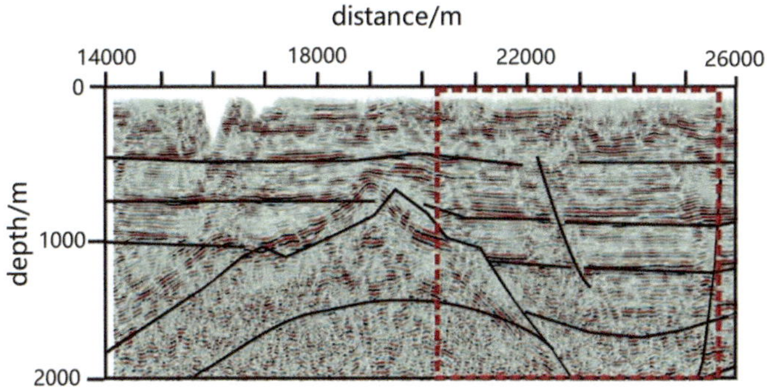

Fig. 4.12 Seismic layer analysis of the survey line

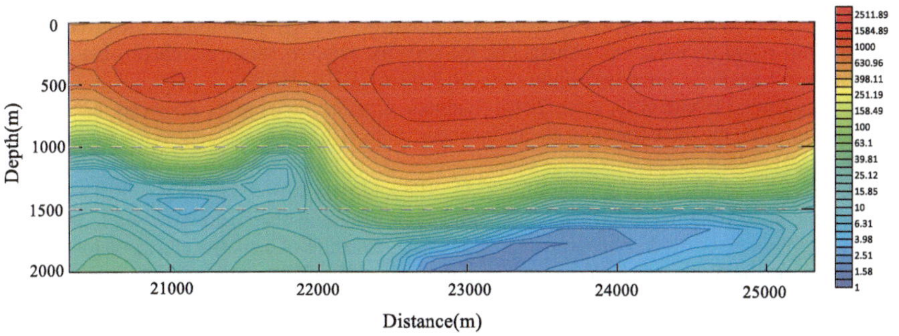

(a)

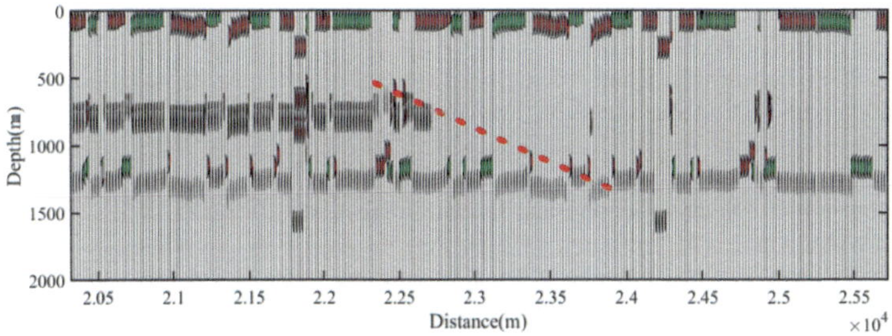

(b)

Fig. 4.13 Inversion results of 20,320–25,720 m. **a** Transient electromagnetic resistivity inversion result, **b** simulated seismic offset imaging result

Fig. 4.12 (horizontal 22,000–23,000 m, depth 500–1200 m), the fault interpretation of the simulated seismic wave field in Fig. 4.13b is better than the fault interpretation of the resistivity contour line in Fig. 4.13a.

References

Avdeev, D., & Avdeeva, A., 2009. 3D magnetotelluric inversion using a limited-memory quasi-Newton optimization. Geophysics, 74, F45–F57.

Auken, E., & Christiansen, A.V., 2004. Layered and laterally constrained 2D inversion of resistivity data. Geophysics, 69, 752–761.

Cai, J., Qi, Y., & Yin, C., 2014. Weighted laterally-constrained inversion of frequency-domain airborne EM data. Chinese Journal of Geophysics, 57, 953–960.

Cai, H., Liu, M., Zhou, J., Li, J., & Hu, X., 2022. Effective 3D-transient electromagnetic inversion using finite-element method with a parallel direct solver. Geophysics, 87, E377–E392.

Candansayar, M.E., 2008. Two-dimensional inversion of magnetotelluric data with consecutive use of conjugate gradient and least-squares solution with singular value decomposition algorithms. Geophysical Prospecting, 56, 141–157.

Cao Xiaoyue, Yin Changchun, Zhang Bo, et al. Research on finite memory quasi-Newton inversion of 3D magnetotelluric method based on unstructured grid[J]. Applied Geophysics: English Edition, 2018, 3: 556–565.

Chen W Y, Xue G Q, Olatayo A L, et al. A comparison of loop time-domain electromagnetic and short-offset transient electromagnetic methods for mapping water-enriched zones - A case history in Shaanxi, China[J]. Geophysics, 2017, 82(6): B201–B208.

Cheng J L, Chen D, Xue G Q, et al. 2016. Synthetic aperture imaging in advanced detection of roadway using the mine transient electromagnetic method. Chinese J. Geophysics. (in Chinese), 59(2): 731–738

Constable, S.C., Parker, R.L., & Constable, C.G., 1987. Occam's inversion: A practical algorithm for generating smooth models from electromagnetic sounding data. Geophysics, 52, 289–300.

Cox, L.H., Wilson, G.A., & Zhdanov, M.S., 2010. 3D inversion of airborne electromagnetic data using a moving footprint. Exploration Geophysics, 41, 250–259.

Degroot-Hedlin, C., & Constable, S., 1990. Occam's inversion to generate smooth, two-dimensional models from magnetotelluric data. Geophysics, 55, 1613–1624.

Di Qingyun, Zhu Rixiang, Xue Guoqiang, Yin Changchun, Li Xiu. Research progress on new electromagnetic detection technology for deep earth resources in China[J]. Chinese Journal of Geophysics, 2019, 62(06):2128-2138.

Donoho, D.L., 2006. For most large underdetermined systems of linear equations the minimal L1-norm solution is also the sparsest solution. Communications on Pure and Applied Mathematics, 59, 797–829.

Egbert, G.D., 2012. Hybrid conjugate gradient-Occam algorithms for inversion of multifrequency and multitransmitter EM data. Geophysical Journal International, 190, 255–266.

FAN KeRui, WANG Gang, LU KaiLiang, ZHENG Xin, LI Xiu, SUN XingBang. 2023. Inverse wavefield transform of vertical magnetic signal of galvanic source TEM when slowing down pseudo-wavefield velocity. Chinese Journal of Geophysics (in Chinese), 66(4): 1743–1757

Fan Yanan, Li Xiu, Qi Zhipeng, Lu Kailiang. Transient Electromagnetic Virtual Wave Field Second Order Born Approximation Imaging Algorithm. Journal of Geophysics, 2022, 65(3): 1144-1159.

Gallardo, L.A., & Meju, M.A., 2003. Characterization of heterogeneous near-surface materials by joint 2D inversion of dc resistivity and seismic data. Geophysical Research Letters, 30, 2–5.

Gallardo, L.A., & Meju, M.A., 2007. Joint two-dimensional cross-gradient imaging of magnetotelluric and seismic traveltime data for structural and lithological classification. Geophysical Journal International, 169, 1261–1272.

Haber, E., 2005. Quasi-Newton methods for large-scale electromagnetic inverse problems. Inverse Problems, 21, 305–323.

Haber, E., Oldenburg, D.W., & Shekhtman, R., 2007. Inversion of time domain three-dimensional electromagnetic data. Geophysical Journal International, 171, 550–564.

Haber, E., & Schwarzbach, C., 2014. Parallel inversion of large-scale airborne time-domain electromagnetic data with multiple OcTree meshes. Inverse Problems, 30, 1–28.

He Y M, Chen W Y, Lei K X, et al. Semi-Airborne electromagnetic 2.5D inversion based on a PSO–LCI strategy[J]. Journal of applied geophysics, 2022, 197(6): 104541.

Kamm, J., & Pedersen, L.B., 2014. Inversion of airborne tensor VLF data using integral equations. Geophysical Journal International, 198, 775–794.

Kelbert, A., Egbert, G.D., & Schultz, A., 2008. Non-linear conjugate gradient inversion for global EM induction: Resolution studies. Geophysical Journal International, 173, 365–381.

Key, K., 2009. 1D inversion of multicomponent, multifrequency marine CSEM data: Methodology and synthetic studies for resolving thin resistive layers. Geophysics, 74(2), F9–20.

Kirkegaard, C., & Auken, E., 2015. A parallel, scalable and memory efficient inversion code for very large-scale airborne electromagnetics surveys. Geophysical Prospecting, 63, 495–507.

Lavoué, F., Brossier, R., Métivier, L., Garambois, S., & Virieux, J., 2014. Two-dimensional permittivity and conductivity imaging by full waveform inversion of multioffset GPR data: A frequency-domain quasi-Newton approach. Geophysical Journal International, 197, 248–268.

Lee, S., McMechan, G.A., & Aiken, C.L.V., 1987. Phase-field imaging: the electromagnetic equivalent of seismic migration. Geophysics, 52, 678–693.

Li, H., Xue, G. qiang, Zhou, N. nan, & Chen, W. ying., 2015. Appraisal of an array TEM method in detecting a mined-out area beneath a conductive Layer. Pure and Applied Geophysics, 172, 2917–2929.

Liu Yunhe, Yin Changchun. Research on three-dimensional frequency domain airborne electromagnetic inversion [J]. Chinese Journal of Geophysics, 2013, 56(12): 4278-4287.

Liu, G., Kovacs, A., & Becker, A., 1991. Inversion of airborne electromagnetic survey data for sea-ice keel shape. Geophysics, 56, 1986–1991.

Liu, Y.H., Yin, C.C., Ren, X.Y., & Qiu, C.K., 2016. 3D parallel inversion of time-domain airborne EM data. Applied Geophysics, 13, 701–711.

LU KaiLiang, FAN YaNan, LI Xiu, ZHOU JianMei. 2023. Finite difference transient electromagnetic migration imaging for grounded sources. Chinese Journal of Geophysics (in Chinese), 66(2): 854–869,

Luan, X., Di, Q., Xue, G., & Chen, B., 2020. Ground-wire source TEM 3D full time multinary inversion using adaptive regulation. Journal of Environmental and Engineering Geophysics, 25, 403–413.

Luan, X., Becken, M., Thiede, A., & Kotowski, P.O., 2023. 3D tomographic inversion for frequency semi-airborne EM under multinary constraints. Exploration Geophysics, 1–18.

Meju, M.A., Saleh, A.S., Mackie, R.L., Miorelli, F., Miller, R. V., & Mansor, N.K.S., 2018. Workflow for improvement of 3D anisotropic CSEM resistivity inversion and integration with seismic using cross-gradient constraint to reduce exploration risk in a complex fold-thrust belt in offshore northwest Borneo. Interpretation, 6, SG49–SG57.

Meju, M.A., MacKie, R.L., Miorelli, F., Saleh, A.S., & Miller, R. V., 2019. Structurally tailored 3D anisotropic controlled-source electromagnetic resistivity inversion with cross-gradient criterion and simultaneous model calibration. Geophysics, 84, E387–E402.

Mittet, R., 2010. High-order finite-difference simulations of marine CSEM surveys using a correspondence principle for wave and diffusion fields. Geophysics, 75, F33–F50.

Moghadas, D., Engels, M., & Schwalenberg, K., 2015. 1D joint multi-offset inversion of time-domain marine controlled source electromagnetic data. Geophysical Prospecting, 63, 1334–1354.

Nittinger, C.G., & Becken, M., 2016. Inversion of magnetotelluric data in a sparse model domain. Geophysical Journal International, 206, 1398–1409.

Newman, G.A., & Alumbaugh, D.L., 2000. Three-dimensional magnetotelluric inversion using non-linear conjugate gradients. Geophysical Journal International, 140, 410–424.

Newman, G.A., & Commer, M., 2008. New advances in three-dimensional controlled-source electromagnetic inversion. Geophysical Journal International, 172, 513–535.

Oldenburg, D.W., Haber, E., & Shekhtman, R., 2013. Three dimensional inversion of multisource time domain electromagnetic data. Geophysics, 78, E47–E57.

Pedersen, L.B., 2004. Determination of the regularization level of truncated singular-value decomposition inversion: The case of 1D inversion of MT data. Geophysical Prospecting, 52, 261–270.

Puzyrev, V., 2019. Deep learning electromagnetic inversion with convolutional neural networks. Geophysical Journal International, 218, 817–832.

Qi, Z., Zhang, Y., Li, X., & Li, H., 2022. S-inversion of electrical source semi-airborne TEM data to determine the electric interface underground. Journal of Applied Geophysics, 204.

Ren, X., Lai, M., Wang, L., Yin, C., Liu, Y., Su, Y., Zhang, B., Ben, F., & Huang, W., 2023. A fast 3D inversion for airborne EM data using preconditioned stochastic gradient descent. Geophysical Journal International. 234, 737–754.

Rodríguez, J., Esparza, F.J., & Gómez-Treviño, E., 2010. 2-D Niblett-Bostick magnetotelluric inversion. Geologica Acta, 8, 15–30.

Sattel, D., 2005. Inverting airborne electromagnetic (AEM) data with Zohdy's method. Geophysics, 70, G77–G85.

Su, Y., Yin, C., Liu, Y., Zhang, B., & Ren, X., 2019. Inversions of time-domain airborne EM based on generalized model constraints. Chinese Journal of Geophysics, 62, 743–751.

Su, Y., Yin, C., Liu, Y., Ren, X., Zhang, B., & Xiong, B., 2021. Three-dimensional anisotropic inversions for time domain airborne electromagnetic data. Minerals, 11, 1–17.

Simons, F.J., Loris, I., Nolet, G., Daubechies, I.C., Voronin, S., Judd, J.S., Vetter, P.A., Charléty, J., & Vonesch, C., 2011. Solving or resolving global tomographic models with spherical wavelets, and the scale and sparsity of seismic heterogeneity. Geophysical Journal International, 187, 969–988.

Tartaras, E., Zhdanov, M.S., Wada, K., Saito, A., & Hara, T., 2000. Fast imaging of TDEM data based on Sinversion. Journal of Applied Geophysics, 43, 15–32.

Vallée, M.A., & Smith, R.S., 2009. Inversion of airborne time-domain electromagnetic data to a 1D structure using lateral constraints. Near Surface Geophysics, 7, 63–71.

Viezzoli, A., Christiansen, Anders, V., Auken, E., & Sørensen, K., 2008. Quasi-3D modeling of airborne TEM data by spatially constrained inversion. Geophysics, 73, F105–F113.

Vignoli, G., Fiandaca, G., Christiansen, A.V., Kirkegaard, C., & Auken, E., 2015. Sharp spatially constrained inversion with applications to transient electromagnetic data. Geophysical Prospecting, 63, 243–255.

Vignoli, G., Sapia, V., Menghini, A., & Viezzoli, A., 2017. Examples of improved inversion of different airborne electromagnetic datasets via sharp regularization. Journal of Environmental and Engineering Geophysics, 22, 51–61.

Wang, R., Yin, C., Wang, M., & Wang, G., 2012. Simulated annealing for controlled-source audio-frequency magnetotelluric data inversion. Geophysics, 77, E127–E133.

Wang, F., Morten, J.P., & Spitzer, K., 2018. Anisotropic three-dimensional inversion of CSEM data using finite-element techniques on unstructured grids. Geophysical Journal International, 213, 1056–1072.

Wang, X., You, N., Di, Q., Deng, J., & Chang, Y., 2021. 3-D parallel inversion of multichannel transient electromagnetic data using a moving footprint. Geophysical Journal International, 226, 1783–1799.

Wu X, Xue G Q, He Y M, et al. Removal of the multi-source noise in airborne electromagnetic data based on deep learning[J]. Geophysics, 2020, 85(6):1-72.

Wu, X., Xue, G., Zhao, Y., Lv, P., Zhou, Z., & Shi, J., 2022. A deep learning estimation of the earth resistivity model for the airborne transient electromagnetic observation. Journal of Geophysical Research: Solid Earth, 127, 1–19.

Xue Guoqiang, Deng Xiang. Detection ability of transient electromagnetic method for thin layer [J]. Petroleum Geophysical Exploration, 2007, 42(6):709-713.

Xue Guoqiang, Li Yong, Yang Jingdong. Application of transient electromagnetic method in highway geological survey [J]. Petroleum Instrument, 2006, 20(2): 41-43.

Xue Guoqiang, Li Xiu, Di Qingyun. Progress in the theory and application of transient electromagnetic method[J]. Progress in Geophysics, 2008, 22(4): 1195-1200.

Xue Guoqiang, Yan Shu, Zhou Nannan. Analysis of the deviation of large loop source transient electromagnetic response caused by dipole hypothesis[J]. Chinese Journal of Geophysics, 2011, 54(9): 2389-2396.

Xue Guoqiang, Zhang Linbo, Zhou Nannan, Chen Weiying. Developments measurements of TEM sounding in China. Geological Journal, 2020a, 55(3):1636–1643.

Xue, G., Li, H., He, Y., Xue, J., & Wu, X., 2020b. Development of the inversion method for transient electromagnetic data. IEEE Access, 8, 146172–146181.

Xue G Q, He Y M, Chen W Y, Wu X, Song W T. 3-D inversion based on the particle swarm optimization-quasi-newton hybrid algorithm for SOTEM[J]. IEEE Transactions on Geoscience and Remote Sensing, 2023, 61: 1–11.

XUE JunJie, ZHONG HuaSen, LI Hai, LI Xiu, DI QingYun, WANG HeYuan, TONG ZhaoCheng. 2018. An improvement to the transformation algorithm for the transient electromagnetic field. Chinese Journal of Geophysics (in Chinese), 61(12): 5077-5083

Yang F S, Ma J W. Deep-learning inversion: a next generation seismic velocity model building method[J]. Geophysics, 2019, 84: R583–R599.

Yu, C., Liu, H., Zhang, X., Yang, D., & Li, Z., 2013. The analysis on IP signals in TEM response based on SVD. Applied Geophysics, 10, 79–87.

Zhang, J., Liu, Y., Yin, C., Qiu, C., Wang, Q., & Li, J., 2019. Three-dimensional regularized inversion of marine controlled-source EM data based on unstructured tetrahedral meshes. Chinese Journal of Geophysics(in Chinese)), 62, 4451–4461.

Zhang, B., Engebretsen, K.W., Fiandaca, G., Cai, H., & Auken, E., 2021. 3D inversion of time-domain electromagnetic data using finite elements and a triple mesh formulation. Geophysics, 86, E257–E267.

Zhdanov, M.S., 2002. Geophysical inverse theory and regularization problems. Elsevier.

Zhdanov, M.S., Gribenko, A., & Wilson, G., 2012. Generalized joint inversion of multimodal geophysical data using Gramian constraints. Geophysical Research Letters, 39, 1–7.

Zhdanov, M.S., & Cox, L.H., 2013. Multinary inversion for tunnel detection. IEEE Geoscience and Remote Sensing Letters, 10, 1100–1103.

Zhdanov, M.S., Jorgensen, M., & Cox, L., 2021. Advanced methods of joint inversion of multiphysics data for mineral exploration. Geosciences (Switzerland), 11, 1–15.

Chapter 5
IP Additional Field Characteristics and Information Extraction

The induced polarization (IP) effect is a fundamental electrical characteristic of rocks and minerals. As an essential component in various electromagnetic exploration methods—such as frequency-domain magnetotellurics (MT), controlled-source audio-frequency magnetotellurics (CSAMT), and time-domain magnetic source systems—the IP effect plays a crucial role. Notably, the short-offset transient electromagnetic (SOTEM) method, with its unique near-source observation capabilities, exhibits a stronger induced electric effect in its electromagnetic response. Therefore, it is essential to investigate the SOTEM method while considering the presence of the induced electric additional field.

When polarized bodies—such as massive sulfide ore, groundwater, graphite, or carbonaceous layers—exist underground, the observed SOTEM electromagnetic field consists of both the conventional electromagnetic response and the induced electric response resulting from the IP effect. This interaction alters the decay characteristics of the electromagnetic field compared to traditional transient electromagnetic methods. Notably, when the induced electric response and the electromagnetic response have opposite signs, a polarity reversal phenomenon may occur. While early research regarded this influence as interference, the induced electric response embedded in the SOTEM electromagnetic field provides critical information, such as the polarization rate of subsurface materials. This is particularly significant when exploring polarized bodies like massive sulfide ore. In such cases, the induced electric effect should not be dismissed as noise but rather leveraged to enhance SOTEM data interpretation. Effectively extracting induced electric information from the SOTEM electromagnetic field can yield valuable insights for data processing and improve the accuracy of deep mineral resource exploration.

Current research on the induced electric effect in time-domain electromagnetism, both domestically and internationally, has primarily focused on loop-source transient electromagnetic (TEM) systems. However, the SOTEM system, which employs near-source observation, contains stronger induced electric signals within its electromagnetic response. Thus, investigating the IP effect in SOTEM is of greater

G. Xue et al., *Exploration with Short-Offset Grounded-Wire Transient Electromagnetic Method*, https://doi.org/10.1007/978-981-96-8846-3_5

significance, yet this research area remains largely unexplored. Unlike loop-source systems, which predominantly measure magnetic field components, SOTEM relies heavily on the collection and processing of electric field components to achieve high-resolution detection of both high- and low-resistivity targets. Consequently, investigating the induced electric effect in SOTEM necessitates a comprehensive consideration of several factors: the unique excitation mode of the grounded-wire source, the complexity of three-dimensional forward modeling incorporating the IP effect, and the challenges associated with extracting useful information. The lack of research in these areas has hindered the further development and application of the SOTEM detection system.

To address these research gaps and overcome existing challenges, this study aims to analyze the response characteristics of SOTEM with induced electric effects and develop effective methods for extracting this information. To enhance the theoretical framework of SOTEM detection and improve data processing and interpretation accuracy, we will focus on the following research areas: (1) one-dimensional forward modeling of the SOTEM response incorporating the IP effect, (2) three-dimensional forward simulation and characteristic analysis, (3) methodologies for extracting induced electric information, and (4) case studies demonstrating the practical application of these findings.

5.1 IP Effect in Transient Electromagnetic Method

5.1.1 Mechanism of Rock and Mineral IP Effect

The induced polarization (IP) effect in rock minerals was first discovered in the nineteenth century. Since the 1940s, this phenomenon has been widely applied across various geophysical fields, including metal and graphite ore exploration, as well as hydrogeological investigations (Liangkui 1982; Dias 1972). By the 1960s, it had become a primary method for detecting metal deposits. Today, IP methods are extensively used in mineral exploration, oil and gas prospecting, and hydro-environmental geological surveys (Jinming 2004).

Induced polarization methods are generally categorized into two types: the time-domain direct current (DC) induced polarization method and the frequency-domain alternating current (AC) induced polarization method. While both share similar fundamental principles, their response characteristics and decay rates differ. As illustrated in Fig. 5.1, the time-domain induced polarization response exhibits a characteristic charging and discharging curve, where ΔV_1 represents the primary field, and ΔV_2 represents the polarized secondary field.

At a macroscopic level, the induced polarization effect can be described as follows: during the current injection phase, or within a short time after the current is switched off, a transient additional electric field emerges due to electrochemical interactions between the artificially applied electric field and the rock or mineral. This

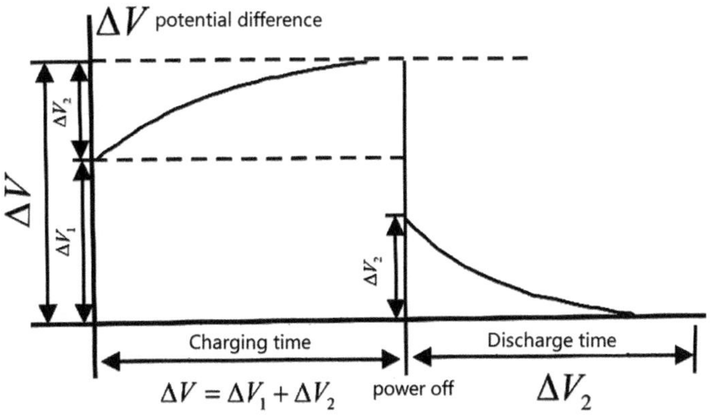

Fig. 5.1 Schematic diagram of time-domain induced electric phenomenon

phenomenon, which follows a regular time-dependent behavior, is referred to as induced polarization.

At the microscopic level, the IP effect of rock minerals can be classified into two primary types: electron conductor polarization and ion conductor polarization. Domestic and international experts share a relatively unified understanding of the IP effect mechanism in electron conductors (which include most metal ores, graphite, and carbonaceous rocks). It is generally attributed to overpotential occurring at the interface between the electron conductor and the surrounding solution (Liangkui 1982). Specifically, under an applied external current, the double electrical layer at this interface undergoes directional movement of electrons and ions, leading to charge accumulation. Once the current is switched off, the stored charge discharges, generating an observable secondary polarization response.

Figure 5.2 illustrates the excitation polarization process of an electron conductor. When the excitation current is applied, charge redistribution occurs within the electron conductor particles, leading to the formation of cathodic and anodic regions, which in turn increase the potential difference across the double layers. This potential difference, known as overpotential, is shown in Fig. 5.2a. As the charging period continues, charge accumulation increases, causing the overpotential to rise until it reaches saturation, at which point no further charge accumulation occurs (Fig. 5.2b). When the external current is turned off, the accumulated opposite charges at the interface discharge through the interface itself. At this stage, both the surrounding solution and the electron conductor undergo a discharge process until the system returns to its normal potential state (Fig. 5.2c). This process represents the discharge phase of overpotential.

However, general rock-forming minerals and clay minerals function as ion conductors, and their polarization behavior is best explained by the Double-Layer Deformation Hypothesis, proposed by Liangkui (1983). This widely accepted hypothesis is based on the distribution characteristics of mobile cations within the electrical double layer at the interface between rock particles and the surrounding

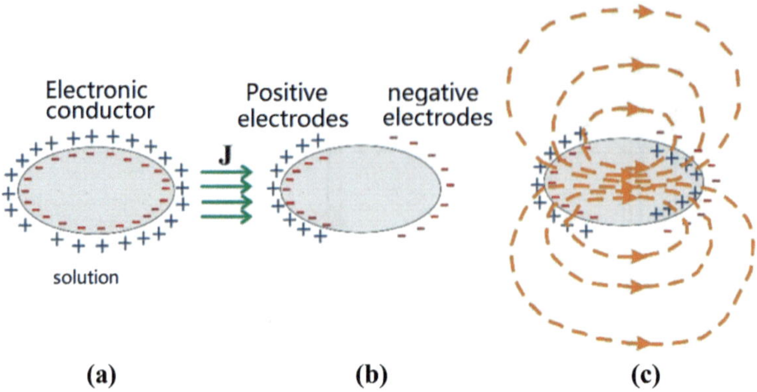

Fig. 5.2 Excitation polarization process of electron conductor

solution. Under an external current, deformation of the double layer occurs due to ion displacement. When the current is turned off, the accumulated ions discharge, resulting in the observed polarization effect.

As shown in Fig. 5.3, when an external current is applied, cations in the diffuse layer of the double layer are displaced (Fig. 5.3a), leading to double-layer deformation (Fig. 5.3b). Once the current is removed, the accumulated ions are released, restoring the system to equilibrium (Fig. 5.3c), thereby generating the excitation polarization field.

This differentiation between electron conductor and ion conductor polarization is crucial for accurately interpreting induced polarization data in geophysical surveys. By understanding these mechanisms, researchers can enhance the effectiveness of IP methods in various geological applications, ranging from mineral prospecting to hydrogeological investigations.

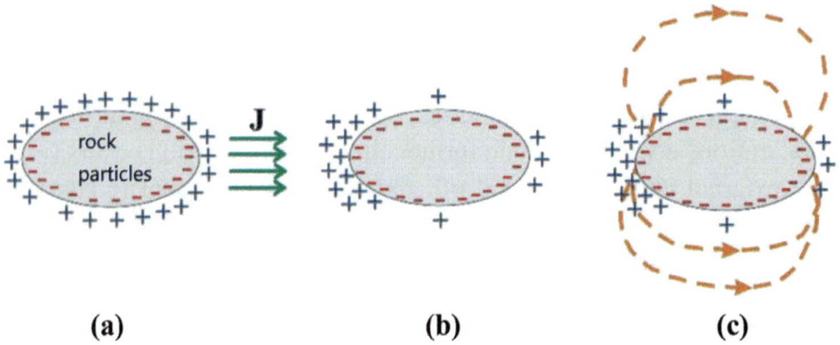

Fig. 5.3 Excitation polarization process of ion conductor

5.1.2 IP Effect Mechanism in Transient Electromagnetic Method

However, traditional IP technology is not the only electromagnetic method sensitive to polarizable bodies. Any electromagnetic measurement conducted in the presence of a polarizable body will be influenced by its effects. The transient electromagnetic (TEM) method is an effective tool for imaging the electrical properties of the near-surface region, typically from tens to hundreds of meters in depth (Xue et al. 2007; Guo et al. 2020). This method utilizes grounded wires or return wires to transmit the source signal into the ground and records the decay curve of the electromagnetic response after the source is switched off (Xue et al. 2012; Chen et al. 2017).

In conventional TEM measurements, the resistivity of each subsurface grid cell is assumed to be constant, meaning that the excitation polarization effect of polarizable media is not considered. However, when a polarizable target body is present, an electric phenomenon is triggered, generating an additional electromagnetic field that superimposes onto the conventional TEM response (Kang and Oldenburg 2018). In such cases, conventional inversion methods assuming constant resistivity fail to accurately fit the observed data, making it difficult to obtain reliable subsurface resistivity estimates.

In some instances, the TEM response with an IP effect exhibits a sign reversal phenomenon. Whether this reversal occurs, and the timing of its occurrence, depends on the relative magnitudes of the electric field polarization current (caused by the IP effect) and the induced eddy current at different response times. Flis et al. (1989) provided a physical explanation for how the IP effect in TEM can lead to response sign reversal by analyzing the process of ion migration within rock voids under an applied electric field.

As illustrated in Fig. 5.4, the physical mechanism of this process is as follows:

(a) Before the current is switched off, metal particles in the rock remain in electrical equilibrium with the negatively charged clay particles.
(b) In the early stage after the current is switched off, induced eddy currents develop underground, forming a secondary electric field. Cations within the polarized body are stimulated by these eddy currents, generating a polarization current in the same direction as the eddy currents. At this stage, the polarized body is undergoing the charging process, and the polarization current contributes positively to the overall response. However, due to the dominance of the strong electromagnetic effect, this influence is often masked.
(c) As charging continues, before the polarized body reaches a saturated charge state, cation accumulation continues. At this point, the polarization current reaches zero, and the induced eddy current dominates the response.
(d) Over time, the induced eddy current gradually weakens. Once it becomes too weak to sustain the saturated charge state of the polarized body, cations begin to return to their original positions, initiating the discharge process. This discharge produces a polarization current in the opposite direction to the induced eddy current. If, at a certain moment, the eddy current has sufficiently decayed, or if the IP effect of the polarized body is particularly strong, the two opposing fields

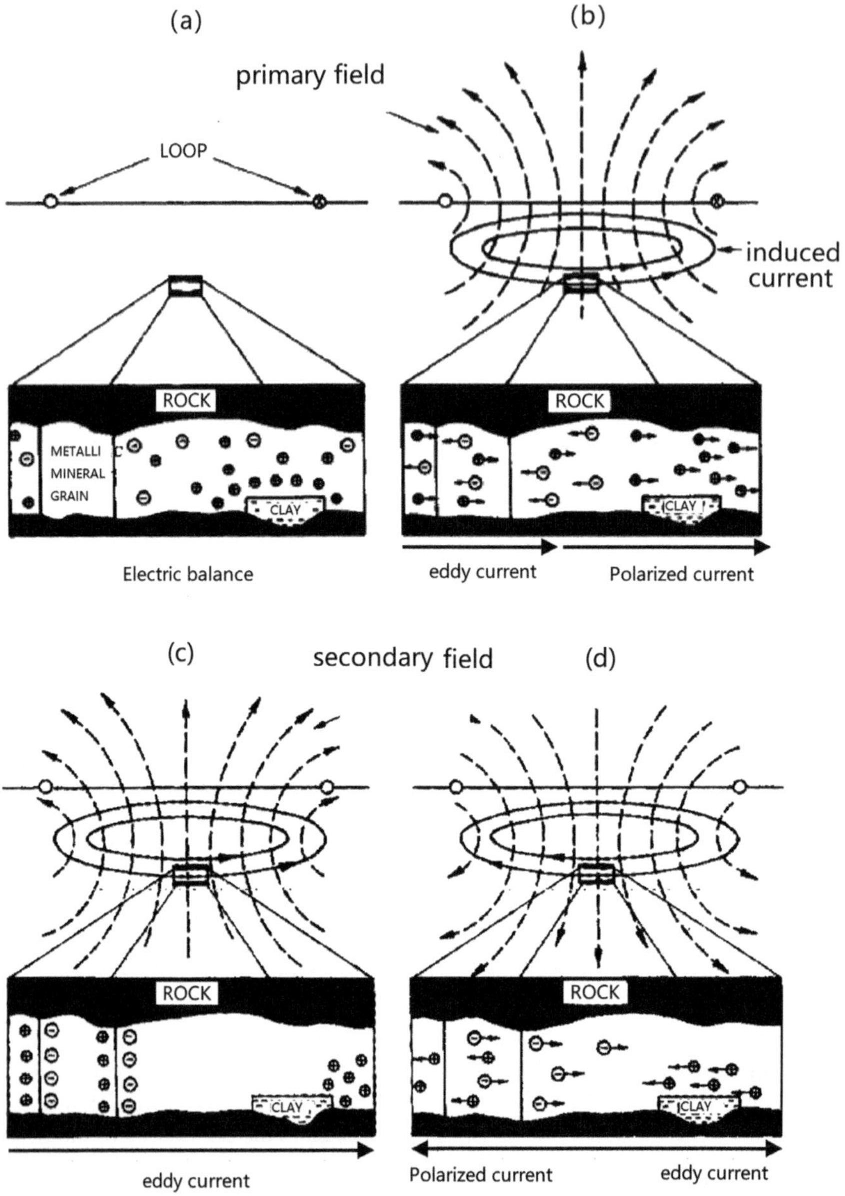

Fig. 5.4 Physical mechanism of IP effect causing TEM response symbol reversal (Flis 1987)

may couple together. When the polarization current exceeds the decaying eddy current, the sign reversal phenomenon occurs in the total measured field.

The macroscopic manifestation of this mechanism is illustrated in Fig. 5.5. The excitation effect of subsurface polarizable materials—such as porphyry copper,

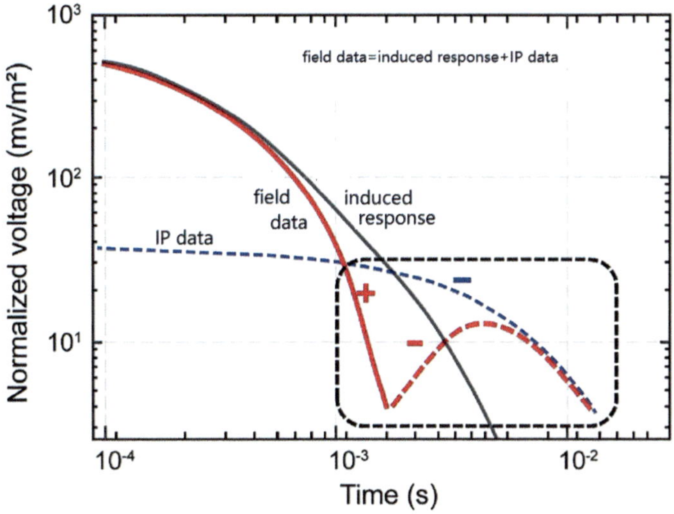

Fig. 5.5 Transient electromagnetic response with IP effect

magnetite, pyrite, sulfides, water bodies, carbonaceous layers, etc.—induces an additional field that is superimposed on the TEM response generated by electromagnetic induction. Because this additional field decays more slowly and is opposite in direction to the inductive field, it leads to distortion and sign reversal in the late-time TEM response. This explanation integrates both microscopic and macroscopic perspectives to elucidate the role of the IP effect in transient electromagnetic responses.

5.1.3 Cole–Cole Complex Resistivity Model

The IP effect refers to the reversible charge storage phenomenon in porous materials at low frequencies (usually < 100 Hz) (Revil et al. 2015). When an artificial source is applied to a polarizable medium, boundary charges and free charges will move, causing displacement current $\mathbf{J_d}$ and conduction current $\mathbf{J_c}$. According to Ohm's law, the current generated by the source can be divided into:

$$\begin{aligned} \mathbf{J_c} &= \sigma \mathbf{E}, \\ \mathbf{J_d} &= \frac{\partial \mathbf{D}}{\partial t} = i\omega\varepsilon_a\varepsilon\mathbf{E}, \end{aligned} \tag{5.1}$$

where **E** is the electric field strength, **D** is the electric displacement vector, σ is the conductivity, ε_a is the dielectric constant in vacuum, the current density is the sum of the displacement current and conduction, and the complex conductivity form of Ohm's law is obtained:

$$\mathbf{J} = \sigma \cdot \mathbf{E} + i\omega\varepsilon\varepsilon_a\mathbf{E} = \sigma^*\mathbf{E}, \tag{5.2}$$

where $\sigma^* = \sigma + i\omega\varepsilon\varepsilon_a \mathbf{E} = \sigma^* \mathbf{E}$ can be used to describe the electrical characteristics of polarizable media.

The excitation polarization effect is essentially a complex electrochemical reaction of rocks and their contained aqueous solutions under the action of an external current, and a mathematical model is needed to describe it. The characteristic of complex conductivity is that the conductivity of the medium changes with frequency or time. In order to establish the mathematical formula of the excitation polarization effect, excitation polarization models have been established to describe the dispersion characteristics of conductivity, including the Dias model (Dias 1972), the Cole–Cole model (Cole and Cole 1941) and the GEMIP model (Zhdanov 2008). Although there is currently no unified physicochemical model to describe dispersive media, the Cole–Cole model, including some modified parameterizations by Pelton et al. (1978) and Fiandaca et al. (2018), is the most widely used model in geophysical forward and inverse modeling.

Use the most accurate and classic Cole–Cole model to describe the IP effect:

$$\rho(i\omega) = \rho_0 \left[1 - m\left(1 - \frac{1}{1 + (i\omega\tau)^c} \right) \right] \tag{5.3}$$

where $\rho(i\omega)$ is the dispersive resistivity, m is the charging rate of the medium, the charging rate affects the intensity of the spectrum but does not affect the shape of the spectrum, it is mainly affected by the content of metal minerals, the degree of mineral compaction and impregnation; t is the time constant, in some cases, the degree of compaction of the ore body, depth and scale of the ore body have the greatest impact on the peak phase of the spectrum; c is the frequency-related coefficient, the depth of the ore body is an important factor determining the degree of spectrum characteristics.

These parameters characterize the dispersive resistivity of the Cole–Cole model, and their physical meanings are also very clear, for example, the time constant determines the time length required for time decay. Laboratory studies of the IP effect show that the frequency-related coefficient is not equal to 1, but is usually in the range of 0.1–0.6 (Pelton et al. 1978). For typical sulfide ore deposits, the value of the frequency-related coefficient is 0.3 (Macnae and Hine 2016). The general range of these parameters is: $m = 0 \sim 0.98$, $\mathrm{t} = 10^{-3} \sim 10^5$ s, $c = 0.1 \sim 0.6$.

Considering the dispersion characteristics introduced by the excitation effect, it is necessary to incorporate dispersive conductivity into the SOTEM response calculation equation. The calculation process first involves determining the frequency-domain response, followed by obtaining the transient electromagnetic (TEM) response through digital filtering using the fast Hankel transform for sine and cosine transformations.

The subsequent sections will provide a detailed explanation of the solution method and the calculation formulas for computing the transient electromagnetic response in a layered medium. Additionally, the approach for integrating complex conductivity from Eq. (5.2) into the calculation framework will be outlined, demonstrating how to effectively model the transient electromagnetic response while incorporating the IP effect.

5.2 Analysis of the Characteristics of SOTEM Three-Dimensional Forward Response with Excitation Effect

In order to fully utilize the response characteristics of different components in the SOTEM system for a three-dimensional polarization model, we separately investigate the impact of IP effects, the underlying physical mechanisms, and the analysis of typical polarization models. This approach facilitates the construction of complex real-world models, the selection of optimal field observation parameters, and ultimately enhances the extraction of valuable geological information.

We conduct a comprehensive analysis of the three-dimensional SOTEM response with the induced electric effect and compare responses with and without a three-dimensional polarizable body in both the frequency domain and time domain. These comparisons highlight the importance of considering the IP effect and demonstrate the feasibility of extracting IP information using SOTEM. Additionally, we examine the impact of polarization effects on the diffusion of transient electromagnetic fields, elucidate the charging and discharging mechanisms of induced polarization, and provide a rational explanation for the sign reversal phenomenon observed in SOTEM measurements. Furthermore, we perform three-dimensional forward modeling of thin plate-shaped polarizable bodies in horizontal, vertical, and inclined orientations, analyzing their electromagnetic field response characteristics. The findings in this chapter lay a solid foundation for subsequent case analyses and IP information extraction studies.

5.2.1 The Impact of IP Effects on FDEM and SOTEM in the Three-Dimensional Polarization Model

To evaluate the impact of IP effects in both the frequency-domain electromagnetic method (FDEM) and the transient electromagnetic method (TEM), we use the same three-dimensional model and grid partitioning method for comparative simulations. A half-space model is selected for both frequency-domain and time-domain simulations, with an earth electrical source as the excitation device.

In the simulation, a 1000 m-long wire source is placed at the interface between air (conductivity = 10^{-6} S/m) and half-space (conductivity = 0.01 S/m). The injected current is 10 A. The computational domain spans 4000 × 4000 × 1000 m, with an infinite source layer of 100 m thickness included for simulation. The transmitter is positioned along the x-axis at x = 200 m, and the transmitter grid is refined from a minimum size of 10 m to a maximum size of 50 m. The observation area is set to 40 m, gradually increasing to 500 m at the model boundary, with a maximum grid growth rate of 1.4, ensuring high-quality elements.

The middle target layer is the polarization layer, with a thickness of 300 m. The Cole–Cole model of the polarizable layer is m = 0.3, c = 0.3, t = 1 s, $\rho = 100\Omega\cdot$m, and its grid partitioning adopts local dense partitioning technology, with the smallest size of 10 m.

A local dense partitioning strategy is applied to the polarization layer, using a minimum element size of 10 m, as illustrated in Fig. 5.6. These Cole–Cole parameters are typical for porous media containing metallic particles, and a strong IP effect is expected to emerge (Revil et al. 2015).

For the half-space model, a linear shape function is selected as the basis function, resulting in a computational mesh containing 265,670 elements and 296,837 unknowns. The solution is computed across 81 frequencies ranging from 0.001 to 10^4 Hz, using 10 frequency points per decade. The total computation time is 248 min, with a physical memory usage of 15.7 GB.

(1) Three-dimensional modeling of FDEM with IP effects

Since the same model and grid are used for modeling FDEM with and without IP effects, the results obtained at the same frequency are nearly identical. To quantitatively assess this difference, the relative IP effect can be defined as follows:

$$\gamma = \left|\frac{f_{\mathrm{IP}} - f_{\mathrm{noIP}}}{f_{\mathrm{noIP}}}\right| \times 100\% \tag{5.4}$$

where f_{IP} and f_{noIP} represent the responses with and without IP effects respectively.

The observation point coordinates selected for this analysis are (−300, 0, 0), with an offset distance of 500 m. Figure 5.8 compares the response of the horizontal electric field (Ex) component in the frequency-domain electromagnetic method (FDEM) for the half-space model with and without a polarization layer. As shown in Fig. 5.8, when the frequency is below 1000 Hz($\gamma < 5\%$), the real part of the Ex component

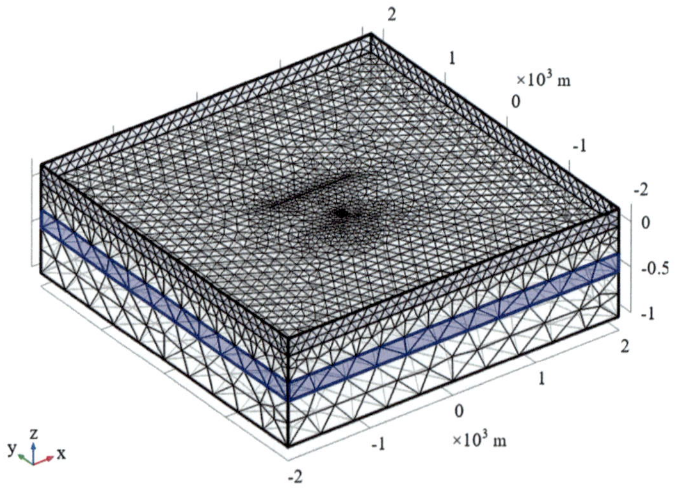

Fig. 5.6 Schematic diagram of the three-dimensional forward model (the middle is the polarizable layer)

response in the half-space with the polarization layer is nearly identical to that of the non-polarized half-space. However, for frequencies above 1000 Hz, the relative IP effect increases significantly, indicating that in shallow regions, the real part of the Ex component is strongly influenced by the IP effect.

The imaginary part of the Ex component is more significantly affected by the IP effect, particularly in the frequency range of 1 Hz to 10^3 Hz ($\gamma > 20\%$). However, it is also observed that the total measured horizontal electric field Ex does not undergo a sign reversal, with errors exceeding 20% only at the highest frequencies. This suggests that across the entire frequency range, the relative IP effect of the Ex component is greater than the normal field noise in the medium- to high-frequency range (Fig. 5.7)

Figure 5.8 compares the vertical magnetic field (Hz) component responses in FDEM for half-space models with and without a polarization layer, using the same observation point as above. As shown in Fig. 5.8, when the frequency is below 100 Hz ($\gamma < 5\%$), the real part of the Hz component response in the polarized and non-polarized half-space models remains nearly identical. However, as the frequency increases beyond 100 Hz, the relative IP effect gradually increases, reaching its maximum impact at 1000 Hz.

The imaginary part of the Hz component response is highly influenced by the IP effect, particularly in the 1–10^3 Hz ($\gamma > 10\%$), and even more so at higher frequencies ($\gamma > 50\%$). However, the effect is less pronounced in the several hundred hertz range. Despite this, the total observed vertical magnetic field component Hz does not exhibit a sign reversal, with errors exceeding 40% only at the highest frequencies.

Overall, the relative IP effect is significantly larger than normal field noise across the entire frequency range, indicating that the IP effect has a major impact on both the electric and magnetic field components of FDEM responses. Consequently,

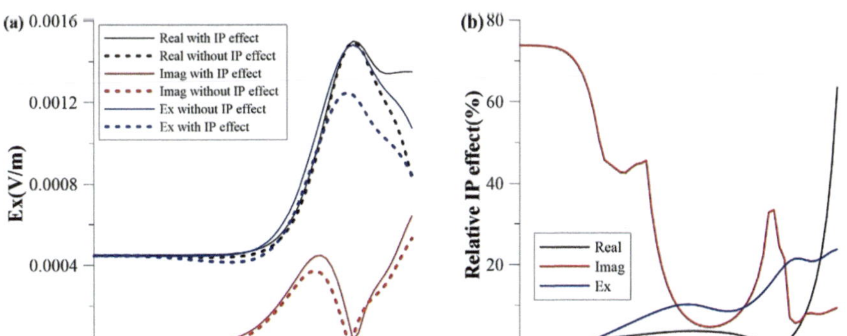

Fig. 5.7 Comparison of FDEM horizontal electric field component responses of half-space models with and without polarization layers. **a** and **b** represent the response values and the corresponding relative IP effects respectively

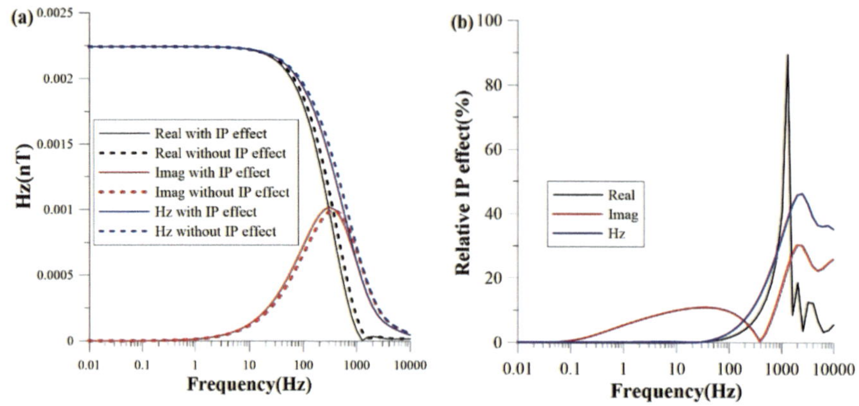

Fig. 5.8 Comparison of FDEM vertical magnetic field component responses for half-space models with and without a polarization layer. **a** and **b** represent the response values and the corresponding relative IP effects, respectively

neglecting the IP effect in this context may introduce significant errors in data inversion and interpretation.

To further illustrate how the induced polarization effect influences the ground observation field, Figs. 5.9 and 5.10 present plan-view maps at a single frequency, including the transmitter location within the observed ground field area, for cases with and without the IP effect. For ground observation field mapping, we analyze the horizontal electric field component (Ex) and the vertical magnetic field component (Hz) at two representative frequencies: Ex at 1 Hz, Hz at 1000 Hz. Results indicate that for the horizontal electric field Ex at 1 Hz, the polarizable layer significantly distorts the amplitude of the imaginary part of the measured field. Conversely, for the vertical magnetic field Hz at 1000 Hz, the real part of the measured field exhibits minimal distortion due to the polarizable layer.

(2) 3D Modeling of SOTEM with IP Effect

The same model and grid as described in the previous section are used for modeling SOTEM with and without IP effects, as illustrated in Fig. 5.6. Using the frequency-time transformation method, the previously computed frequency-domain response without IP effect is first transformed into the time domain. The response error of the FCT method remains mostly below the field noise threshold of 5%, meeting the requirements for scientific accuracy. Following the same approach, the frequency-domain response incorporating the IP effect is also transformed into the time domain. The time response of the induced polarization effect is then obtained, as shown in the figures.

A sign reversal in the late-time response is observed in both the electric field and magnetic field, a phenomenon that has also been identified in TEM field data and is widely attributed to the IP effect (Weidelt 1982; Zhou et al. 2020; Li et al. 2021).

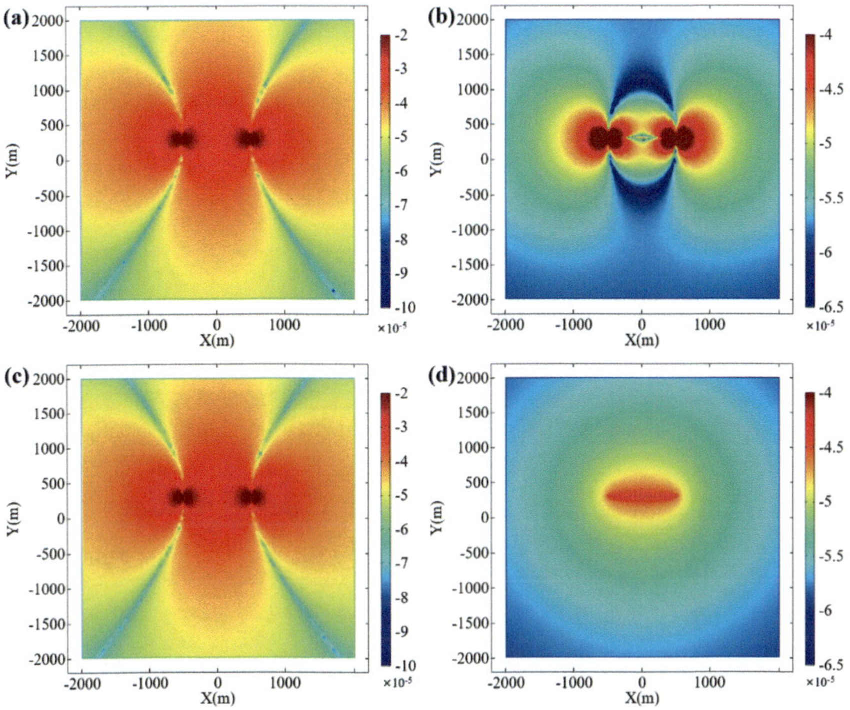

Fig. 5.9 At 1 Hz, the response of the horizontal component of the surface electric field in half-space models with and without a polarization layer. **a** and **b** correspond to the real and imaginary parts of the half-space with a polarization layer, respectively, **c** and **d** are the real and imaginary parts of the non-polarized half-space, respectively

As shown in Figs. 5.11 and 5.12, under the given model parameters, the horizontal electric field component (Ex) exhibits a sign change within tens of milliseconds, whereas the vertical magnetic field component (Hz) undergoes sign reversal at a slightly later stage, typically within hundreds of milliseconds. Overall, the impact of the induced polarization effect is relatively minor in the early stages of SOTEM but becomes more pronounced in the middle and late stages.

Despite exhibiting the smallest peak amplitude, the relative IP effect gradually increases over time in the early-stage response. In this example, the early-stage Ex response exhibits a relative anomaly exceeding 5%, which could potentially impact data inversion. Thus, it cannot be assumed that the early response is unaffected by the IP effect.

Additionally, thc timing of sign reversal cannot be directly determined from the relative anomaly curve, implying that using the sign change phenomenon as a criterion to distinguish whether the data is affected by the IP effect is not appropriate.

Similarly, to further illustrate how the induced polarization effect influences the SOTEM ground observation field, Figs. 5.13 and 5.14 present planar maps of the

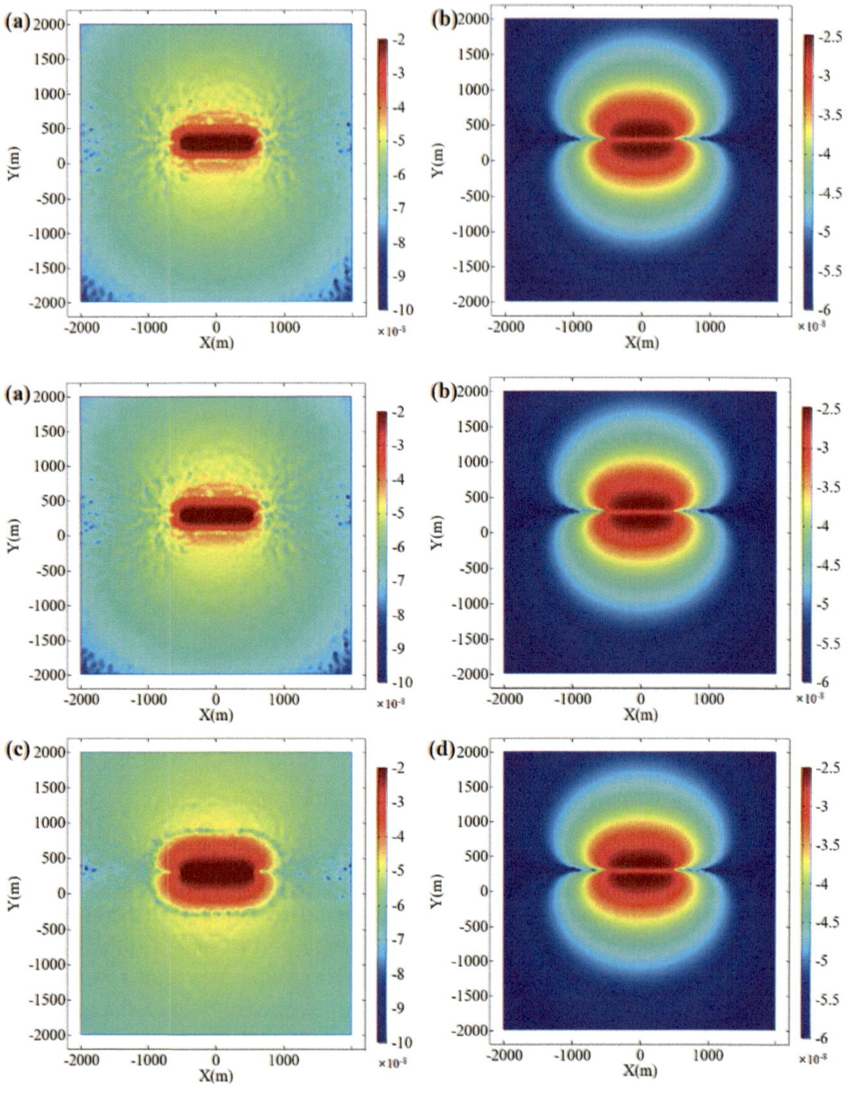

Fig. 5.10 At 1000 Hz, the response of the vertical component Hz of the magnetic induction on the surface in half-space models with and without a polarization layer. **a** and **b** correspond to the real and imaginary parts of the half-space with a polarization layer, respectively, **c** and **d** are the real and imaginary parts of the non-polarized half-space, respectively

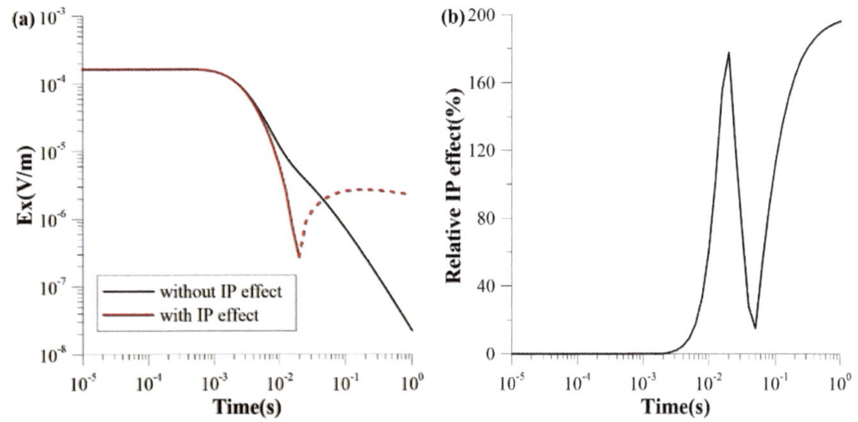

Fig. 5.11 Comparison of SOTEM horizontal electric field component responses for models with and without a polarization layer in half-space. **a** and **b** represent the response values and the corresponding relative IP effects, respectively

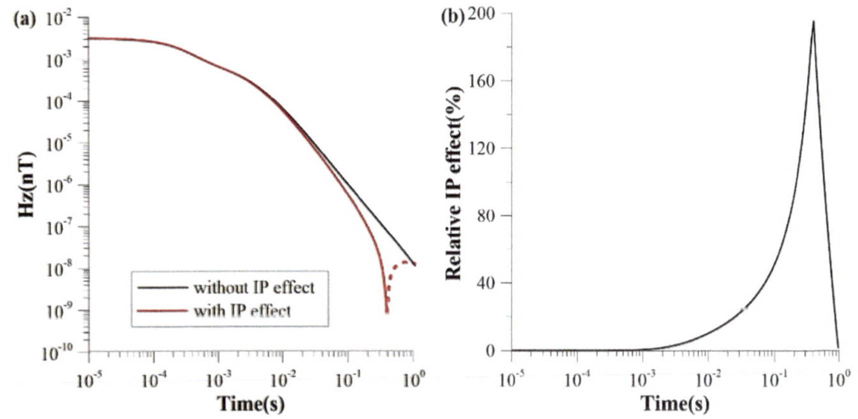

Fig. 5.12 Comparison of SOTEM vertical magnetic field component responses for models with and without a polarization layer in half-space. **a** and **b** represent the response values and the corresponding relative IP effects, respectively

field response (displayed in logarithmic scale, without unit labels), comparing results with and without the IP effect across different time channels in the ground field area, including the transmitter location.

For analysis, the horizontal electric field component (Ex) and the vertical magnetic field component (Hz) are examined at 0.1 ms and 100 ms, and 0.1 ms and 500 ms, respectively.

As shown in Figs. 5.13 and 5.14, at 0.1 ms, the polarization layer has negligible impact on the ground measurement field. However, at 100–500 ms, in contrast to the

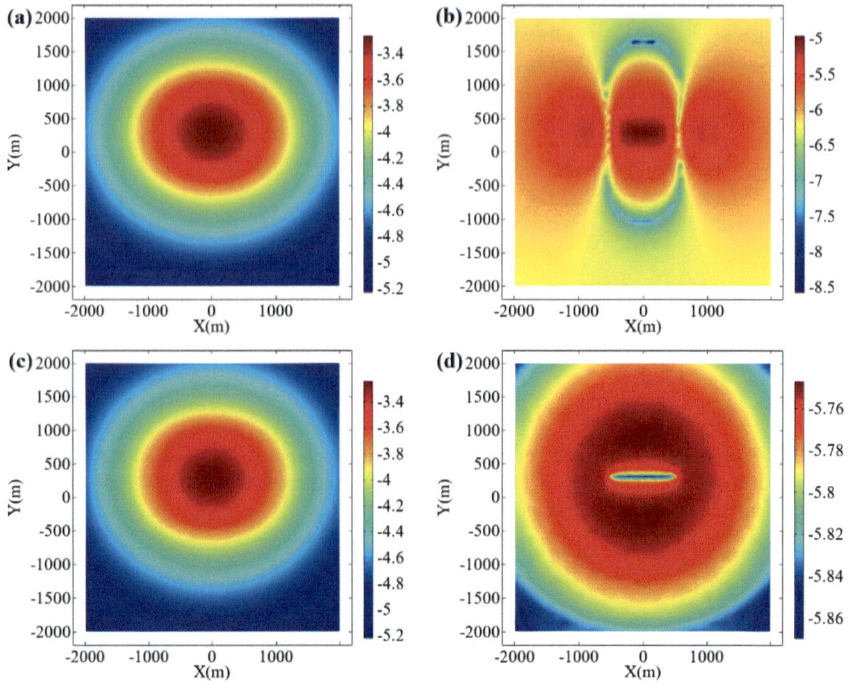

Fig. 5.13 Ground horizontal electric field components at different times for half-space with and without a polarization layer. **a** and **b** are the fields of the half-space with a polarization layer at 0.1 ms and 100 ms, respectively. **c** and **d** are the fields of the non-polarized half-space at 0.1 ms and 100 ms, respectively

non-polarized half-space, the polarization layer introduces significant distortions in the horizontal electric field component (Ex) and vertical magnetic field component (Hz). Furthermore, components influenced by the IP effect exhibit distinct behaviors, further highlighting the critical role of the IP effect in SOTEM data interpretation.

(3) Summary of the Impact of Induced Polarization Effect on SOTEM and FDEM

The above study separately examined the effects of excitation on the frequency-domain and time-domain electromagnetic responses of both the electric field and magnetic field. Based on the figures presented in this section, it can be observed that whether using the transient electromagnetic method (TEM) in the time domain or the controlled-source electromagnetic (CSEM) method in the frequency domain, the horizontal component of the electric field is strongly influenced by the IP effect, whereas the vertical magnetic field component is less affected.

For the electric field component in FDEM, the IP effect leads to an increase in field values across the entire frequency band. In contrast, for the magnetic field component, the IP effect has a relatively minor influence in the low- and mid-frequency bands, while its effect becomes more pronounced at high frequencies. Compared to the

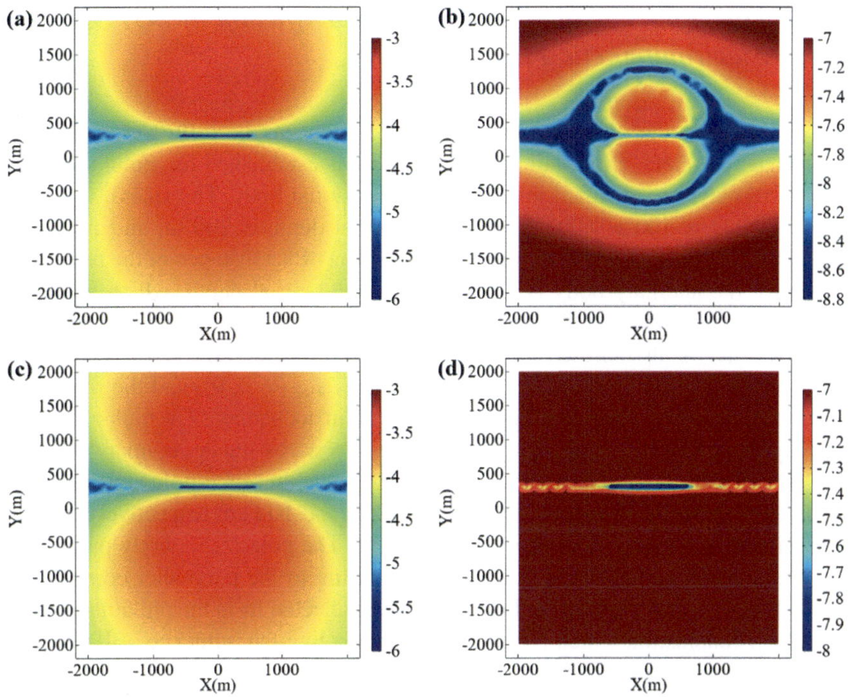

Fig. 5.14 Ground vertical magnetic field components at different times for half-space with and without a polarization layer. **a** and **b** are the fields of the half-space with a polarization layer at 0.1 ms and 500 ms, respectively. **c** and **d** are the fields of the non-polarized half-space at 0.1 ms and 500 ms, respectively

frequency-domain electromagnetic method, the transient electromagnetic method is capable of short-offset measurements near the source.

Unlike the IP effect's influence on frequency-domain electromagnetic responses, in the time domain, the IP effect not only alters the transient electromagnetic response but also induces sign reversal, making its presence stronger and more distinct. Furthermore, it is known that near-field detection technology using electrical sources provides greater observation depth and higher resolution for deep subsurface structures compared to far-field detection (Weiying 2016). The SOTEM method offers the advantage of greater detection depth compared to magnetic-source devices and enables electric field component observation, making it a promising approach for IP information extraction.

5.2.2 The Influence of IP Effect on the Diffusion of SOTEM Electromagnetic Field

In order to study the influence of polarization effect on the diffusion of SOTEM electromagnetic field, a three-dimensional model is established for simulation. The polarization layer is modified into a three-dimensional body. The size of the target anomaly body is 400 × 400 × 400 m, the burial depth is 200 m, and the center coordinates of the anomaly body are (0, −300). The offset distance of the measuring line above the anomaly body is less than twice the burial depth, which meets the definition of SOTEM for the receiving and transmitting distance. The length of the transmitter is also chosen as 1km, the transmitting current is 10 A, and it is deployed along the y-axis. The center coordinates are (0, 0, 0), and the rest of the grid division and boundary condition settings remain unchanged. The high-resistance anomaly body without excitation effect and the low-resistance anomaly body, as well as the high-resistance polarized body and low-resistance polarized body are calculated separately. The resistivity is 1000 Ω·m and 10 Ω·m respectively, and the uniform ground resistivity is chosen as 100Ω·m. The polarizable body parameters are chosen as m = 0.6, c = 0.6, t = 1 s, according to the parameter combination that will produce the strongest polarization effect selected from the previous forward simulation conclusion, which is conducive to analyzing the influence of IP effect in electromagnetic diffusion.

The three-dimensional geometric model of the anomalous response is shown in Fig. 5.15. The simulation results include 765,670 elements and 296,837 unknowns. A total of 81 frequencies are computed in 248 min, with a physical memory usage of 15.5 GB. The electromagnetic diffusion of multiple components is analyzed separately for both high- and low-resistivity anomalous bodies, with and without the excitation polarization effect.

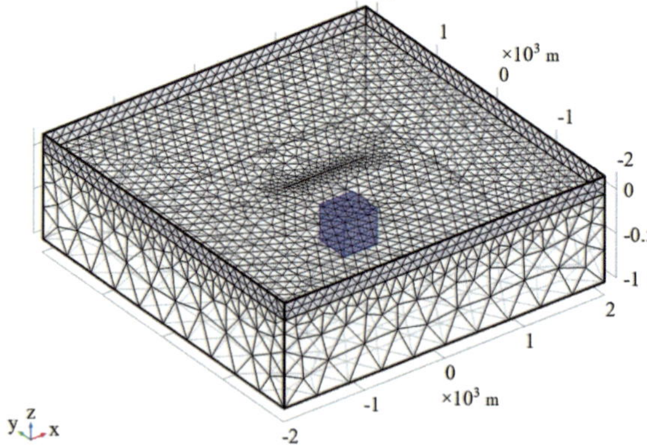

Fig. 5.15 SOTEM Three-dimensional anomaly electromagnetic field diffusion geometric model

To visualize the magnetic field diffusion in SOTEM with the IP effect, this study computes the induced current values at all positions within the entire three-dimensional anomaly body at 41 time steps, ranging from 0.1 ms to 1000 ms after power-off, under step-wave excitation. The distribution characteristics and diffusion processes of the horizontal electric field component Ex and vertical magnetic field component Hz in the subsurface are analyzed sequentially.

According to the relative IP effect error analysis of the single point response of the model with or without IP effect, several different moments are given, and two types of figures, the horizontal distribution map of the ground and the distribution map of the XOZ section, are drawn respectively. It is worth noting that the sign-changing response that appears in various figures in the text has been processed by taking the absolute value.

(1) The influence of IP effect on the electromagnetic field diffusion of three-dimensional low-resistance body

a. Horizontal electric field component

First, the comparison of the single point response of the horizontal electric field component of the low-resistance anomaly body of SOTEM with or without excitation effect is given, so as to judge the influence of the polarization effect in the diffusion process at each time channel. The measuring point is selected directly above the anomaly body, with coordinates (0, −400, 0), as shown in Fig. 5.16.

As shown in Fig. 5.16, the IP effect of the low-resistivity three-dimensional polarized body in this model on the horizontal electric field component (Ex) of the SOTEM electromagnetic field is analyzed. After the current is cut off, the influence of the IP effect gradually increases over time. Due to the strong excitation effect of the three-dimensional polarized body, by approximately 3 ms, the reverse polarization current

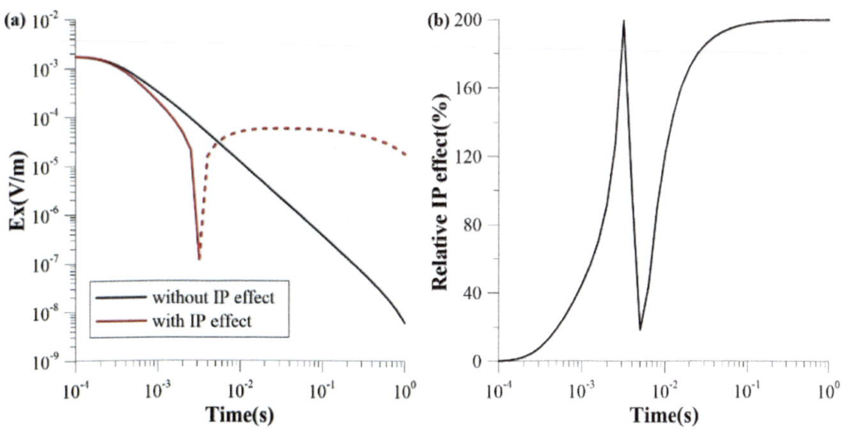

Fig. 5.16 Comparison of the horizontal electric field component of the three-dimensional low-resistance anomaly body of SOTEM with and without IP effect. **a** and **b** represent the response value and the corresponding relative IP effect, respectively

exceeds the eddy current secondary field, which has decayed to a certain extent. At this moment, a sign reversal occurs.

To better illustrate and analyze the diffusion and distribution of the horizontal component response, four time channels—0.1, 3, 100, and 500 ms—are selected for visualization. Figures 5.17 and 5.18 separately display the horizontal distribution diagrams and XOZ cross-section (vertical to the ground) distribution diagrams at each time step, comparing cases with and without the excitation effect to examine the differences in diffusion characteristics. For optimal visualization, the legend range is adjusted based on the actual response decay behavior.

From Figs. 5.17a, e and 5.18a, e, it can be observed that at 0.1 ms after power-off (early response stage), the electromagnetic field influence is strong, while the IP effect on the horizontal electric field component is not yet prominent. As a result, the electric field responses on both the left and right sides appear nearly identical.

At t = 3 ms, the electromagnetic field without the excitation effect continues to propagate downward and outward following the "smoke ring" effect (Figs. 5.17f and 5.18f) and continues to decay. However, in the presence of the IP effect, the

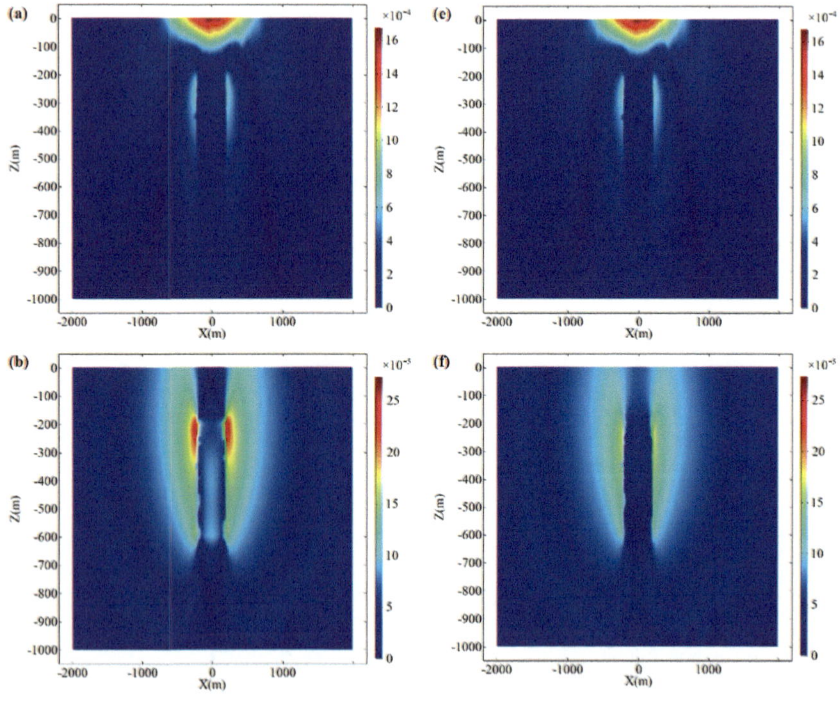

Fig. 5.17 Distribution diagram of the XOZ cross-section of the horizontal electric field component response at different times (the left side is the result of the low-resistance polarized body, the moments of **a**, **b**, **c**, **d** are 0.1 ms, 3 ms, 100 ms, 500 ms respectively; the right side is the result of the low-resistance non-polarized body, the moments of **e**, **f**, **g**, **h** are 0.1 ms, 3 ms, 100 ms, 500 ms respectively

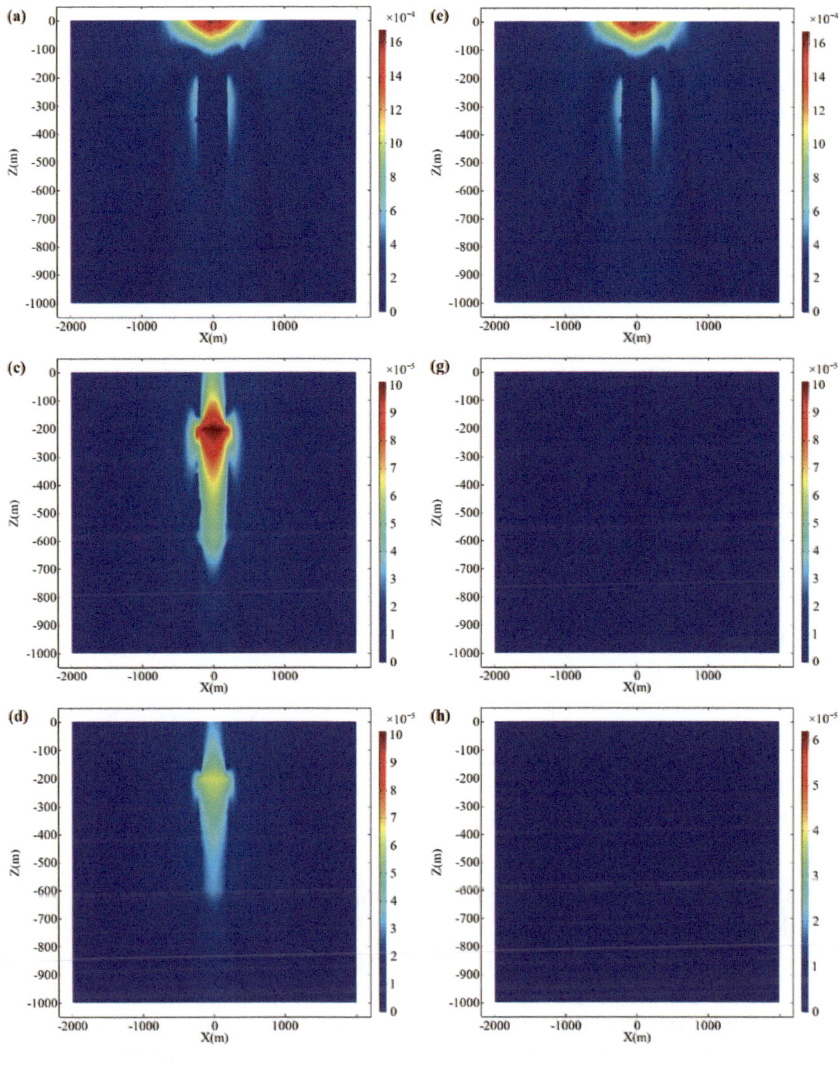

Fig. 5.17 (continued)

polarization current magnitude of the three-dimensional low-resistivity polarized body becomes greater than the induced current, leading to a sign reversal (Figs. 5.17b and 5.18b).

As time progresses, the late-time polarization current becomes dominant, further enhancing the excitation effect. The electric field response of the non-excitation model continues to decay on the right side (Figs. 5.17g, h and 5.18g, h). Meanwhile, in the excitation model with a low-resistivity polarized body, the electric field response undergoes a decay process driven by the reverse polarization current (Figs. 5.17c, d

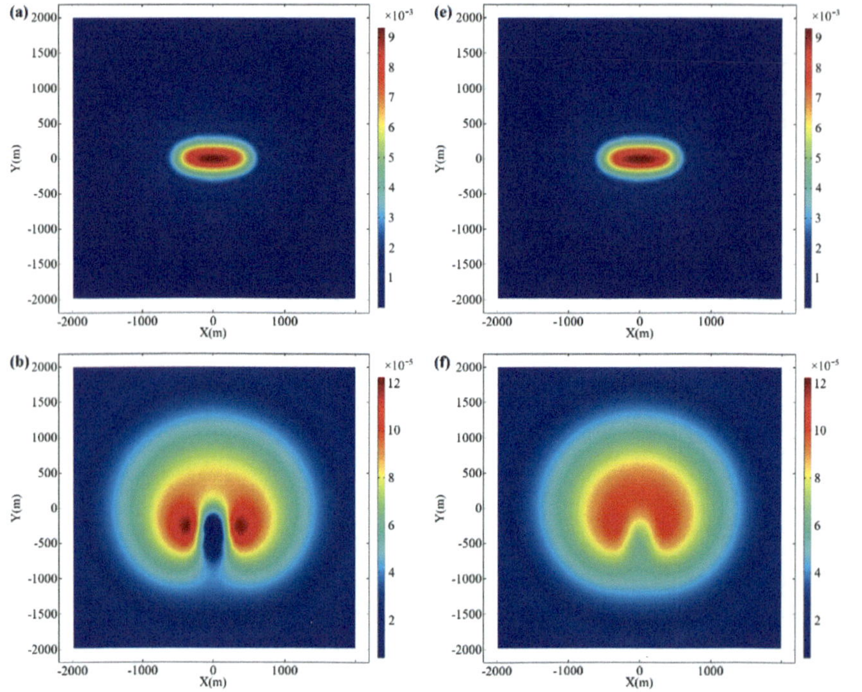

Fig. 5.18 Ground distribution diagram of the horizontal electric field component response at different times (the left side is the result of the low-resistance polarized body, the moments of **a**, **b**, **c**, **d** are 0.1 ms, 3 ms, 100 ms, 500 ms respectively; the right side is the result of the low-resistance non-polarized body, the moments of **e**, **f**, **g**, **h** are 0.1 ms, 3 ms, 100 ms, 500 ms respectively

and 5.18c, d), during which the decay rate of the negative Ex component response slows down.

By continuously comparing the left-side images (a–d, e–g) and right-side images (a–d, e–g) in Figs. 5.17 and 5.18, the distinct diffusion characteristics of the electromagnetic field in cases with and without the excitation effect can be more intuitively observed. Additionally, the process of current sign reversal within this model is clearly demonstrated.

b. Vertical magnetic field component

Similarly, a comparison of the single-point response of the Hz vertical magnetic field component in SOTEM for the low-resistivity anomalous body is provided with and without the excitation effect. This allows for an evaluation of the influence of the polarization effect on the diffusion process at various time channels. The measurement point is selected directly above the anomaly, with coordinates (0, –400, 0).

As shown in Fig. 5.19, the IP effect of the low-resistivity three-dimensional polarized body on the SOTEM vertical magnetic field component is demonstrated. A

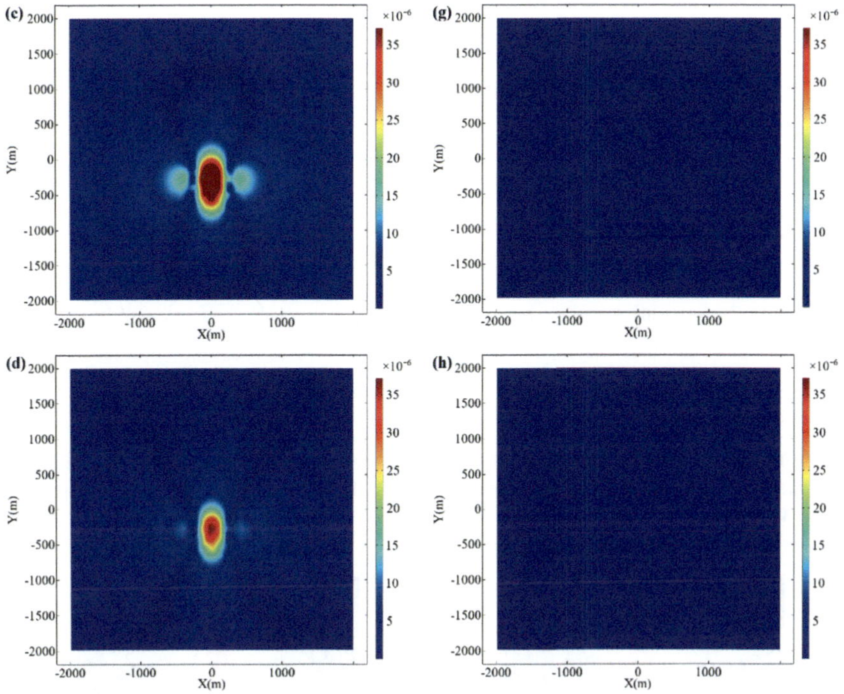

Fig. 5.18 (continued)

significant double sign reversal phenomenon is observed under this model. The first sign reversal occurs at approximately 30 ms, and the second sign reversal occurs around 580 ms.

To analyze the diffusion and distribution of the vertical magnetic field component response, four time channels—0.3 ms, 10 ms, 30 ms, and 300 ms—are selected for visualization. The ground horizontal distribution maps and XOZ profile distribution maps for each time step are presented under conditions with and without the excitation effect, allowing for a direct comparison of their diffusion behavior. The results are shown in Figs. 5.20 and 5.21. Similar to the analysis of the horizontal electric field component, the legend range is adjusted based on the actual response attenuation to ensure optimal visualization.

From Figs. 5.20a, e and 5.21a, e, it is evident that at 0.3 ms after power-off, the excitation effect on the vertical magnetic field component is not very pronounced. However, in the XOZ profile results, a local high-value region is observed above the anomaly, caused by the slower decay of the response due to the polarization effect.

At t = 10 ms, the electromagnetic field without the excitation effect continues to propagate downward and outward, following the well-known "smoke ring" effect (Figs. 5.20f and 5.21f). The field continuously decays, forming a strip-like distribution on the ground. However, in the polarized medium, the polarization current

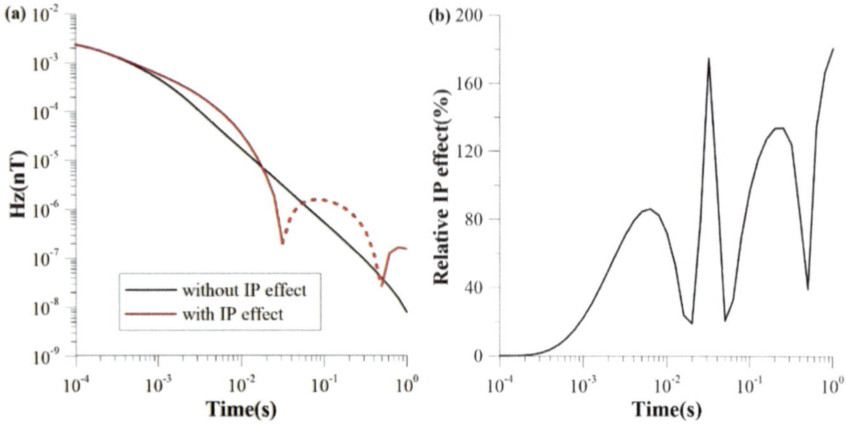

Fig. 5.19 compares the SOTEM vertical magnetic field of three-dimensional low-resistance anomalies with and without IP effects. **a** and **b** represent the response values and the corresponding relative IP effects, respectively

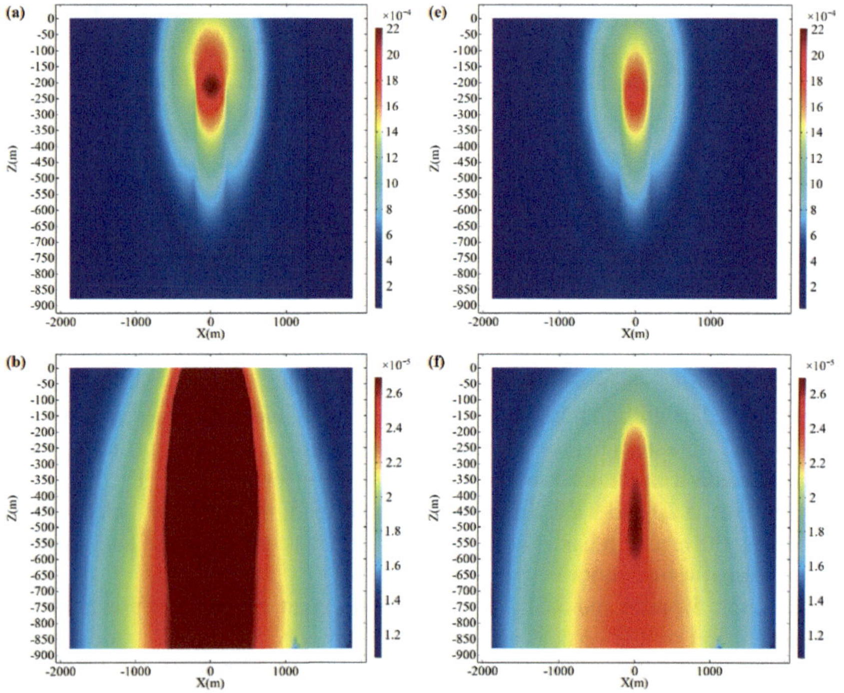

Fig. 5.20 shows the XOZ profile distribution of the magnetic field vertical component response at different times (the left side is the result of the low-resistance polarized body, **a**, **b**, **c**, **d** are 0.1 ms, 3 ms, 100 ms, 500 ms respectively; the right side is the result of the low-resistance non-polarized body, **e**, **f**, **g**, **h** are 0.3 ms, 10 ms, 30 ms, 300 ms respectively

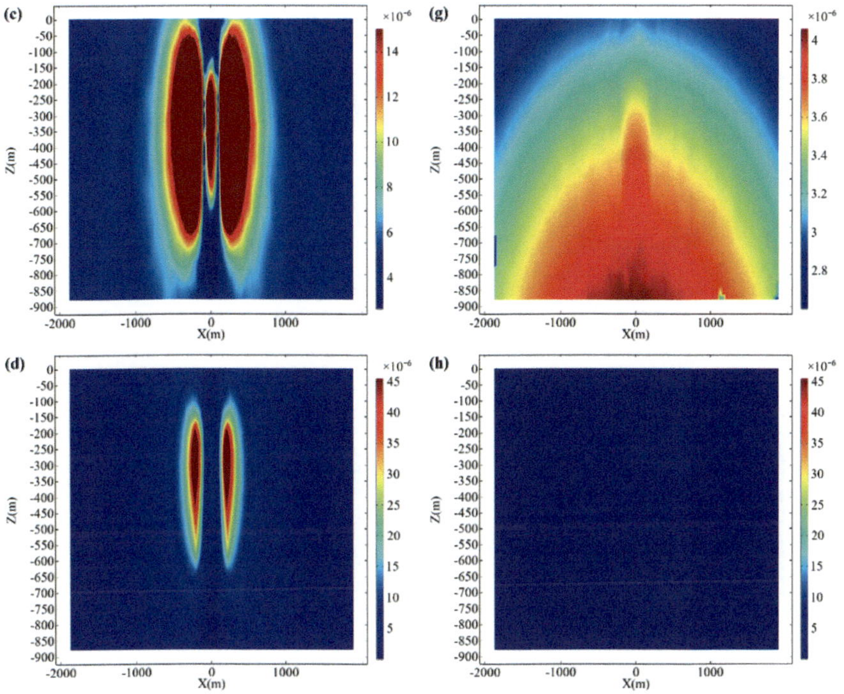

Fig. 5.20 (continued)

exceeds the induced current, resulting in higher response values in the low-resistivity polarized body compared to the non-polarized medium (Figs. 5.20b and 5.21b).

As time progresses, the late-stage polarization current becomes dominant, making the excitation effect more evident. The magnetic field response of the non-excitation model on the right side continues to decay (Figs. 5.20g, h and 5.21g, h). However, in the excitation model, the magnetic field response reaches a critical sign reversal point at 30 ms (Figs. 5.20c, 5.21c). This occurs because the reverse polarization current at this moment exceeds the forward decaying transient field.

Compared to the horizontal electric field component, the sign reversal time of the vertical magnetic field component is delayed. At t = 300 ms, the Hz response of the three-dimensional low-resistivity body with excitation effect is significantly greater than that of the non-excitation model. At this point, the polarization current is also undergoing decay (Figs. 5.20d and 5.21d).

By continuously comparing the left-side images (a–d, e–g) of Figs. 5.20 and 5.21, the diffusion characteristics of the Hz response under the excitation effect and the sign reversal process become more intuitively apparent. The secondary sign change phenomenon observed here is consistent with findings from Smith and Walker (1988), who identified a similar phenomenon in central-loop TEM observation data. They attributed the first sign change to the polarization current of a near-surface,

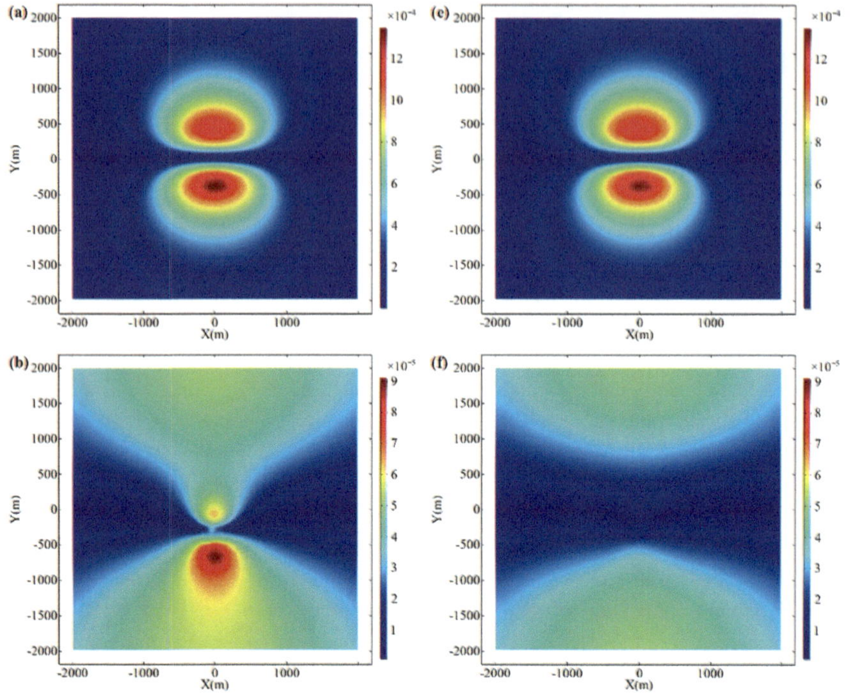

Fig. 5.21 Ground distribution of the vertical component response at different times (the left side is the result of low resistance polarization, the moments of **a**, **b**, **c**, **d** are 0.1 ms, 3 ms, 100 ms, 500 ms respectively; the right side is the result of low resistance non-polarization, the moments of **e**, **f**, **g**, **h** are 0.3 ms, 10 ms, 30 ms, 300 ms respectively

low-resistivity conductor, and suggested that, because the Hz component decays more slowly in a low-resistivity model, the transient secondary induced current decays slower than the polarization current, resulting in a secondary sign change phenomenon.

However, it is important to note that, based on existing SOTEM detection equipment, most observations are typically conducted within 100 ms, making it rare to detect this type of transient signal secondary reversal phenomenon.

In addition, it can be seen that the response of the vertical magnetic field component is different from the diffusion form of the horizontal electric field component. The maximum value of the vertical magnetic field response is located under the two endpoints of the transmitter, gradually diffusing outward. There is no vertical induced current under the midpoint of the transmitter, and its response intensity decays faster than the electric field response over time.

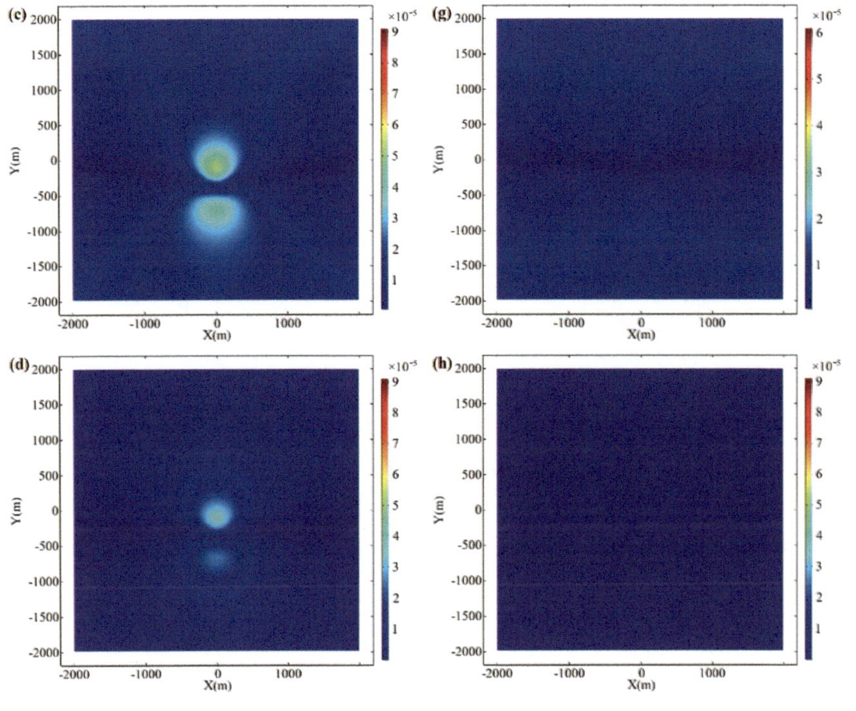

Fig. 5.21 (continued)

(2) The influence of IP effect on the electromagnetic field diffusion of 3D high resistance body

a. Horizontal electric field component

As with the low-resistivity model, we first present a comparison of the single-point response of the SOTEM horizontal electric field component (Ex) for a high-resistivity anomaly body, both with and without the excitation effect. The measurement point is selected directly above the anomaly body, with coordinates (0, −400, 0). Additionally, a comparison of the single-point response of the horizontal electric field component between the 3D low-resistivity polarization model and the 3D high-resistivity polarization model is also provided. The results are shown in Fig. 5.22.

As shown in Fig. 5.22a, b, the IP effect of the high-resistivity 3D polarized body in this model on the SOTEM electromagnetic field horizontal electric field component (Ex) is illustrated. After the current is cut off, the influence of the IP effect increases over time. At approximately 6 ms, the reverse polarization current exceeds the eddy current secondary field, which has decayed to a certain extent, resulting in a sign change.

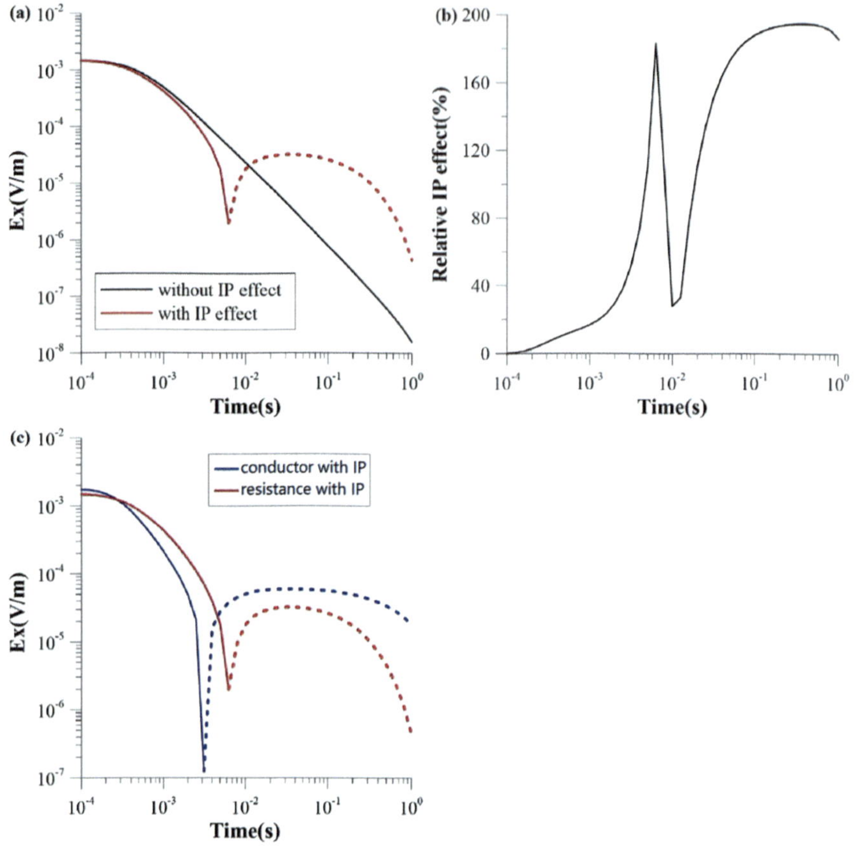

Fig. 5.22 Comparison of SOTEM horizontal electric field components of 3D high resistance anomalies with and without IP effects. **a** and **b** represent the response value and the corresponding relative IP effect, respectively, and **c** represents the comparison of the single point response of the horizontal electric field component of the 3D low resistance polarization model and the 3D high resistance polarization model

To examine the diffusion and distribution of the horizontal component response, four time channels—1 ms, 6 ms, 100 ms, and 500 ms—are selected for visualization. The ground horizontal distribution maps and XOZ profile distribution maps for each time step are presented under conditions with and without the excitation effect, enabling direct diffusion comparisons. The results are shown in Figs. 5.23 and 5.24.

Additionally, in Fig. 5.22c, it can be observed that the sign reversal time of the 3D low-resistivity polarized body occurs earlier, and the excitation effect after sign reversal is stronger. This indicates that the horizontal electric field component (Ex) in SOTEM is more sensitive to the low-resistivity polarized body.

As seen in Figs. 5.23a, e and 5.24a, e, at 1 ms after power-off, the influence of the horizontal electric field excitation effect has already emerged. From the XOZ profile

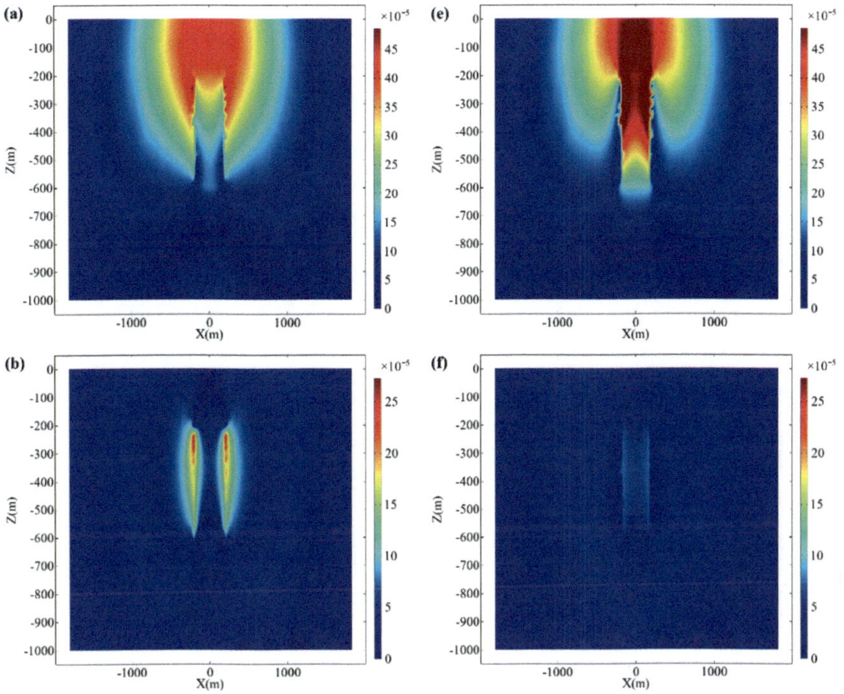

Fig. 5.23 Distribution map of the horizontal component of the electric field response at different times in the XOZ section (the left side shows the results of the low-resistance polarized body, the moments **a**, **b**, **c**, **d** are 1 ms, 6 ms, 100 ms, 500 ms respectively; the right side shows the results of the low-resistance non-polarized body, the moments **e**, **f**, **g**, **h** are 1 ms, 6 ms, 100 ms, 500 ms respectively

results, it is evident that the response decay rate above the high-resistivity polarized anomaly body is slower due to the polarization effect, leading to the formation of a local low-value region.

At t = 6 ms, the electromagnetic field without the excitation effect continues to propagate downward and outward, following the "smoke ring" effect (Figs. 5.23f, 5.24f), and it continues to decay. However, at this moment, in the presence of the IP effect, the polarization current in the 3D high-resistivity polarized body exceeds the induced current, resulting in a sign change (Figs. 5.23b, 5.24b).

Since low-resistivity anomalies have a greater impact on the transient response horizontal induced current, the sign reversal occurs later in the high-resistivity model compared to the low-resistivity model.

Similarly, as time progresses, the late-stage polarization current becomes dominant, further enhancing the excitation effect. The electric field response of the non-excitation model on the right side continues to decay (Figs. 5.23g, h, 5.24g, h). Meanwhile, in the low-resistivity model with excitation, the electric field response

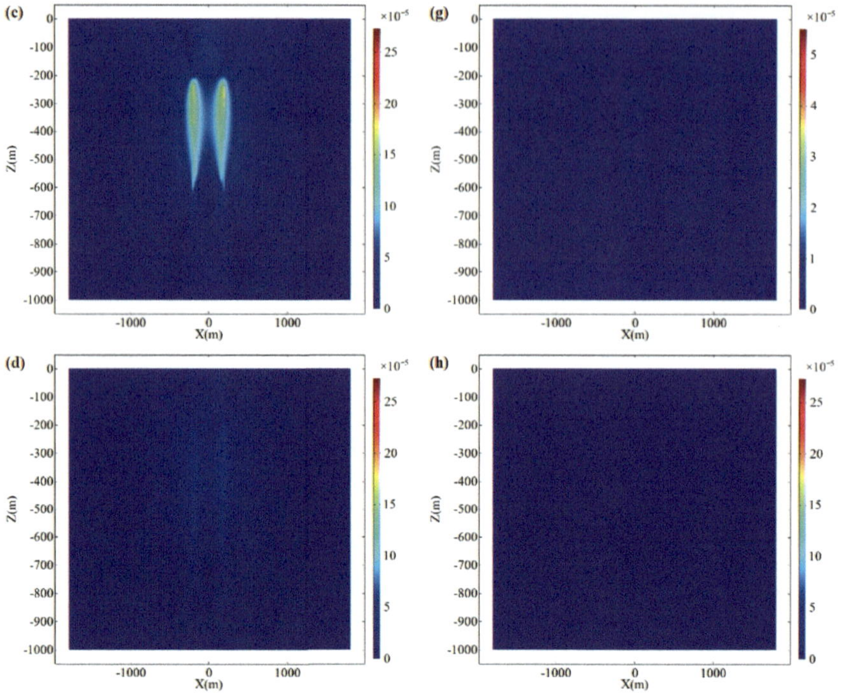

Fig. 5.23 (continued)

undergoes a decay process of the reverse polarization current (Figs. 5.23c, d, 5.24c, d). At this stage, the decay rate of the negative Ex component response is slower.

b. Vertical magnetic field component

Similarly, a comparison of the single-point response of the SOTEM vertical magnetic field component (Hz) for the three-dimensional high-resistivity anomaly body, both with and without the IP effect, is provided to assess the impact of polarization effects during the diffusion process at various time channels. The observation point remains the same, and an additional comparison is made between the horizontal electric field component responses of the three-dimensional low-resistivity polarization model and the three-dimensional high-resistivity polarization model. The results are shown in Fig. 5.25.

As shown in Fig. 5.25a and b, the IP effect of the three-dimensional high-resistivity polarized body on the SOTEM vertical magnetic field component (Hz) is demonstrated. It is observed that under this model, the Hz component response does not exhibit a sign reversal phenomenon. Additionally, the relative IP effect only reaches several tens in the late stage, which is significantly lower than the Ex component result for the high-resistivity polarization model, and also less than the relative IP effect of the Hz component in the low-resistivity polarization model.

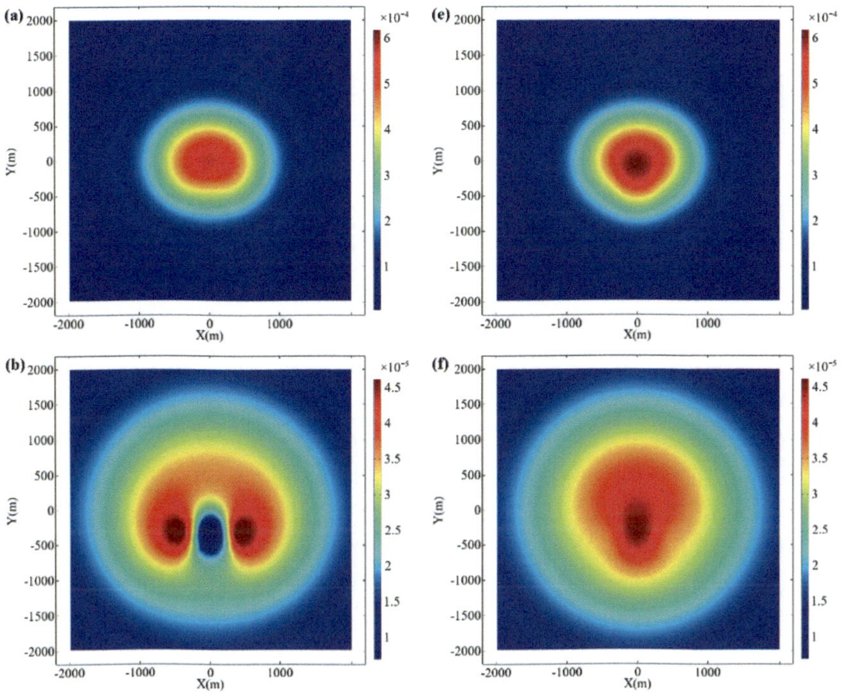

Fig. 5.24 Distribution map of the horizontal component of the electric field response at different times on the ground (the left side shows the results of the high-resistance polarized body, the moments **a**, **b**, **c**, **d** are 1 ms, 6 ms, 100 ms, 500 ms respectively; the right side shows the results of the high-resistance non-polarized body, the moments **e**, **f**, **g**, **h** are 1 ms, 6 ms, 100 ms, 500 ms respectively

To analyze the diffusion and distribution of the vertical magnetic field component response, three time channels—1 ms, 30 ms, and 300 ms—are selected for visualization. The ground horizontal distribution maps and XOZ section distribution maps for each time step are presented under conditions with and without the IP effect, allowing for a direct diffusion comparison. The results are shown in Figs. 5.26 and 5.27.

Additionally, in Fig. 5.25c, it can be observed that in the three-dimensional low-resistivity polarized body, the vertical magnetic field response undergoes a sign reversal, and a secondary sign change phenomenon occurs under this model. From this, it can be concluded that the Hz component in SOTEM is more sensitive to the low-resistivity polarized body. This implies that when detecting a low-resistivity polarized body, both Ex and Hz components exhibit a strong response to the IP effect. However, when detecting a high-resistivity polarized body, the Ex component provides a more pronounced response to the IP effect compared to the Hz component.

As seen in Figs. 5.26a, d and 5.27a, e, at 1 ms after power-off, the excitation effect on the vertical magnetic field component is not prominent. However, in the

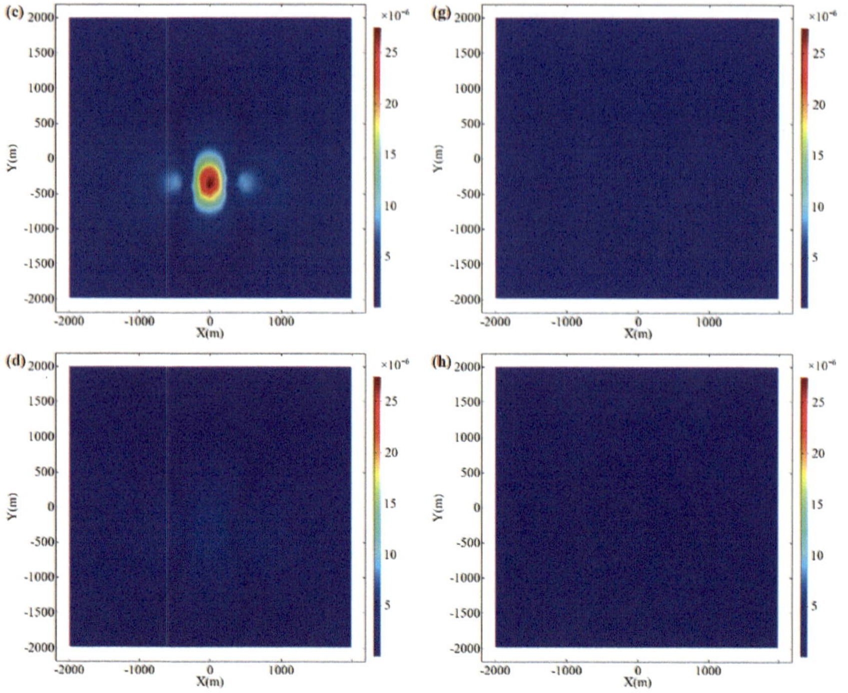

Fig. 5.24 (continued)

XOZ cross-section results, a slower response decay rate due to the polarization effect above the high-resistivity anomaly can be observed, leading to the formation of a local high-value region in the Hz component.

At t = 30 ms, the Hz field without the excitation effect continues to propagate downward and outward, following the diffusion effect (Figs. 5.26e, 5.27e). The field continuously decays, forming a strip-like distribution on the ground. However, in the presence of the IP effect, the polarization current value in the three-dimensional high-resistivity polarized body exceeds the induced current, leading to a higher response value in the polarized medium compared to the non-polarized medium (Figs. 5.26b, 5.27b).

Similarly, as time progresses, the late-stage polarization current becomes dominant, further enhancing the excitation effect. The magnetic field response of the non-excitation model on the right continues to decay (Figs. 5.26g, 5.27g). However, the response under the high-resistivity polarization model is not sufficient to cause a sign reversal in the transient response.

At t = 300 ms, the Hz response of the three-dimensional high-resistivity body with the excitation effect is significantly larger than that of the non-excitation model. At this stage, the polarization current is also undergoing a decay process (Figs. 5.26c, 5.27c).

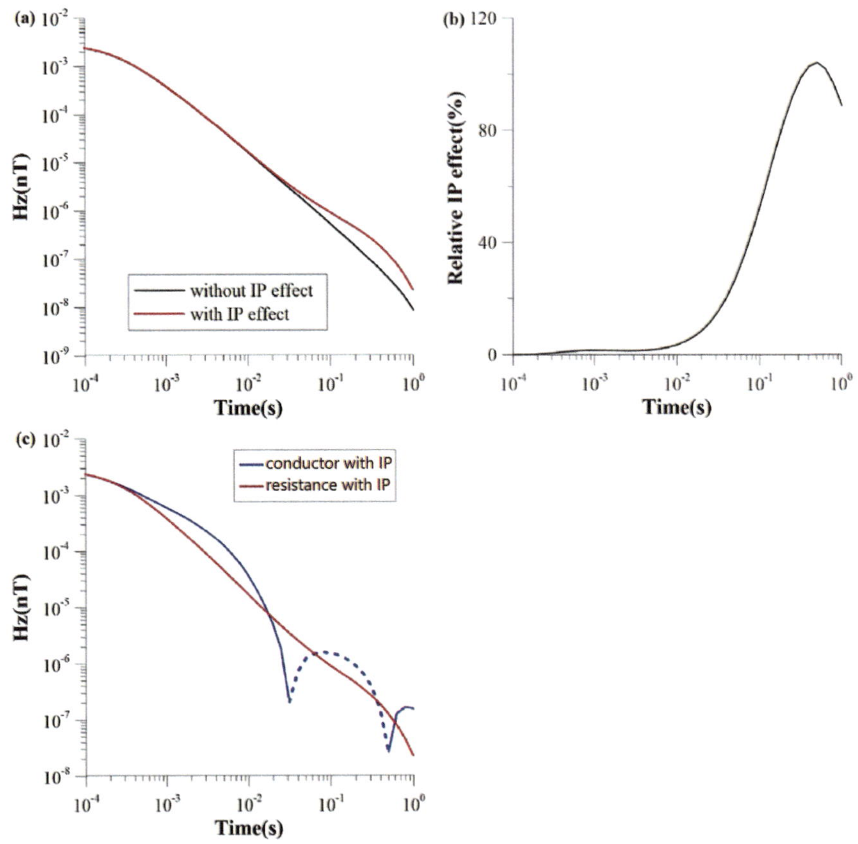

Fig. 5.25 Comparison of the SOTEM vertical magnetic field of the three-dimensional high-resistance anomaly body with and without the IP effect. **a** and **b** represent the response values and the corresponding relative IP effects, respectively. **c** shows the comparison of the single-point response of the horizontal electric field component of the low-resistance polarization model and the high-resistance polarization model

(3) Summary of the impact of IP effect on electromagnetic diffusion

From the distribution of SOTEM responses at different times, with and without polarization effects, it can be observed that the induced current always flows in the positive direction, diffusing downward and outward over time, with its intensity gradually decaying. This behavior aligns with the typical characteristics of transient electromagnetic (TEM) coil diffusion. However, in contrast to the induced current, the diffusion characteristics of the polarization current exhibit distinct differences.

In the early stage, the polarization current flows in the positive direction, with its intensity initially increasing and then weakening, representing the charging process. The charging process continues over time until it reaches saturation, at which point

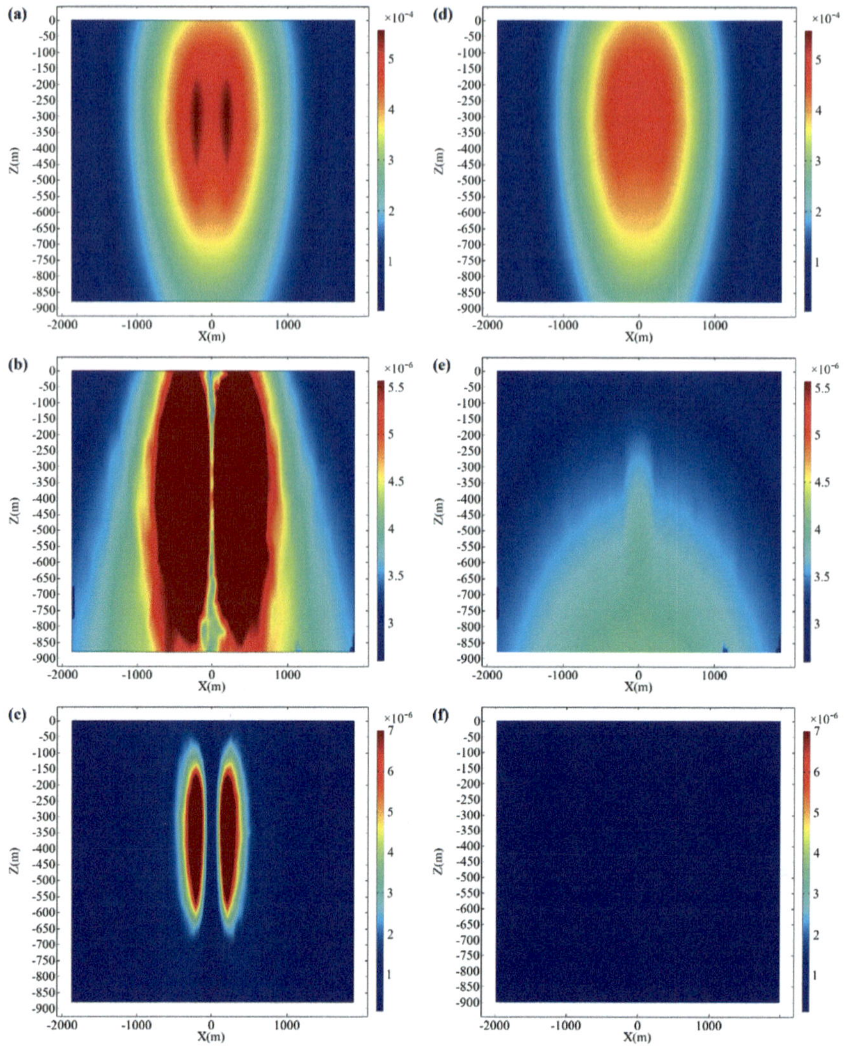

Fig. 5.26 Distribution of the vertical component of the magnetic field response at different times in the XOZ cross-section (the left side shows the results of the high-resistance polarized body, **a**, **b**, **c** times are 1 ms, 30 ms, 300 ms respectively; the right side shows the results of the high-resistance non-polarized body, **d**, **e**, **f** times are 1 ms, 30 ms, 300 ms respectively)

the polarization current becomes zero. After saturation, the discharge process begins, causing the polarization current to reverse direction.

Throughout the entire discharge period, the polarization current diffusion exhibits characteristics similar to those of the induced current, following a coil diffusion pattern. However, during the discharge process, the polarization current does not adhere to the coil diffusion behavior, and the reverse polarization current intensity

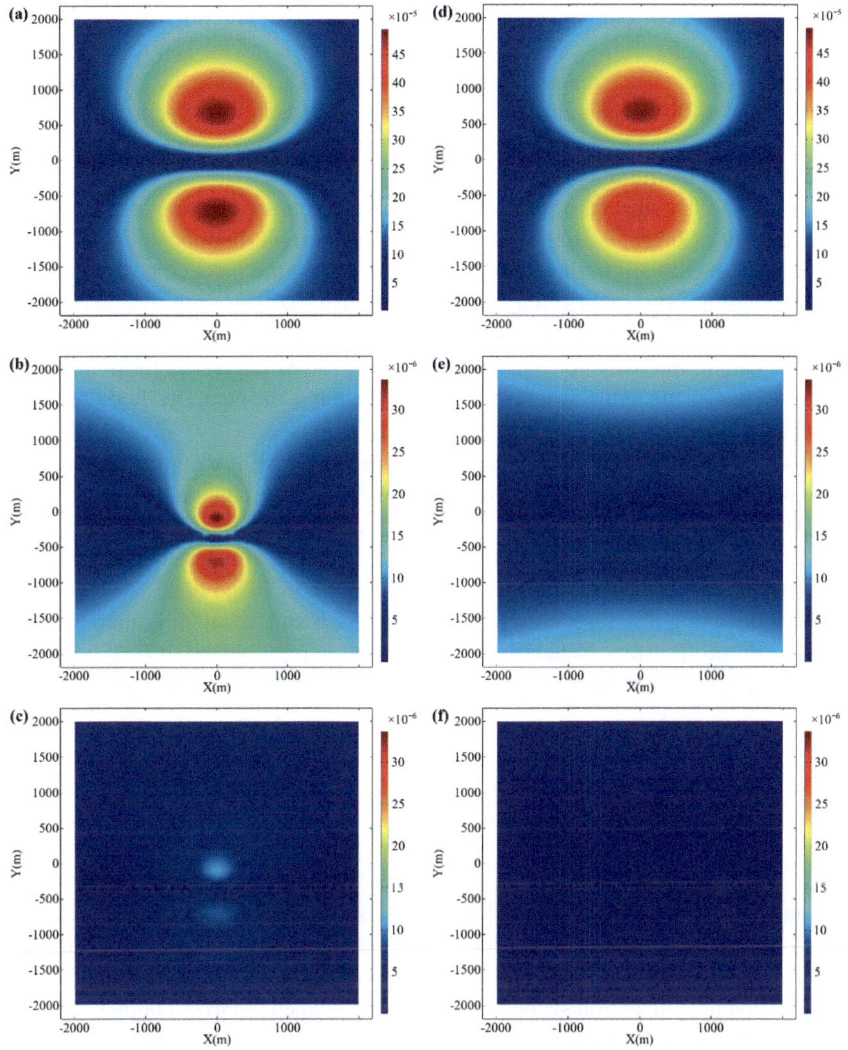

Fig. 5.27 Distribution of the vertical component of the electric response on the ground at different times (the left side shows the results of the high-resistance polarized body, **a**, **b**, **c** times are 1 ms, 30 ms, 300 ms respectively; the right side shows the results of the high-resistance non-polarized body, **d**, **e**, **f** times are 1 ms, 30 ms, 300 ms respectively)

increases over time. Its maximum value is always located at the projection position of the source center on the surface.

In the early charging stage, the induced electric field dominates, and the polarization current follows coil diffusion characteristics. In the middle and late stages, the polarization current becomes dominant, exhibiting non-coil diffusion characteristics.

Since the total current consists of both induced current and polarization current, the initial diffusion follows the same characteristics as the induced current. However, at later times, the reverse polarization current dominates, causing the overall current to exhibit the diffusion characteristics of a non-helical polarization current field.

A physical explanation can be provided for the negative response of the current in the polarized medium:

During the early stage, the polarized medium undergoes a charging process, where the polarization current is aligned with the induced current (both positive), diffusing downward and outward in the form of smoke rings. As the induced current decays, the polarization body reaches charge saturation, and the polarization current becomes zero. After saturation, the discharge process begins, making the polarization current negative, with its intensity first increasing and then decreasing. When the polarization current surpasses the induced current, the total current in the polarized body becomes negative, leading to sign reversal in the SOTEM response.

This explains the diffusion characteristics of the SOTEM electromagnetic field under polarization effects and the process of response sign change.

Additionally, by comparing the impact of the IP effect on the vertical magnetic field component (Hz) and the horizontal electric field component (Ex) in low-resistivity and high-resistivity polarized bodies, it is concluded that:

When detecting low-resistivity polarized bodies, both Ex and Hz components are sensitive to the IP effect. When detecting high-resistivity polarized bodies, the Ex component is more effective for IP effect analysis than the Hz component.

5.2.3 Three-Dimensional Forward Response Characteristics of Thin Plate-Shaped Polarized Bodies

To better understand the response characteristics of the three-dimensional polarization model and to provide guidance for data processing, interpretation, and fieldwork, this section presents three-dimensional forward modeling for a typical polarized body found in practical detection scenarios—thin plate-shaped polarized bodies.

In actual mineral exploration, such as the detection of impregnated brass ore and blocky polymetallic ore, large low-resistivity ore bodies are often difficult to detect directly. Many ore bodies exist in the form of thin plate-shaped structures.

Another common example of thin plate-shaped polarized bodies in mineral exploration is the carbonaceous layer, which typically exhibits relatively low resistivity and high magnetic susceptibility. When a carbonaceous layer is located in shallow subsurface layers, it may interfere with the identification of underlying high-resistivity or low-resistivity anomalous bodies.

Given the widespread occurrence of these two geological types in the form of thin plate-shaped polarized bodies, and in preparation for future complex three-dimensional body modeling in field studies, it is essential to analyze their responses under different conditions.

Thus, this section presents the three-dimensional forward modeling of SOTEM in three forms of thin plate-shaped polarized bodies: upright, horizontal and inclined (inclination angle 45°).

The forward model is adjusted based on the semi-space model containing polarized bodies shown in Fig. 5.15. The parameters of the grounded electrical source device remain unchanged, the current size is 10 A, the source length is 1000 m, it is arranged along the y-axis direction, the center point coordinates are (0, 0, 0), and the rest of the grid division, the setting of boundary conditions remain unchanged. A finite boundary thin plate-shaped three-dimensional polarized body model is designed in a uniform semi-space with a conductivity of 0.003 S/m. The size of the abnormal body is $200 \times 300 \times 50$ m, the top surface is buried at a depth of 500 m, and the center point coordinates are horizontal form: (0, −600, −525); vertical form: (0, −600, −650), inclined form: (0, −800, −650), the polarizable body parameters are chosen as m = 0.3, c = 0.3, t = 1 s, resistivity is 200 $\Omega \cdot$m, the maximum size is set to 5 m, the maximum cell growth rate is 0.8, maintaining the high quality of the grid division. The three-dimensional geometric model is shown in Fig. 5.28. Finally, 79,821 cells were generated, and the degrees of freedom to be solved were 1,808,154. The results of 81 frequencies were obtained in 623 min, and the physical memory usage was 15.8 Gbytes.

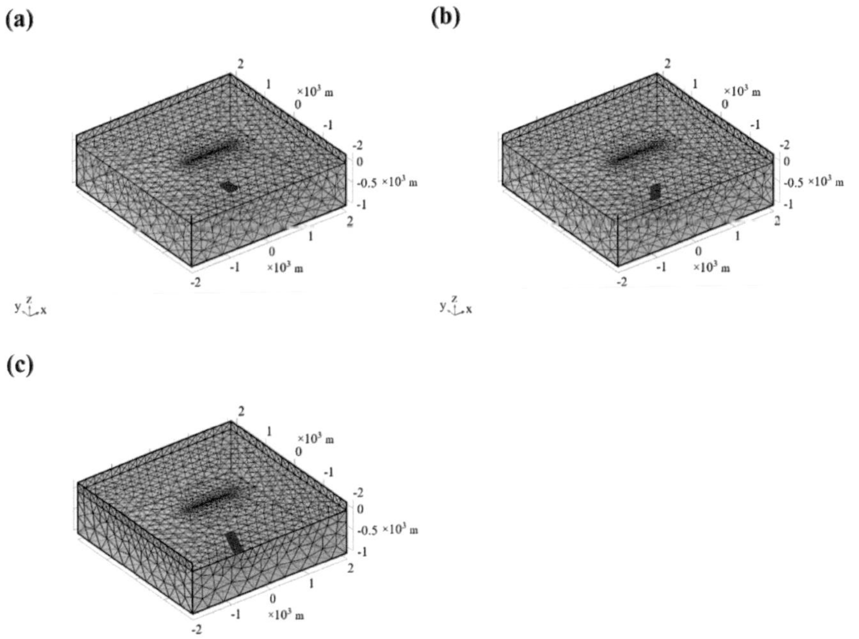

Fig. 5.28 Grid division diagram of SOTEM three-dimensional forward modeling of thin plate-shaped polarized bodies. **a** Horizontal form **b** Upright form **c** Inclined form

Three example observation points are selected in the SOTEM observation area, the offset distance perpendicular to the transmitter direction gradually increases, and the XYZ coordinates are observation point 1 (0, −500, 0), observation point 2 (0, −600, 0) observation point 3 (0, −700, 0), these three observation points are selected for response analysis in all three cases.

(1) Horizontal thin plate-shaped polarized body three-dimensional forward response characteristics

As shown in Fig. 5.29a, the horizontal electric field components at the three calculation points of the horizontal thin plate polarization model all undergo a sign change, which occurs before 100 ms. Within the transient electromagnetic field observation time range, as the offset distance increases, the early-stage transient electromagnetic response decreases. However, since the offset distance variation is relatively small, and the thickness of the horizontal thin plate polarized body remains constant, the time of the negative response phenomenon remains the same across all calculation points. Additionally, the magnitude of the reverse transient current is also very similar across the three points.

As shown in Fig. 5.29b, for the vertical magnetic field component, a sign change is observed only at the smallest offset distance among the three calculation points. At observation point 3, the late-time response is higher than at the other two points. This is because, for the vertical magnetic field component, a larger offset distance results in a slower decay rate. Given that the polarized layer thickness remains constant, the decay rate of the induced current is the primary factor affecting this behavior.

Overall, the horizontal electric field component exhibits a higher sensitivity in detecting the polarized body compared to the vertical magnetic field component. Since the horizontal thin plate polarized body is relatively thin (only 50 m), the ability to detect such structures depends on the charging rate of common carbonaceous layers

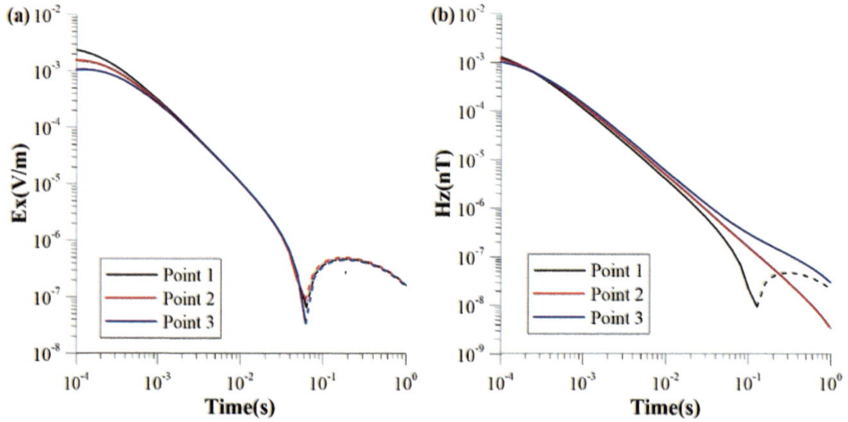

Fig. 5.29 SOTEM response of three example points of the horizontal thin plate-shaped polarized body model. **a** Horizontal electric field component **b** Vertical magnetic field component

or ore bodies, typically ranging from 0.2 to 0.3. Consequently, this model is most effectively identified when the offset distance is small.

(2) Response characteristics of three-dimensional forward modeling of inclined thin plate polarized body

As shown in Fig. 5.30a, the horizontal electric field components at the three calculation points of the inclined thin plate polarized body model all undergo a sign change, which occurs before 100 ms. Within the transient electromagnetic field observation time range, as the offset distance increases, the early-stage transient electromagnetic response decreases.

Although observation point 1 is not directly above the polarized body, it is still influenced by the polarized structure located adjacent to it, demonstrating that the influence range of the polarized body extends beyond the area directly above it. Even with a slight change in offset distance, a clear distinction can be observed in the late-stage negative response magnitudes, with observation point 3 exhibiting a stronger negative response than observation point 2, which in turn is stronger than observation point 1.

Although the offset distance increases and the depth of the polarized body beneath the observation points also increases, the volume of the polarized body at greater depths increases as well. The enhancement of the galvanic effect indicates that at this stage, the horizontal electric field component is predominantly influenced by the polarized body rather than the offset distance alone.

As shown in Fig. 5.30b, the vertical magnetic field components at the three calculation points undergo a sign change, except at the largest offset distance. The sign reversal in the vertical magnetic field component at observation point 1 is similar to that of the electric field component, as the transient field decays more rapidly, leading to an earlier sign change and a larger polarization current.

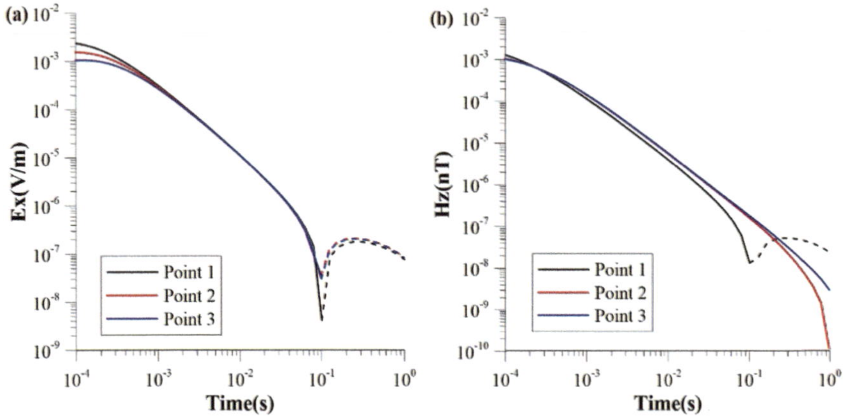

Fig. 5.30 SOTEM response of three example points of the inclined thin plate polarized body model. **a** Horizontal electric field component **b** Vertical magnetic field component

For observation point 3, the vertical magnetic field component does not undergo a sign change, similar to what was observed in the horizontal model. This phenomenon is attributed to the slow decay of the transient field at this position and the weaker galvanic effect of the magnetic field at greater depths.

In the inclined model, the response at observation point 2 closely resembles that of the horizontal model. At this stage, the primary variable affecting the response is the thickness of the polarized body, which appears to be a minor factor influencing the magnetic field component in this scenario.

(3) Response characteristics of three-dimensional forward modeling of vertical thin plate polarized body

As shown in Fig. 5.31a, the horizontal electric field components at the three calculation points of the upright thin plate polarized body model all undergo sign changes, with the sign change occurring before 100 ms. Within the transient electromagnetic field observation time range, as the offset distance increases, the early-stage transient electromagnetic response decreases.

The sign change phenomenon observed at observation point 1 further confirms that the influence range of the polarized body extends beyond the region directly above it. Similar to the inclined model, even with small variations in offset distance, it remains distinguishable that the late-stage response at observation point 2 is the strongest. This is because the thickness of the polarized body beneath observation point 2 is the greatest, and the enhancement of the electric effect suggests that at this time, the horizontal electric field component is predominantly influenced by the polarized body.

As shown in Fig. 5.31b, the vertical magnetic field components at the three calculation points all undergo sign changes, except at the largest offset distance. The overall response behavior is similar to that observed in the inclined model. The key

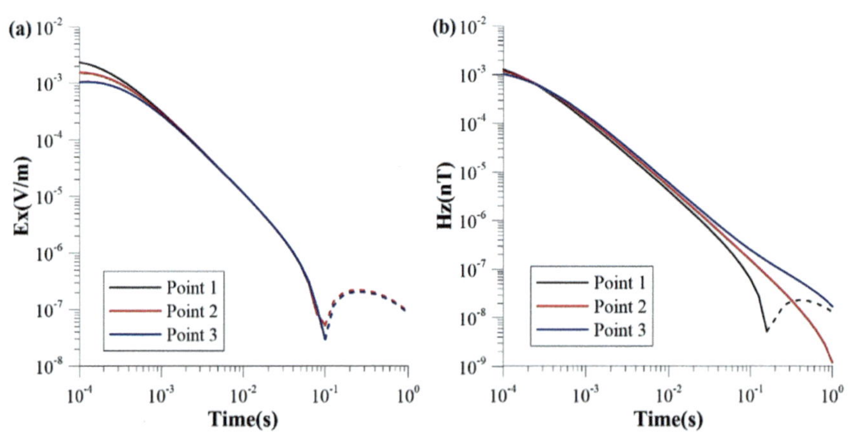

Fig. 5.31 SOTEM response of three example points of the vertical thin plate polarized body model. **a** Horizontal electric field component **b** Vertical magnetic field component

difference lies in the thickness of the polarized body beneath observation point 2, which ranges from 25 to 300 m. It can be inferred that under this offset distance, the magnetic field components do not exhibit a negative response, indicating that the vertical resolution capability of the magnetic field components is relatively poor when detecting thin plate-shaped polarized bodies.

Additionally, the forward modeling results for the horizontal thin plate polarized body are largely consistent with those for the vertical thin plate polarized body, but with some notable differences. The similarity arises because the offset distance is the primary factor influencing the electric effect. By comparing the results of both models, it is evident that the response values at the three observation points in the horizontal model are higher than those in the upright model at the same time, with the largest difference observed at observation point 1. This suggests that, under equal offset distances, a horizontal polarized body located directly beneath an observation point produces a stronger electric effect.

In summary, for common carbonaceous layers or ore bodies, given their actual size and polarization parameters, the sensitivity of the horizontal electric field component (Ex) to thin plate-shaped polarized bodies is higher than that of the vertical magnetic field component (Hz). Additionally, the electric effect of the vertical magnetic field component is most significantly influenced by offset distance. Therefore, selecting appropriate observation parameters based on different detection objectives and information extraction needs is crucial.

5.3 Research on SOTEM Response Information Extraction with Electric Effect

If the electric signal can be effectively extracted from the SOTEM response, it would provide valuable additional information for data processing and interpretation, ultimately improving interpretation accuracy and broadening its application scope, particularly in the exploration of metal minerals and geothermal resources. Thus, the effective extraction of electric information holds both theoretical and practical significance.

In light of this, domestic and international researchers have conducted extensive studies on electric information extraction. As introduced earlier, the electric effect in long-offset transient electromagnetic (TEM) fields from electrical sources is fundamentally similar to that observed in the natural-source electrical method. In both cases, electric field observations are performed in the far zone, where the electromagnetic effect dominates, leading to limited application prospects. In contrast, using SOTEM to extract electric information is much more promising.

Furthermore, while one-dimensional inversion provides only an approximate representation of real exploration data, its computational demand is significantly lower than that of three-dimensional inversion. Despite its simplified approach, 1D inversion can effectively reflect the electrical properties and variations of the subsurface medium. Since resistivity and chargeability are key indicators of the underground electrical structure, it is necessary to invert these two parameters from the measured magnetic field data.

5.3.1 Introduction to Adaptive Regularization Inversion Method

In practice, electromagnetic inversions are typically ill-posed problems. To achieve stable inversion results, high-precision observation data alone is insufficient; instead, reasonable parameter settings and appropriate inversion techniques must be employed.

The SOTEM-observed magnetic field information represents the induced electromotive force, and most current data processing techniques convert this induced electromotive force into early-time, late-time, or full-period apparent resistivity (Changchun and Hwarong 1991; Baiyao 1999; Yao 2002; Liangjun et al. 2002; Denghai et al. 2003; Qingyan et al. 2014). However, traditional real resistivity models struggle to produce accurate inversion results when processing data containing electric information.

A promising solution is the Adaptive Regularization Inversion Algorithm (ARIA), which has proven effective in TEM inversion (Qingyan et al. 2014; Yucong et al. 2015; Chaofeng 2020). This method enhances inversion stability and reduces ambiguity in inversion solutions. Specifically, the adaptive regularization inversion approach is applied to estimate the two key polarization parameters—resistivity and polarization rate—within the inversion model.

Following the regularization inversion framework, the goal is to identify a model that best fits the observed data while simultaneously satisfying the required model constraints defined by the objective function. Based on this principle, the regularization method objective function can be expressed as:

$$\phi(\mathbf{m}) = \phi_d(\mathbf{m}) + \lambda\phi_m(\mathbf{m}) \tag{5.5}$$

where $\phi(\mathbf{m})$ is the total objective function; $\phi_d(\mathbf{m})$ is the observation data objective function; $\phi_m(\mathbf{m})$ is the objective function of the prior constraint model; λ is the regularization factor; $\mathbf{m}$ is the model vector, which is the resistivity parameter ρ and the charge rate parameter m. Where $\mathbf{m} = [\rho_1, \rho_2, \ldots, \rho_n;\ m_1, m_2, \ldots, m_n]$.

In Eq. (5.5), $\phi_d(\mathbf{m})$ is given by the following formula

$$\phi_d(\mathbf{m}) = \Delta d^T \mathbf{W}_d \Delta d \tag{5.6}$$

where Δd is the difference vector between the measured response and the theoretical data; $\mathbf{W}_d$ is the data variance matrix, represented as below, where σ_j is the data standard deviation

$$\mathbf{W}_d = \mathrm{diag}\{1/\sigma_1, 1/\sigma_2, \ldots, 1/\sigma_j, \ldots, 1/\sigma_m\} \tag{5.7}$$

For different constraint conditions, various methods can be used to construct the objective function of model constraints. The three most commonly used are the minimum square sum of model parameters (minimum model constraint), the minimum square sum of the second derivative of model parameters (smoothest model constraint), and the minimum square sum of the derivative of model parameters (flattest model constraint). Since this paper uses a layered medium model, the discrete expressions of these three constraints are as follows

$$\begin{cases} \phi_m(\mathbf{m}) = \mathbf{m}^T\mathbf{m} \\ \phi_m(\mathbf{m}) = (\mathbf{Rm})^T(\mathbf{Rm}) \\ \phi_m(\mathbf{m}) = \left(\mathbf{R}^2\mathbf{m}\right)^T\left(\mathbf{R}^2\mathbf{m}\right) \end{cases} \tag{5.8}$$

In the formula, **R** represents roughness, which is

$$\mathbf{R} = \begin{Bmatrix} 0 & & & & 0 \\ -1 & 1 & & & \\ & -1 & 1 & & \\ & & \cdots & \cdots & \\ & & & -1 & 1 \end{Bmatrix} \tag{5.9}$$

The regularization factor λ is the weight between the objective functions of the prior model and the data. Its function is to balance the two objective functions and determine the fitting object. $\lambda \to \infty$ indicates that the inversion target is entirely prior information, and the data fitting degree is poor at this time; $\lambda = 0$ indicates that the inversion is entirely aimed at fitting the data, and the model's constraints are poor at this time. Therefore, choosing an appropriate regularization coefficient has a significant impact on the effect of inversion. On the one hand, it requires sufficient data fitting, and on the other hand, it requires the obtained model to conform to existing information.

Referencing Chen Xiaobin's adaptive regularization method (Xiaobin et al. 2005), there are two schemes to adjust the regularization factor λ:

$$\begin{cases} \lambda^1 = \lambda^0 \\ \lambda^k = \frac{\phi_d^{k-1}}{\phi_m^{k-1}} \end{cases} \text{(MD scheme)} \tag{5.10}$$

$$\lambda^k = \frac{\phi_d^{k-1}}{\phi_d^{k-1} + \phi_m^{k-1}} \text{(CMD scheme)} \tag{5.11}$$

In the formula, k indicates that the inversion iteration number is k. The initial model of one-dimensional electromagnetic inversion is usually a uniform half-space. When the first iteration ϕ_m is zero, this makes the regularization factor of the MD scheme meaningless, so the CMD scheme is used.

Derive the model in Eq. (5.5). Due to the principle of minimization, the inversion equation can be obtained

$$\left(\mathbf{J}_k^T \mathbf{W}_d^T \mathbf{W}_d \mathbf{J}_k + \lambda \mathbf{R}^T \mathbf{R}\right) \Delta \mathbf{m} = \mathbf{J}_k^T \mathbf{W}_d^T \mathbf{W}_d \Delta \mathbf{d}_k - \lambda \mathbf{R}^T \mathbf{R} m_k \tag{5.12}$$

Let $\mathbf{A} = \mathbf{J}_k^T \mathbf{W}_d^T \mathbf{W}_d \Delta \mathbf{d}_k + \lambda \mathbf{R}^T \mathbf{R}$, $\mathbf{b} = \mathbf{J}_k^T \mathbf{W}_d^T \mathbf{W}_d \Delta \mathbf{d}_k - \lambda \mathbf{R}^T \mathbf{R}$, Eq. (5.12) can be simplified to $\mathbf{A} \Delta \mathbf{m} = \mathbf{b}$. In the formula, $\Delta \mathbf{d}_k$ is the vector difference between the current model theoretical response and the observed data when the iteration number is k, $\mathbf{J}_k^T$ is the sensitivity matrix of the model at this time, and the differential method is used to solve the $\mathbf{J}_k^T$ of one-dimensional inversion. When $\mathbf{A}$ is non-singular, the vector used to correct the model is represented as $\Delta \mathbf{m} = \mathbf{A}^{-1} \mathbf{b}$, and the model used for the next iteration can be obtained as $\mathbf{m} = \mathbf{m}_k + \Delta \mathbf{m}$.

The end condition of inversion is: the relative fitting error (rms) is lower than the specified value.

$$rms = \sqrt{\frac{\sum_{i=1}^{N} \left[\left(d_i^{\text{Observations}} - d_i^{\text{Theoretical value}}\right) / d_i^{\text{Observations}}\right]^2}{N}} \tag{5.13}$$

5.3.2 *Inversion Extraction of IP Information*

According to the Cole–Cole model, the transient electromagnetic response of an underground polarized medium affected by the induced polarization (IP) effect is described by four parameters: (1) Polarization resistivity; (2) Charge rate; (3) Frequency-related coefficient; (4) Time constant.

Simultaneous inversion of multiple parameters often leads to local minima, making the iteration process unstable, slowing inversion speed, and resulting in unsatisfactory inversion outcomes. Among these parameters, the time constant and frequency-related coefficient have a relatively minor impact on the SOTEM electromagnetic response.

Moreover, in the absence of additional influencing parameters, the individual inversion of the frequency-related coefficient and time constant produces suboptimal results. When multiple parameters are inverted simultaneously, restoring these values to their actual physical properties becomes even more challenging (Kaifeng 2021).

Given these challenges, this study focuses on inverting only the two parameters that exert the greatest influence on the SOTEM response: Polarization resistivity and Charge rate.

The frequency-related coefficient and time constant are set to typical values. The following sections validate the effectiveness of the inversion method and present inversion results for different types of geoelectric models.

(1) Verification of the effectiveness of the inversion method

The H model is selected for verification, the basic parameters of the model are set as shown in the table, the device parameters are set to an offset distance of 500 m, a transmission current of 10 A, and a transmission source length of 1000 m. The verification of the effectiveness of the inversion method carried out in this section includes two parts: the impact of different initial model parameters on the inversion results; the impact of different noise on the inversion results (Table 5.1).

a. **Impact of different initial models**

The inversion method used in this study involves configuring several key parameters, including the top layer depth, depth change coefficient, depth to be calculated, and the number of iterations. However, the primary factor that may significantly influence the inversion results is the excitation parameter values of the half-space background field. The adaptive regularization inversion method assumes an initial model based entirely on half-space models. The theoretical forward model consists of 32 layers, with a top layer depth of 10 m, a depth change coefficient of 1.1, and a maximum depth of 1516.31 m. To assess the impact of different initial model parameters, four inversion schemes were designed. Scheme 1 and Scheme 2 evaluate the impact of the frequency correlation coefficient and time constant on inversion results, Scheme 2 and Scheme 3 compare the influence of the initial charging rate, while Scheme 3 and Scheme 4 examine the effect of initial resistivity. The specific parameter settings for these four schemes are detailed in Table 5.2.

Table 5.1 Parameters of the model for inversion calculation

Model	Resistivity (ρ /Ω·m)	Layer thickness (h/m)	Charging rate (m)	Frequency correlation coefficient (c)	Time constant (t/s)
H	$\rho_1 = 100$ $\rho_2 = 20$ $\rho_3 = 100$	$h_1 = 500$ $h_2 = 200$	$m_1 = 0$ $m_2 = 0.5$ $m_3 = 0$	$c_1 = 0.3$ $c_2 = 0.3$ $c_3 = 0.3$	$t_1 = 0.1$ $t_2 = 0.1$ $t_3 = 0.1$

Table 5.2 Parameters of the initial model scheme for inversion

	Resistivity ρ (Ω·m)	Charging rate m	Frequency correlation coefficient c	Time constant t(s)
Scheme one	$\rho = 100$	0.5	c = 0.6	t = 100
Scheme two	$\rho = 100$	0.5	c = 0.3	t = 0.1
Scheme three	$\rho = 100$	0.1	c = 0.3	t = 0.1
Scheme four	$\rho = 10$	0.1	c = 0.3	t = 0.1

The inversion results are shown in Figs. 5.32 and 5.33. As can be seen from the figures, all four sets of initial value models achieve good inversion results, with inversion errors of less than 1% across all methods. The resistivity and chargeability models obtained from the inversion closely approximate the actual model, and while they do not perfectly match, they effectively capture the variations in resistivity and chargeability along the vertical direction. A comparison with the second method confirms that the frequency correlation coefficient and time constant have minimal influence on the inversion results. Similarly, the comparison between Scheme One and Scheme Two further substantiates that these two parameters have negligible effects on the inversion process. The comparison between Scheme Two and Scheme Three indicates that the initial chargeability setting does not affect the final inversion results, while the results of Scheme Three and Scheme Four demonstrate that the initial resistivity setting does not impact the stability of inversion convergence.

b. **The impact of different noise data**

Further analysis was conducted to examine the impact of different levels of noise in the inversion process. Gaussian noise at levels of 3%, 8%, and 15% was added to the forward results, and the corresponding inversion outcomes are shown in Figs. 5.34 and 5.35. As depicted in the figures, when 3% Gaussian noise is introduced, the fitting error of the magnetic field response remains below 5%, and the inversion results remain satisfactory, demonstrating the inversion method's strong anti-interference capability in real-world exploration scenarios. The inverted resistivity and chargeability models continue to effectively reflect the model's structural characteristics. However, with the introduction of 8% and 15% Gaussian noise, the magnetic field response fitting error increases, with the former remaining below 10% and the latter approaching 50%. Although the resistivity model inversion results deviate significantly from the actual model parameters, they still capture the general "high-low–high" H-type electrical characteristics. The chargeability inversion results, while

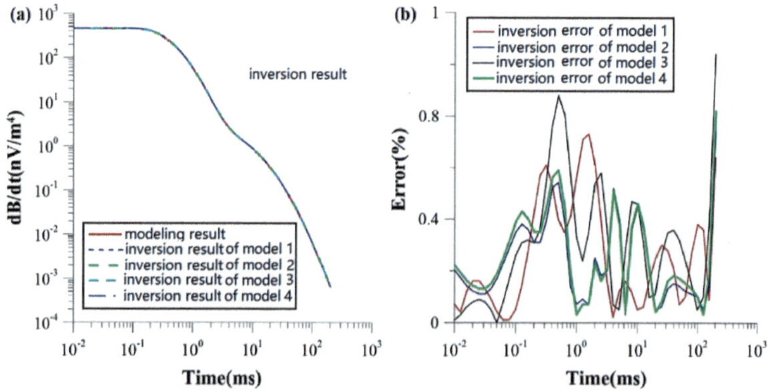

Fig. 5.32 Inversion results of different initial models. **a** and **b** represent the fitting curves of forward response and inversion results, and the relative error curves of fitting, respectively

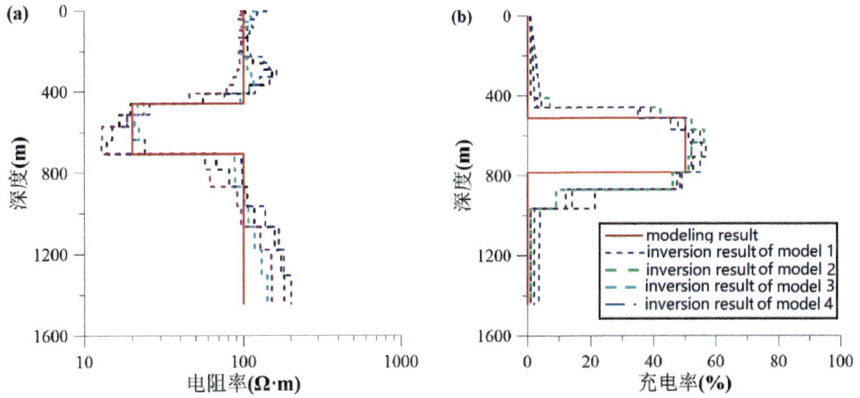

Fig. 5.33 Extraction results of stimulated electrical information of different initial models. **a** and **b** represent the resistivity results and charging rate results, respectively

capable of indicating chargeability variations across different layers, exhibit substantial discrepancies from the actual model parameters. These findings suggest that as interference noise levels increase, their effects propagate into the inversion process, leading to deviations in the inversion results.

(2) Inversion Model Calculation

After verifying the effectiveness of the inversion method, we proceed with the inversion calculations for different geoelectric models in SOTEM, maintaining the same basic device parameters as in the previous section. The selected parameters

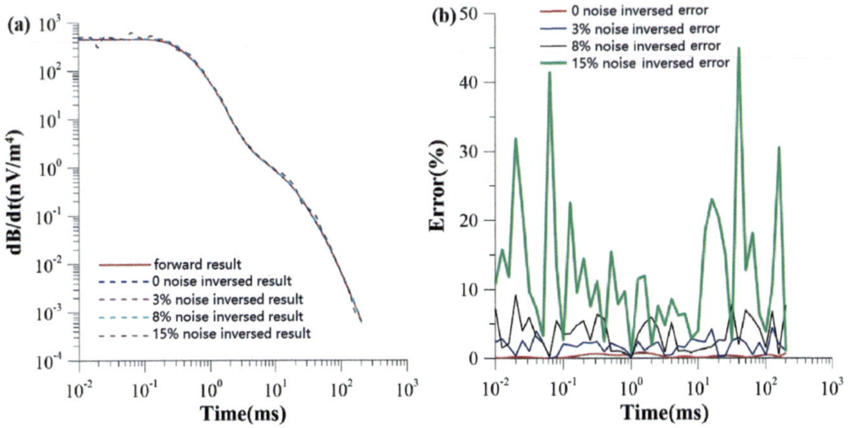

Fig. 5.34 Responses to different Gaussian noise models. **a** and **b** respectively represent the fitting curves of the forward response and the inversion result, as well as the relative error curve of the fitting

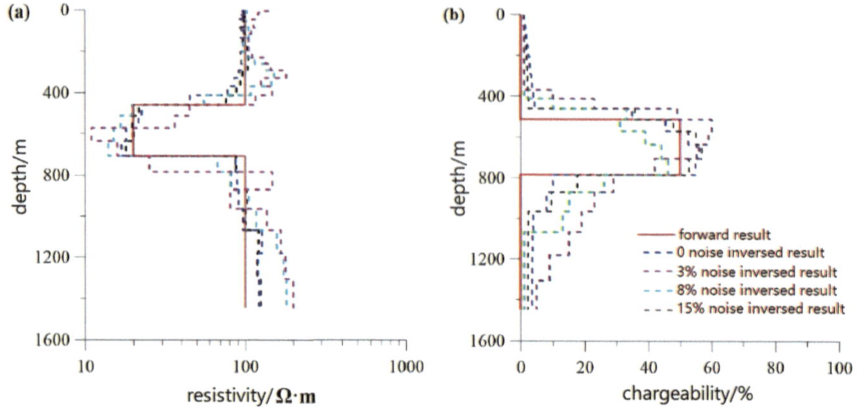

Fig. 5.35 Results of electric information extraction from different Gaussian noise models. **a** and **b** respectively represent the resistivity results and the charge rate results

include an offset distance of 500 m, a transmission current of 10 A, and a source length of 1000 m. The inversion results for various geoelectric models will be analyzed to further evaluate the method's applicability and stability in different subsurface conditions.

a. **Uniform Half-Space Model**

First, using the uniform half-space as an example, the geoelectric model parameters and initial parameters are listed in the corresponding table, and the inversion results are shown in the accompanying figure. From the results, it is evident that the inversion of the half-space model achieves highly satisfactory outcomes, with the fitting error of the magnetic field time derivative response remaining below 0.1%. Additionally, the resistivity and chargeability information in the model is effectively extracted, demonstrating the robustness and accuracy of the inversion method in a uniform half-space scenario (Figs. 5.36, 5.37 and Table 5.3).

b. Two-Layer Geoelectric Model

The parameters and initial parameters of the two-layer geoelectric model (G and D models) are provided in the corresponding table, and the inversion results are illustrated in the accompanying figure. From the results, it can be observed that the inversion of the two-layer model achieves relatively ideal results, with the fitting difference for the G model remaining below 0.8%, while the D model exhibits a slightly higher fitting difference of less than 3%. Furthermore, the resistivity and chargeability information in the two-layer model has been successfully extracted, and the inversion results closely approximate the actual model, demonstrating the reliability of the method in handling layered geoelectric structures (Figs. 5.38, 5.39, 5.40, 5.41 and Table 5.4).

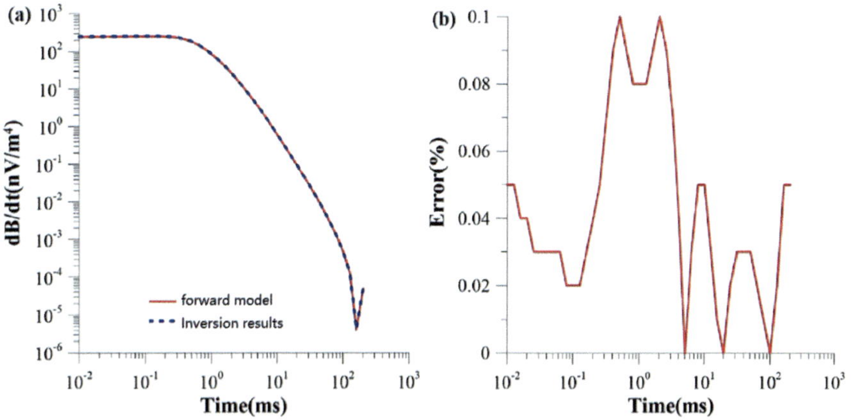

Fig. 5.36 Half-space model response inversion results. **a** and **b** respectively represent the fitting curves of the forward response and the inversion result, as well as the relative error curve of the fitting

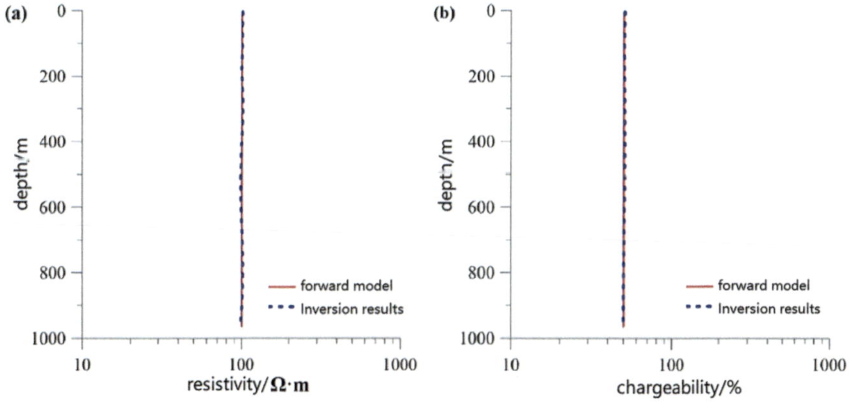

Fig. 5.37 Half-space model electric information extraction results. **a** and **b** respectively represent the resistivity results and the charge rate results

Table 5.3 Half-space model and initial model parameters

	Model	ρ (Ω·m)	m	c	$t(s)$
Model parameters	Half-space	100	0.5	0.3	0.1
Initial values	Half-space	20	0.1	0.3	0.1

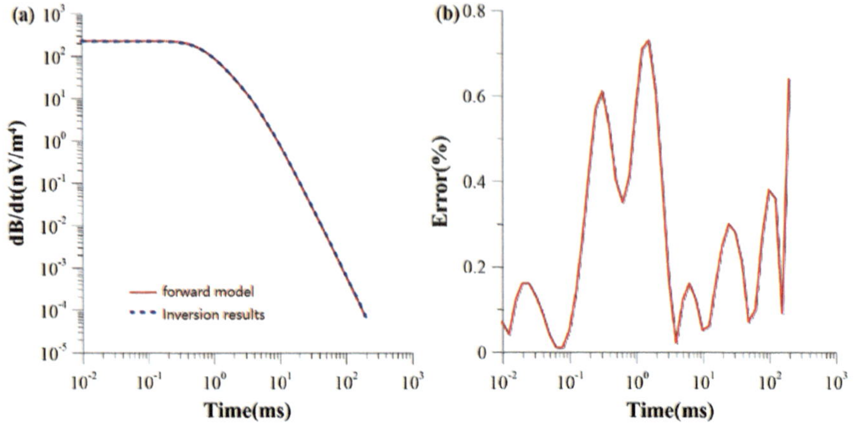

Fig. 5.38 Response inversion results of the G model. **a** and **b** respectively represent the fitting curve of the forward response and the inversion result, and the relative error curve of the fitting

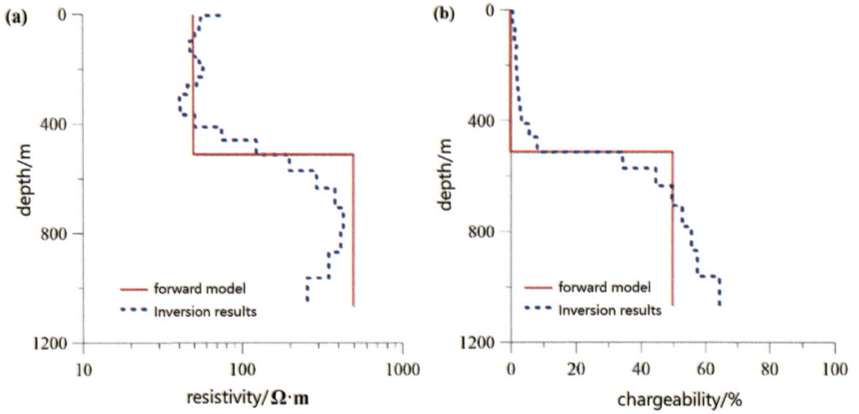

Fig. 5.39 The results of the electric information extraction of the G model. **a** and **b** respectively represent the resistivity results and the charge rate results

c. Three-layer Geoelectric Model

The parameters of the three-layer geoelectric model (A, Q, H, and K models) are listed in Table 5.5. The initial parameters remain the same as those used in the two-layer model, with a half-space background resistivity of 100 Ω·m and polarization parameters set as $m = 0.1$, $c = 0.3$, $t = 0.1$ s. The inversion results are presented in the accompanying figure.

From the results, it is evident that the inversion of the three-layer model achieves relatively ideal results, with the inversion fitting error for the G model response

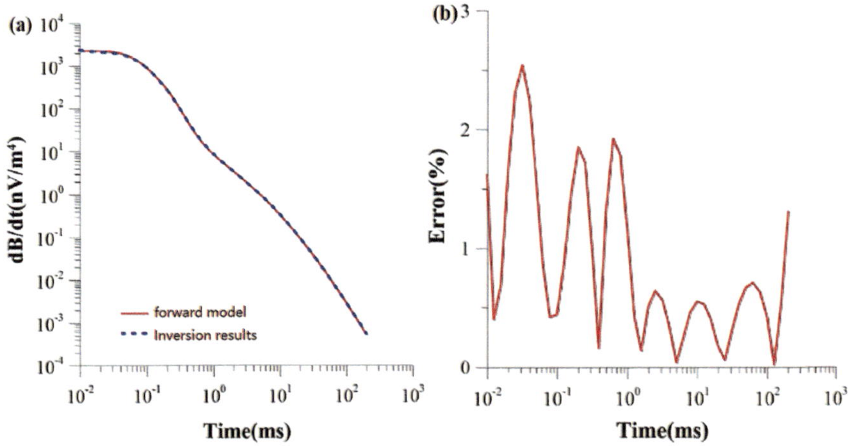

Fig. 5.40 Response inversion results of the D model. **a** and **b** respectively represent the fitting curve of the forward response and the inversion result, and the relative error curve of the fitting

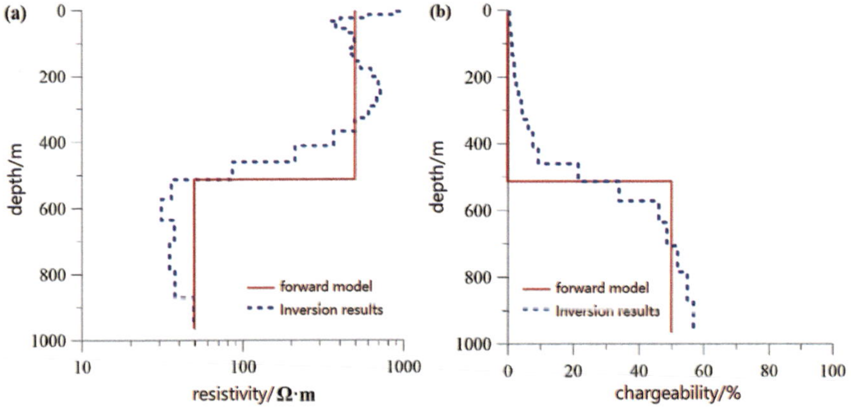

Fig. 5.41 The results of the electric information extraction of the D model. **a** and **b** respectively represent the resistivity results and the charge rate results

Table 5.4 Two-layer model and initial model parameters

	Model	$\rho_1(\Omega \cdot m)$	$\rho_2(\Omega \cdot m)$	m_1	m_2	$h(m)$	c_1–c_2	$t_1(s)$–$t_2(s)$
Model parameters	G	50	500	0	0.5	500	0.3	0.1
	D	500	50	0	0.5	500	0.3	0.1
Initial value	G/D	100		0.1			0.3	0.1

Table 5.5 Three-layer model parameters

	Model	$\rho_1(\Omega\cdot \mathrm{m})$	$\rho_2(\Omega\cdot \mathrm{m})$	$\rho_3(\Omega\cdot \mathrm{m})$	$h_1(m)$	$h_2(m)$	m_1	m_2	m_3	c_1–c_3	$t_1(s)$–$t_3(s)$
Model Parameters	A	20	100	500	500	200	0	0.5	0	0.3	0.1
	Q	500	100	20	500	200	0	0.5	0	0.3	0.1
	H	100	20	100	500	200	0	0.5	0	0.3	0.1
	K	100	500	100	500	200	0	0.5	0	0.3	0.1

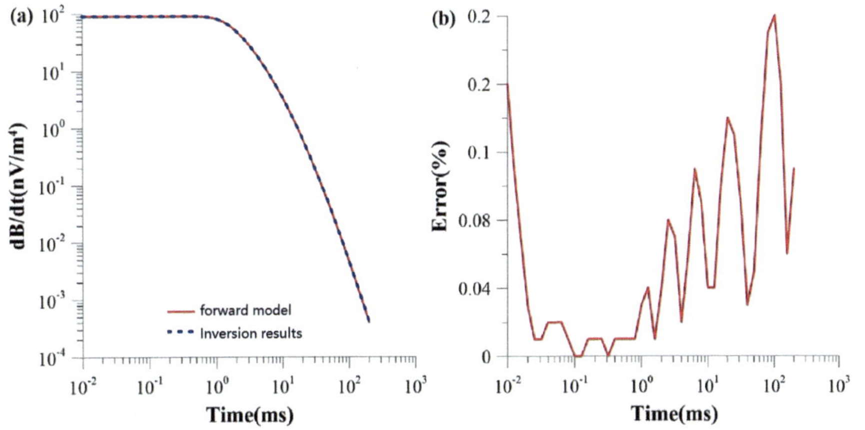

Fig. 5.42 Inversion result of A model response. **a** and **b** respectively represent the fitting curve of forward response and inversion result and the relative error curve of fitting

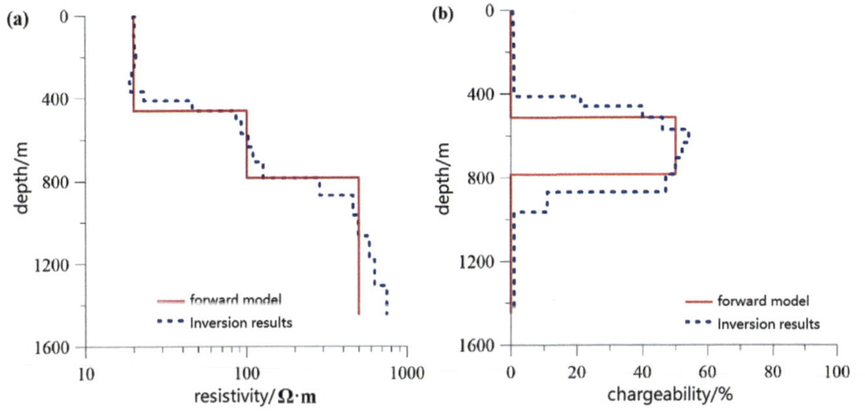

Fig. 5.43 Results of electric information extraction of Model A. **a** and **b** represent the resistivity results and charge rate results, respectively

value remaining below 0.8% and the fitting error for the D model below 3%. Additionally, the resistivity and chargeability information in the three-layer model has been successfully extracted, and the inversion results closely approximate the real model. While the retrieved parameters are not perfectly consistent with the actual model, the horizontal and vertical distribution ranges appear slightly broader, with a certain degree of depth extension. Nevertheless, the inversion effectively captures the overall resistivity and chargeability variations in the vertical direction, demonstrating the feasibility and accuracy of the method in multi-layer geoelectric structures (Figs. 5.42, 5.43, 5.44, 5.45, 5.46, 5.47).

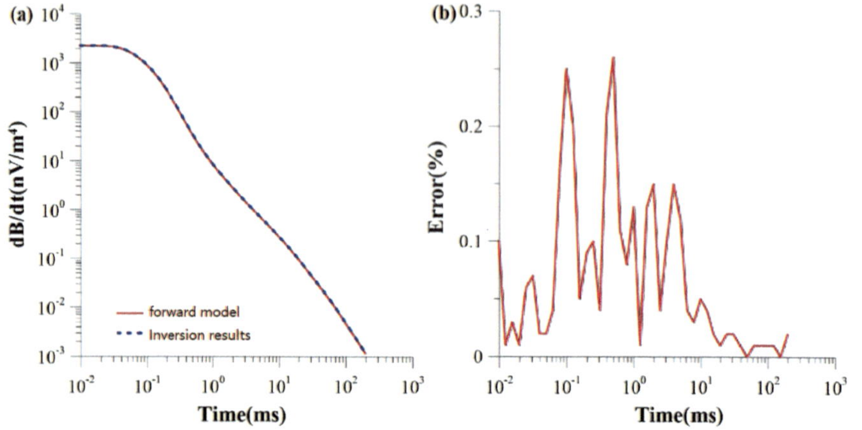

Fig. 5.44 Q Model response inversion results. **a** and **b** represent the forward response and the fitting curve of the inversion results, as well as the relative error curve of the fitting

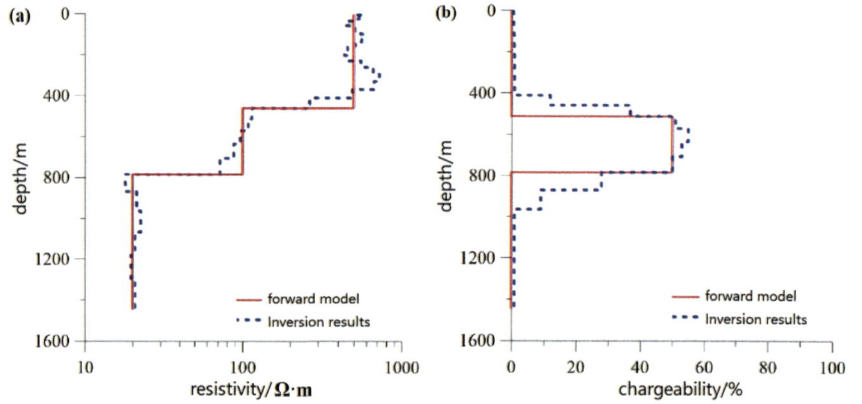

Fig. 5.45 Q Model electric information extraction results. **a** and **b** represent the resistivity results and charge rate results, respectively

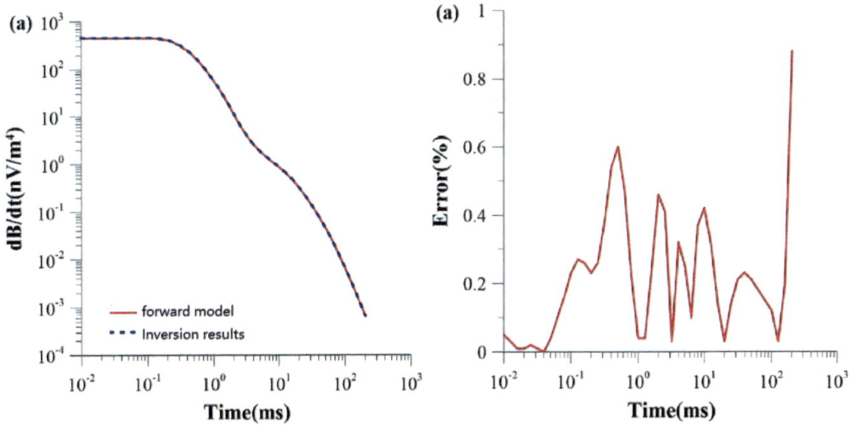

Fig. 5.46 H Model response inversion results. **a** and **b** represent the forward response and the fitting curve of the inversion results, as well as the relative error curve of the fitting

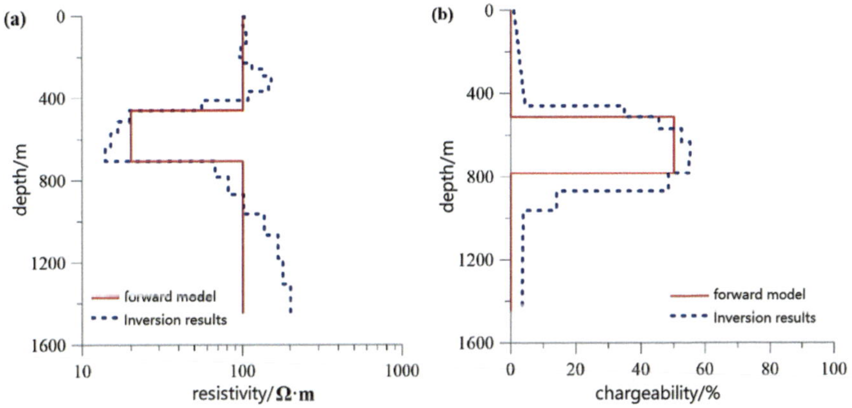

Fig. 5.47 H Model electric information extraction results. **a** and **b** represent the resistivity results and charge rate results, respectively

References

Guo, Z. W., Xue, G. Q., Liu, J., Wu, X. Electromagnetic methods for mineral exploration in China: A review[J]. Ore Geology Reviews, 2020,118, 103357.

Xue Guoqiang, Deng Xiang. Detection ability of transient electromagnetic method for thin layer [J]. Petroleum Geophysical Exploration, 2007, 42(6):709–713

Xue G Q, Bai C Y, Li X. Extracting the virtual reflected wavelet from TEM data based on regularizing method[J]. Pure and Applied Geophysics, 2012, 169(7): 1269–1282

Chen Weiyin, Li Hai, Xue Guoqiang, et al. 1D OCCAM inversion of SOTEM data and its application to 3D model effects[J]. Chinese Journal of Geophysics, 2017, 60(9): 3667–3676.

Chen W Y, Xue G Q, Khan M Y. Quasi MT inversion of short-offset transient electromagnetic data[J]. Pure and Applied Geophysics, 2016, 173(7): pp.2413–2422

Li G, He Z, Deng J, et al. Robust CSEM data processing by unsupervised machine learning[J]. Journal of Applied Geophysics, 2021, 186: 104262.

Revil, A., Florsch, N., Mao, D. Induced polarization response of porous media with metallic particles—Part 1: A theory for disseminated semiconductors[J]. Geophysics, 2015,80(5), D525–D538.

Zhdanov, M. Generalized effective-medium theory of induced polarization[J]. Geophysics, 2008, 73: F197–F211.

Zhou, N. N., Lei, K.X., Xue, G.Q., and Chen, W. Induced polarization effect on grounded-wire transient electromagnetic data from transverse electric and magnetic fields[J]. Geophysics,2020, 85, 111–120.

Chapter 6
Progress and Application of Machine Learning in Electromagnetic Detection Data Processing

6.1 Overview of Artificial Intelligence in Electromagnetic Exploration

The electromagnetic method is essential for exploring metal minerals, oil and gas reservoirs, coal mine voids, and geothermal resources. However, due to the complex observation environment and significant noise interference, achieving precise measurements of underground targets with a single geophysical method remains challenging. Currently, the comprehensive interpretation of multi-physical data by experts is the primary approach for delineating subsurface rock and ore bodies, but this method heavily relies on human experience.

With the rapid advancement of machine learning, this chapter emphasizes how training large datasets can establish the internal relationships between multi-source physical data and rock and ore bodies, ultimately enhancing target delineation. Recent progress in machine learning applications for electromagnetic data denoising, inversion, and multi-source geophysical information fusion is analyzed. It is demonstrated that deep network-based denoising methods effectively eliminate multi-source noise, while clustering-based approaches improve the correlation between multi-source geophysical data and target ore bodies, facilitating the 3D interpretation of ore body distributions.

The research presented in this chapter offers valuable insights for the comprehensive use of multi-source data to address practical subsurface detection challenges.

6.1.1 Artificial Intelligence Electromagnetic Exploration

Geophysics is a fundamental method for studying the Earth, typically inferring subsurface target properties from observed data. The electromagnetic method, as a key branch of geophysics, leverages the differences in conductivity, polarization

G. Xue et al., *Exploration with Short-Offset Grounded-Wire Transient Electromagnetic Method*, https://doi.org/10.1007/978-981-96-8846-3_6

rate, and other physical properties of subsurface rock and ore bodies for effective target detection. It plays an essential role in mineral, oil and gas, and geothermal resource exploration.

Due to the complexity of geological environments, multiple geophysical methods are often employed to obtain multi-source geophysical data, which allows for a more comprehensive understanding of underground structures. This approach enhances the reliability of target rock (mineral) detection, but it relies heavily on the expertise of data interpreters.

In recent years, machine learning has been increasingly applied to geophysical data processing and analysis. As an integrated method based on multivariate statistical analysis, machine learning enables models to identify underlying patterns in large datasets, making it possible to automate classification and prediction of new data. This technique is particularly suited for processing massive geophysical datasets. For instance, clustering analysis facilitates data-driven, automated integration, and enhances the efficiency of data processing while generating value-added geophysical maps.

With the rapid advancement of machine learning, research has focused on establishing correlations between multi-source physical parameters and rock/ore bodies through extensive data training. Machine learning applications in electromagnetic exploration primarily focus on three key areas:

Noise removal—Electromagnetic observations are often affected by various types of noise, including atmospheric electricity, anthropogenic interference, and background noise, each with complex characteristics. Traditional denoising methods involve multiple processing steps and suffer from subjectivity and poor effectiveness. Machine learning has introduced integrated multi-source noise removal techniques, significantly improving data quality.

Data inversion—Machine learning techniques enhance inversion accuracy by incorporating nonlinear relationships between physical properties and geophysical measurements.

Information fusion—Clustering and deep learning methods enable multi-source data integration, allowing for more precise interpretations of subsurface geological structures.

This section explores the latest research on machine learning applications in electromagnetic exploration, focusing on noise removal, data inversion, and multi-source data fusion. It further examines the challenges and transformations that machine learning brings to geophysical methods and discusses future trends and potential applications in electromagnetic exploration.

6.1.2 Overview of Machine Learning

Traditional geophysical data inversion follows a model-driven approach, where actual observations are used to infer past or future geophysical phenomena based on mathematical models. In contrast, machine learning-based inversion follows a data-driven approach, where models are trained on large datasets to learn the hidden relationships within the data, enabling classification and prediction of new observations. This approach is particularly effective in handling large and complex datasets.

The model-driven method focuses on the physical cause-and-effect relationships, using models such as wave equations, diffusion equations, and heat conduction equations to infer subsurface structures. Meanwhile, the data-driven approach relies on statistical relationships within the data, using techniques such as dictionary learning, deep learning, and neural networks to classify input datasets and extract meaningful geological information.

Figure 6.1 compares the model-driven and data-driven approaches. The model-driven method infers geophysical phenomena through established physical models, whereas the data-driven method applies advanced machine learning algorithms to classify and extract information from geophysical datasets, facilitating a more efficient and automated approach to subsurface exploration.

Figure 6.2 illustrates the relationship between artificial intelligence (AI), machine learning (ML), neural networks, and deep learning, highlighting the hierarchical structure of these technologies. Machine learning methods are increasingly applied in earth sciences, effectively addressing the challenge of subjective decision-making that previously relied on expert experience. In the broader field of artificial intelligence, machine learning is a key approach, while neural networks constitute one of the fundamental algorithms within machine learning, typically consisting of input layers, hidden layers, and output layers. By utilizing multilayer perceptrons, neural networks can identify and represent distributed features within data.

Deep learning is a subset of artificial neural networks, encompassing various learning paradigms, including supervised learning, unsupervised learning, semi-supervised learning, transfer learning, and reinforcement learning. Supervised learning involves training an intelligent algorithm with labeled data, mapping input data to specific labels. In contrast, unsupervised learning does not require manually labeled datasets, with clustering being a typical algorithm in this category. Semi-supervised learning combines elements of both supervised and unsupervised learning to leverage partially labeled datasets. Transfer learning allows a model trained on one task to transfer knowledge to a new task, enabling further training and adaptation. Reinforcement learning, on the other hand, is based on learning through interaction with the environment, optimizing decision-making strategies over time.

6.1.3 Progress in the Application of Machine Learning in Electromagnetic Detection

Geophysical data is characterized by large volumes, rapid updates, and high uncertainty in classification, making it well-suited for machine learning applications. Techniques such as clustering analysis facilitate data-driven, automated integration and value-added map generation, enabling more efficient interpretation of geophysical

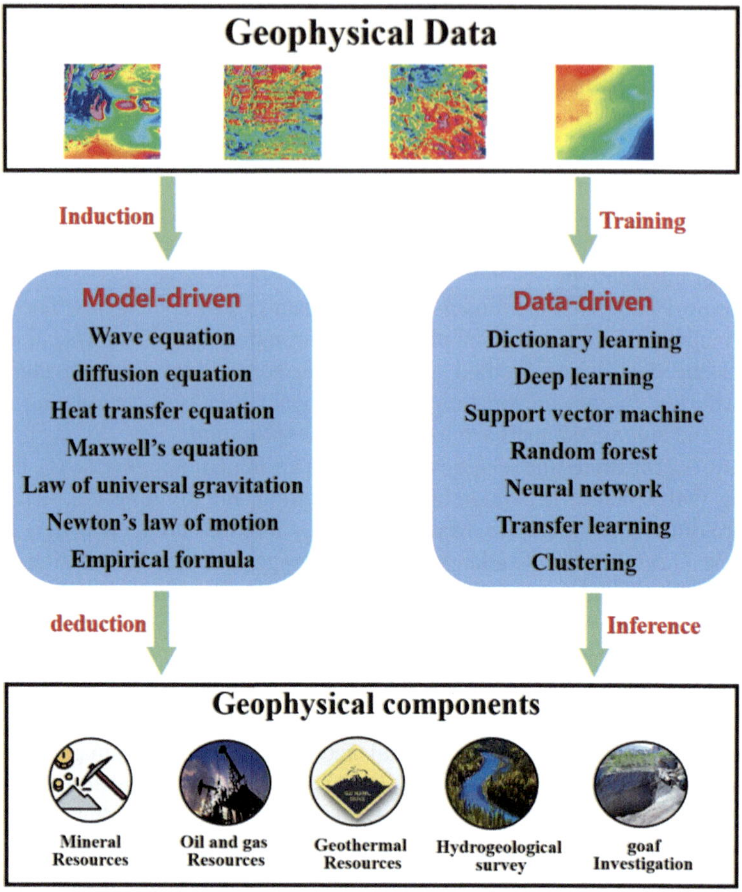

Fig. 6.1 Relationship diagram of model-driven and data-driven methods

data. As a result, the adoption of machine learning technologies for denoising, inversion modeling, and information fusion in electromagnetic exploration has become an inevitable trend. This section introduces recent advancements in machine learning applications in electromagnetic exploration, focusing on electromagnetic data denoising, inversion modeling, and multi-source data fusion.

(1) Application of Machine Learning in Electromagnetic Noise Removal.

The sources of electromagnetic detection noise can be categorized into internal noise and external noise. Motion-induced noise is a specific type of noise that arises in airborne or semi-airborne electromagnetic detection methods, where the stable geomagnetic field is significantly stronger than the excitation field source used in electromagnetic detection. In such cases, motion-induced noise must be carefully addressed to ensure accurate data acquisition. A major research challenge is

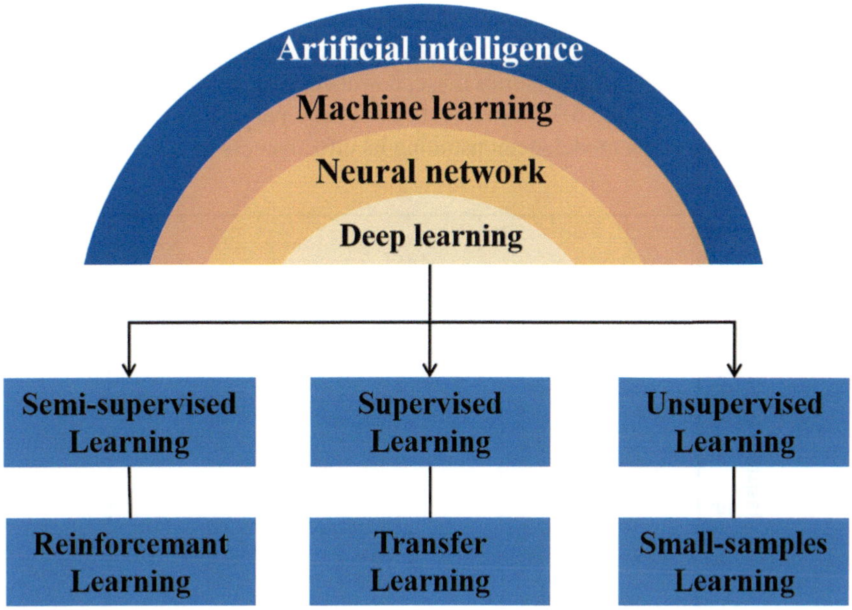

Fig. 6.2 The inclusion relationship diagram of artificial intelligence, machine learning, neural networks, and deep learning

improving the efficiency of electromagnetic data noise processing while minimizing the impact of subjective operator judgment on data processing results.

In recent years, the application of machine learning in electromagnetic data denoising has become increasingly widespread. Lin et al. used an autoencoder to build an unsupervised learning model (SFSDSA) to process the secondary field signal (SFS) in TEM. SFSDSA is based on the deep features of the signal, mapping the noise-interfered signal points to the high-probability points with the noise-free signal as a reference, thereby achieving signal denoising. Chen and Xue (2021) proposed a new denoising framework based on deep convolutional neural networks (CNN), converting the TEM signal denoising task into an image denoising task (i.e., TEMDnet). This method has the advantages of strong flexibility, easy operation, and fast computation speed. Wang et al. proposed a new noise learning-inspired deep denoising network (i.e., TEM-NLnet) to solve the problem of severe uneven noise distribution in measured TEM signals. When working in urban areas with strong noise, due to the extremely low signal-to-noise ratio (SNR), Sun et al. proposed a method combining the minimum noise fraction (MNF) algorithm and machine learning to deal with this problem.

Electromagnetic exploration methods based on flight platforms require specialized approaches to address motion-induced noise. Wu and colleagues proposed a wavelet neural network to filter out high-frequency motion noise from helicopter transient electromagnetic data. Building on this, Wu et al. (2020) introduced a machine learning-based denoising autoencoder (DAE) to comprehensively remove multiple sources of noise in airborne electromagnetic data, including motion-induced noise, nearby and moderately distant sferics noise, power-line noise, and background electromagnetic noise. Figure 6.3 illustrates the denoising performance of this method in a 5-layer theoretical model, demonstrating its effectiveness in noise suppression.

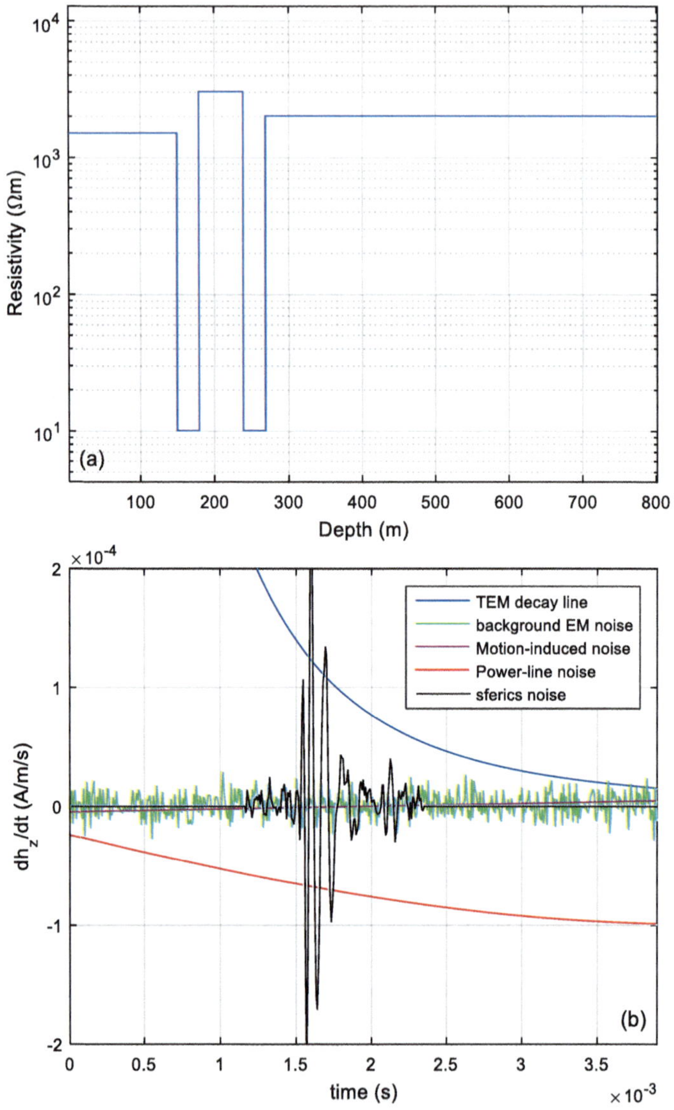

Fig. 6.3 Test results of the return performance of the multi-source noise suppression method for airborne electromagnetic detection based on DAE (Wu et al., 2020)

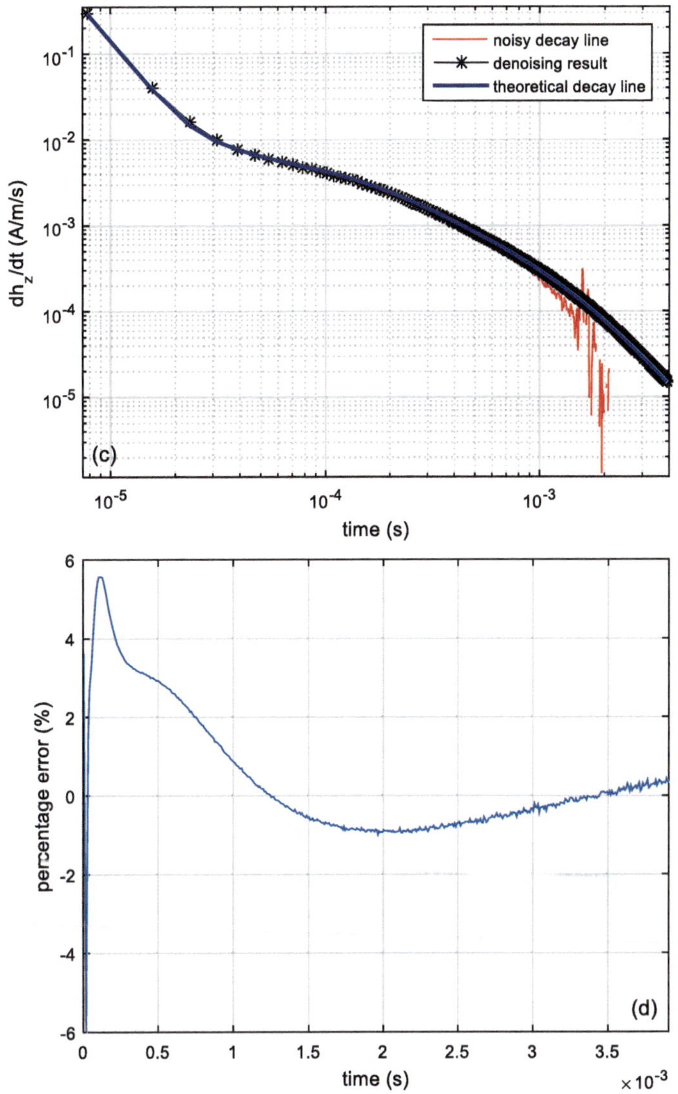

Fig. 6.3 (continued)

Beyond airborne systems, significant research progress has been made in machine learning applications for electromagnetic data denoising in ground-based artificial source methods. Given the severe interference from anthropogenic noise, Li et al. developed an unsupervised machine learning algorithm under a data-driven framework—the shift-invariant sparse coding (SISC) method. This approach effectively removes large-amplitude power-line noise, baseline drift, white noise, pulse noise, and periodic/non-periodic square wave noise. Additionally, Li et al. utilized normalized cross-correlation (NCC), absolute maximum amplitude (Max), and detrended fluctuation analysis (DFA) to accurately identify and extract high-quality time series data from noisy measurements.

In practical applications, the distinction between weak useful signals and noise remains a critical challenge, often leading to two primary issues in data processing: (1) incomplete noise removal, and (2) loss of valuable signals. These limitations directly impact the reliability of resistivity models obtained after inversion. To address this, Wu et al. proposed the manifold hypothesis and established a feedback mechanism between signal–noise distinction and signal-model mapping (Fig. 6.4). This approach allows the entire neural network to simultaneously perform traditional denoising and inversion, leading to more objective and reliable resistivity model estimations, significantly improving the accuracy and robustness of electromagnetic inversion results.

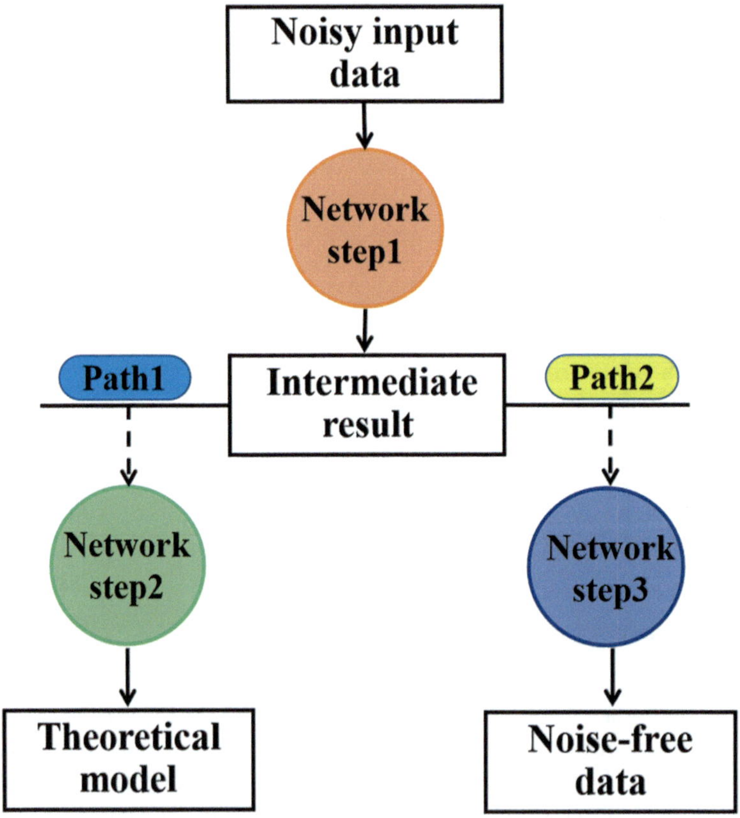

Fig. 6.4 The network structure designed based on the proposed manifold hypothesis. The network structure includes two paths, and the intermediate results shared by the two paths represent the common low-dimensional manifold of the model and data

(2) Application of machine learning in electromagnetic data inversion modeling.

The goal of electromagnetic inversion is to quantitatively interpret the internal structure of the Earth, which can be broadly categorized into model-driven inversion and data-driven inversion. Recently, machine learning-based data-driven inversion has gained significant attention, as it enables real-time prediction of subsurface structures by directly processing input data through pre-trained models.

Traditionally, the interpretation of electromagnetic data relies on solving inverse problems, where the objective is to minimize the discrepancy between synthetic data (computed using Maxwell's equations) and observed data. Puzyrev demonstrated the feasibility of machine learning techniques for electromagnetic interpretation through numerical experiments. Later, Puzyrev and Swidinsky introduced a 1D electromagnetic inversion model based on convolutional neural networks (CNNs), applying CNNs to controlled-source electromagnetic (CSEM) and time-domain electromagnetic (TEM) data. Meanwhile, Bang et al. developed an efficient 2D machine learning inversion technique, which improves upon 1D methods by incorporating spatial continuity constraints between adjacent data points during training.

Wang et al. introduced two inversion modeling approaches: model-driven inversion and data-driven inversion. The data-driven inversion technique employs two neural networks to approximate random model parameters and solutions within TgNN-geo. Figure 6.5 illustrates the TgNN proxy network structure, where inputs include time (t), spatial coordinates (x, y), and random model parameters (ξi) parameterized by Karhunen-Loève expansion (KLE) (highlighted by the yellow dashed box). The TgNN proxy incorporates physical/engineering constraints and domain knowledge as prior information during training, while the output represents the quantity of interest, such as hydraulic head (green dashed box).

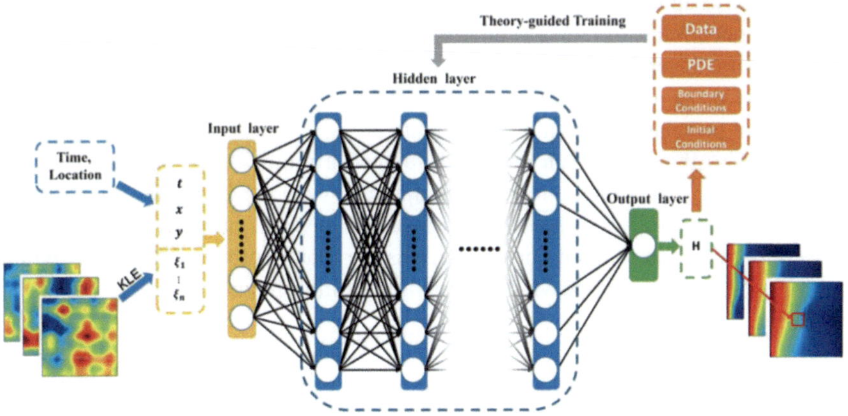

Fig. 6.5 Theoretical guidance neural network proxy structure

Inspired by Google's Neural Machine Translation System, Wu et al. developed a fast inversion operator, which directly converts time-domain airborne electromagnetic (AEM) measurements into subsurface resistivity structures using deep learning guidance. The inversion results provide a clear depiction of geological features such as lakes, surrounding mountains, and fault structures, enabling real-time underground resistivity reconstruction from AEM data.

To perform the inversion of airborne electromagnetic data, Wu established a three-stage inversion framework that transforms airborne electromagnetic (AEM) signals into resistivity models. In the first stage, a large number of synthetic sample pairs are randomly generated for training and validating the inversion system. Each data sample includes a 1D resistivity model, the height of the transmitting and receiving antennas, and the corresponding synthetic AEM signal. In the second stage, an instant inversion system is established based on the Long Short-Term Memory (LSTM) network and trained with the aforementioned dataset, capturing the intrinsic mapping relationship between the AEM signal and the resistivity structure, and considering the height of the transmitting and receiving antennas. Finally, in the third stage, the newly acquired AEM signal is directly input into the trained inversion system, outputting the resistivity model in real time and reconstructing the pseudo 3D resistivity structure.

Joint inversion is a widely used approach to enhance the reliability of geophysical exploration by integrating multiple geophysical datasets (e.g., electromagnetic, gravity, magnetic, and seismic data). Unlike single-property inversion, joint inversion incorporates additional constraints, reducing non-uniqueness in subsurface models and providing a more comprehensive geological interpretation.

Recently, machine learning has been increasingly employed in geophysical joint inversion, where it automatically integrates prior information and multimodal datasets, significantly improving inversion accuracy. Zhou et al. proposed a deep learning-assisted joint inversion method based on attribute fusion constraints, which automatically captures structural similarities and rock physics relationships between Audio Magnetotellurics (AMT) and seismic first-arrival time data. One major challenge in joint inversion is the multimodal nature of rock physical property data, as different properties may exhibit various relationships (e.g., linear, quadratic, or exponential correlations) within the study area.

To address this issue, Sun and Li developed a multi-source physical property joint clustering inversion method, which applies Fuzzy C-Means (FCM) clustering to constrain the statistical behavior of inverted physical property values. Unlike conventional spatial-domain constraints, this approach enforces constraints within the parameter domain, improving the usability of multimodal geophysical data.

(3) Application of machine learning in multi-source information fusion.

Traditionally, geoscience research experts interpret the inverted physical property models based on their experience, which involves many human factors and is unreliable. With the continuous increase of current geoscience data, which is characterized by large volume, diverse types and sources, fast speed, and high value, information fusion technology based on machine learning is rapidly developing in the field of geoscience. Depending on the standards that information fusion technology is based on, it can be classified. For example, based on the attributes of the input data of information fusion, it can be divided into data fusion and image fusion; based on the fusion level, it can be divided into pixel-level fusion, feature-level fusion, and decision-level fusion.

Firstly, introducing information fusion based on data fusion, Wang et al. used unsupervised clustering algorithms to fuse multi-source geophysical data to generate clustering maps. Each category in the generated clustering map has multiple geophysical features. Then, based on the existing geological maps, field data, and rock sample analysis in the survey area, the clustering map with multiple physical property features can be associated with certain geological attributes or rock types. Multi-geophysical data includes airborne magnetic, gravity, frequency electromagnetic, and radiation measurements. Magnetic measurements capture the magnetization contrast from the surface to the Curie isotherm depth; gravity measurements reflect the rock density contrast at all depths below the surface; radiation data captures the K, Th, and U concentration contrast in the upper 50 cm of the ground; AFEM (Airborne Frequency Electromagnetic) data reflects the lateral or vertical changes in resistivity/conductivity of the earth at 50–100 m depth. Due to the nature of these input data, the generated cluster mapping carries multi-depth information. When associated with existing geological information, the cluster map can not only help interpret bedrock outcrops but also help interpret rocks under sedimentary cover layers. Figure 6.6 overlays the cluster map with a simplified bedrock map to find connections between clusters and specific geological units or lithologies.

Fu et al. proposed using K-nearest neighbor, back-propagation neural network, and support vector machine, three machine learning methods, respectively, for 3D mineral prediction, all of which are supervised learning algorithms. The entire information fusion process is mainly divided into three parts (Fig. 6.7): The first step is to establish a three-dimensional geological model based on a large amount of geological cross-sections, drilling data, profile data, rock sample data, etc., and then invert gravity, magnetic, electromagnetic, seismic geophysical data to establish a three-dimensional physical property model (all models are within the same depth range). The second step is to extract the ore-controlling feature samples according to the known mining area locations. The third step is to use the known samples to train the model, and then use the trained model to predict the minerals in the entire area.

Figure 6.8 presents a comparison of 3D mineral prediction results obtained using different machine learning algorithms. Figure 6.8a, c, and e incorporate lithological information as feature parameters in the model training process, while Fig. 6.8b, d, and f exclude lithological information. The comparison of prediction results demonstrates that the models incorporating lithological data exhibit superior predictive performance compared to those that do not. These results suggest that considering lithological information significantly enhances exploration accuracy, allowing for the delineation of potential mineralized zones, thereby improving exploration efficiency and reducing drilling risks.

Information fusion techniques based on image fusion have gained increasing applications in geosciences, as geophysical images provide a more direct representation of geological information and offer advantages such as low computational

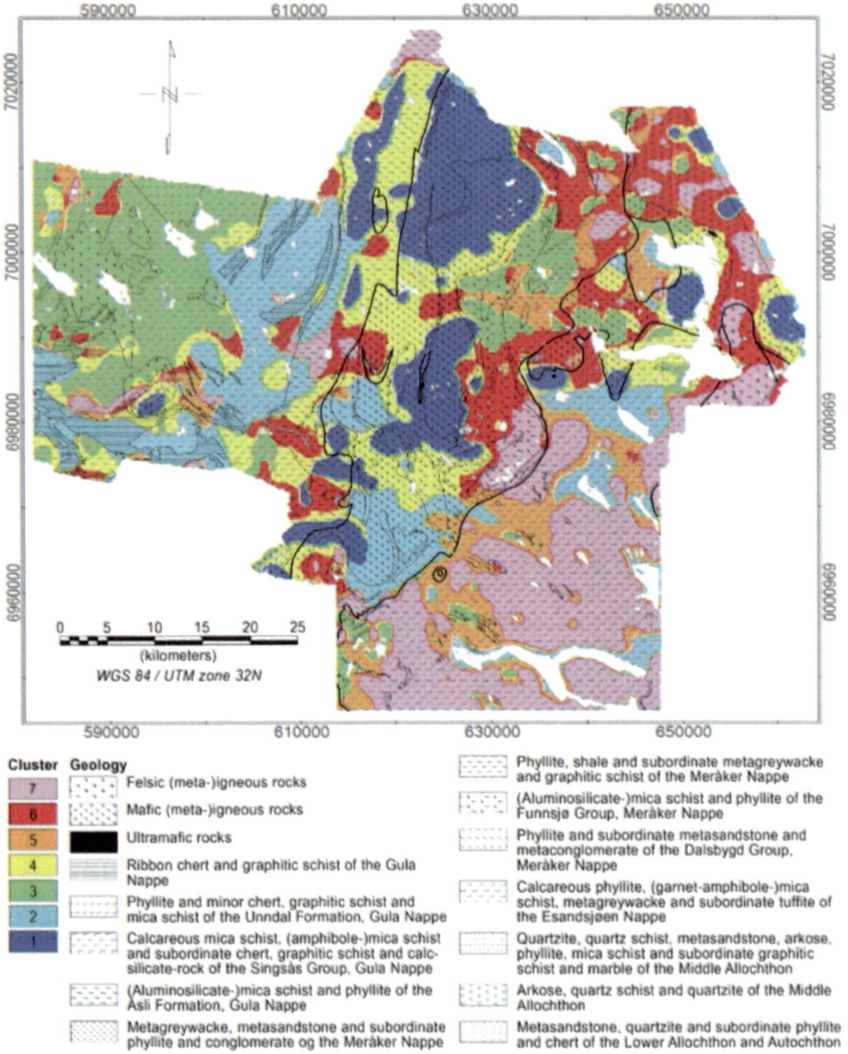

Fig. 6.6 Clustering map overlaid on a simplified geological bedrock map based on a 1:250,000 unified bedrock map

complexity, easy operation, and the elimination of complex physical parameter relationships between multiple datasets. By integrating two or more geophysical images, machine learning-based image fusion techniques generate composite results that enhance the localization of geological targets.

Figure 6.9 illustrates a strategy flowchart for constructing a unique geological model of the target body using the three-dimensional wavelet fusion method, as proposed by Nasri et al. This method has been effectively applied to reconstruct the

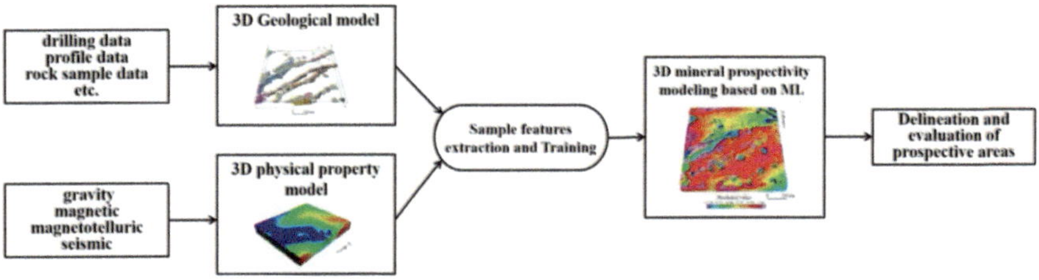

Fig. 6.7 Workflow of 3D mineral prospecting modeling based on machine learning

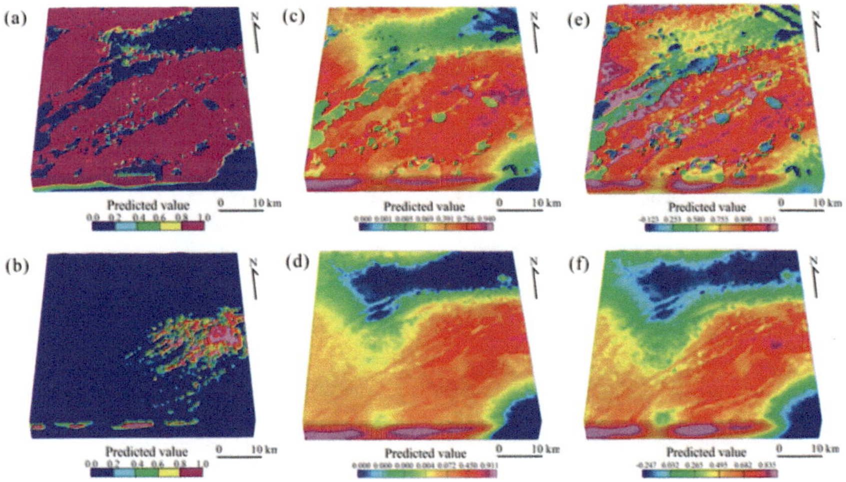

Fig. 6.8 3D mineral prospecting prediction results. The results include: **a** KNN with lithological information, **b** KNN without lithological information, **c** BPNN with lithological information, **d** BPNN without lithological information, **e** SVM with lithological information, **f** SVM without lithological information

geological model of the Makran subduction zone in southwestern Iran, providing crucial insights into the location of faults, subduction zones, oceanic crust geometry, and geological boundaries. Since the early 2010s, several researchers have advanced geophysical image fusion techniques to improve data interpretation. Leblanc et al. introduced an image information fusion method to enhance multi-source geophysical data interpretation, effectively reducing errors in interpretation results. Mishra and Palkar further refined image fusion approaches, leading to significant reductions in storage requirements and computation time. Nasri et al. applied 3D wavelet-based image fusion to the interpretation of complex three-dimensional geological models, improving subsurface structural mapping.

Karamitrou et al. utilized curvelet transforms to fuse resistivity and magnetization images, transforming different geophysical datasets into raster images to enhance the visualization of subsurface archaeological features. Karamitrou and Bogiatzis

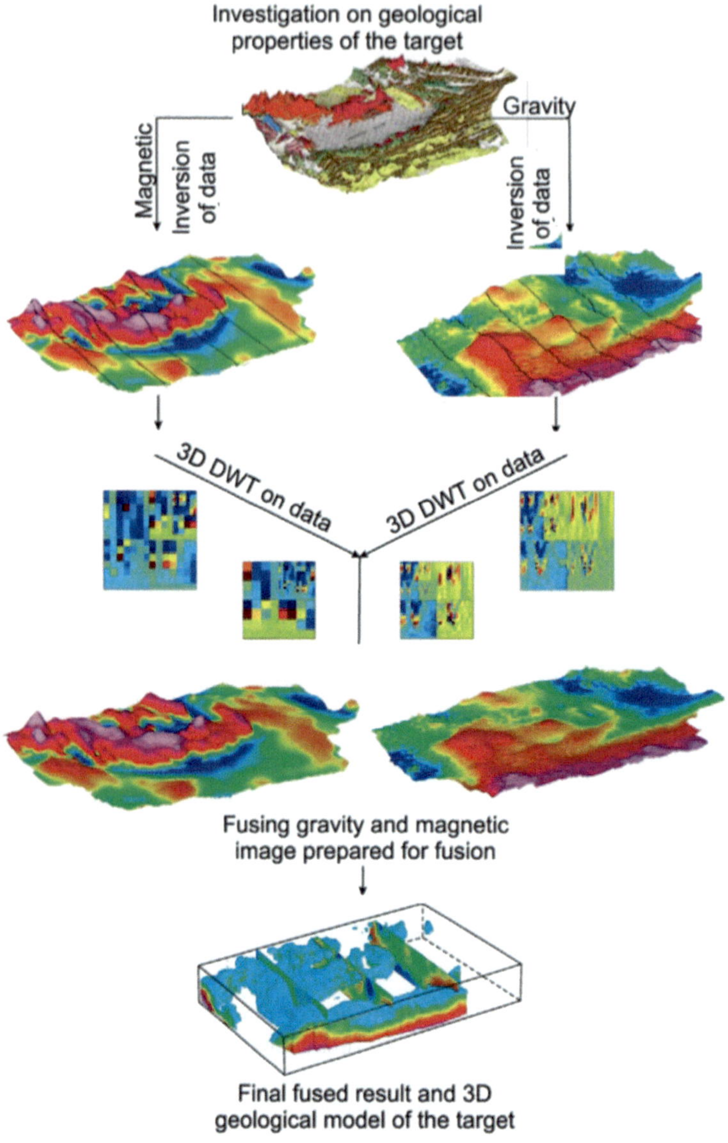

Fig. 6.9 Strategy flowchart for establishing a unique geological model of the target body using the three-dimensional wavelet fusion method

integrated multiple geophysical images from different exploration instruments, such as apparent resistivity, magnetic field gradient, and electromagnetic (EM) reflection data, thereby improving the identification of archaeological targets.

Given the diverse nature of geophysical imaging data, the primary challenge in image fusion lies in feature extraction. Different geophysical survey techniques capture distinct geological properties, necessitating specialized image processing techniques for effective fusion. In seismic imaging, reflection horizons in seismic sections are first converted into 2D digital images, where special attention is given to edge structure information to preserve stratigraphic details. For gravity and magnetic exploration, image processing methods translate abstract field characteristics such as field strength, rate of change, and morphological variations into intuitive visual representations, including brightness, chroma, roughness, and spatial geometry. In the case of resistivity and magnetization contour maps, feature extraction focuses on image texture details and gray-scale variations, which play a crucial role in fine-scale geological analysis, particularly in identifying sedimentary structures. Currently, deep learning-based feature extraction techniques in geophysics are primarily applied in remote sensing and seismic studies. However, as image fusion technology advances, machine learning is expected to play an increasingly significant role in the integration of multi-source geophysical data, ultimately enhancing subsurface geological interpretation and improving resource exploration efficiency.

6.1.4 Research Prospects of Machine Learning Technology in Electromagnetic Detection Applications

(1) Challenges faced by machine learning technology in electromagnetic detection applications.

Machine learning has demonstrated its ability to capture deep features in geophysical data through neural networks trained on large-scale datasets, leading to successful applications in multiple domains. However, the practical use of machine learning in geophysics faces significant challenges, primarily due to the lack of labeled datasets. Many current studies rely on numerical simulations, which, while useful, do not fully reflect the complexities of real-world geophysical environments. Another fundamental limitation is that machine learning models require the test dataset to exhibit feature similarity with the training dataset; otherwise, the generalization capability of the model is significantly reduced. This limitation hinders the direct application of machine learning methods to actual geophysical data, where geological conditions and electromagnetic responses vary considerably.

Similarly, multi-information fusion techniques encounter challenges such as disparities in accuracy and scale between different geophysical instruments and detection methods, as well as inconsistencies in multi-source datasets. When applying machine learning to predict the geometric shape and spatial extent of geological targets, different machine learning algorithms may yield different results even when analyzing the same geological structure, target body, or detection dataset. On the one hand, three-dimensional visualization and quantitative predictions can significantly enhance mineral exploration efficiency by integrating multi-source geophysical data. On the other hand, if machine learning is not properly handled, it may introduce erroneous results, including over-extrapolation, leading to unrealistic models; over-interpolation, creating artificial data that does not reflect physical reality; overfitting, generating misleading knowledge from limited datasets; and over-simplification, causing models to overlook critical geological features. Therefore, choosing an appropriate information fusion algorithm based on the detection objectives and survey conditions in different regions is crucial to ensuring reliable results.

(2) The development direction of machine learning in electromagnetic detection.

As electromagnetic detection methods continue to evolve, geophysical data is becoming increasingly diverse and complex, requiring the application of machine learning techniques for the comprehensive analysis of multi-source electromagnetic datasets (Fig. 6.10). Machine learning can aid in summarizing ore-forming patterns, identifying diagenetic environments, and predicting mineralization zones. However, due to the unique nature of electromagnetic data, the modeling process must account for multi-source, multi-scale, high-dimensional, and complex spatio-temporal dependencies. The main challenges in applying machine learning to electromagnetic exploration include the shortage of labeled training data, low signal-to-noise ratios in collected datasets, and the strong non-linearity of geophysical responses. Among these, the lack of labeled electromagnetic data remains the most significant hurdle.

Transfer learning is emerging as a potential solution to address the scarcity of labeled data, allowing models trained on one dataset to be adapted for a different but related task with minimal additional training. In addition, semi-supervised learning, unsupervised learning, multimodal machine learning, federated learning, and active learning offer promising directions for improving geophysical data analysis. Furthermore, enhancing equipment performance by reducing instrumental noise levels and improving data acquisition quality will provide a stronger foundation for subsequent data processing and interpretation.

Future research will focus on four key aspects: (1) integrating physical laws with machine learning to improve the interpretability and reliability of electromagnetic exploration results; (2) collecting relevant geological priors to enhance uncertainty estimation in machine learning models, thereby improving model credibility; (3) addressing the discrepancies between measured and simulated data, as geological structures in the field often differ from theoretical models, which can lead to invalid training results; and (4) advancing database construction, particularly through cloud computing platforms, to support large-scale data integration and promote innovation across the geoscience field.

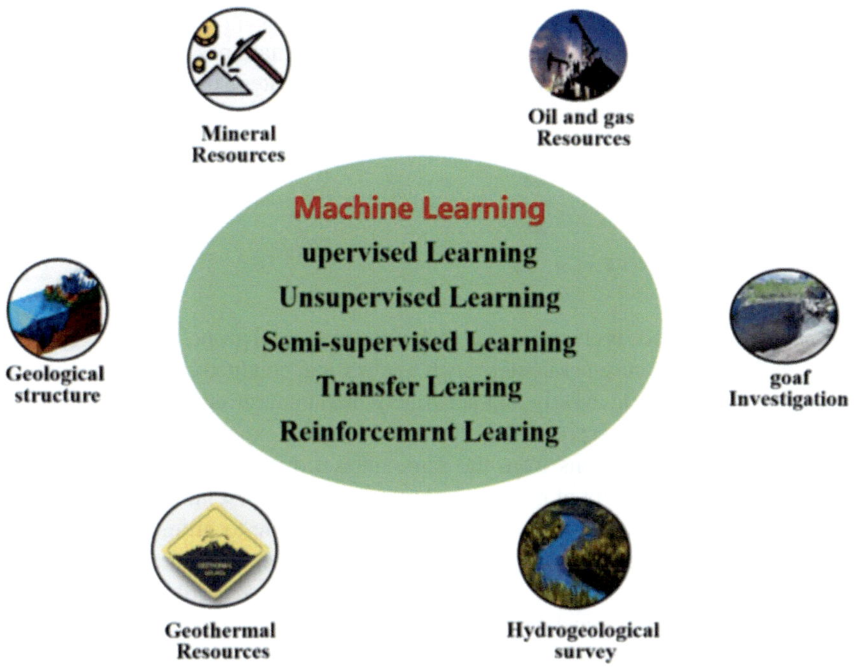

Fig. 6.10 The application direction of machine learning technology in the electromagnetic field and the trend of modern technology development

6.2 Using Multi-source Geophysical Image Fusion Method to Identify Physical Property Anomalies—Taking Airborne Electromagnetic and Aeromagnetic Data as Examples

The comprehensive interpretation of multi-source geophysical data enables accurate identification and localization of target bodies in complex terrain environments. However, when dealing with a single geological model, different geophysical datasets are often inverted separately, resulting in multiple inversion results that do not fully utilize the advantages of integrating multi-source data. To address this, a multi-source geophysical information fusion method based on non-subsampled contourlet transform (NSCT) is applied, using images as carriers for a comprehensive analysis of multi-source geophysical data to enhance target body identification and localization accuracy. This study applies the method to publicly available geophysical data from the Iliamna volcanic area in southwest Cook Bay, Alaska, USA, demonstrating the fusion of magnetization and resistivity models. The process involves using NSCT to decompose multi-source images from different strata into low-frequency and high-frequency sub-band images, followed by a fusion rule based on weighted averaging and a new metric parameter (NMP) to fuse the low- and high-frequency sub-band images separately. Finally, the NSCT inverse transform is applied to obtain a fused image that integrates magnetization and resistivity information. A comparison of the fusion results with prior geological information demonstrates a high degree of consistency, confirming the effectiveness and reliability of this approach and providing a new solution for the accurate and rapid identification of target bodies.

6.2.1 Case Preface

The Earth's subsurface is characterized by multiple physical properties, and geophysical detection aims to use parameters such as electrical conductivity, magnetization, and density to comprehensively and accurately map underground structures. Traditionally, geophysical interpretation relies on subjective expert analysis, which can lead to different conclusions from the same dataset due to variations in expertise and judgment. The concept of geophysical joint inversion, first introduced by Vozoff and Jupp, has since evolved into a widely used technique to mitigate inconsistencies in inversion structures and increase the utilization of multi-source data. Various joint inversion approaches have been proposed, including Gallardo et al. who integrated gravity and magnetic data with geometric constraints to reduce internal ambiguity. However, this method is mainly applicable to simple layered models, and its reliability decreases in complex geological settings. Mojica and Bassrei introduced parameter regularization and cross-validation techniques to improve the resolution of 3D gravity inversion at greater depths. Gallardo and Meju proposed the cross-gradient function to establish structural coupling between different physical properties, offering a flexible and effective joint inversion method independent of specific petrophysical correlations. Zhdanov et al. developed a generalized joint inversion method based on Gramian constraints, transforming models into a Gramian space domain to enhance inversion stability. Yin et al. (2018) further refined joint inversion techniques by incorporating local Pearson correlation coefficient constraints, effectively improving convergence and addressing non-uniqueness in single geophysical inversions.

Image fusion technology has emerged as a rapidly advancing method for multi-source geophysical data interpretation. The principle of geophysical image fusion involves integrating or extracting relevant information from multiple geophysical datasets, converting them into a single composite image with higher visual and digital quality than the original input images. From a conceptual perspective, image fusion can be regarded as a subset of data fusion, commonly applied in deep mining exploration where multiple datasets from the same location are mapped into pixel-based cluster analyses. Algorithmically, image fusion extends conventional image analysis techniques, encompassing image compression, feature extraction, registration, pattern recognition, and segmentation.

Image fusion techniques can be classified into three levels: pixel-level fusion, which directly combines information at the pixel scale to generate a final composite image; feature-level fusion, which employs multi-scale and signal decomposition techniques such as pyramid decomposition and wavelet decomposition for advanced feature extraction; and decision-level fusion, which integrates high-level data interpretation techniques such as Dempster–Shafer (D–S) evidence theory, expert systems, and Bayesian estimation. Various geophysical image fusion algorithms have been explored, including simple averaging, spatial frequency matching, Brovey transformation, intensity-hue-saturation transformation, principal component analysis (PCA), and discrete wavelet transform (DWT). Leblanc et al. demonstrated that fusing multiple geophysical datasets reduces false anomalies in final interpretations while retaining the advantages of each dataset in evaluating the subsurface model. Erkan et al. applied information fusion techniques to improve gravity and magnetic model integration, enhancing the accuracy of geological models. Mishra and Palkar highlighted that image-based information fusion significantly reduces storage requirements and computation time, improving spatial resolution, classification accuracy, feature representation, geometric precision, and computational efficiency. Nasri et al. applied a 3D wavelet-based fusion method to analyze and interpret geological models, successfully improving the accuracy of fault mapping, subduction zone delineation, and oceanic crust boundary detection.

While wavelet-based image fusion techniques are widely used in geophysical applications, they do not effectively capture directional image features. In geophysical datasets, spatial relationships between geological units must be preserved to ensure accurate feature extraction. Identifying the geometric shape of subsurface targets requires multi-directional feature extraction, while localized feature information is critical for detecting specific target bodies. Therefore, selecting a fusion algorithm with strong multi-directionality and local feature extraction capabilities is essential for producing reliable fusion results. The Non-Subsampled Contourlet Transform (NSCT) algorithm addresses these challenges by providing multi-scale, multi-directional, translation-invariant transformations with superior localization capabilities, making it well-suited for extracting high-dimensional image features in geophysical datasets.

Based on the advantages of NSCT, this study applies an image fusion method using NSCT, utilizing images as the primary fusion medium. The methodology is demonstrated using measured geophysical data from the Iliamna volcanic area in southwest Cook Bay, Alaska, USA, effectively enhancing target body identification and positioning in a big data environment while maintaining accuracy and improving interpretation efficiency.

6.2.2 NSCT Algorithm

The NSCT algorithm effectively captures high-dimensional features in geophysical images, providing multi-scale, multi-directional, and translation-invariant transformations. The NSCT framework consists of two main components: the Non-Subsampled Pyramid Filter Bank (NSPFB) and the Non-Subsampled Directional

Filter Bank (NSDFB). The NSPFB is responsible for multi-scale decomposition, breaking down the source image into low-frequency and high-frequency sub-band images at each decomposition level. Following NSPFB decomposition, a total of K + 1 sub-images (where K represents the number of decomposition levels) are obtained, including K high-frequency images and one low-frequency image of the same size as the original source image. The NSDFB is then applied to the high-frequency sub-bands to achieve directional decomposition, allowing for enhanced feature extraction and improved detection of geological structures.

The NSCT inverse transform is finally used to reconstruct a fused image, integrating information from multiple geophysical sources to enhance interpretability and accuracy in subsurface exploration. The NSCT-based fusion technique is particularly effective in regions with complex geological formations, such as the Iliamna volcanic region, where magnetization and resistivity anomalies must be accurately mapped. The resulting fused image provides enhanced spatial resolution, improved geological feature differentiation, and more reliable target identification, making it a valuable tool for modern geophysical interpretation.

$$H_n^{eq}(z) = \begin{cases} H_1\left(z^{2^{n-1}I}\right)\prod_{k=0}^{n-2} H_0\left(z^{2^k I}\right), & 1 \le n \le K \\ \prod_{k=0}^{n-2} H_0\left(z^{2^k I}\right), & n = K+1 \end{cases} \tag{6.1}$$

where, $H_0(z)$ and $H_1(z)$ respectively represent the corresponding low-pass and high-pass filters. Figure 6.11 is an NSPFB diagram with three decomposition levels. Then, NSDFB performs level directional decomposition on the high-frequency sub-band images on NSPFB, and obtains directional sub-images of the same size as the source image, so that NSCT can obtain multi-directional characteristics and more accurate directional information.

Figure 6.12 is a display of a four-channel NSDFB. When the output of the filter is combined with the output of the sector filter in the first step, a frequency decomposition in four directions can be obtained. The equivalent filter for each direction can be given by (6.2). The synthesized filter is obtained in a similar way. All filter groups in the NSDFB tree structure are selected from a single non-downsampling sector filter group. In order to achieve multi-directional decomposition, the high-pass image obtained by NSDFB decomposition iteration needs to be processed by NSPFB again.

$$U_k(z) = U_i(z)U_j\left(z^D\right) \tag{6.2}$$

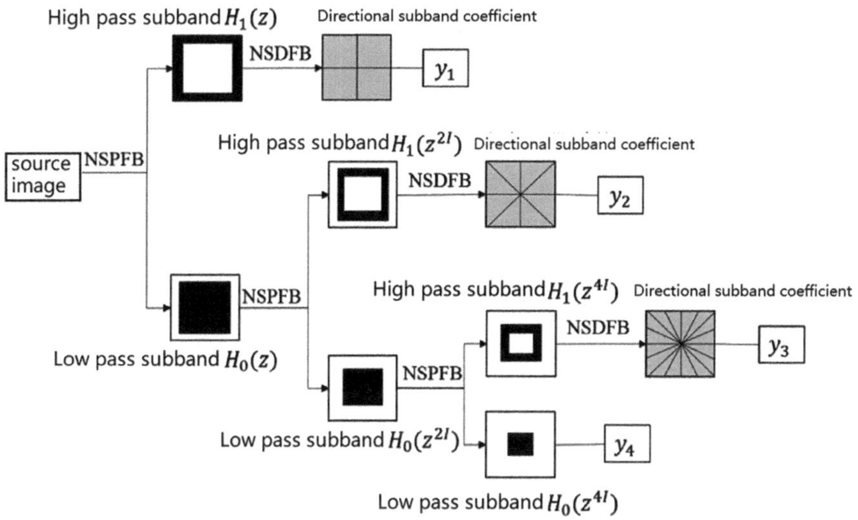

Fig. 6.11 Non-subsampled Contourlet Transform (NSCT)

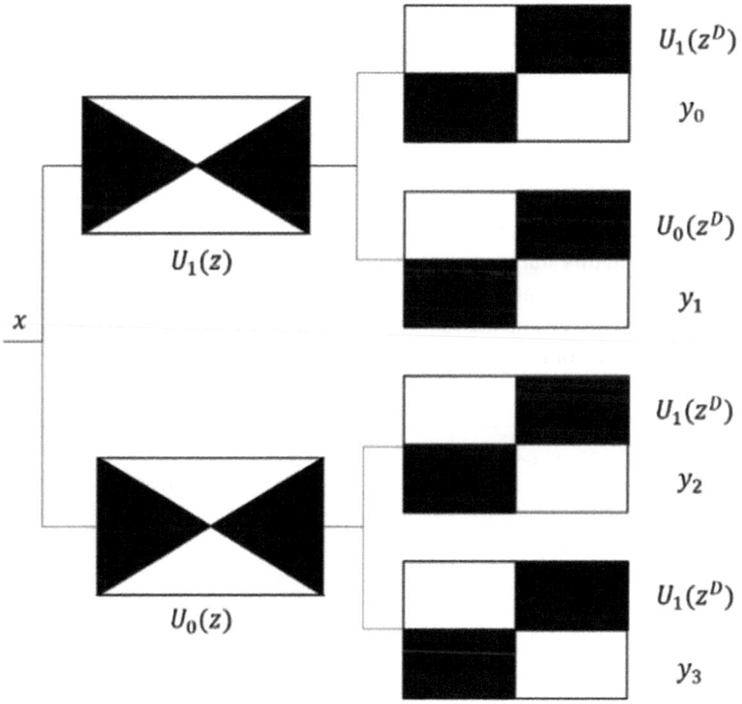

Fig. 6.12 NSDFB diagram with four channels

6.2.3 Image Fusion Algorithm Based on NSCT

After the above analysis, the NSCT algorithm is a fast transform with translation invariance, multi-directionality, and multi-scale capabilities. It possesses neighbor sampling and local positioning abilities, allowing it to effectively capture high-dimensional information features and geometric characteristics in images. These properties make NSCT highly suitable for geophysical image fusion. This case study applies an NSCT-based image fusion algorithm, which consists of three key steps: NSCT decomposition, fusion process, and NSCT inverse transformation (Fig. 6.13).

NSCT decomposition: The Non-Subsampled Pyramid Filter Bank (NSPFB) is used to decompose the source image into low-frequency and high-frequency sub-band images at multiple decomposition levels. Directional decomposition: The Non-Subsampled Directional Filter Bank (NSDFB) then performs directional decomposition on the high-frequency sub-bands obtained from NSPFB, extracting multi-directional features and enhancing directional information accuracy. Fusion and reconstruction: Based on detection requirements and image characteristics, specific fusion rules are designed separately for low-frequency and high-frequency sub-band images. Finally, an appropriate reconstruction tool is selected to apply the NSCT inverse transformation, producing the final fused geophysical image.

In the fusion process, in addition to considering the number of multi-scale decomposition layers and reconstruction tools, the fusion rules play a crucial role in determining the quality of the fused image. In practical applications, fusion rules must be carefully designed based on data characteristics and exploration objectives, making them the most influential parameter in fusion performance.

From the perspective of geological interpretation, the final fused model should preserve all useful information from different geophysical datasets. The fusion strategy adopted in this study ensures that no artifacts are introduced, prevents degradation of image accuracy, and avoids the generation of misleading or erroneous information. The fused image and the fusion process must be reliable and robust, ensuring that no information from any geophysical dataset is discarded.

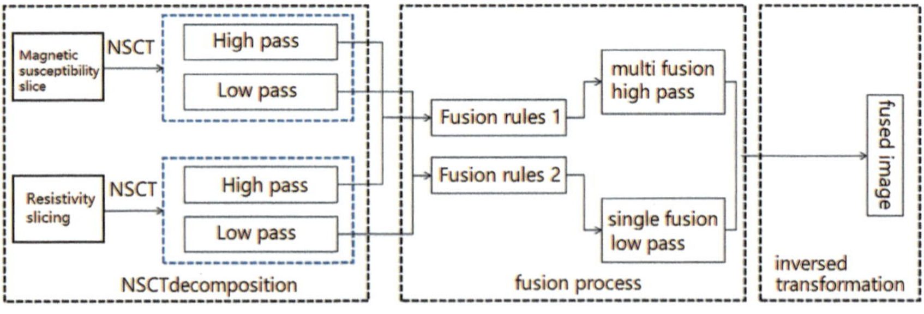

Fig. 6.13 Flowchart of NSCT fusion algorithm

This section provides a detailed explanation of the fusion strategy, focusing on the resistivity and magnetization models from the Iliamna volcanic area and the detection target of low resistivity and low magnetism.

(1) Fusion rules of low-frequency sub-band images.

The primary goal of the fusion process in this study is to integrate resistivity and magnetization models into a single unified information model, which can then be used to identify the location of altered rock bodies characterized by low resistivity and low magnetism.

After NSCT decomposition, the original source images are transformed into low-frequency sub-band images and multiple high-frequency sub-band images. The low-frequency sub-band images primarily represent the structural components of the geophysical images, reflecting the average characteristics of subsurface geological structures. These sub-band images contain spectral information and most of the energy information from the original images.

Given the significance of these low-frequency components in maintaining geophysical structure integrity, this study applies the weighted average fusion rule to merge the low-frequency sub-band images, as described in Eq. (6.3). This fusion rule is designed to effectively combine the average characteristics of both source images, ensuring that both resistivity and magnetization information contribute equally to the final fused model.

$$L(x, y) = \alpha * L_{Mag}(x, y) + \beta * L_{\mathrm{Res}}(x, y), \alpha + \beta = 1 \tag{6.3}$$

where, α, β is the positive weighting coefficient, $L(x, y)$ is the approximate component of the final model. $L_{Mag}(x, y)$ and $L_{Res}(x, y)$ are the approximate components of the first and second models respectively. The weighting coefficients α and β determine the proportion of information occupied by different models. In order to extract the best information from the data, these parameters need to be analyzed to determine the appropriate coefficients. Considering the accuracy of the resistivity and magnetization inversion models, and that this case has conducted a large amount of fusion analysis, it is believed that setting these two parameters to 0.25 and 0.75 respectively will yield the most favorable results.

(2) Fusion rules of high-frequency sub-band images.

The high-frequency sub-band images decomposed by NSCT primarily represent detailed information in the source image, such as edges, lines, and region boundaries. These high-frequency components play a crucial role in geophysical interpretation, as the existence, shape, occurrence, and spatial position of underground target bodies depend on detailed structural features. Therefore, the key objective of high-frequency sub-band fusion is to enhance the detailed features present in each source image.

In this study, the detection target in the Iliamna volcanic area consists of low-resistivity, low-magnetism altered rocks. Thus, the fusion strategy for high-frequency sub-band images prioritizes enhancing the feature representation of low-resistivity and low-magnetism regions in the source images.

To achieve this, Phase Congruency (PC) is utilized as a dimensionless metric to quantify the importance of image features. In high-frequency sub-band images, PC values correspond to the sharpness of geological targets, meaning that a higher PC value indicates a more pronounced edge or boundary in the image.

Additionally, to address the unique characteristics of geophysical anomaly bodies, this study incorporates a Local Sharpness Change Measure (LSCM) to capture local contrast variations in the image. The LSCM helps highlight regions with abrupt changes in resistivity or magnetization, which are critical for delineating geological structures.

However, PC and LSCM alone are not sufficient to fully describe local brightness variations in a two-dimensional geophysical image. To compensate for this, the study introduces Local Energy (LE) as an additional metric to represent localized brightness intensity in the high-frequency sub-band images.

Based on the above analysis, to enhance the information content carried by high-frequency sub-band images, this study introduces a New Metric Parameter (NMP) as shown in Eq. (6.4). The NMP is a composite metric that integrates:

Phase Congruency (PC): Describes the sharpness of geological targets in the image. Local Sharpness Change Measure (LSCM): Captures local contrast variations within the high-frequency sub-band images. Local Energy (LE): Represents local brightness information, ensuring that all critical features are retained.

By combining these three metrics, the NMP enhances the visibility of low-resistivity and low-magnetism anomalies, making it an effective fusion strategy for high-frequency sub-band images in geophysical data interpretation.

$$\mathrm{NMP}(x, y) = (PC(x, y))^{\alpha_1} \cdot (LSCM\,(x, y))^{\beta_1} \cdot (LE(x, y))^{\gamma_1} \tag{6.4}$$

where, $\alpha_1, \beta_1, \gamma_1$ is used to adjust the parameters of PC, LSCM, and LE in NAM. In this section, the parameters $\alpha_1, \beta_1, \gamma_1$ are set to 1, 2, and 2 respectively. For the specific content of the NMP formula, please refer to the next section.

After obtaining the fusion coefficients of the high-frequency sub-band image and the low-frequency sub-band image, the fused image can be obtained through NSCT inverse transformation. In order to make the proposed fusion strategy have better performance, it is necessary to analyze and set parameters. The decomposition layers, the choice of filters, the weighting coefficients of the fusion strategy, etc. are all parameters that must be precisely defined. Generally, by increasing the number of decomposition layers, the role of detail components will increase in the final fusion model, but the contribution of structural components to the final fusion model is relatively small. At present, there is no good theory to introduce how to choose the decomposition level of the model for the fusion of geophysical data, so a large number of experiments need to be verified.

(3) NMP calculation details

$$\mathrm{NMP}(x, y) = (PC(x, y))^{\alpha_1} \cdot (LSCM\,(x, y))^{\beta_1} \cdot (LE(x, y))^{\gamma_1} \tag{6.5}$$

In formula (6.5), PC is a dimensionless method, which is used to describe the importance of image features. In the high-frequency sub-band image, the PC value corresponds to the sharpness of the target in the image. Because the image can be regarded as a two-dimensional signal, the PC value at (x, y) can be calculated according to formula (6.6):

$$\mathrm{PC}(x, y) = \frac{\sum_k E_{\theta_k}(x, y)}{\varepsilon + \sum_n \sum_k A_{n,\theta_k}(x, y)} \tag{6.6}$$

Here θ_k represents the direction angle at k, A_{n,θ_k} represents the nth Fourier component and direction angle θ_k. And ε is the normal number to remove the DC component of the image signal. In this fusion framework, the parameter ε is set to 0.001. $E_{\theta_k}(x, y)$ can be calculated by Eq. (6.7).

$$E_{\theta_k}(x, y) = \sqrt{F_{\theta_k}^2(x, y) + H_{\theta_k}^2(x, y)}$$
$$\begin{cases} F_{\theta_k}(x, y) = \sum\limits_n e_{n,\theta_k}(x, y) \\ H_{\theta_k}(x, y) = \sum\limits_n o_{n,\theta_k}(x, y) \end{cases} \tag{6.7}$$

where, $e_{n,\theta_k}(x, y)$, $o_{n,\theta_k}(x, y)$ is the convolution value of the input image at (x, y), which can be calculated by the following formula:

$$\left[e_{n,\theta_k}(x, y), o_{n,\theta_k}(x, y)\right] = \left[I(x, y) * M_n^e, I(x, y) * M_n^0\right] \tag{6.8}$$

where, $I(x, y)$ is the pixel value at (x, y), M_n^e, M_n^0 is the two-dimensional log-Gabor symmetric filter at the nth level.

However, PC is a contrast invariant and does not reflect local contrast changes. To make up for the shortcomings of PC, sharpness change (SCM) is given in Eq. (6.9).

$$\mathrm{SCM}(x, y) = \sum_{(x_0, y_0)\in\Omega_0} (I(x, y) - I(x_0, y_0)) \tag{6.9}$$

where,

Ω_0 represents the local area of size 3×3 input at (x, y). (x_0, y_0) represents a pixel point in the local area at Ω_0. At the same time, in order to calculate the neighborhood contrast of (x, y), the local sharpness change measure (LSCM) is introduced, see (6.10):

$$LSCM(x, y) = \sum_{a=-M}^{M} \sum_{b=-N}^{N} SCM(x + a, y + b) \tag{6.10}$$

The formula (2M+ 1) × (2N + 1) represents the size of the neighborhood.

Since LSCM and PC cannot fully reflect local brightness information, local energy (LE) is introduced, which can be calculated by Eq. (6.11).

$$LE(x, y) = \sum_{a=-M}^{M} \sum_{b=-N}^{N} (I(x + a, y + b))^2 \tag{6.11}$$

After obtaining NAM, the fused high-frequency sub-band image can be calculated using the rule proposed in formula (6.12).

$$H_F(x, y) = \begin{cases} H_A(x, y), \ if \ Lamp_A(x, y) = 1 \\ H_A(x, y), \ otherwise \end{cases} \tag{6.12}$$

where, $H_F(x, y)$, $H_A(x, y)$ and $H_B(x, y)$ are respectively the fused image and the source images I_A and I_B of the high-frequency sub-band images. $Lamp_i(x, y)$ represents the decision map of the high-frequency sub-band image fusion, which can be calculated by formula (6.13).

$$Lamp_i(x, y) = \begin{cases} 1, \ if \ \left[S_i(x, y)\right] > \frac{\tilde{M} \times \tilde{N}}{2} \\ 0, \ otherwise \end{cases} \tag{6.13}$$

In formula (6.13), Ω_1 represents a sliding window of size $\tilde{M} \times \tilde{N}$ centered on (x, y), K is the number of source images.

6.2.4 *Fusion Results of Actual Data*

(1) Data Source.

Iliamna Volcano is an andesitic stratovolcano located in the southwest of Cook Inlet, Alaska, USA. Hydrothermal alteration weakens the stability of the volcano, leading to frequent landslides. Therefore, understanding the three-dimensional distribution of altered rocks is crucial for assessing the volcano's slope stability. Figure 6.14 presents the geological map of the study area.

The volcano's surface is extensively covered with ice and snow, making it difficult to directly identify alteration zones through conventional field surveys. However, intense hydrothermal alteration significantly reduces the resistivity and magnetization of volcanic rocks. As a result, geophysical exploration methods can be used to detect subsurface alteration zones that remain hidden beneath the surface.

In June 2012, various geophysical datasets were collected in the Iliamna Volcano area (Table 6.1), with the most critical being airborne electromagnetic (AEM) data and airborne magnetic (AM) data. Peterson et al. detailed the preprocessing procedures for these datasets and conducted separate inversions to obtain resistivity and magnetization models.

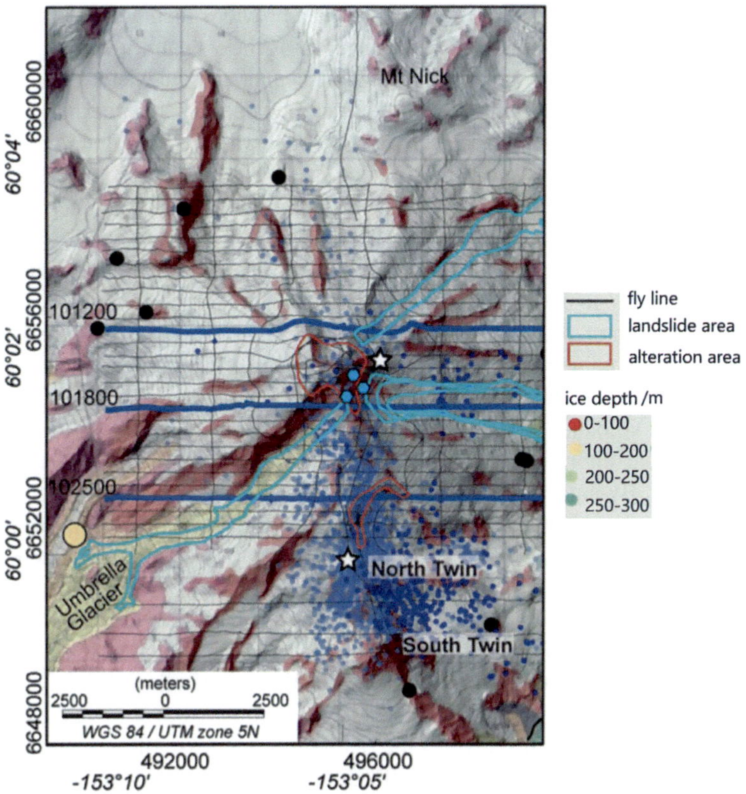

Fig. 6.14 Geological map of the Iliamna Volcano study area

Table 6.1 Data of the Iliamna Volcano area

Data type	Function
Airborne electromagnetic data	Inversion modeling, determining shallow electrical properties and horizontal stratification
Airborne magnetic data	Inversion modeling, determining lateral boundaries and larger sampling depths of magnetization
Magnetization data	Detecting the magnetization of altered samples
Synthetic aperture radar data	Determining the thickness of the ice layer in a certain area
Aerial photos	Observing the existence of altered rocks, new rocks and ice and snow near the peak
Satellite images	Drawing the polygonal shape of new rocks

The resistivity model derived from AEM data provides insights into the shallow electrical properties and horizontal stratification of the volcano, resolving features down to ~450 m. However, this depth limitation is inherent to the detection capability of AEM methods, making it insufficient for mapping deeper hydrothermal alteration zones. To complement this, a magnetization model was incorporated. The magnetization inversion model is more sensitive to lateral boundaries and enables greater depth penetration, making it more effective in delineating alteration zones within the volcano.

Peterson et al. utilized both resistivity and magnetization models, along with additional datasets, to determine the thickness and extent of ice cover and to map the spatial distribution of altered rocks. This study aims to fuse the resistivity and magnetization models obtained from actual data, generating a comprehensive information model to accurately depict low-resistivity, low-magnetism altered rock formations within Iliamna Volcano.

Hydrothermal alteration will reduce the resistivity and magnetization rate of volcanic rocks. By analyzing the physical changes of rocks, the area where altered rocks are located can be found. Based on the research goal of this section, Iliamna Volcano is approximately divided into three materials: ice, new rock and altered rock. Table 6.2 is the physical parameters of the three materials.

The magnetization of ice and metamorphic rock can both be considered negligible. As a result, relying solely on the magnetization inversion model makes it challenging to identify hidden alteration zones on the volcanic surface. To accurately simulate the magnetization of the volcano's surface, it is essential to first remove the ice layer.

Ice is typically highly resistive (>300,000 $\Omega{\cdot}m$), whereas volcanic metamorphic rock exhibits low resistivity. This significant contrast in electrical properties allows for the differentiation between ice and metamorphic rock based on resistivity values. However, in dry environments, the resistivity of fresh volcanic rock may be comparable to or even greater than that of ice, making it impossible to completely distinguish between the two using resistivity inversion alone.

By analyzing Table 6.2, it is evident that the three media (ice, metamorphic rock, and altered rock) exhibit distinct physical properties. This suggests that integrating resistivity and magnetization data through image fusion is a feasible approach for identifying low-resistivity, low-magnetization metamorphic rocks.

After preprocessing the data, resistivity and magnetization cross-sections at the same depth are obtained. Considering the accuracy limitations of the resistivity model, the depth range for fusion in this study is 0–425 m (details of the data preprocessing process are provided in the next section). These cross-sections serve as input images for the image fusion process, where each input image is referred to as a

Table 6.2 Physical parameters of three media

	Ice	Metamorphic rock	New rock
Resistivity (Res)	High	Low	High
Magnetization (Mag)	0	Low (→0)	High

source image. Figure 6.15 presents a cross-section at a depth of 25 m, illustrating the distribution of resistivity and magnetization properties in the study area.

(2) Data preprocessing.

Before image fusion, the data preprocessing process consists of the following steps:

1. Generate digital cross-sections from the inverted resistivity model and magnetization model.
2. Standardize data resolution and coordinate systems by applying data interpolation, sorting, and coordinate transformation. This step ensures that all data points, regardless of their type or initial geometric distribution, are converted into grid values with the same resolution and aligned within a common coordinate system, strictly registered according to grid positions.
3. Convert standardized grid values into color images, which are then grayscaled to prepare for further processing.
4. Normalize the grayscaled images, ensuring that the normalized cross-sections (source images) serve as a consistent input for image fusion.

To maintain consistency with the detection accuracy of airborne electromagnetic exploration, the inverted model is segmented into nine depth cross-sections at 25, 75, 125, 175, 225, 275, 325, 375, and 425 m below the surface. Each cross-section has a pixel size of 660 × 630. Following segmentation, the cross-sections undergo grayscale conversion and normalization. The purpose of this step is to digitize the image information and eliminate the inherent physical differences between the two datasets, thereby enabling a seamless fusion of the resistivity and magnetization models. Figure 6.16 presents an example of the source images at 25 m depth, showcasing the processed magnetization and resistivity cross-sections used for further fusion analysis.

(3) NSCT-based fusion process.

After completing the data preprocessing, the source images undergo fusion using the NSCT method. Figure 6.17 illustrates the processing workflow, which consists of three main steps: decomposition, fusion, and inverse transformation.

Currently, no established theoretical framework dictates the optimal number of decomposition layers for NSCT. In this study, four decomposition layers were selected as the optimal configuration, determined through extensive experimentation. The NSCT framework consists of two key components: the Non-Subsampled Pyramid Filter Bank (NSPFB) and the Non-Subsampled Directional Filter Bank (NSDFB).

A decomposition filter, denoted as "vk", is applied to the cross-sections, producing low-frequency sub-band images and high-frequency sub-band images at each decomposition level.

The NSDFB further decomposes the high-frequency sub-band images from the NSPFB, extracting directional details using the "pyrexc" directional filter.

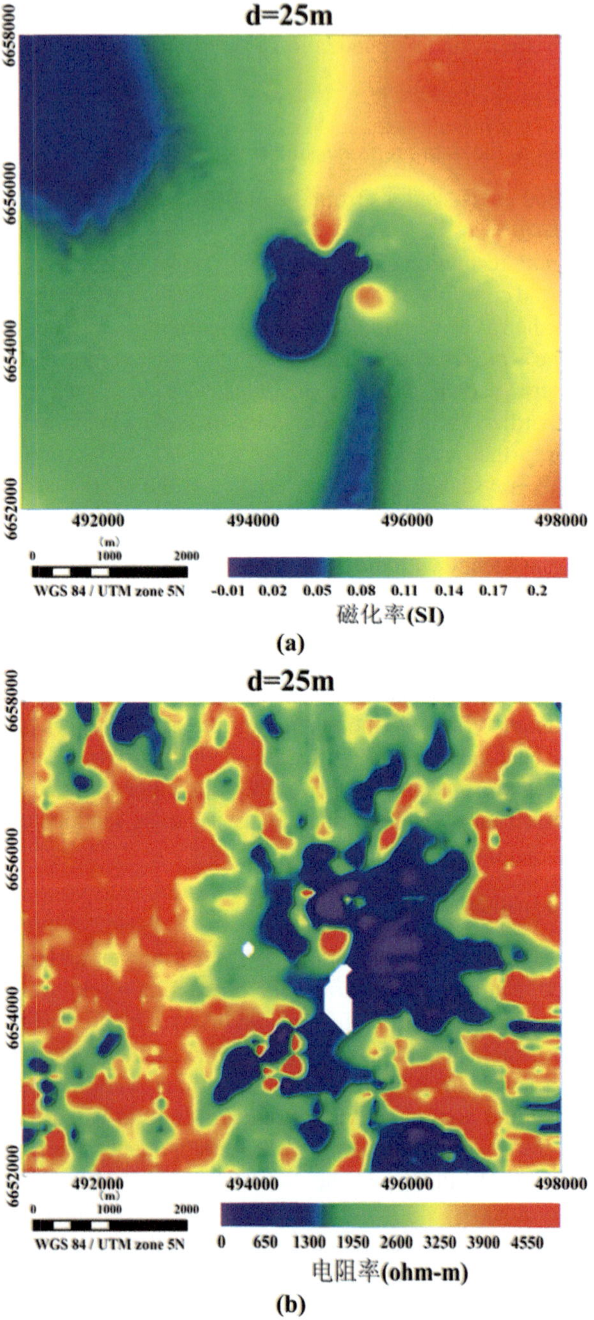

Fig. 6.15 **a** Magnetization cross-section (d = 25 m); **b** Resistivity cross-section (d = 25 m)

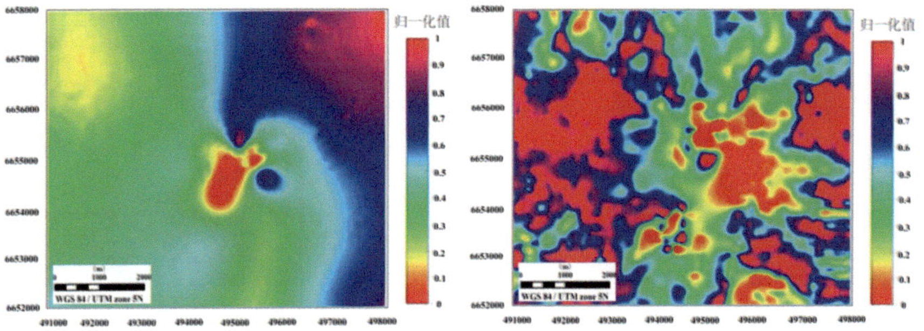

Fig. 6.16 **a** Normalized magnetization cross-section (25 m), **b** normalized resistivity cross-section (25 m)

Since the source images are decomposed into four layers, the process generates five sub-images of the same size as the original image, comprising four high-frequency images and one low-frequency image (Fig. 6.17).

The NSCT algorithm is applied to perform multi-scale decomposition on the cross-sectional images, followed by fusion of high- and low-frequency components using predefined fusion rules based on weighted averaging and New Metric Parameter (NMP).

For low-frequency sub-band images, the fusion rule weighting coefficients are set to 0.25 and 0.75. For high-frequency sub-band images, the weighting coefficients are set to 1, 2, and 2, respectively. Using the corresponding formulas, the fused low-frequency and high-frequency sub-band images are obtained.

After fusion, the NSCT inverse transformation is applied to reconstruct the final fused image from the fused low- and high-frequency sub-band images. The decomposition filters and directional filters used in this step remain consistent with those in the decomposition process. The final fused image, obtained through NSCT inverse transformation, provides an enhanced visualization of the target features, integrating resistivity and magnetization characteristics more effectively (Fig. 6.17).

For fusion of resistivity and magnetization maps at different depths, the same processing method is applied uniformly. This results in a 3D fusion model map (Fig. 6.18), enabling an in-depth analysis of altered rock depth and distribution. The detailed interpretation of these results is presented in Sect. 6.2.5, which focuses on the actual data fusion results.

6.2.5 *Actual Data Fusion Results*

The study in this section focuses on the identification of low-resistance, low-magnetism altered rock. According to the judgment coefficient in Fig. 6.7, different geological materials are classified as follows: when the judgment coefficient <0.18, it corresponds to the altered rock zone; when $0.18 <$ judgment coefficient < 0.88,

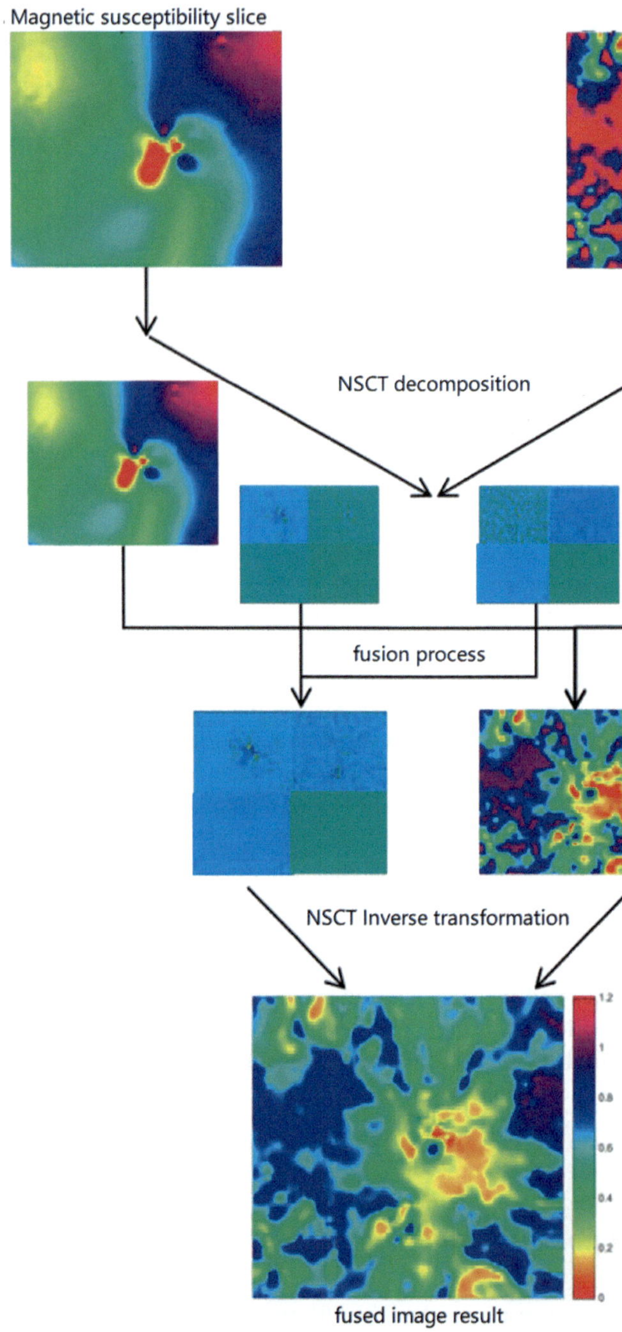

Fig. 6.17 Principle flowchart of NSCT method fusion

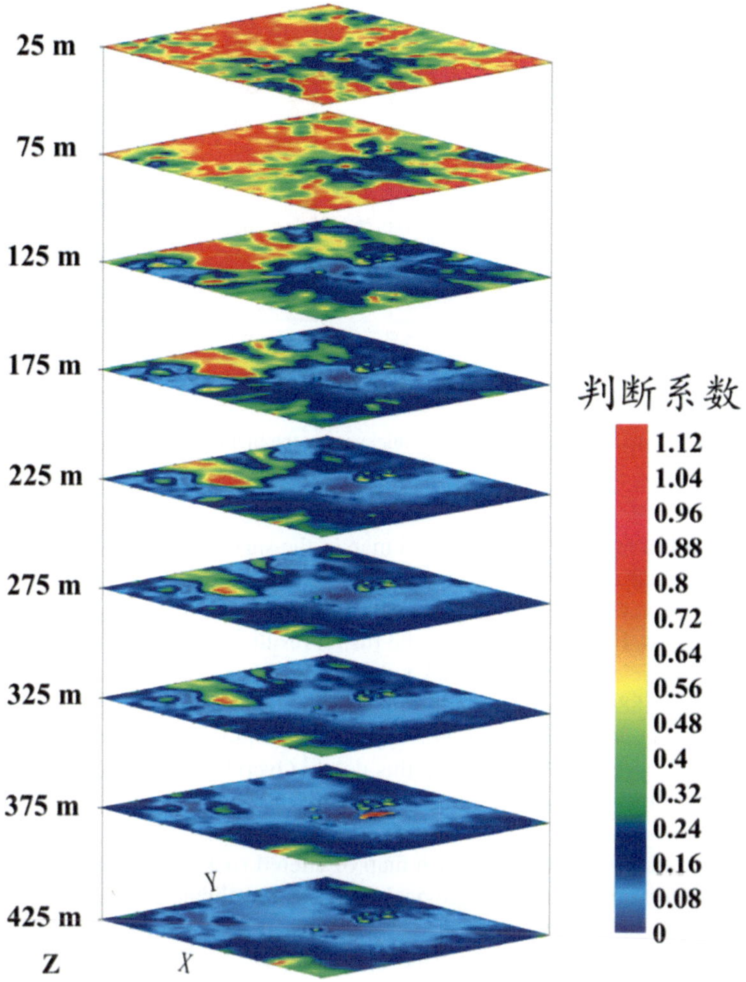

Fig. 6.18 Three-dimensional fusion model map obtained based on NSCT

it represents the ice layer; and when the judgment coefficient >0.88, it belongs to the new rock zone. The fusion results obtained in this study are compared with the comprehensive analysis results reported in the literature.

According to the literature, the maximum thickness of the ice layer is located in the northwest of Iliamna Volcano's peak, reaching 390 m, with a deviation of approximately ± 30 m (Fig. 6.19a). The three-dimensional fusion model obtained in this study (Fig. 6.18) provides a similar estimate. As observed in Fig. 6.18, when the judgment coefficient falls within 0.18–0.88, the maximum ice thickness is also concentrated northwest of the volcano's peak, measuring ~375 m, demonstrating a high level of consistency with previous findings.

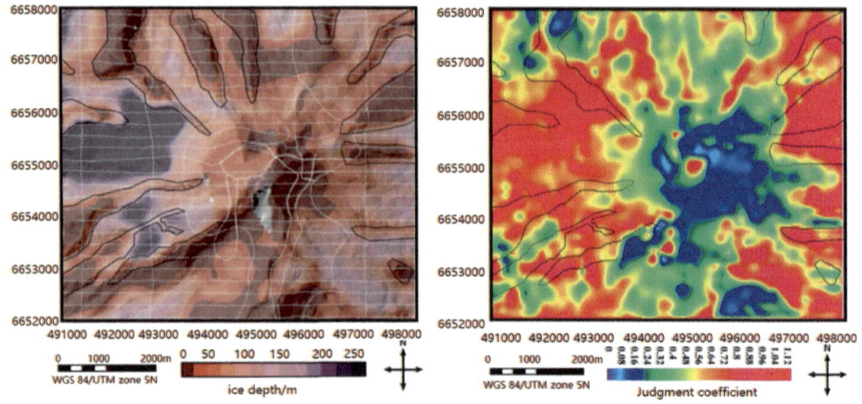

Fig. 6.19 **a** Schematic diagram of ice layer thickness in the literature, **b** area map of new rock at 25 m

Figure 6.19a presents the geological map of the survey area, where the new rock is extensively covered by ice and snow. The black polygon in Fig. 6.19a represents the estimated new rock boundary beneath the ice, delineated using Google Earth satellite images. Figure 6.19b shows the fusion result at a depth of 25 m, where the judgment coefficient within the black polygon area is primarily concentrated within the range of 0.88–1.2, corresponding to the new rock zone in the fusion model. However, some areas within the polygon exhibit a judgment coefficient of 0.48–0.87, indicating residual ice cover at this depth. Overall, the new rock distribution inferred from the fusion model aligns well with the geological interpretation from the literature.

The literature provides an isopach map of altered rock thickness (Fig. 6.20, red line) based on comprehensive geological analysis. To validate this, an isopach thickness map of altered rock was generated in this study and compared with the literature's results. The mapping process consisted of two steps:

1. Within the 425 m depth range, the image threshold segmentation method was applied to identify regions where the judgment coefficient <0.18, corresponding to altered rock.
2. Between 425 and 825 m depth, only the magnetization model was used, generating cross-sectional maps at 50 m intervals. The image threshold segmentation method was again applied to delineate the altered rock zone.

The comparison between the two maps indicates that, while minor discrepancies exist, the general distribution and thickness of altered rock are consistent with the literature's interpretation. It is important to note that the accuracy of information fusion is closely dependent on the quality of the inversion model, underscoring the importance of precise inversion results in achieving high-accuracy fusion.

From the above analyses, the NSCT-based image fusion method in this study successfully differentiates the three geological materials (altered rock, ice, and new

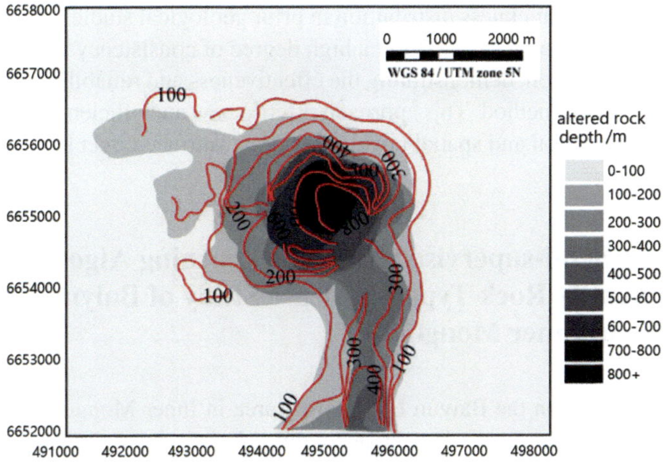

Fig. 6.20 Isopach map of altered rock thickness

rock) and provides a comprehensive interpretation of the resistivity and magnetization models through image fusion. The entire fusion process is completed in just a few seconds, significantly improving computational efficiency. Comparing the fusion results with the literature's comprehensive interpretation, we find a high degree of consistency, demonstrating that the NSCT-based image fusion method enables accurate identification and precise positioning of target bodies while enhancing computational efficiency.

6.2.6 *Case Summary*

In this study, an NSCT-based image fusion method is employed to integrate and analyze multiple geophysical datasets using images as carriers. The NSCT image fusion process consists of three main steps: NSCT decomposition, fusion processing, and NSCT inverse transformation. A crucial aspect of this approach is the selection of an appropriate fusion strategy, which must be flexibly adjusted based on detection requirements and the characteristics of the source images to achieve optimal results. The method is validated using publicly available measurement data from the Iliamna volcano area, covering the entire workflow from data preprocessing and image fusion to result comparison and analysis. By comparing key geological features—ice layer coverage and thickness, new rock locations, and the spatial distribution of altered rocks—with prior geological data, the study demonstrates that the maximum ice thickness in the detection area is found northwest of the volcano peak, measuring ~375 m, which aligns with previous findings. The new rock locations identified in the fusion model correspond closely to those determined by satellite imagery. The three-dimensional distribution of altered rocks in the fusion results matches the overall contour and thickness distribution in prior geological studies. These findings confirm that the fusion results exhibit a high degree of consistency with the existing geological information, demonstrating the effectiveness and reliability of the NSCT-based image fusion method. This approach offers a new and efficient solution for the accurate identification and spatial positioning of subsurface target bodies.

6.3 Using Semi-supervised Machine Learning Algorithm to Identify Rock Types—A Case Study of Baiyun Ebo Mine in Inner Mongolia

This study focuses on the Baiyun Ebo mining area in Inner Mongolia, which hosts the world's largest rare earth deposit, accounting for nearly one-third of global rare earth resources. Despite extensive geophysical exploration and drilling, the complex morphology of the ore body—characterized by steep dip angles and rapid lateral lithological variations—poses significant challenges for interpretation. The multiple solutions inherent in geophysical imaging, combined with the limited availability of deep drilling data, have left the distribution of ore-bearing dolomite in the deeper sections of the mining area unclear, restricting further resource development.

To address these challenges, this research applies a machine learning-based approach that integrates gravity, magnetic, and seismic inversion imaging results with the limited available drilling data. Specifically, a semi-supervised dual-constraint variational Gaussian mixture model algorithm is utilized. This algorithm incorporates geological constraints via a penalty function and distance constraints from the KMeanspp clustering method, ensuring high precision and recall rates even with scarce and unbalanced labeled data. Moreover, it has the capability to identify new lithologies, a crucial advantage for deep mineral exploration where labeled training data are often lacking.

Using this method, the spatial distribution of four primary lithologies was successfully delineated along a representative section between the main mine and the east mine in the Baiyun Ebo mining area. The study integrates four types of geophysical inversion results with lithological data from six drill holes, enabling an accurate classification of subsurface formations down to a depth of ~2 km. The results indicate that the ore-bearing dolomite is predominantly distributed between depths of 200 and 1600 m, exhibiting a sloping, plate-like structure that deepens toward the south and becomes shallower in the north. These predictions align closely with actual drilling observations, verifying the effectiveness and practicality of the proposed algorithm.

This research highlights the significant application potential of machine learning-based multi-source geophysical data fusion in mineral exploration, particularly in geologically complex mining regions. By providing a novel technical approach and theoretical framework for deep prospecting in the Baiyun Ebo mining area, this study is expected to facilitate further advancements in rare earth resource exploration and development.

6.3.1 Case Background and Introduction

In the Baiyun Ebo mining area, understanding the distribution of deep ore-bearing dolomite has always been a challenge. Despite extensive geophysical exploration and drilling, the limitations of geophysical imaging and restrictions on drilling locations and depths have made it difficult to fully understand the deep dolomite distribution. To overcome this challenge, this study developed a new machine learning technology capable of accurately identifying the deep underground distribution of dolomite by integrating various geophysical exploration results, including gravity, magnetism, and seismic imaging, with limited drilling data. This technology is particularly valuable in areas where it is difficult to obtain large amounts of labeled data, offering high precision and recall rates even with scarce and unbalanced data, and providing the ability to identify new lithologies.

The research results of this study align closely with actual drilling observations of the dolomite distribution, marking a significant breakthrough. This new approach provides both technical and theoretical support for deep mineral exploration in the Baiyun Ebo mining area, demonstrating the vast potential of machine learning in integrating multi-source geophysical data and exploring complex mining areas. These findings are expected to contribute to the continued advancement of mineral exploration in the region.

Rare earths are known as "industrial vitamins". Due to their excellent optical, electrical, and magnetic physical properties, they are widely used in cutting-edge scientific and technological fields such as national defense, aerospace, low-carbon energy, and special materials. They have a "pivotal" strategic significance for national security, national economy, and high-tech development. China is a major country in rare earth resources, with a full range of types of rare earth deposits, huge reserves, and excellent endowment. The Bayan Obo mine in Inner Mongolia is located on the northern edge of the North China Craton, (Fig. 6.21), it is the largest rare earth deposit in the world, with its rare earth resources accounting for nearly 1/3 of the world's total rare earth resources, and it has important economic and scientific research value.

Researchers have conducted geological, mineralogical, ore-forming chronology, and ore genesis studies on the Bayan Obo deposit. More and more research results show that the dolomite that hosts the Bayan Obo deposit is a carbonate of mantle-derived magmatic origin, with rare earths mainly occurring in the dolomite, and iron ore bodies mainly distributed in the dolomite within the contact zone between the dolomite and the slate. The deep part of the mining area has good prospects for ore prospecting.

A series of gravity-magnetic-electric-seismic comprehensive geophysical exploration work has been carried out in the Bayan Obo mining area, revealing the deep geophysical field response characteristics of the medium in the mining area. For example, gravity and magnetic anomalies show that the main, east, and west mining areas as a whole appear as high magnetic and high gravity anomaly areas, which basically coincide with the iron ore enrichment areas that have been identified on the surface of the mining area; controlled source audio-frequency magnetotelluric

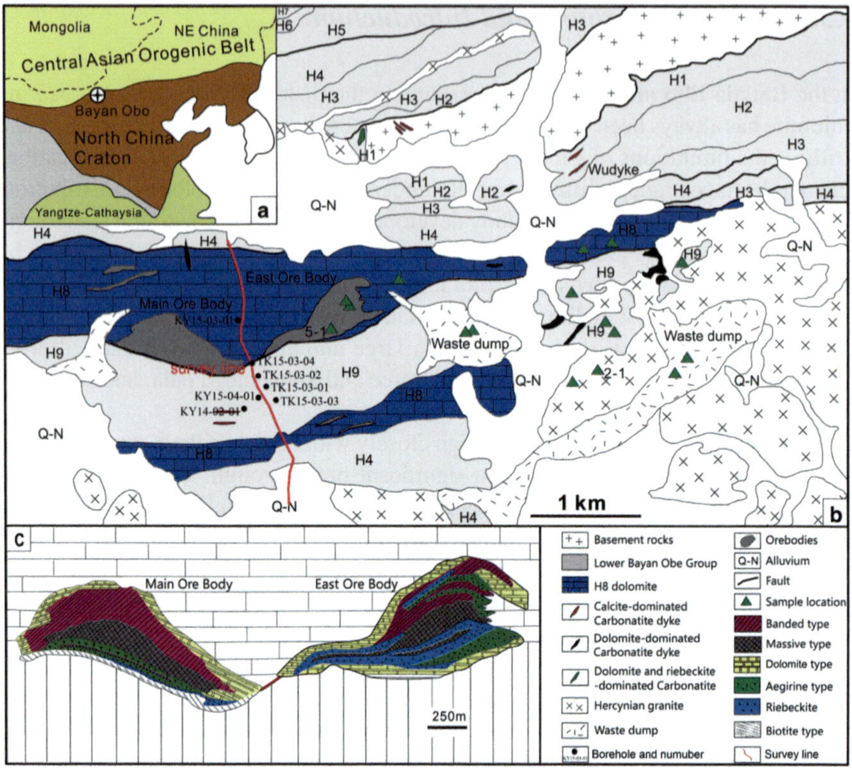

Fig. 6.21 **a** Geological overview of the Baiyun Ebo deposit. **b** Structural location map of the Bayan Obo deposit (located in the North China Craton). **c** Distribution map of different types of rare earth ores in the main ore body and the east ore body

exploration reveals a low resistivity anomaly body extending to the deep part (2 km deep) between the main mine and the east mine, which becomes shallower to the east and west, and is speculated to be closely related to the spatial distribution of carbonates; using short-period dense array continuous waveform observation, the shear wave velocity structure of the entire mining area 2.5 km shallow obtained by the background noise imaging method shows that there is a change from a relatively high-speed area to a relatively low-speed area under the main, east, and west mines, which may imply a significant decrease in the content of the ore-bearing dolomite at this depth or a fundamental change in the rock type.

However, due to the complex geological conditions and the limited horizontal distribution and drilling depth coverage of existing drilling holes, and the complexity of the detection environment such as ongoing construction activities, high-voltage lines and other supporting equipment, and complex terrain and landforms in the mining area, there is serious uncertainty about the reliability, accuracy, and interpretation of the results of single geophysical exploration. Therefore, it severely restricts the current research on the controlling structures of Bayan Obo and the prediction of deep resources.

The multi-source geoscience information fusion method based on machine learning can extract useful information from multi-dimensional and multi-scale geoscience data, thereby improving the interpretation accuracy of the results. Such algorithms have achieved significant results in fields such as mineral prediction, geological mapping, geological body delineation and classification, geological structure analysis. Recently, Xue et al. proposed a model fusion algorithm based on transfer learning (using the ResNet50 network trained by the ImageNet image dataset as the transfer model) and local energy weighted average fusion rule (TL-LA), and carried out fusion research on the inversion results of gravity, magnetic, electric, and seismic of the representative section between the main and east mines in the Bayan Obo mining area. They obtained the possible spatial distribution of the ore-bearing dolomite with generally high resistance, high magnetism, high density, and high speed characteristics. The results are in good agreement with the known drilling data, and the accuracy of the detection results of any single geophysical method has been improved to a certain extent. However, this algorithm did not use the rock type observation data obtained from drilling as label constraints in the fusion process, which weakened the accuracy of this algorithm in rock type identification and positioning and its ability to depict rock type boundaries.

Given the scarcity of deep drilling data in the Baiyun Ebo mining area, the amount of lithology distribution label data obtained in space is far less than the unlabeled data, and various lithology labels are obviously unbalanced. In this case, a semi-supervised dual-constraint variational Gaussian mixture model algorithm (TCS-VGMM) is applied. This algorithm combines geological constraints based on penalty functions and KMeanspp distance constraints, which can effectively solve the problems of prediction accuracy brought about by the scarcity and imbalance of label data, and also has the ability to identify new lithologies. This algorithm is used to fuse the multi-source geophysical imaging results of the cross-section between the main and east mines of Baiyun Ebo and the small amount of lithology distribution data obtained from deep drilling. It successfully identified the spatial distribution of four lithologies in the cross-section and compared it with the latest scientific deep drilling data. The results showed a high degree of consistency, verifying the effectiveness of this fusion method and its application potential in complex mining area mineral exploration.

6.3.2 Case Data Support

Although numerous studies have applied geophysical detection technology across the entire Baiyun Ebo mining area, this case specifically focuses on a cross-section between the main and east mining areas, as illustrated by the red line in Fig. 6.21b. This cross-section is particularly significant, as it includes six completed drill holes, two of which reach a depth of 2 km. These drill holes provide crucial data labels and serve as the validation basis for the algorithm used in this case. Furthermore, this area is highly important for studying and developing the Baiyun Ebo deposit, making it a key target for detecting the deep ore body morphology and potential resources.

In this study, inversion physical results from various geophysical methods, including transient electromagnetic, gravity, aeromagnetic, and background noise imaging, were collected, along with lithology distribution data from the six drill holes along the cross-section shown in Fig. 6.21b. The data set includes resistivity, density, magnetization rate, and shear wave velocity cross-sections (Fig. 6.22a–d), as well as the lithology distribution from the six drill holes (Fig. 6.22e). A comprehensive dataset containing 4747 data points was constructed, with intervals of 100 m in horizontal depth and 20 m in vertical depth. This dataset encompasses the geophysical structures from gravity, magnetism, electricity, and seismic methods along this cross-section, along with lithology distribution data obtained from the drill holes.

Among the total dataset, only 401 data points have lithology labels, making up 8.4% of the total, while 4346 data points lack lithology labels (Fig. 6.23). This results in a significant imbalance ratio between labeled and unlabeled data, approximately 1:11. Additionally, the labeled data encompasses four types of rocks: slate, dolomite, mineralized dolomite, and sandstone, with proportions of 6:2:5:1. This further highlights the imbalance between categories, with some rock types being much more prevalent than others.

6.3.3 Semi-supervised Machine Learning Method

According to the amount of labeled data used in training, machine learning can be categorized into supervised learning, unsupervised learning, and semi-supervised learning. Supervised learning is the preferred method when there is a sufficient amount of labeled data, as it uses the trained model to classify or predict the test set. Conversely, when labeled data is scarce, unsupervised learning is applied, which involves clustering by analyzing the inherent structure and relationships within the data. When a small amount of labeled data exists alongside a large volume of unlabeled data, semi-supervised learning is typically used. This approach guides the learning process of unlabeled data using the limited labeled data available. Given the characteristics of multi-source geophysical observations and lithology labels in the Baiyun Ebo mining area (Figs. 6.2 and 6.3, and Table 6.1), this case applies a semi-supervised dual-constraint variational Gaussian mixture model algorithm (TCS-VGMM).

(1) Unsupervised Variational Gaussian Mixture Model (VGMM).

The traditional variational Gaussian mixture model (GMM) assumes that the data is generated from multiple Gaussian distributions, each representing a potential category or cluster within the data space. This mixture model is capable of approximating complex, nonlinear, and non-convex data distributions with high accuracy (Fig. 6.24a). The probability density function of a GMM is a weighted sum of each

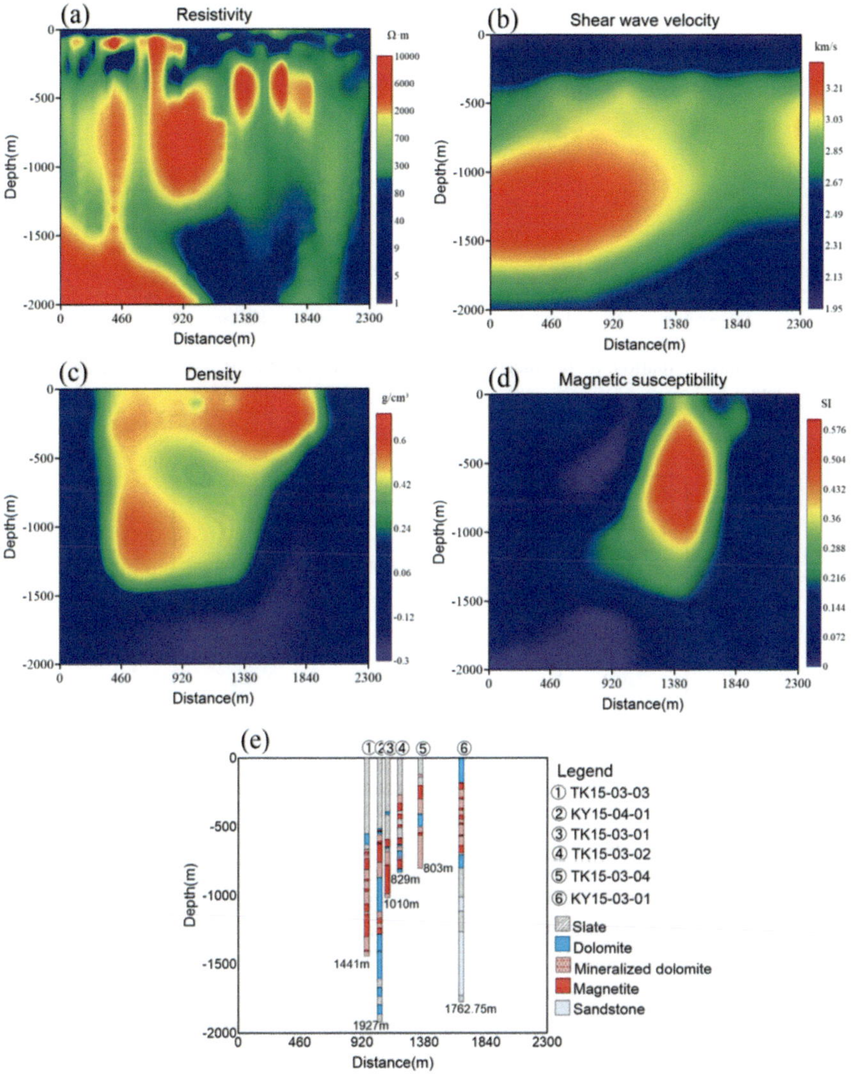

Fig. 6.22 Four different geophysical parameter inversion cross-sections within a depth of 2000 m. **a** Resistivity; **b** velocity; **c** density; **d** magnetization rate; **e** drilling data

Gaussian distribution, as described by Eq. (6.14), where each distribution contributes to a potential data cluster.

$$p(x) = \sum_{k=1}^{K} \pi_k \mathcal{N}(x|\mu_k, \Lambda_k) \tag{6.14}$$

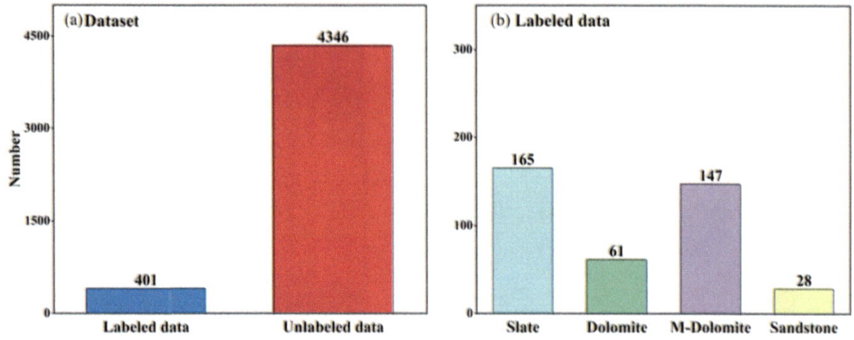

Fig. 6.23 Dataset situation: **a** classification of labeled data and unlabeled data; **b** classification of labeled data

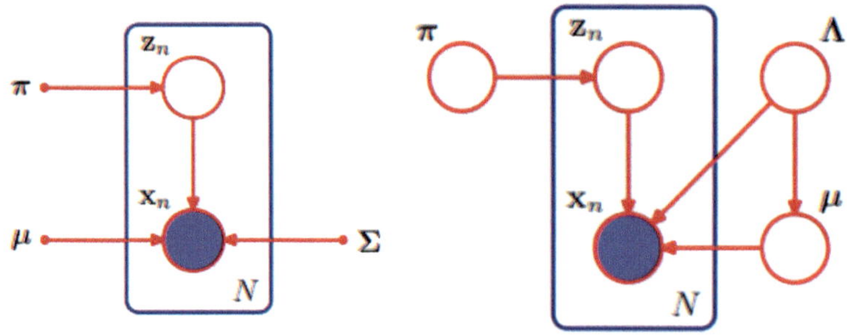

Fig. 6.24 Directed acyclic graph: **a** Gaussian mixture model, **b** Bayesian mixture model representing the Gaussian model

where, x is the observed sample, π_k is the mixing coefficient of the k Gaussian component, μ_k and Λ_k are the mean and variance of the k Gaussian component, $\mathcal{N}(x|\mu_k, \Lambda_k)$ It is a Gaussian distribution with mean μ_k and matrix Λ_k. To simplify calculations, matrix Λ_k is a precision matrix rather than a covariance matrix.

The parameter estimation of GMM usually uses the Expectation–Maximization (EM) algorithm, which is an iterative algorithm that mainly includes two steps: the Expectation step (E step) and the Maximization step (M step). In the E step, the probability of each data point belonging to each Gaussian component is calculated. These probabilities are called "responsibilities", represented by. Specifically, given the current estimate of the Gaussian distribution parameters, the posterior probability of each data point being generated by each Gaussian distribution is calculated using Bayes' theorem. In the M step, the model parameters (mean μ_k, covariance Λ_k, and mixture weight π_k of each component) need to be updated to maximize the likelihood probability of the model for the observed data. These two steps are alternated, and each iteration is expected to improve the overall data likelihood until it converges to a local maximum.

The Variational Gaussian Mixture Model (VGMM) is an extension of GMM (Fig. 6.24b), a probabilistic model based on variational inference. It simplifies the model parameters in the form of variational inference to approximate the posterior distribution of complex models, effectively improving the convergence speed of the clustering process. Variational inference involves finding a simplified distribution $q(\mathbf{Z})$ to approximate the true posterior distribution $(Z|x)$. The goal is to find the variational distribution that minimizes the Kullback–Leibler (KL) divergence between $q(\mathbf{Z})$ and the true posterior, which is equivalent to maximizing the Evidence Lower Bound (ELBO), defined as:

$$ELBO = \sum_{\mathbf{Z}} \iiint q(\mathbf{Z}, \boldsymbol{\pi}, \boldsymbol{\mu}, \boldsymbol{\Lambda}) \ln\left\{\frac{p(\mathbf{X}, \mathbf{Z}, \boldsymbol{\pi}, \boldsymbol{\mu}, \boldsymbol{\Lambda})}{q(\mathbf{Z}, \boldsymbol{\pi}, \boldsymbol{\mu}, \boldsymbol{\Lambda})}\right\} d\boldsymbol{\pi} d\boldsymbol{\mu} d\boldsymbol{\Lambda}$$

$$q(\mathbf{Z}) = \sum_{n=1}^{N}\sum_{k=1}^{K} r_{nk}^{z_{nk}} \tag{6.15}$$

$$\gamma_{nk} = E\left[\ln \pi_k\right] + \frac{1}{2}E\left[\ln|\boldsymbol{\Lambda}_k|\right] - \frac{D}{2}\ln(2\pi) - \frac{1}{2}E_{\boldsymbol{\mu}_k, \boldsymbol{\Lambda}_k}\left[(x_n - \boldsymbol{\mu}_k)^T \boldsymbol{\Lambda}_k (x_n - \boldsymbol{\mu}_k)\right] \tag{6.16}$$

$$\pi_k = \frac{1}{N}\sum_{n=1}^{N} r_{nk} \tag{6.17}$$

$$\mu_k = \frac{\sum_{n=1}^{N} r_{nk} x_n}{\sum_{n=1}^{N} r_{nk}}, \ \Lambda_k = \frac{\sum_{n=1}^{N} r_{nk}(x_n - \mu_k)(x_n - \mu_k)^T}{\sum_{n=1}^{N} r_{nk}} \tag{6.18}$$

where N is the number of data points, Z_{nk} is an indicator variable indicating whether data point n belongs to distribution, and r_{nk} is the responsibility of data point n belonging to distribution k, π_k is the mixture coefficient, μ_k and Λ_k are the mean and covariance respectively.

VGMM accurately depicts the category attributes and clustering structure of data points similar to GMM, which is achieved by continuously iterating the variational E step [formula (6.16)] and variational M step [formulas (6.17) and (6.18)] during model training until convergence. The iteration of these steps is monitored by observing the lower bound of the model (i.e., the lower bound of the log-likelihood) to ensure that the value of the variational lower bound does not decrease after each iteration, thereby optimizing the overall performance of the model.

(2) Twin-constraint Semi-supervised Variational Gaussian Mixture Model (TCS-VGMM).

Although the VGMM algorithm provides a mathematical framework and optimization mechanism for data modeling, it is mainly used for unsupervised learning scenarios and cannot directly handle labeled data. In this case, a small amount of lithology label data is provided in the mining area, which only accounts for 8.4% of the total data, and the data ratio of different lithology labels is extremely unbalanced, so the VGMM algorithm cannot be directly used for effective lithology prediction.

Faced with the problem of scarce lithology label data in the mining area and the imbalance of various lithology labels, the VGMM algorithm has been improved. In this case, a twin-constraint semi-supervised learning method (TCS-VGMM) is applied, which includes distance constraints (KMeanspp) and geological constraints (penalty function).

Distance constraint is to assign cluster labels by minimizing the distance between data points and their nearest cluster center through kmeanspp, and to select the initial cluster center to optimize this assignment. The basic principle is: randomly (uniform distribution) select a sample point from the dataset **X** as the first initial cluster center c_i. Then calculate the shortest distance between each sample and the current existing cluster center, use $D(x)$ It indicates; then calculate the probability of each sample point being selected as the next cluster center $P(x)$, and finally select the sample point corresponding to the maximum probability value (or probability distribution) as the next cluster center:

$$P(x) = \frac{D(x)^2}{\sum_{x \in X} D(x)^2} \tag{6.19}$$

Then, the cluster centers and data point cluster assignments are iteratively updated through the standard k-means algorithm until the cluster assignments no longer change. Subsequently, the most likely category of known data points is calculated by comparing the posterior probabilities of each labeled data point on each cluster and selecting the category with the highest probability.

On this basis, geological constraints based on penalty functions are further imposed, using the provided lithology labels as constraint conditions, thereby guiding the KMeanspp algorithm to achieve more appropriate data partitioning. Its basic principle is: for each class, samples with the same label (lithology) can be assigned to the same or different Gaussian components; samples with different labels cannot be assigned to the same Gaussian component. That is to identify all labeled data points with labels different from the current labeled data points. Apply a penalty function to the probability of these data points on the current cluster:

$$r_{nk} = \begin{cases} \alpha \cdot r_{nk}, & if \ n \in r_{label\_ind} \\ r_{nk}, & otherwise \end{cases} \tag{6.20}$$

where, r_{nk} is the responsibility of data point n belonging to distribution k, r_{label_ind} is a set of data point indices, these data points have labels different from the label of the data point with the highest probability of belonging to the k cluster, α is the penalty factor, in this case $\alpha = 10^{-8}$. This is equivalent to avoiding assigning these data points to this cluster. This step is key to implementing the constraint conditions, it directly uses known label information to influence the cluster assignment process.

Finally, normalize the adjusted posterior probability (R) to ensure that the sum of the cluster probabilities of each data point is 1. Ensure that the adjusted R can be correctly interpreted as a probability distribution.

In general, TCS-VGMM analyzes data through the VGMM algorithm, obtains patterns in the data through the dual-constraint semi-supervised (TCS) learning method, enforces dual constraints during the clustering process by modifying the variational E-step learning method in VGMM, and finally makes classification decisions based on the learned information. At this time, the constructed TCS-VGMM has solved the most difficult part of lithology identification, that is, accurate prediction of data under the condition of few labels and unbalanced labels, and has the ability to identify new classes.

6.3.4 TCS-VGMM Algorithm Process

Figure 6.25 shows the complete process of target body identification and positioning using this algorithm, which mainly includes four key steps: data generation and preprocessing, feature dimension reduction, model training and performance evaluation, and lithology classification and prediction.

The dataset $X = [X_1, X_2, \ldots X_n]^T$ is composed of a feature set and labels, and the features and labels correspond one-to-one in spatial position. Given the continuity of geological structures, it is particularly important to maintain the local neighbor information of the data. Therefore, when constructing the dataset matrix, not only the traditional geoscience features are considered, but also the spatial position information X_i (horizontal position coordinates) and D_i (depth position coordinates) are included as important features in the model. The feature set $(x_{1i}, x_{2i}, x_{3i}, \ldots, x_{mi})$ is composed of different geophysical data, where $= 1, 2, \ldots n$, m represents physical parameters, such as resistivity, density, n represents the number of samples. The label refers to the lithology category, which is provided by borehole geological data in this case. Samples with known lithology are called labeled data or labeled data, and samples with unknown lithology are called unlabeled data. The representation of the dataset is as follows (refer to Fig. 6.25):

$$X = \begin{bmatrix} x_1 \\ x_2 \\ x_3 \\ \vdots \\ x_n \end{bmatrix} = \begin{bmatrix} X_1 & D_1 & x_{11} & x_{21} & \cdots & x_{m1} & b_1 \\ X_2 & D_2 & x_{12} & x_{22} & \cdots & x_{m2} & b_2 \\ X_3 & D_3 & x_{13} & x_{23} & \cdots & x_{m3} & b_3 \\ \vdots & \vdots & \vdots & \vdots & \vdots & \vdots & \vdots \\ X_n & D_n & x_{1n} & x_{2n} & \cdots & x_{mn} & b_n \end{bmatrix} \tag{6.21}$$

After generating the dataset, the features are preprocessed first, mainly including "pseudo-Gaussian" transformation and data standardization. "Pseudo-Gaussian" transformation is a data preprocessing technique, the purpose is to make non-Gaussian distributed data closer to Gaussian distribution, thereby improving the

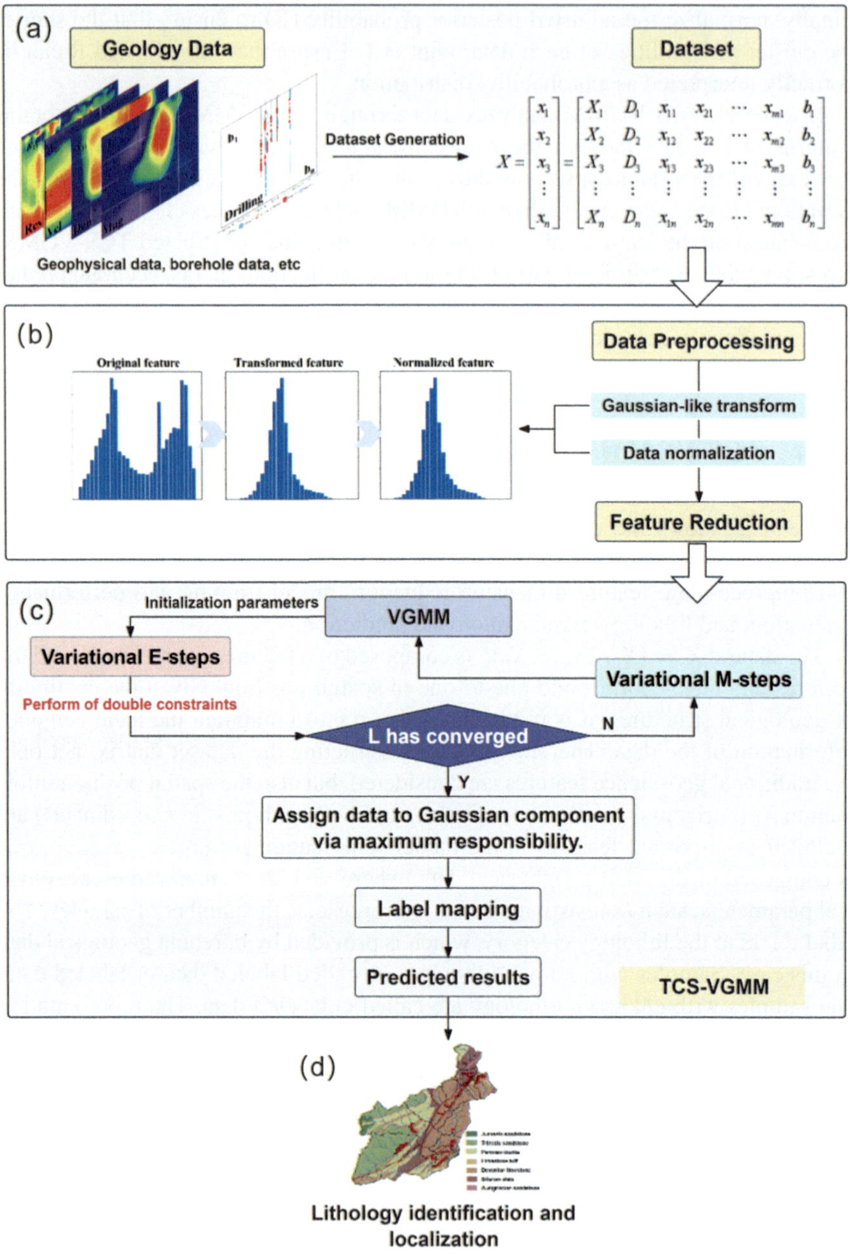

Fig. 6.25 TCS-VGMM for target body identification and positioning Workflow. **a** Generation of datasets from multi-source geoscience data; **b** data processing, including various transformations and feature dimension reduction; **c** TCS-VGMM model training process; **d** lithology identification and positioning

performance of statistical methods that depend on the Gaussian distribution assumption. Common transformation methods include logarithmic transformation, Box and Cox (1964) transformation, and Yeo and Johnson transformation. In this case, a combination of these transformations is used to achieve pseudo-Gaussian transformation of the data. After this process is completed, the data is standardized to eliminate the differences in dimensions and physical properties between different data, ensuring the accuracy and effectiveness of subsequent analysis.

Because the above dataset considers the spatial information of the data, in this case, the Locality Preserving Projections (LPP) technique is used, which specifically achieves dimensionality reduction by preserving the local structure of the data. The processed dataset not only has fewer features but also retains key structural attributes, which helps to simplify the model learning process.

In the LPP method, the key step is to construct a weight matrix, which captures the local adjacency relationships between data points. This weight matrix is usually defined by the Gaussian kernel function, where two data points that are close in position x_i and x_j will be assigned a larger weight. Then, the goal of LPP is to find a projection matrix so that the data still maintains its original local structure in the new low-dimensional space. The optimization objective function of this process is:

$$\min_{A} \sum_{i,j} \|A^T x_i - A^T x_j\|^2 W_{ij}$$
$$A^T XDX^T A = I \tag{6.22}$$

where, X is a number matrix, D is a diagonal matrix, $D_{ii} = \sum_j W_{ij}$, I is a unit matrix.

In this case, accuracy, confusion matrix, and F1 Score are used as three common model performance metrics to evaluate the lithology classification and predictive performance of the trained model. Accuracy is the ratio of the number of correctly predicted instances to the total number of instances. The confusion matrix is a matrix that specifically describes the performance of the model and is used to evaluate the accuracy of the classification model. For the confusion matrix (taking a binary classification problem as an example), it usually includes four parts: True Positive (TP), False Positive (FP), True Negative (TN), False Negative (FN), which respectively represent the number of correctly predicted positive cases, the number of incorrectly predicted positive cases (actually negative), the number of correctly predicted negative cases, and the number of incorrectly predicted negative cases (actually positive). The F1 score is a measure that comprehensively considers precision and recall. Their calculation formulas are as follows:

$$Accuracy = \frac{N_c}{N_T} \tag{6.23}$$

$$Precision = \frac{TP}{TP + FP} \tag{6.24}$$

$$Recall = \frac{TP}{TP + FN} \tag{6.25}$$

$$F1 = 2 \times \frac{Precision \times Recall}{Precision + Recall} \tag{6.26}$$

In the formula, N_c and N_T represent the number of samples that are accurately classified or predicted and the total number of samples, respectively. The higher the precision, the more accurate the model is in predicting the positive class, and the fewer false alarms; the higher the recall, the stronger the model's ability to identify actual positives, and the fewer missed positives; The F1 score ranges from 0 to 1, where 1 represents the best possibility and 0 represents the worst performance. The higher its value, the better the model performs in considering both precision and comprehensiveness (i.e., recall).

6.3.5 Case Application Results Analysis and Discussion

In order to evaluate and verify the effectiveness of the TCS-VGMM method, a theoretical synthetic dataset was designed based on the characteristics of the measured data (Figs. 6.22 and 6.23). This dataset was generated using the multivariate normal distribution function, containing four-dimensional features and a column of labels, totaling 4747 Gaussian-distributed data points. These four-dimensional features simulate the four physical properties of rocks, and the label column represents the lithology distribution in the measured data. The dataset is divided into five categories: Class1 (949 entries, corresponding to the slate in the measured data), Class2 (475 entries, corresponding to the dolomite in the measured data), Class3 (949 entries, representing mineralized dolomite in the measured data), Class4 (475 entries, corresponding to the slate sandstone in the measured data), and Class5 (1899 entries, representing possible unknown lithology).

For ease of visualization, only the first three features are shown in panel a. Position coordinates were assigned to the dataset, which includes the four features and one label column, as seen in Fig. 6.26b and c. Based on the actual borehole positions, lithology categories, and quantities, "pseudo boreholes" were assigned to the synthetic data, as shown in Fig. 6.26d. It is assumed that the lithology categories in the non-borehole areas are unknown, which makes the labeling of the synthetic data consistent with the measured data.

Next, five sets of experiments were designed (Fig. 6.27). Each experiment uses lithology data obtained from a different number of drill holes as the training set, while the areas without drill holes serve as the test set. The performance of the model was evaluated by calculating the classification or prediction accuracy, confusion matrix, and F1 score of the test set obtained.

As shown in the algorithm flowchart in Fig. 6.25, before inputting data into the TCS-VGMM model, feature normalization and dimensionality reduction are

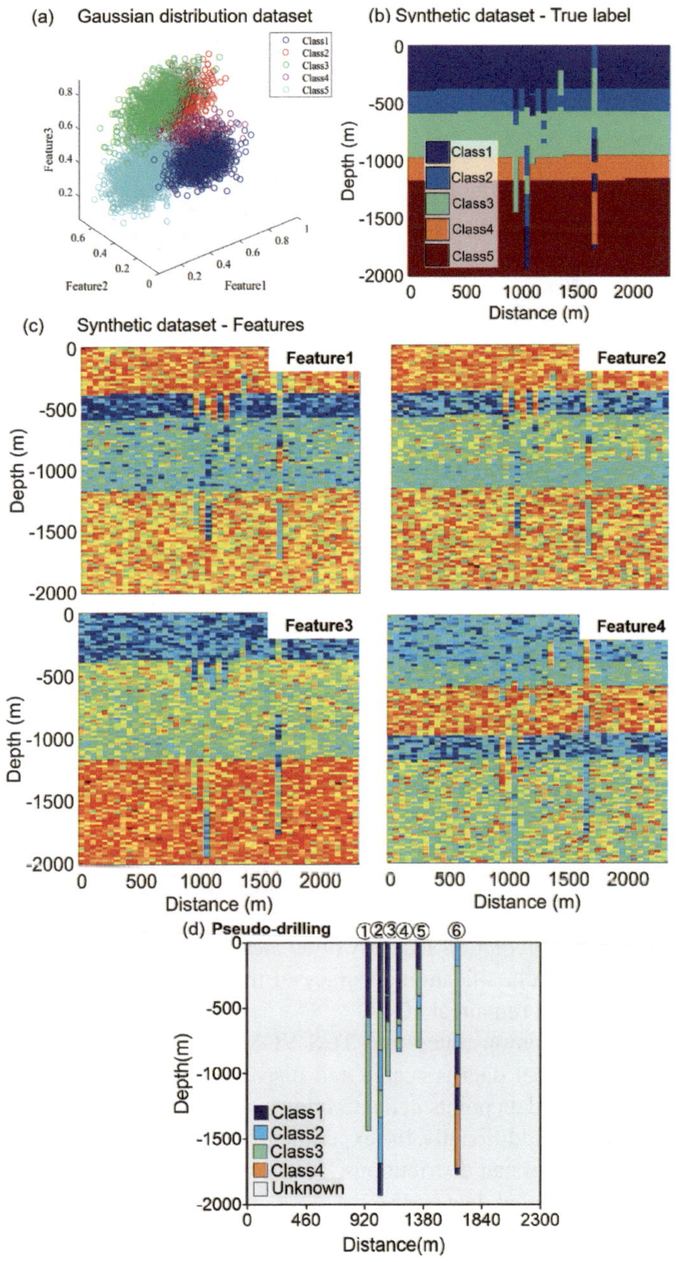

Fig. 6.26 Distribution and feature information of theoretical synthetic data: **a** A Gaussian distributed dataset with 4-dimensional features generated using the multivariate normal distribution function, the dataset includes 5 classes; **b** after assigning position coordinates to the dataset, the distribution of the 5 classes on the plane; **c** after assigning position coordinates to the dataset, the distribution of the 4 features on the plane; **d** pseudo borehole, analogous to the drilling data in the measured data. It is assumed that the classes in the borehole area are known, and the classes in other areas are unknown

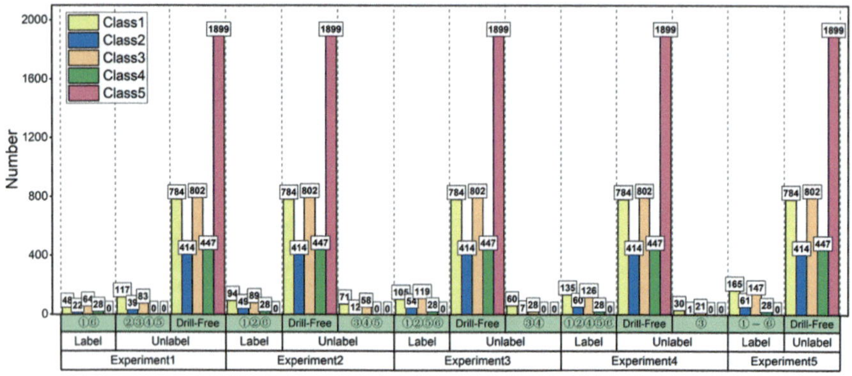

Fig. 6.27 Five different experimental designs: different experiments use different numbers of drill holes as the training set, and the remaining drill holes and areas without drill holes as the test set. For example, in Experiment 1, the labels provided by drill holes ① and ⑥ are used as the training set, and the labels for drill holes ②–⑤ and areas without drill holes are assumed to be unknown and used as the test set. The purpose of this experimental setup is to verify the impact of different numbers of drill holes (number of labels) on model training. In Experiments 1–5, the proportions of label data are 3.4, 5.5, 6.4, 7.4, and 8.4%, respectively

performed. Since the theoretical synthetic data already conforms to a Gaussian distribution, the "quasi-Gaussian" transformation is not used as a preprocessing step, and the processed data is directly used for TCS-VGMM model training. Based on numerous experiments, all model initial parameters π_k and μ_k are set to 0.1, the model iterates 1000 times, and the tolerance for variational inference is set to 10^{-8}.

Once the model converges, the classification and prediction accuracy for different numbers of drill hole data are calculated. According to the results, in Experiment 1, the classification accuracy for the training set is 100%, but there are errors in predicting the test set. Specifically, in Class 3, one data point is misclassified into Class 4, resulting in a prediction accuracy of 99.58%, with an F1 score of 0.9980. For Experiments 2–4, the classification accuracy of the training set and the prediction accuracy of the test set remain at 100%.

This experiment demonstrates that TCS-VGMM can effectively classify and predict even when label data is scarce and unevenly distributed. Furthermore, as the number of labeled data points in the training set increases, the accuracy of model prediction improves. Additionally, the experiment shows that for datasets that follow Gaussian or quasi-Gaussian distributions, good predictive performance can still be achieved when the labeled data is <5%.

Subsequently, the predictive ability of different experimental scenarios for the entire area is compared. As shown in Fig. 6.28, the prediction accuracy for the entire area in each experiment is >92%, which demonstrates the characteristic that the more labeled data available, the higher the prediction accuracy. Moreover, the TCS-VGMM model can successfully predict categories (Class 5) that were not included in the training set.

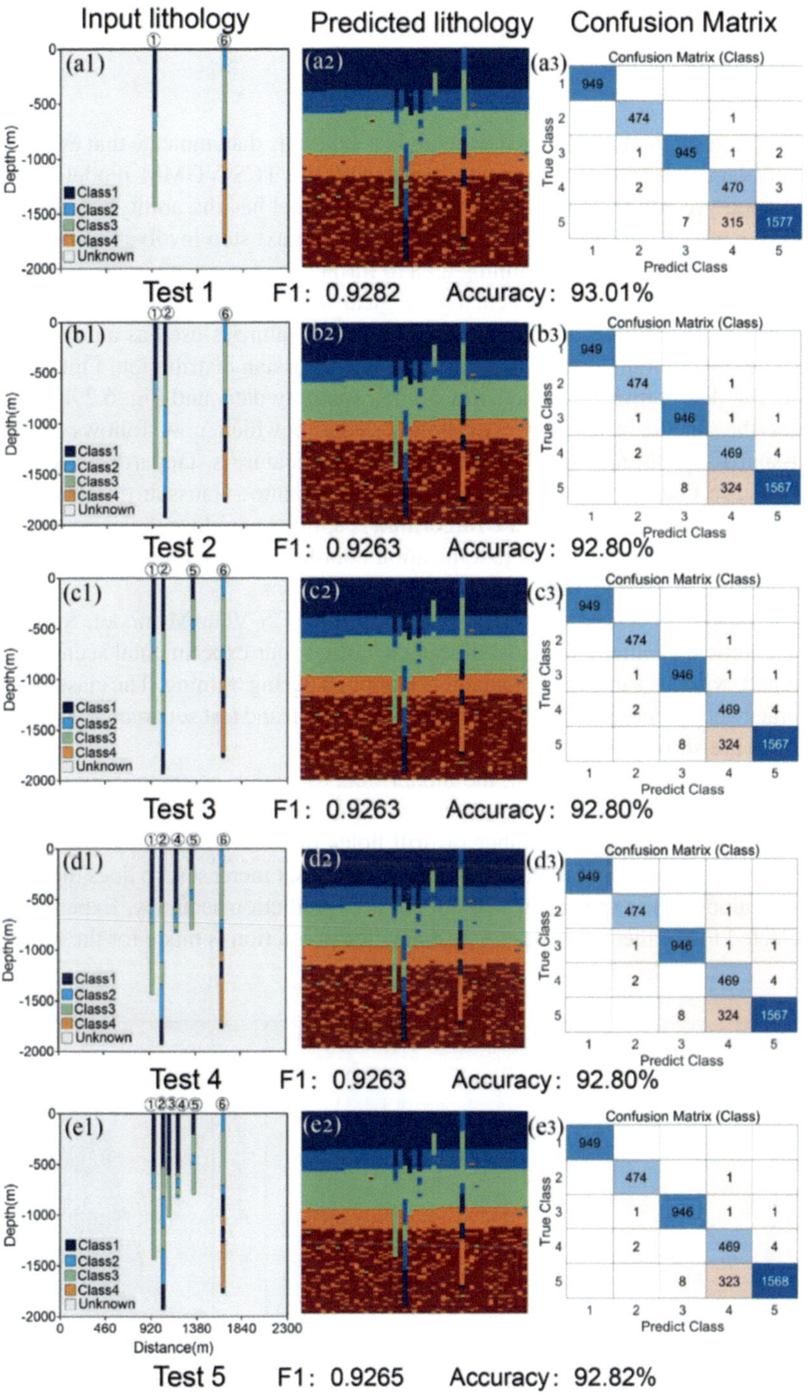
Input lithology
Predicted lithology
Confusion Matrix
(a1)
(a2)
(a3)
(b1)
(b2)
(b3)
(c1)
(c2)
(c3)
(d1)
(d2)
(d3)
(e1)
(e2)
(e3)
Depth(m)
Distance(m)
Class1
Class2
Class3
Class4
Unknown
Confusion Matrix (Class)
True Class
Predict Class
Test 1 F1: 0.9282 Accuracy: 93.01%
Test 2 F1: 0.9263 Accuracy: 92.80%
Test 3 F1: 0.9263 Accuracy: 92.80%
Test 4 F1: 0.9263 Accuracy: 92.80%
Test 5 F1: 0.9265 Accuracy: 92.82%

◂**Fig. 6.28** The impact of different numbers of drill hole data on the prediction results for the entire area: The Confusion Matrix, F1 score, and Accuracy here are comprehensive evaluations of the classification and prediction of the training and test sets

The experimental results on the theoretical synthetic data indicate that even when labeled data is scarce and unevenly distributed, the TCS-VGMM model can still yield good prediction results. Additionally, the model has the ability to effectively identify new categories present in the test set. The next step involves applying this algorithm to the Baiyun Ebo mining area to further verify and assess its application performance and reliability in actual geological scenarios.

When preprocessing the dataset, the resistivity feature is used as an example to illustrate how to transform the data into a quasi-Gaussian distribution. Figure 6.29a shows the distribution characteristics of the resistivity data, and Fig. 6.29b demonstrates the feature after Box–Cox transformation, which now follows a quasi-Gaussian distribution. After this transformation, the feature is standardized, as shown in Fig. 6.29c. Once all features have been transformed into a Gaussian distribution, the Local Preserving Projection (LPP) algorithm is applied to reduce the dimensionality of the data features, which include location coordinates, resistivity, magnetization rate, density, and speed, reducing them to 3 dimensions.

The newly processed dataset is input into the TCS-VGMM model. Similar to the theoretical synthetic data test described earlier, four experimental scenarios are designed, with the same model parameters applied during training. The classification accuracy and F1 score of the training set classification and test set prediction for each scenario are shown in Fig. 6.30.

As observed from Fig. 6.30, the model's classification accuracy is quite strong when applied to actual data, and generally, the model's prediction accuracy is positively correlated with the number of drill holes used in the training set. In other words, as the amount of labeled data in the training set increases, so does the prediction accuracy. The experiment with the highest prediction accuracy, Experiment 4, is selected for further analysis, where a detailed prediction is made for the lithology

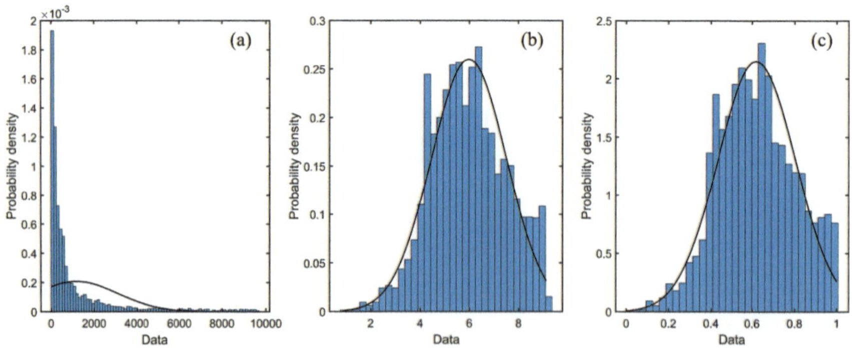

Fig. 6.29 Feature preprocessing, using resistivity feature as an example: **a** original data distribution; **b** data distribution after quasi-Gaussian transformation; **c** data distribution after normalization

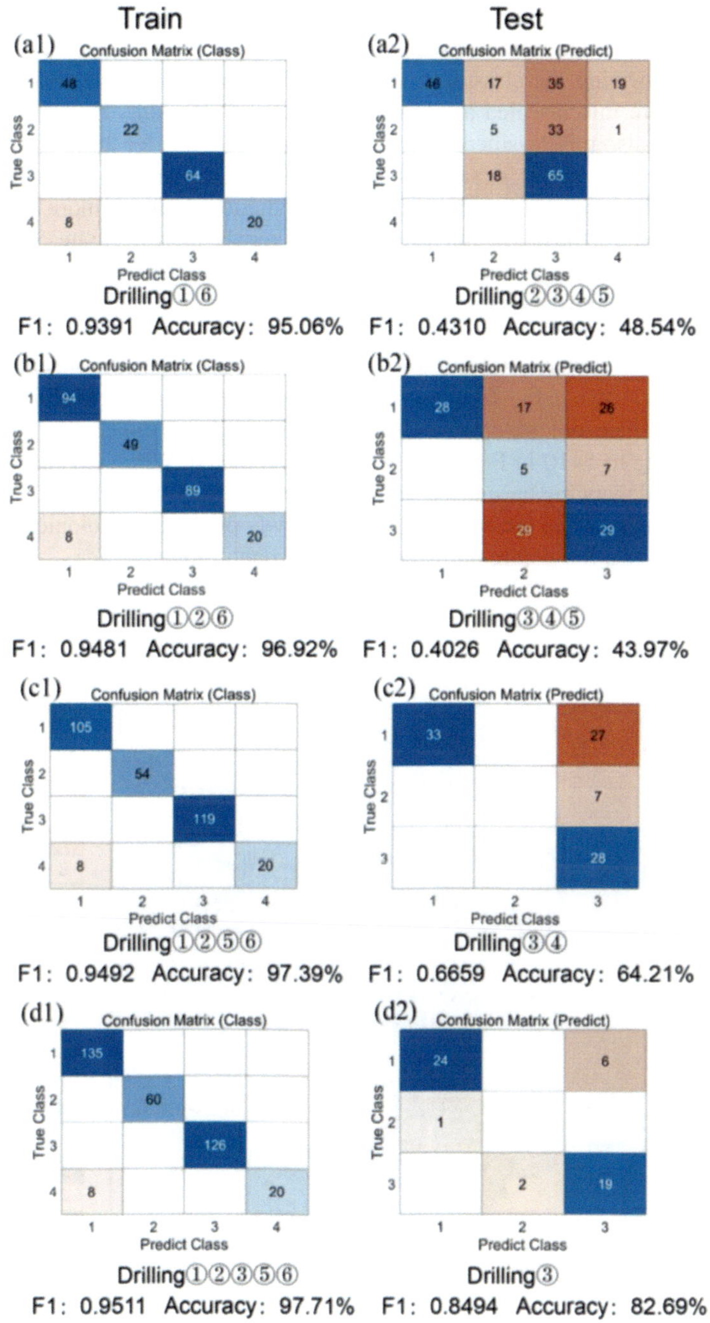

Fig. 6.30 Prediction results under different experimental conditions: In the confusion matrix, 1, 2, 3, 4 respectively represent Slate, Dolomite, Mineralized dolomite, Sandstone

distribution between the main and east mines of the Baiyun Ebo mining area, based on the results obtained from this scenario.

Firstly, reviewing the classification results from the training set, 8 samples of sandstone were misclassified as shale. In the prediction results from the test set, dolomite was also misclassified. There are two main reasons for these errors in this case study: first, the labels for sandstone and dolomite are relatively fewer compared to shale, which causes a noticeable skew in the prediction results. If more labels were added in the future, the prediction results would likely be more accurate. Second, the labels for sandstone are primarily found at depths below 1300 m, where the resolution of the geophysical inversion imaging dataset decreases with depth, which could also introduce some error.

Subsequently, further analysis was conducted on the prediction results for the entire area. As shown in Fig. 6.31, new categories are present in the prediction results. In addition to shale, dolomite, mineralized dolomite, and sandstone, other lithologies (corresponding to 5–10 in Fig. 6.11) were also predicted. While existing research and geological mapping have identified rocks such as gneiss and granodiorite in and around the mining area, the accuracy of these newly predicted lithologies and their precise categorization will require verification through future drilling.

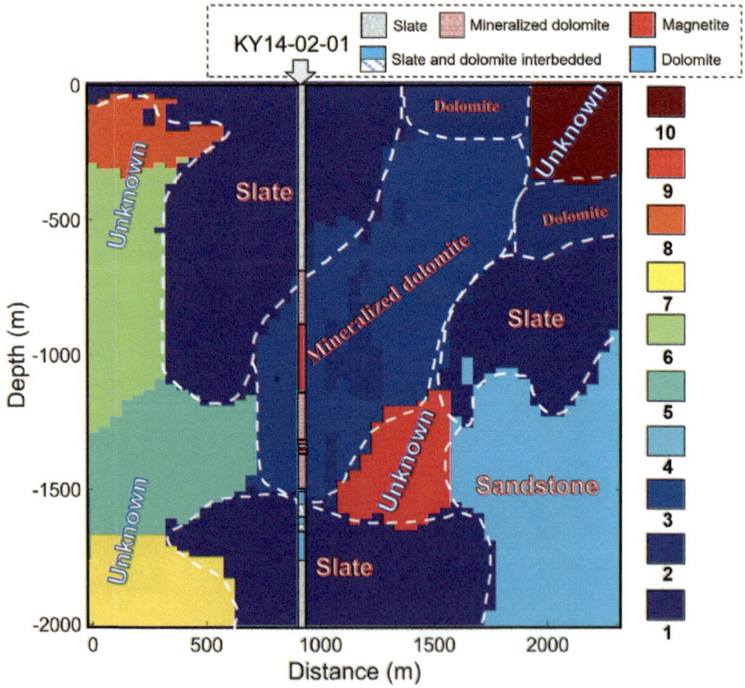

Fig. 6.31 Prediction results. The labels 1, 2, 3, 4 in the figure correspond to shale, dolomite, mineralized dolomite, and sandstone, respectively, and 5–10 are the new lithologies predicted. But the lithology category is unknown

As shown in Fig. 6.31, the predicted positions of the dolomite and mineralized dolomite are approximately within the depth range of 200–1600 m, exhibiting a plate-like distribution that tilts to the south. According to assay results from drill cores near the east mine, the rare earth grade in the iron ore body and the underlying dolomite in the eastern part of the east mine ore body is extremely low. In contrast, the rare earth grade in the iron ore body and the underlying dolomite in the western part gradually increases from east to west, from top to bottom, and from north to south. This observation led to the speculation that the mineralized dolomite was formed by magma channel upwelling from the southwest to the northeast. Therefore, the ore formation in the main and east mining areas is believed to dip southward, which aligns with the model's prediction results in this case.

To further verify the accuracy of the model's prediction results, an additional verification drill hole (KY14-02-01, shown in Fig. 6.1) was drilled, located ~140 m west of the 950 measurement point. By projecting the lithology distribution of this drill hole onto the model's prediction results, the comparison reveals a high degree of consistency in the distribution of shale, dolomite, and mineralized dolomite. This further demonstrates that the algorithm used in this case is highly reliable.

6.3.6 Case Summary

The study and resource prediction of the Bayan Obo super-large rare earth deposit is a current major demand, with the spatial distribution of mineralized carbonates being a key basis for understanding the ore-forming mechanism and assessing resource potential. However, due to the complexity of the detection object and environment, it is challenging to accurately locate the mineralized carbonates using a single geophysical method. How to effectively combine the results of comprehen sive geophysical detection with drilling data to improve the accuracy of underground structure interpretation has become a key scientific issue in mining exploration. This case study applies a semi-supervised variational Gaussian mixture learning model to address the spatial heterogeneity of geoscience data and the imbalance of limited label distribution.

The study analyzed the representative cross-section between the main mine and the east mine in the Bayan Obo mining area, utilizing the inversion results of four types of heavy magnetic seismic geophysical data within 2 km of the surface and the lithology observation data from six drill holes. The study successfully distinguished the spatial distribution of four main lithologies. The results show that mineralized dolomite is mainly distributed between the depth intervals of 200–1600 m underground, with a south-deep and north-shallow inclined plate-like distribution feature. The prediction results align closely with the actual distribution of dolomite observed in the drill holes, verifying the effectiveness and practicality of the algorithm.

With the rapid development of computer technology, multi-source information fusion methods based on machine learning have proven effective in extracting useful information from multi-dimensional, multi-scale geoscience data, thus improving interpretation accuracy. Given the scarcity of lithology label data obtained from drilling and the uneven distribution of various lithologies, a constrained semi-supervised variational Gaussian mixture model algorithm (TCS-VGMM) was developed in this case. By integrating the heavy magnetic seismic imaging results with a small amount of lithology data from drilling, the algorithm achieved accurate identification and positioning of underground lithology structures.

The results demonstrate that the stability of the model increases with the amount of labeled data, meaning that more drilling data enhances the predictive power of the model. However, the accuracy of lithology identification and positioning is significantly influenced by the resolution of the geophysical imaging dataset, which limits the efficiency of machine learning fusion algorithms in geoscience applications. As more drilling data becomes available, and with advancements in data processing technologies in comprehensive geophysical exploration and improved imaging resolution, the prediction accuracy of the algorithm is expected to improve further.

The TCS-VGMM algorithm developed in this study not only provides a new tool and perspective for the detection and resource assessment of the deep structure of the Bayan Obo deposit but also offers a novel solution to other complex problems in the field of earth science, highlighting the vast potential of machine learning in processing and interpreting geophysical data.

6.4 Using Deep Learning for Electromagnetic Data Inversion Imaging—Underground Structure Investigation of a Coal Mine in Xuejiawan, Inner Mongolia

The SOTEM detection method is advantageous due to its small volume effect and strong signal strength, making it highly suitable for achieving deep, high-precision detection. In contrast, deep learning methods, which are known for their fast computation speed and high inversion accuracy, have been widely applied in solving inverse problems in geophysics. In this case, an improved 3D U-Net model is used for performing three-dimensional inversion of SOTEM data based on deep learning. Numerical simulation experiments demonstrate that this algorithm exhibits stable noise resistance and strong generalization ability, significantly reducing computational time while enhancing imaging accuracy.

The three-dimensional inversion algorithm is applied to image the resistivity structure of SOTEM data from an underground structure investigation conducted in a coal mine in Xuejiawan, Inner Mongolia. The results from this application confirm that the algorithm not only improves the efficiency of data processing but also provides more accurate and reliable imaging of the resistivity structure.

6.4.1 Case Introduction

The Transient Electromagnetic Method (TEM) is a widely used electromagnetic exploration technology, which can be categorized into electrical source TEM and magnetic source TEM based on the excitation source. Within the electrical source TEM, Long Offset Transient Electromagnetic (LOTEM) and Short Offset Transient Electromagnetic (SOTEM) methods each have distinct characteristics. LOTEM is more suitable for large-scale exploration but is prone to volume effects in deep exploration, particularly in complex geological environments. On the other hand, SOTEM effectively suppresses the volume effect, providing enhanced resolution for thin-layer structures, making it highly advantageous for detecting complex geological bodies.

Despite the clear advantages of SOTEM technology, the current data interpretation methods are primarily limited to one-dimensional and quasi-two-dimensional inversion approaches. This restriction hinders its application in complex geological environments. Electromagnetic inversion technology is broadly divided into two categories: deterministic inversion and stochastic inversion. Deterministic inversion methods, such as gradient descent and conjugate gradient methods, are computationally efficient but highly sensitive to the choice of the initial model. These methods are prone to falling into local minima, and the resulting error is often significant when there is insufficient prior information. In contrast, stochastic inversion methods, such as genetic algorithms, simulated annealing, and Bayesian algorithms, can achieve global optimization, making them suitable for cases where prior information is lacking. However, these methods come with high computational demands, particularly in three-dimensional inversion, where the requirements for.

In recent years, deep learning, especially neural networks, has made significant progress in the application of geophysical data processing. Since the 1990s, neural networks have been applied to the field of geophysics, and with the development of computer technology, they have gradually been widely used in inversion problems. For example, neural networks have been used for electromagnetic data inversion and denoising, becoming an important tool for solving complex electromagnetic inversion problems (Song et al., 2018; Xiao et al., 2020). Especially the Convolutional Neural Network (CNN) due to its powerful feature extraction ability, performs excellently in dealing with inverse problems in geophysics, and the U-Net architecture as a derivative structure of CNN, has been widely used in tasks such as seismic data interpolation, full waveform inversion (Wu et al., 2019; Yang and Ma, 2019; Qu et al., 2021).

In recent years, deep learning, particularly neural networks, has made substantial advancements in the field of geophysical data processing. Since the 1990s, neural networks have been applied to geophysics, and with the continuous development of computer technology, they have gained significant traction in solving inversion problems. Neural networks have become an essential tool for tackling complex electromagnetic inversion problems, including applications in electromagnetic data inversion and denoising (Song et al., 2018; Xiao et al., 2020). Among various deep learning architectures, the Convolutional Neural Network (CNN) stands out due to its powerful feature extraction capabilities, excelling in addressing inverse problems in geophysics. The U-Net, a derivative structure of CNN, has also been widely applied to tasks such as seismic data interpolation and full waveform inversion (Wu et al., 2019; Yang and Ma, 2019; Qu et al., 2021).

The advantages of U-Net in electromagnetic data applications are particularly prominent. Not only can it efficiently process and analyze temporal features within electromagnetic data, but it also significantly improves data processing accuracy during the inversion process. The introduction of deep learning models, especially the three-dimensional convolutional neural network (3D-CNN), allows for more effective handling of complex three-dimensional geological structures, enabling more accurate underground imaging and resource detection.

In this case, an improved U-Net model is employed for the resistivity structure inversion of SOTEM data. The key improvements in this model include: (1) replacing traditional two-dimensional convolution with three-dimensional convolution to better handle temporal data in three-dimensional space, thereby enhancing imaging accuracy; (2) introducing noisy data for data augmentation to improve the network's resistance to noise; (3) adding full-scale skip connections and deep supervision mechanisms to enhance prediction accuracy; (4) incorporating regularization constraints into the loss function to focus the model on anomalies; and (5) using residual block structures to replace traditional convolutional layers, addressing the gradient vanishing issue.

The verification of both simulated and measured data demonstrates that the deep learning method employed in this case significantly improves the accuracy and efficiency of deep resource detection in SOTEM data inversion. This method offers robust technical support for electromagnetic data processing, particularly in complex geological environments, and shows considerable promise for further development in this field.

6.4.2 Basic U-Net Structure Algorithm

As shown in Fig. 6.32, a Convolutional Neural Network (CNN) consists of convolutional layers, pooling layers, and fully connected layers. Typically, a CNN's structure involves several convolutional layers connected to pooling layers, and this structure is repeated multiple times before being connected to several fully connected layers. The convolutional layer uses a convolution kernel to extract features from the input data, with the size of the area in the input data that corresponds to the elements on the feature map referred to as the "receptive field." The pooling layer serves to reduce the dimensionality of the data, extracting the most significant features and reducing the number of parameters. This also helps in preventing overfitting, a problem where the model becomes too complex and performs poorly on unseen data (Krizhevsky et al., 2012).

Below is a brief description of the convolution operation:

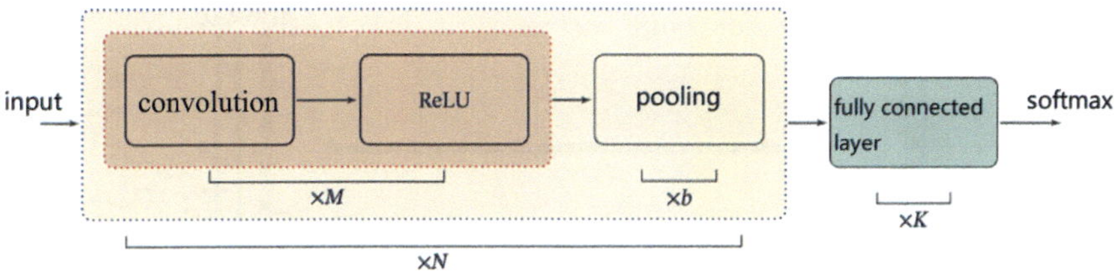

Fig. 6.32 Structure of the convolutional neural network

$$y = f(\omega \otimes x + b) \tag{6.27}$$

where, x represents the input data, ω represents the weight of the convolution kernel, b represents the bias, $\otimes$ represents convolution, y represents the output feature, f represents the activation function.

CNN was initially proposed to solve image classification problems. However, in these networks, the fully connected layer needs to convert the two-dimensional feature map into a one-dimensional vector, which leads to the loss of some spatial information, ultimately reducing the accuracy of the prediction results. This can be problematic in applications such as geophysical imaging where high prediction accuracy is required for regression problems. Additionally, the data size used in the testing process and the training process of the fully connected layer must be the same, making it inconvenient for imaging tasks.

U-Net (Ronneberger et al., 2015) is a variant of the convolutional neural network that adopts an encoder-decoder structure. As shown in Fig. 6.33, U-Net is based on Fully Convolutional Networks (FCN) and enhances them by adding skip connections from the encoder to the decoder, which helps in feature fusion. By injecting detailed information from low-level features into deep high-level features, U-Net can make more accurate predictions. This architecture has been particularly effective in tasks where maintaining spatial information is critical, such as medical image segmentation and geophysical imaging.

6.4.3 Improved 3D U-Net Structure–Full-Scale Skip Connections and Deep Supervision

The 3D U-Net, proposed by Çiçek et al. (2016), is a deep learning architecture designed to handle three-dimensional data directly. However, due to the large number of parameters involved in 3D inversion tasks and the extensive computation required, this necessitates the use of high-performance computing resources and a neural network structure capable of processing three-dimensional inputs. To address the challenges of performing 3D U-Net-based three-dimensional inversion, this case study presents several targeted solutions. These include introducing full-scale skip connections and full-scale deep supervision, which help improve inversion accuracy

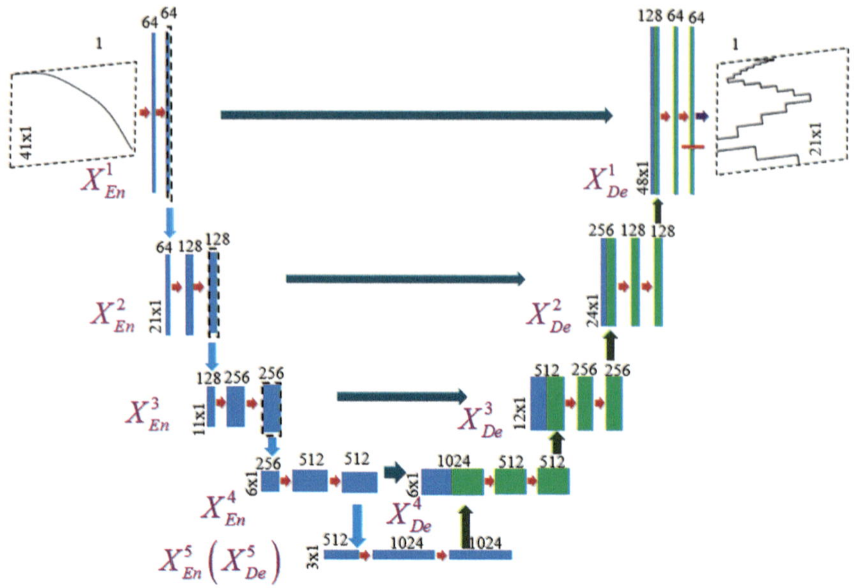

Fig. 6.33 Structure of the 2D U-Net used for electromagnetic data inversion

across all scales while also acting as regularization to effectively avoid overfitting. Additionally, a mixed loss function with three constraints is employed to enhance the algorithm's ability to focus on anomalies. To tackle the gradient vanishing issue, the convolutional layers in the encoder are replaced with a residual block structure that learns feature mapping by fitting residuals.

(1) 3D U-Net network structure with full-scale skip connections and full-scale deep supervision.

In image processing tasks, images are typically divided into different scales, with each feature map capturing various levels of detail about the image. Low-scale feature maps usually contain more fine-grained details, while high-scale feature maps capture more overall structural and semantic information. In the encoder part of the network, the receptive field of the feature map increases as information is passed through the layers. However, this process can lead to the loss of important detailed features. Skip connections are critical in this context because they allow for the fusion of detail features with high-level features. Each time the features from different scales are combined with high-level features, it reduces the loss of fine-grained details and improves the accuracy of the inversion process. Based on this idea, full-scale skip connections are introduced to fuse feature maps from different scales, ensuring that the neural network fully utilizes the feature information extracted at various levels. This approach allows the model to benefit from the richness of details from lower scales while retaining the broader structural and semantic understanding from higher scales, thereby enhancing the model's overall performance.

As shown in Fig. 6.34, in addition to the skip connections between the feature maps of the same scale between the encoder and decoder, the model enhances all skip connections from the large-scale feature maps in the encoder to the small-scale feature maps in the decoder. It also adds skip connections from the small-scale feature maps in the decoder to the large-scale feature maps. When establishing these skip connections, the number of channels in the feature map is first adjusted to 64 through 3D convolution or 3D transpose convolution operations. Specifically, when the large-scale encoding feature map is skip-connected to the small-scale decoding feature map, the data size of the feature map is reduced through 3D max pooling operations. Conversely, when the small-scale decoding feature map is skip-connected to the large-scale decoding feature map, the data size is increased via 3D transpose convolution operations. The data size adjustments are controlled by altering the size and stride of the pooling and transpose convolution kernels.

The advantage of full-scale skip connections lies in their ability to fully utilize features across all scales, which significantly improves inversion accuracy. By integrating feature maps from different scales, the neural network can better capture both the detailed features and the overall structure present in SOTEM 3D data, resulting in more accurate predictions. This approach allows the model to share valuable feature information at various levels, reducing computation time and improving the efficiency of model training.

(2) Optimization of feature extraction structure.

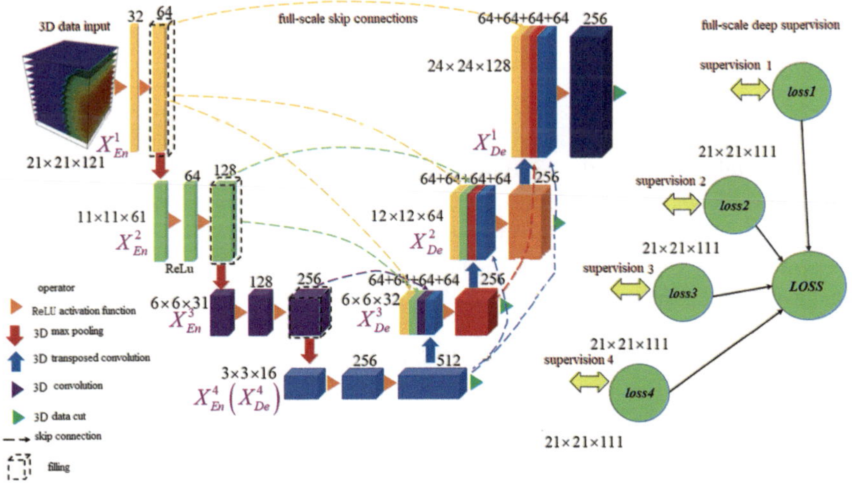

Fig. 6.34 Three-dimensional U-Net structure, with added full-size skip connections and SOTEM inversion depth supervision

The full-scale skip connection technology adds more convolutional layers to the network, which may lead to the problem of gradient vanishing during the backpropagation process, ultimately reducing the learning ability of the network. To address this issue, the convolutional operation is replaced with a residual block structure in the encoding phase. This modification allows the network to learn the mapping more effectively by fitting the residuals, helping to maintain the gradient flow and improve the overall learning efficiency of the model. Residual blocks allow the network to learn the identity function more easily, which mitigates the vanishing gradient problem and enables deeper networks to be trained with better performance.

The structure of the residual block is shown in Fig. 6.35. Taking X_{En}^2 as an example, replace the first convolution operation in X_{En}^2 with residual block structure 1. The input is x. The green mainline path in residual block structure 1 is a three-dimensional convolution operation, the size and stride of the convolution kernel are both 1, and the output obtained after the Batch Normalization (BN) layer is $F(x)$, $F(x)$ and x have the same number of channels. The blue branch path (also called "shortcut connection") is an identity mapping, which can be expressed as:

$$h(x) = x \tag{6.28}$$

Combine $F(x)$ and $h(x)$, that is, add each element to get $H(x)$:

$$H(x) = F(x) + h(x) \tag{6.29}$$

$H(x)$ is passed through the ReLu activation function to get the output of the residual block $O(x)$:

$$O(x) = F(x) + h(x) \tag{6.30}$$

The mainline path in the residual block structure 2 is also a three-dimensional convolution layer, but the output $F(x)$ has twice as many channels as the input x. Therefore, in the branch path, the input x also needs to go through a three-dimensional convolution layer with a kernel size and stride of 1 to double the number of feature channels and get the output $h(x)$:

$$h(x) = Conv\_3D(x) \tag{6.31}$$

Finally, add $F(x)$ and $h(x)$, and pass through the activation function to get the output $O(x)$.

(3) Optimization of loss function.

In order to enhance the algorithm's focusing ability on anomalies, constraints in the x, y, and z directions of the inversion result are added to the loss function. In the four different levels of loss functions, the form of mean square error plus regularization constraint term is used. The constraint term is the average absolute error in the x, y, and z directions:

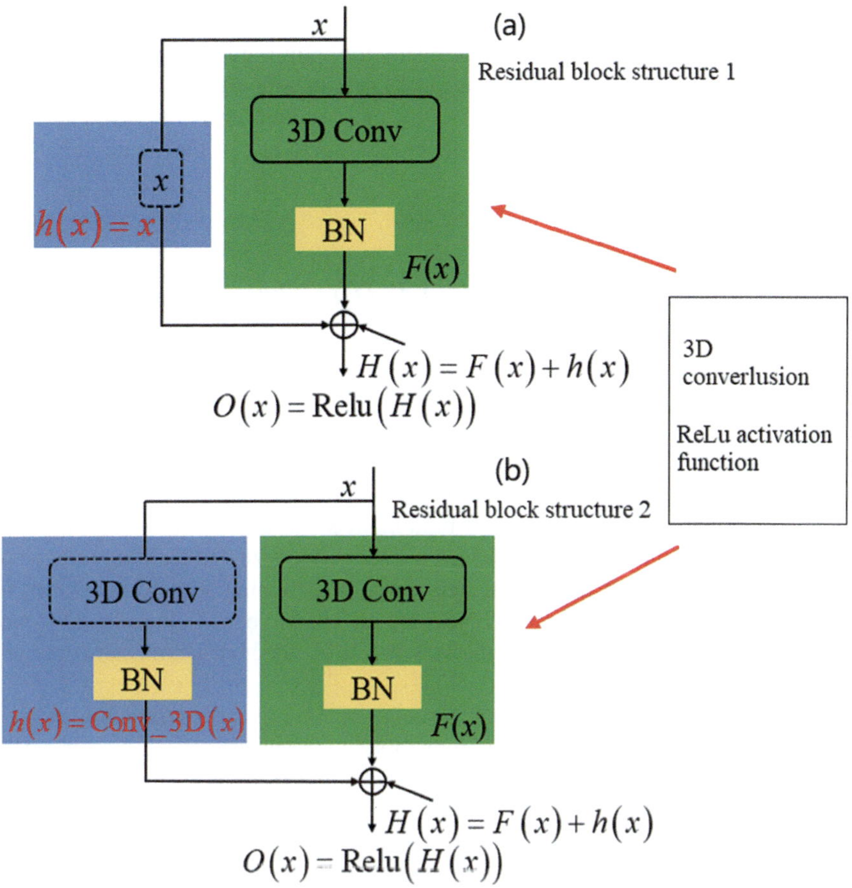

Fig. 6.35 3D U-Net structure with convolution replaced by residual blocks: **a** residual block 1; **b** residual block 2

$$loss_{MSE} = \frac{1}{ph}\sum_{n=1}^{h}\left(\hat{\mathbf{m}}_{n,i,j,k} - \mathbf{m}_{n,i,j,k}\right)^2 \tag{6.32}$$

$$loss_{MAE} = \frac{1}{ph}\sum_{n=1}^{h}\left|\hat{\mathbf{m}}_{n,i,j,k} - \mathbf{m}_{n,i,j,k}\right| \tag{6.33}$$

where p represents the total number of elements in the target output data, h represents the amount of data input into the network in each batch. The regularization factor λ is generally chosen as 0.1, and the average of these four loss functions is the final loss function:

$$Loss = \frac{loss1 + loss2 + loss3 + loss4}{4} \tag{6.34}$$

Through the expansion of the training set, the introduction of full-scale skip connections and deep supervision mechanisms, the optimization of the feature extraction structure, and the optimization of the loss function, a model has been built that has better inversion effects compared to 3D U-Net has a significantly improved network model, which is named as Full-scale 3D U-Net. The effect of Full-scale 3D U-Net is verified with simulation data below.

6.4.4 Full-Scale 3D U-Net SOTEM Data Three-Dimensional Inversion Algorithm Test

(1) Comparison of neural network training.

In order to highlight the superiority of Full-scale 3D U-Net over the traditional 3D U-Net, the training effects of both neural networks on a new noisy dataset were compared. The loss function used for both models was a mixed loss function with three-way constraints. Hyperparameters, such as batch size and learning rate, were kept consistent between the two models during the training process (as shown in Table 6.3). The training of the 3D U-Net took 26.7 h, while the Full-scale 3D U-Net took 25 h. The decline curves of the loss functions for both neural networks are displayed in Fig. 6.36.

As observed from the figure, after 300 rounds of training, the loss functions of both models converged. However, the minimum loss value reached by the Full-scale 3D U-Net was an order of magnitude smaller than that of the 3D U-Net. This indicates that the Full-scale 3D U-Net achieved better learning results. Additionally, the decline of the loss function curve on the test set shows that no overfitting occurred during the training process, confirming the robustness and efficiency of the Full-scale 3D U-Net.

Considering the difference between the actual stratum resistivity distribution and the model parameters designed in this case, three test sets, each containing 1000 models, were generated to evaluate the effectiveness of the algorithm. In the first test set, the resistivity of the first layer was assigned, and the resistivity of each subsequent layer had a 40% probability of being equal to the previous layer; otherwise, it was randomly chosen within the range of [1, 10,000] Ω•m. All layers were assigned resistivity according to this rule. The resistivity generation method in the second test

Table 6.3 Comparison of hyperparameters between Full-scale 3D U-Net and 3D U-Net

Network model	Initial learning rate	Training rounds	Batch size	Optimization algorithm	Dataset capacity	Training duration	Minimum train loss
Full-scale 3D U-Net	0.001	300	100	Adam	40,000	25 h	0.0000158
3D U-Net	0.001	300	100	Adam	40,000	26.7 h	0.000183

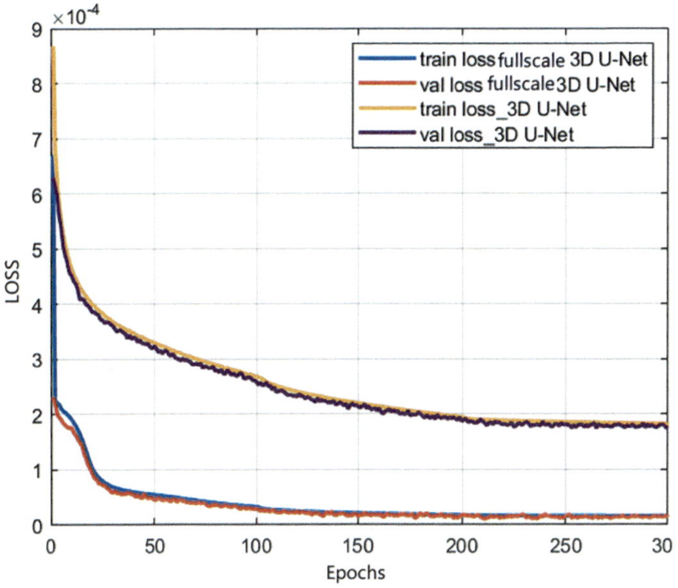

Fig. 6.36 Comparison of loss value decay curves between full-scale 3D U-Net and 3D U-Net

set was the same as in the training set. In the third test set, the number of layers was increased to 28, and the resistivity assignment method remained the same as in the training set.

Figure 6.37 compares an inversion result randomly drawn from each test set with the actual model. As shown in the figure, in all three test cases, the deep learning one-dimensional inversion method achieved high accuracy. Notably, the accuracy of the results obtained from U-Net inversion was higher than that from FCN inversion, and the data fitting of the U-Net inversion results was also better than that of FCN. However, when the number of layers in the test set model differed from that in the training set, while the deep learning one-dimensional inversion method could accurately predict the trend of resistivity changes and the resistivity values of each layer, the depiction of the interface was less precise. This indicates that while the algorithm performs well in predicting resistivity trends, it struggles with accurately identifying interfaces when the model configuration differs significantly from the training set.

(2) Noise resistance test of Full-scale 3D U-Net.

The layout of the transmitting and receiving devices is shown in Fig. 6.38. The grounded long wire source is arranged along the x-axis, the length of the source is 1 km, the midpoint is the coordinate origin, and the positive direction of the z-axis is vertically downward. The shaded part in the figure is the measurement area, and the coordinate range of the observation point R is $\{(x, y)|x \in (-500, 500), y \in (500, 1500)\}$. In this case, d is chosen. B_z/dt is used

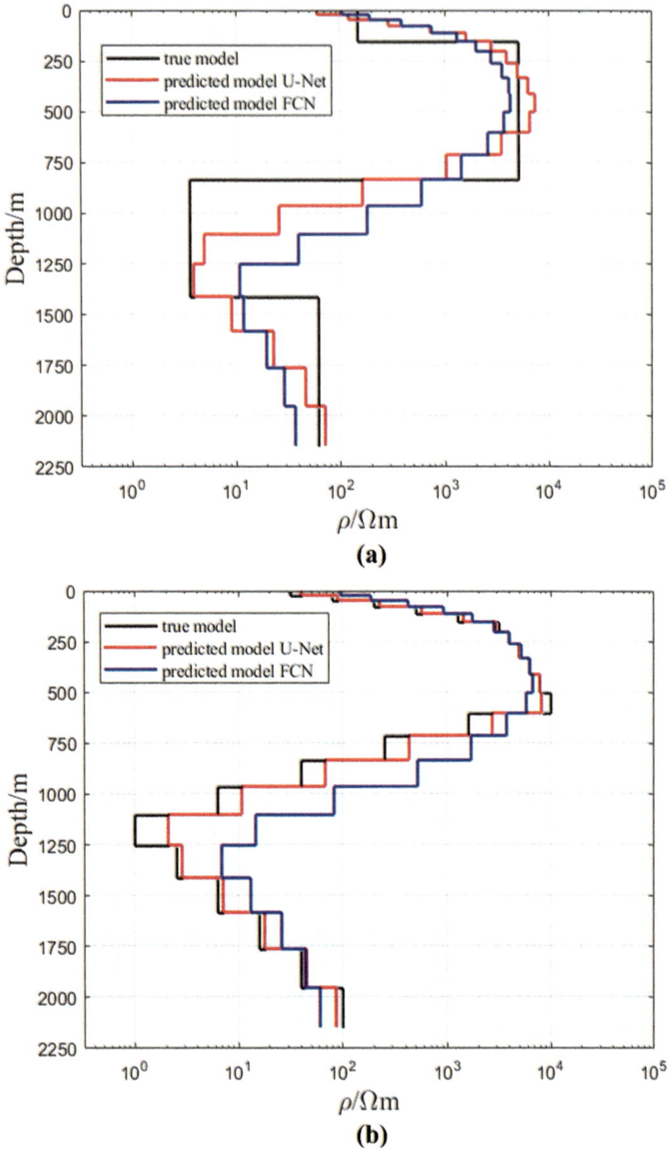

Fig. 6.37 Inversion results of different modes: **a** 4-layer mode; **b** 21-layer mode

as observation data, with sampling times ranging from 10^{-5} to 10^{-1} s, a total of 41 time channels. The size of the input data is 41×1, and the size of the target output is 21×1.

In order to make the objective function easier to converge to the global optimal solution, before the training starts, it is necessary to normalize the training data. Here,

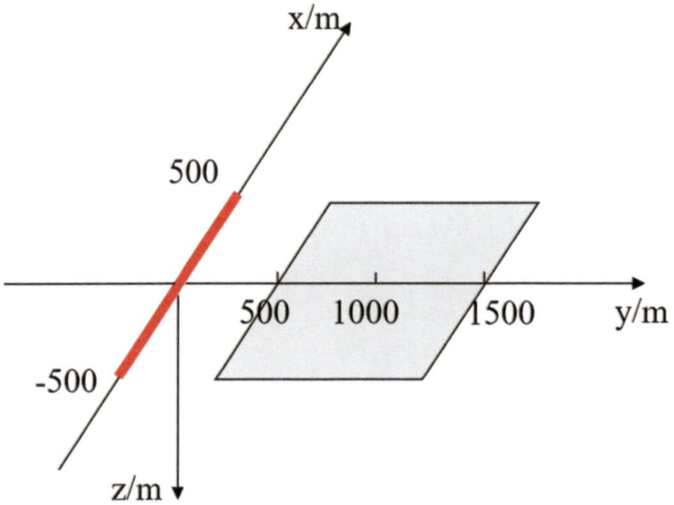

Fig. 6.38 SOTEM data collection schematic

the range transformation method is used to normalize the input data separately. The formula of the range transformation method is as follows:

$$\hat{x}_i = \frac{x_i - \mathbf{X}_{\min}}{\mathbf{X}\min_{\max}} \tag{6.35}$$

where, $\mathbf{X} = [x_1, x_2, ..., x_n]$ represents the SOTEM response collected at a measurement point, n represents the time channel. At the same time, the resistivity value of the geoelectric model is logarithmically taken to reduce the dynamic range of the target output:

$$\hat{\rho}_i = \log_{10}\rho_i \tag{6.36}$$

In the formula, ρ represents the resistivity value.

A total of 100 rounds of training were conducted, with a batch size of 100 pairs of input–output data for each iteration. The initial learning rate is 0.001, and the learning rate is reduced to one-tenth of the original every 30 rounds of training. The Adam algorithm is chosen as the optimization algorithm. The loss function decline curve is shown in Fig. 6.39. After 100 rounds of training, U-Net converges. At the same time, the decline trend of the loss function curve on the test set is consistent with that on the training set, indicating that there is no overfitting.

Next, the noise resistance ability of the full-scale U-Net is tested. Three different levels of Gaussian white noise—3, 8, and 15%—were added to the forward response of the test data, and then the full-scale U-Net was used to invert both the noise-free and noisy data. The results were compared. As shown in Fig. 6.40, when the SOTEM response contains 3% Gaussian white noise, the impact on the inversion

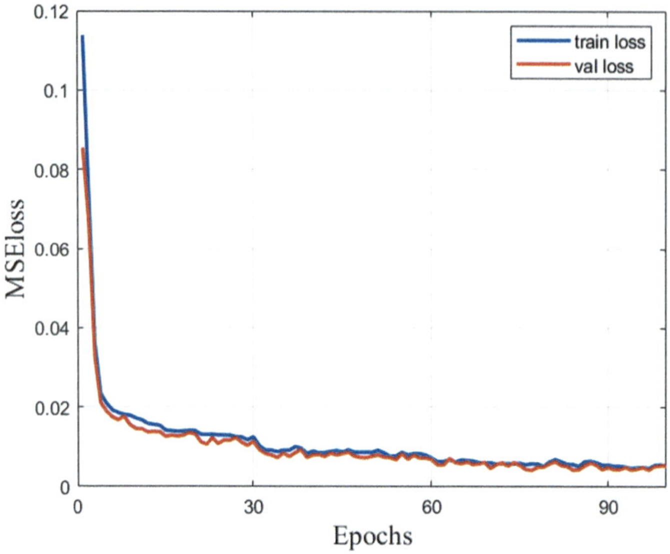

Fig. 6.39 Full-size U-Net loss value decay curve

results is minimal, and the inversion results for both the noise-free data and the 3%-noisy data are in good agreement with the real model. However, as the noise level increases, the gap between the U-Net's inversion results for the noisy data and the real model gradually widens. Despite this, the overall structure and trend of the geoelectric model in the inversion results remain consistent with the real geoelectric model. These results demonstrate that the U-Net inversion strategy in this case can effectively resist low levels of noise interference. However, when faced with higher levels of electromagnetic interference, the inversion accuracy of this method tends to decrease.

In this case, a numerical simulation experiment was designed to demonstrate the superiority of the new method. The test results indicate that the new method exhibits stable noise resistance, strong generalization ability, and high inversion efficiency. These results prove the feasibility of using deep learning methods to achieve SOTEM resistivity structure imaging, laying a solid foundation for processing SOTEM field data.

6.4.5 Underground Structure Investigation of a Coal Mine in Xuejiawan, Inner Mongolia

The work area is located in Longwangqu Village, Xuejiawan Town, Zhungeer Banner, Ordos City, Inner Mongolia Autonomous Region. It is part of the Ordos Plateau. The work area is basically covered by the Quaternary, and bedrock is exposed only in

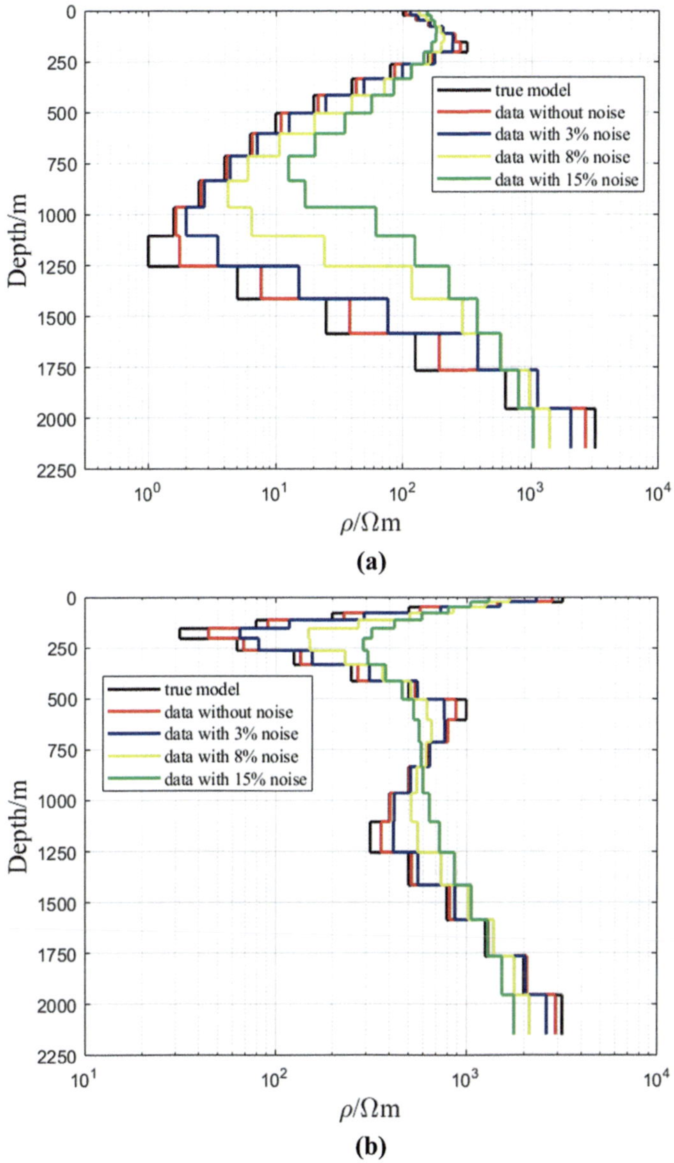

Fig. 6.40 Full-size U-Net noise resistance test

some valleys. According to the stratum exposure and drilling reveal, the strata from old to new are Ordovician Lower Lianjiashan Formation, Carboniferous Middle Benxi Formation, Carboniferous Upper Taiyuan Formation, Permian Lower Shanxi Formation, Lower Shihetzi Formation, Permian Upper Shihetzi Formation, Neogene

Upper New Formation, Quaternary Upper Pleistocene Malan Formation and Quaternary Holocene. The regional structure is simple, and no large folds and faults have been found.

The measurement area has developed gullies, with a vertical depth of several tens of meters. The bedrock layer is fully exposed, and the basic geological structure is clearly revealed. The south side of Longwangqu Station is Longwang Gully, which flows into the Yellow River all year round. The groundwater level in the survey area is low, but the surface runoff water all enters the gully and drains into Longwang Gully through the railway culvert. The main goal of this detection is to investigate the stability of the underground strata in this area and find possible goaf and hidden faults underground.

The layout of the transmitting and receiving devices is shown in Fig. 6.41. The transmitter is 500 m long, powered by a 40 kw generator, with a transmission current of 15 A, a base frequency of 25 Hz, an offset distance range of 310–510 m, a line distance of 20 m, a total of 11 measurement lines were set up, named L1–L11, each measurement line is 400 m long, and the point distance is 10. m, with a single point observation duration of 2 min, using a magnetic rod with an effective receiving area of 10,000 m^2 and a V8 receiver from the Canadian Phoenix Company to measure the vertical induced electromotive force dB_z/dt, as shown in Fig. 6.42.

After completing the data collection, the U-Net model trained in Chap. 3 was employed to perform pseudo-seismic imaging processing on all the SOTEM data from the 11 survey lines. The results are displayed as a three-dimensional composite image of the pseudo-seismic structural surface, as shown in Fig. 6.43. In this figure, the blue interface indicates a resistivity decrease in the lower layers compared to

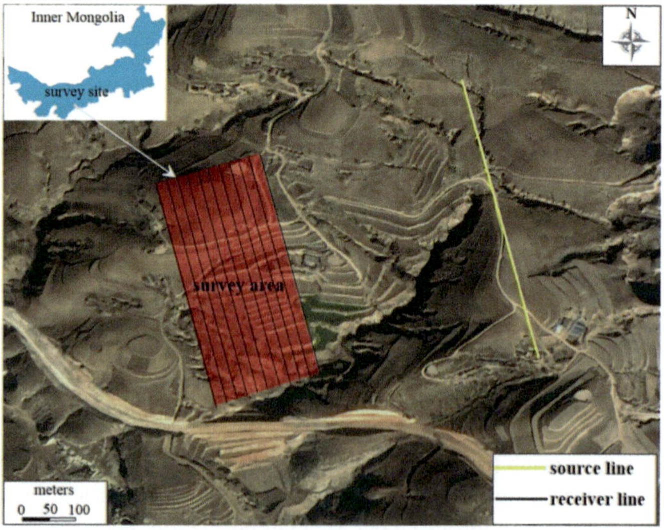

Fig. 6.41 Overview of the work area and schematic diagram of the transmitter and receiver layout

Fig. 6.42 Receiver system **a** V8 receiver, **b** SB-7K magnetic probe

the upper layers, while the red interface represents an increase in resistivity in the lower layers compared to the upper ones. From the imaging results, it is evident that within the 500-m depth range, there are four primary electrical interfaces, dividing the region into five electrical layers. The resistivity pattern exhibits a fluctuation from high to low to high again, then low, and back to high, indicating relatively stable strata with no significant faults.

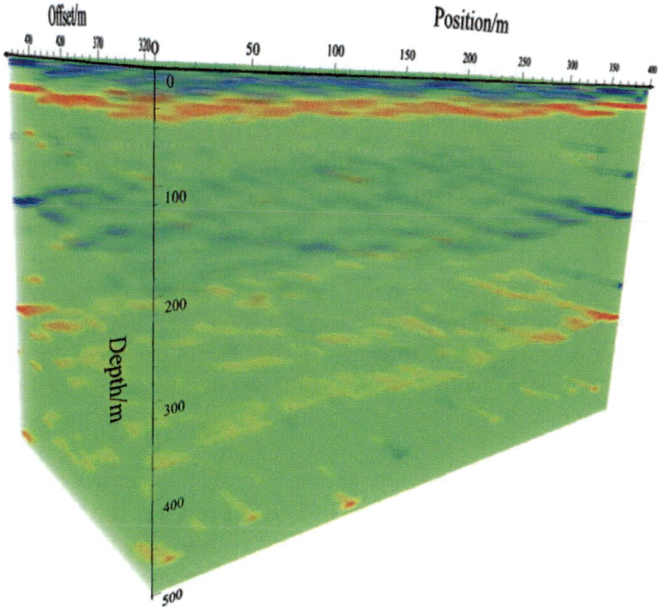

Fig. 6.43 Three-dimensional composite image of the imaging results of the 11 survey lines

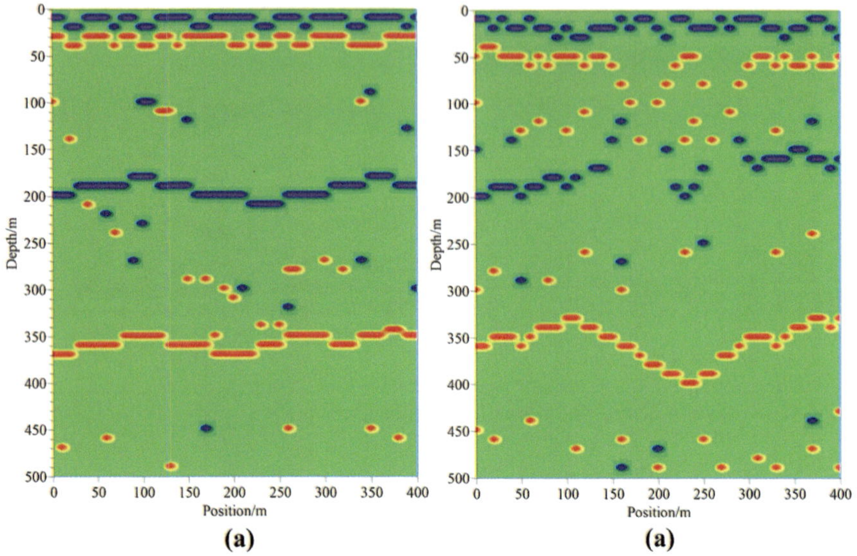

Fig. 6.44 **a** Pseudo-seismic imaging results of Line L1, **b** pseudo-seismic imaging results of Line L6

To further investigate the stability of the strata, the pseudo-seismic imaging results of the L1 and L6 survey lines were selected for detailed analysis, as shown in Fig. 6.44. In Fig. 6.44a, four continuous interfaces are visible at depths of ~20, 40, 200, and 370 m. Similarly, Fig. 6.44b also shows four interfaces at the aforementioned depths. However, at points L6-180 and L6-270, between depths of 40 and 200 m, the middle two layers show discontinuities in the horizontal direction, creating two low-resistance anomaly areas that are connected vertically. Additionally, the interfaces below a depth of 300 m exhibit a distinct step-like change pattern, further indicating the complexity of the subsurface structure.

Next, the SOTEM data from Line L6 were subjected to one-dimensional inversion using both the trained U-Net model and the Occam one-dimensional inversion module of SOTEMSoft. The results from both methods were compared and mutually verified through pseudo-seismic imaging. In the Occam inversion, the initial layer thickness was set to 10 m, with an increment factor of 1.02, and a total of 40 layers were used. Figure 6.45 illustrates the data fitting results for both inversions at point L6-160.

The fitting of the data is good in both cases, but the Occam inversion provides a better fit compared to the U-Net inversion. This is expected, as Occam is a model-driven inversion method, which achieves target accuracy by iteratively fitting the measured data through multiple rounds of forward modeling. While the U-Net inversion does not involve a fitting process, its accuracy is still sufficient to meet the

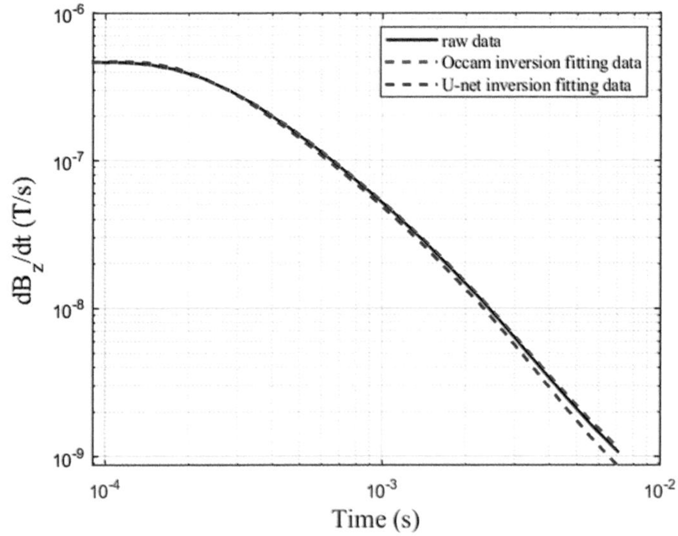

Fig. 6.45 Data fitting situation of the inversion results at L6-160

required standards, demonstrating that the deep learning approach can produce reliable results even without the detailed fitting that model-driven methods like Occam employ.

Figure 6.46a presents the one-dimensional inversion result obtained with U-Net, while Fig. 6.46b shows the result obtained with the Occam one-dimensional inversion. To facilitate comparison, the pseudo-seismic imaging results are projected onto the one-dimensional inversion results. From the figures, it is evident that the pseudo-seismic imaging results align well with the one-dimensional inversion results. The pseudo-seismic imaging clearly outlines the position of the interface, revealing five electrical layers that are shallower than 500 m, with resistivity changing in a pattern of high to low, high to low, and then high again.

The low-resistance loop closure depicted in the pseudo-seismic results for Line L6 is confirmed by both one-dimensional inversion methods. After verification with borehole data, it is found that this low-resistance loop closure corresponds to a water-filled mined-out area. Additionally, both one-dimensional inversion results show a clear step-like change in the interface below a depth of 300 m, which is interpreted as indicating a hidden fault structure. This verification further supports the reliability and accuracy of both inversion methods in identifying underground structures.

The above analysis demonstrates the successful application of pseudo-seismic structural surface imaging to reveal the underground structure of a coal mine area in Xuejiawan, Inner Mongolia. The results were mutually verified with one-dimensional inversion findings, and the stratification within 500 m of the underground structure was explored. The analysis indicates that the strata within 500 m of the depth are generally stable and intact. However, at points 180 and 270 of Line L6, potential

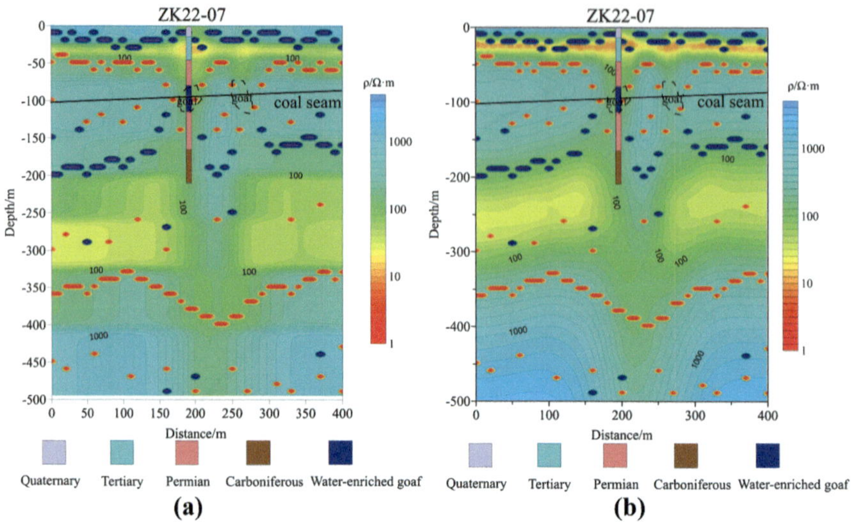

Fig. 6.46 **a** One-dimensional inversion result of Line L6 with U-Net and pseudo-seismic imaging result, **b** one-dimensional inversion result of Line L6 with Occam and pseudo-seismic imaging result

water-filled mined-out areas were detected at depths shallower than 200 m. Additionally, at a depth of 350 m, a hidden fault structure was identified. This fault is not large in scale and has not caused significant damage to the strata.

In comparison to one-dimensional inversion, pseudo-seismic imaging does not provide specific resistivity values for each layer but excels in presenting the interface with greater clarity. Additionally, pseudo-seismic imaging based on deep learning, combined with one-dimensional inversion, can be completed at a remarkably fast speed. Once the data is input into U-Net, the prediction for all data can be made in less than one second, demonstrating the efficiency and practicality of this approach for fast geophysical analysis.

References

Chen Weiyin, Xue Guoqiang. SOTEMsoft, a data processing software system for short-offset transient electromagnetic method with electric source[J]. Journal of Earth Science and Environment, 2021, 43(6): 1050-1056.

Çiçek O, Abdulkadir A, Lienkamp S S, et al. 3D U-Net: Learning Dense Volumetric Segmentation from Sparse Annotation [J], Springer, 2016, Cham.

Krizhevsky A, Sutskever I, Hinton G E. Imagenet classification with deep convolutional neural networks[J]. Advances in Neural Information Processing Systems, 2012, 25: 1097–1105.

Qu S, Verschuur E, Zhang D, et al. Training deep networks with only synthetic data: Deep-learning-based near-offset reconstruction for closed-loop surface-related multiple estimation on shallow-water field data[J]. Geophysics, 2021, 86(3): 1-32.

Ronneberger O, Fischer P, Brox T. U-Net: Convolutional networks for biomedical image segmentation[C]. International Conference on Medical Image Computing and Computer-Assisted Intervention, 2015, 234–241.

Song T, Kuang L, Han L, Wang Y, Liu Q H. Inversion of rough surface parameters from SAR images using simulation-trained convolutional neural networks[J]. IEEE Geoscience and Remote Sensing Letters, 2018, 15: 1130–1134.

Wu H, Zhang B, Lin T F, Cao D P, Lou Y H. Semiautomated seismic horizon interpretation using the encoder-decoder convolutional neural network[J]. Geophysics, 2019, 84(6): B403–B417

Wu X, Xue G Q, He Y M, et al. Removal of the multi-source noise in airborne electromagnetic data based on deep learning[J]. Geophysics, 2020, 85(6):1-72.

Xiao J, Li J, Chen Y, Han F, Liu Q H. Fast electromagnetic inversion of inhomogeneous scatterers embedded in layered media by born approximation and 3-D U-Net[J]. IEEE Geoscience and Remote Sensing Letters, 2020, 17: 1677–1681.

Yang F S, Ma J W. Deep-learning inversion: a next generation seismic velocity model building method[J]. Geophysics, 2019, 84: R583–R599.

Yin Changchun, Zhu Jiao, Qiu Changkai, Cai Jing. Airborne electromagnetic quasi-three-dimensional model space constraint inversion. Chinese Journal of Geophysics, 2018, 61(6): 2537-2547.

Chapter 7
Transient Electromagnetic Detection Equipment System of Electrical Source

7.1 Special Detection Equipment System Requirements

Long ground wire sources typically emit ground wave S_1 and stratum wave S_2. Due to the significant difference in wave speeds between S_1 and S_2, their connection forms an approximately horizontal wavefront S* that propagates almost vertically downward at large transmission and reception distances. Controlled Source Audio Magnetotelluric (CSAMT) and Long Offset Transient Electromagnetic (LOTEM) primarily consider the S* wave as the excitation field. In the far-source region, the S_2 wave is sufficiently attenuated, allowing the S* wave to be regarded as the sole excitation field, thus establishing a "far-source" detection mode.

Due to the increase in transmission and reception distances, the strength and bandwidth of the transmitted signal significantly decrease (Ziolkowski 2009), which, on one hand, results in a considerable degree of transmission energy loss along the propagation path, with only a small portion ultimately utilized for detection. On the other hand, this leads to significant attenuation of the high-frequency components in the signal, thereby diminishing the system's resolution (Hwarong 1990). Consequently, the improvement of detection depth and accuracy in the far-source region is limited.

How to manage the geometric relationship between the observation point and the field source has always been a key issue affecting the detection performance of artificial source electromagnetic methods. He Jishan conducted research on wave field response and developed the Wide Area Electromagnetic Method. Since 2000, the University of Edinburgh in the UK has proposed the MTEM method (Multi-Transient Electromagnetics), based on a near-source in-line arrangement for marine oil and gas exploration. In 2013, Xue Guoqiang proposed the Short Offset Transient Electromagnetic Method based on in-depth research on stratum wave S_2 (Guoqiang et al. 2013). Since 2017, the University of Cologne in Germany, which originated LOTEM, has also conducted research on near-source methods.

A certain range of transmission and reception distances has a significant impact on the response signal and bandwidth. Assuming the Earth is a uniform half-space

G. Xue et al., *Exploration with Short-Offset Grounded-Wire Transient Electromagnetic Method*, https://doi.org/10.1007/978-981-96-8846-3_7

and that the observation point is on the ground, the Z-component response of the frequency domain electromagnetic field is given by.

$$H_z(x, y, 0) = |H_z(x, y, 0)|_P + |H_z(x, y, 0)|_N \tag{7.1}$$

$$|H_z(x, y, 0)|_P = -\frac{iIds}{4\pi^2 k_1^2}\frac{y}{r^5}\left[\left(3 - 3ik_1 r - k_1^2 r^2\right)e^{ik_1 r}\right] \tag{7.2}$$

$$|H_z(x, y, 0)|_N = \frac{iIds}{4\pi^2 k_1^2}\frac{3y}{r^5} \tag{7.3}$$

where $r = (x^2 + y^2)^{1/2}$, that is, the distance between the measurement point and the source point; P and N are Sommerfeld and Foster integrals, respectively, representing the action process of stratum wave and ground wave; *Ids* is the transmission current element.

The primary observation quantity in the SOTEM method is the Z-axis magnetic field. To further investigate the frequency response form of Hz (x, y, 0), the term within the square brackets in formula (2) is extracted separately. Assuming the Earth resistivity is 500 Ω·m, there are four observation points along the Y-axis, with offset distances of 300 m, 900 m, 1800 m, and 3000 m. The term in the square brackets of formula (2) is calculated over the frequency range of 1–100 kHz.

As can be seen intuitively from Fig. 7.1a, the amplitude frequency response exhibits significant low-pass filtering characteristics, with its bandwidth continuing to increase as the offset distance decreases. In Fig. 7.1b, the single response curve in the middle contains three sections, progressing from low frequency to high frequency. Taking the curve with an offset distance of 300 m as an example, from the low frequency end to about 1 kHz, it displays linear phase characteristics. Between 1 kHz and approximately 20 kHz, a non-linear form emerges. When the frequency exceeds 20 kHz, it shows linear characteristics again. The position and extent of the non-linear section result from the joint action of the observation device, which has a certain transmission and reception distance, and the Earth. As the offset distance decreases, the obtainable bandwidth of $H_z(x, y, 0)|_P$ will increase.

In order to estimate the cut-off frequency (Cut-off Frequency, CF) of the amplitude frequency response of $H_z(x, y, 0)|_P$ under different offset distance conditions, we introduce the 1/e amplitude level (black dashed line) in Fig. 7.1a as the cut-off frequency estimation standard. The CF of the amplitude frequency response curve at different offset distances is calculated, and the results are shown in Table 7.1. As the offset distance increases, the CF will significantly decrease.

Through the above analysis, it can be seen that there is an inherent connection between signal attenuation and available bandwidth. As the offset distance decreases, the signal bandwidth increases; when the offset distance is 300 m, the working bandwidth is 25 kHz. Currently, the two systems from Phoenix Company in Canada and Zonge Company in the United States are developed for the traditional "far source area" mode, with the highest base frequency at the 10 kHz level. The shutdown time

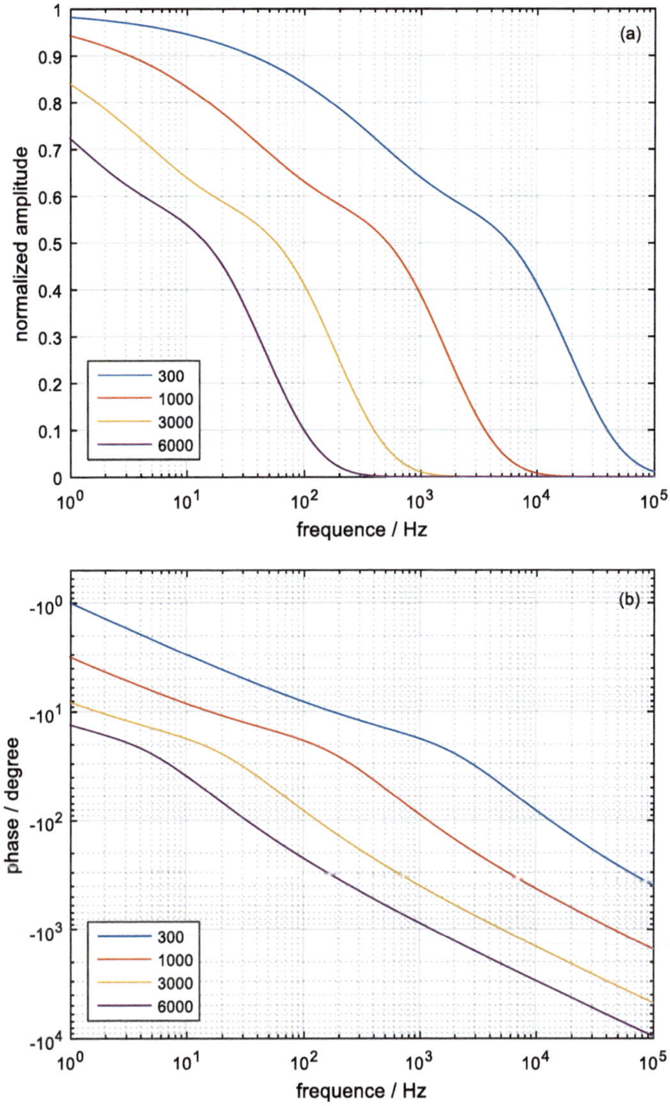

Fig. 7.1 Relationship between amplitude frequency, phase frequency response and offset distance. **a** Amplitude frequency response; **b** Phase frequency response

Table 7.1 Cut-off frequency of amplitude frequency response curve at different offset distances

r_n (m)	300 m	750 m	2100 m	6000 m
1/e (Hz)	12,362.814	1978.0503	252.3023	30.907033

is longer than 100 μs, and the sampling rates of the two systems are 24 kHz and 32 kHz, respectively. The receiver bandwidth is both 8 kHz, which makes it challenging to meet the requirements of the SOTEM method.

7.2 Analysis of MTEM Detection Mechanism

In the previous sections, we have explained the theoretical basis, advantages, and main challenges of the short-offset electromagnetic method. From this section, we will discuss the detection mechanism and detection performance of the Multi-Transient Electromagnetic Method (MTEM) and SOTEM, the two leading short-offset time-domain electromagnetic methods in the world.

MTEM was proposed by researchers at the University of Edinburgh based on the classic long-offset electromagnetic method LOTEM. The main idea of MTEM at its inception was to learn from seismic technology in oil and gas exploration, utilizing multiple emissions of electrical sources, array-type multi-channel reception, and multiple coverage of full wave field information. This approach aims to realize seismic-like processing of data, thereby greatly improving detection accuracy and depth under the same emission intensity, allowing the detection depth to reach more than 2000 m, which is more conducive to the fine detection of deep oil and gas layers and their activities.

The prototype of the MTEM method was formed between 1994 and 1996, when the University of Edinburgh led a project named the THERMIE project OG/0305/92/NL-UK, with CGG Company, DMT Company, and the University of Cologne in Germany participating. A method test was conducted in St. Illiers la Ville, 30 km west of Paris. The purpose of the test was as follows: First, to try a new data processing method that can directly discover gas storage layers from the data; second, by comparing the observation data from 1994 and 1996, to discover the movement of the gas–water interface. The so-called "new data processing method" is the method proposed by Ziolkowski and Hobbs to apply the common midpoint stack processing commonly used in seismic exploration to electromagnetic data processing.

As shown in Fig. 7.2, this is a schematic diagram of the test area for this experiment. The small blue dots on the figure represent the positions of the transmitting electrode piles. Figure 7.3 is a detailed layout diagram of a transmitting position. There are eight circles on the figure, representing eight steel pipes, each three meters long. These eight steel pipes are grouped in pairs, corresponding to a small blue dot in Fig. 7.3. The spacing between the steel pipes in the same group is 3 m, and they are used in conjunction. For each transmitting position, two transmissions are required: one with the electrode direction parallel to the survey line direction (the inline direction in Fig. 7.3), and one perpendicular to the survey line direction (the crossline direction). The distances of the transmitting electrodes in the inline and crossline directions are both 250 m. The system uses Zonge's GGT-30 as the transmitter (equipped with a 30kVA generator), and the transmitting current is adjusted between 5 and 40 A depending on the coupling situation of the transmitting electrodes.

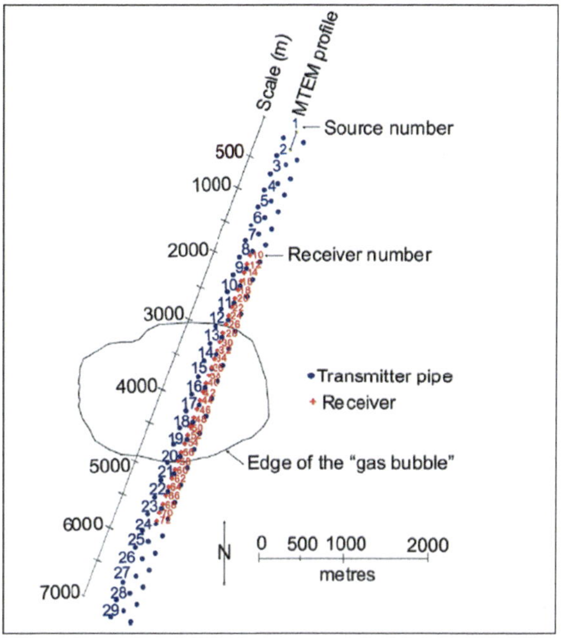

Fig. 7.2 Schematic diagram of the MTEM test area

As shown in Fig. 7.2, the small red crosses on the figure represent the observation point array, and Fig. 7.4 is a detailed layout diagram of this array. As can be seen from Fig. 7.4, a receiving unit is used at each observation position, using the TEAMEX type receiving unit developed by Rüter and Strack (modified from the SEAMEX receiving unit used by DMT for seismic data collection). This receiving unit has two observation channels. For the receiving unit at an odd position, the two channels are used to observe the inline electric field and the crossline electric field; for the receiving unit at an even position, the two channels are used to observe the inline electric field and the vertical magnetic field.

The main components of the receiving unit include an analog low-pass filter, a notch filter, a floating amplifier, and an A/D converter. The sampling rate of the receiving unit is 1 kHz, and the total number of samples is 2048, including 384 pre-trigger samples. The data obtained from the observation is not superimposed on the receiving unit, but the observation data of 16 receiving units is summarized on the receiver (similar to the artificial source seismic method) for data superimposition, storage, and some data preprocessing.

The electrode spacing for observing the electric field signal is 125 m, using standard $Cu/CuSO_4$ electrodes; the return line for receiving the magnetic field attenuation is a square multi-turn return line with a side length of 50 m. The system has a total of 32 receiving unit positions, covering a survey line range of 4 km. During observation, the receiving array is first set up, and then the transmitter is used to perform inline and crossline direction transmissions at each transmitting position.

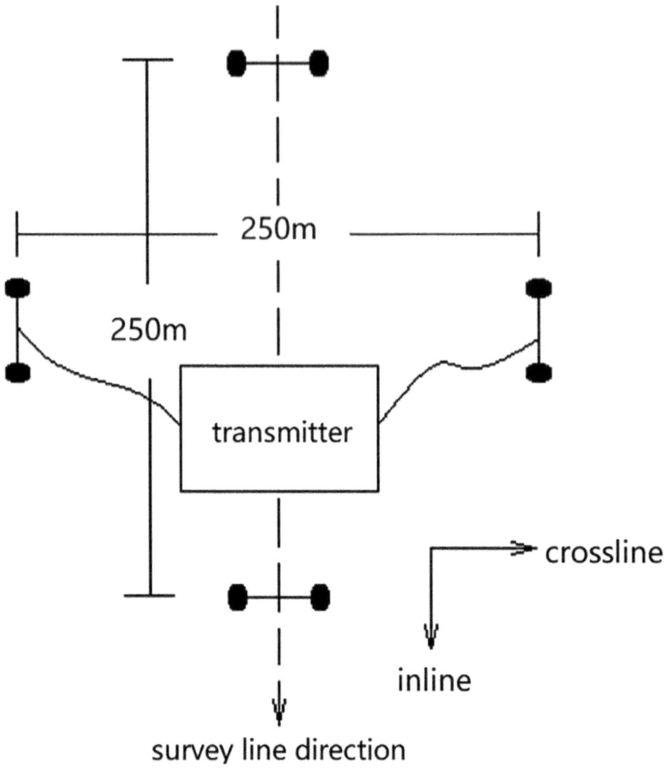

Fig. 7.3 Transmitter layout diagram

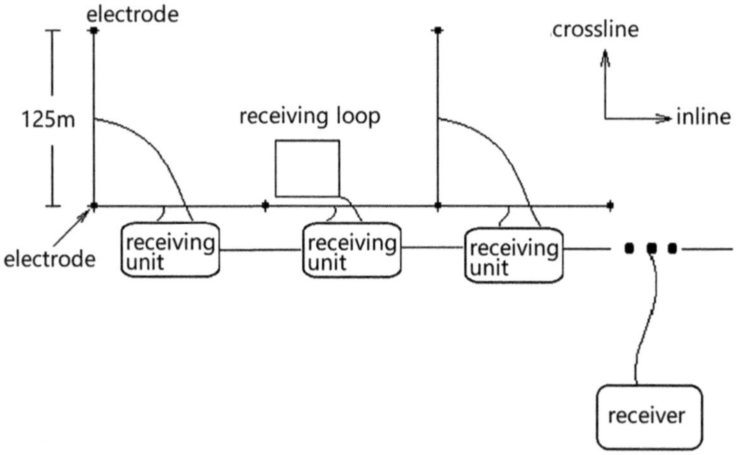

Fig. 7.4 Receiver array layout diagram

In this experiment, based on the experience of LOTEM, it is believed that the system response of the entire system is stable and unchanged. However, combined with experimental data, Wright pointed out that the transmitting current itself (amplitude) is a function of time; its change characteristics will vary with the different positions of the transmitter, the different distances of the transmitting poles, and the current reverse capability of the transmitter itself is also related to the earth load. Therefore, Wright proposed the following four suggestions:

(1) System response observation must be carried out for each transmission.
(2) The recording system used must be completely consistent with the recording system of the receiving unit. If it is not consistent, it is necessary to accurately know the conversion function between the recording system used and the recording system of the receiving unit.
(3) When using a digital recording system, distortion must be prevented. The sampling rate of TEAMEX is 1 kHz, which is not enough for system response observation.
(4) A Hall sensor can be used to directly observe the transmitting current.

In addition, Wright carefully analyzed the data of the St. Illiers la Ville experiment and believed that only inline transmission and inline electric field observation are needed, and other forms of transmission and reception cannot provide more information. Under the support of this theory, the arrangement shown in Fig. 3.10 was proposed. This arrangement has become the typical arrangement of MTEM, whether on land or at sea, whether using bipolar square waves or PRBS coding as the transmission waveform, MTEM observation always uses this arrangement. In the figure, ΔXs represents the transmitting pole distance, ΔXr represents the receiving pole distance, r_1 represents the offset distance of the receiver position closest to the transmitter, so r_1 is also r_{min}, r_n represents the offset distance of the nth receiver position (Fig. 7.5).

Under the guidance of the above experience, TOTAL company, MTEM company (a company of the same name established by the main developers of the MTEM method), and the University of Edinburgh jointly supported a verification experiment of the MTEM method in southern France. Although before this experiment, research had proposed the use of m-sequence (maximal length sequence, the longest linear shift register sequence) pseudo-random binary sequence (Pseudo Random Binary Sequence, PRBS) as the transmission code form, this experiment still used a continuous bipolar square wave (with a frequency of 1–0.5 Hz adjustable). The transmitter still uses Zonge's GGT30, with a transmission voltage of 750–1000 V, to ensure

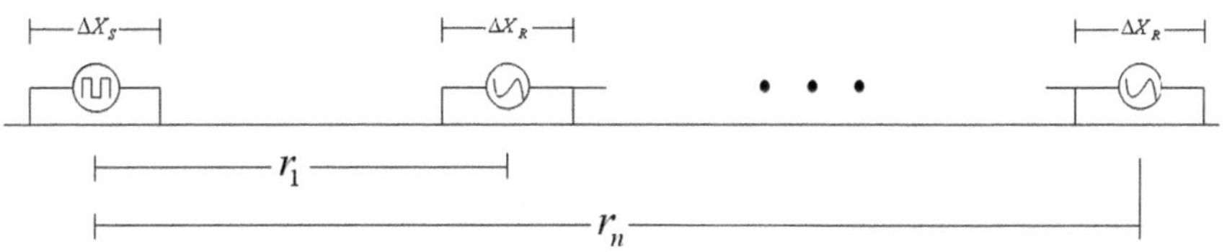

Fig. 7.5 Schematic diagram of the MTEM receiving and transmitting array

that the transmission current can be between 20 and 50 A. The receiving unit (still dual-channel) is independently developed by MTEM, with a sampling frequency of 15 kHz, and is equipped with GPS and a microprocessor. Each receiving unit is synchronized through GPS, independently performs A/D conversion, data recording and storage, and then connects to the receiver using high-speed communication cables. In addition, an online QC system has been introduced into the receiving unit, which can check the system status and data of each channel. A total of 11 receiving units were used in this experiment, 10 of which were used for electric field response observation (only inline direction electric field), and one was used for system response observation. The main progress of this experiment is reflected in: the air wave has been clearly understood, and the data processing process has been improved.

MTEM's first practical application of m-sequence coded transmission waveforms was a marine test conducted from 2007 to 2008 after MTEM was acquired by PGS. The m-sequence is a relatively easy-to-implement "white noise", which is very suitable as an input signal for system identification. The m-sequence has two key parameters: order and code element width. The order represents the number of shift registers, and the code element width represents the duration of the shortest logical state (shortest run) in the sequence. The concept of code element frequency (or clock frequency) is also used in general discussions, which is the reciprocal of the code element width. As shown in Fig. 7.6, it is a 4th order m-sequence with a code element frequency of 400Hz.

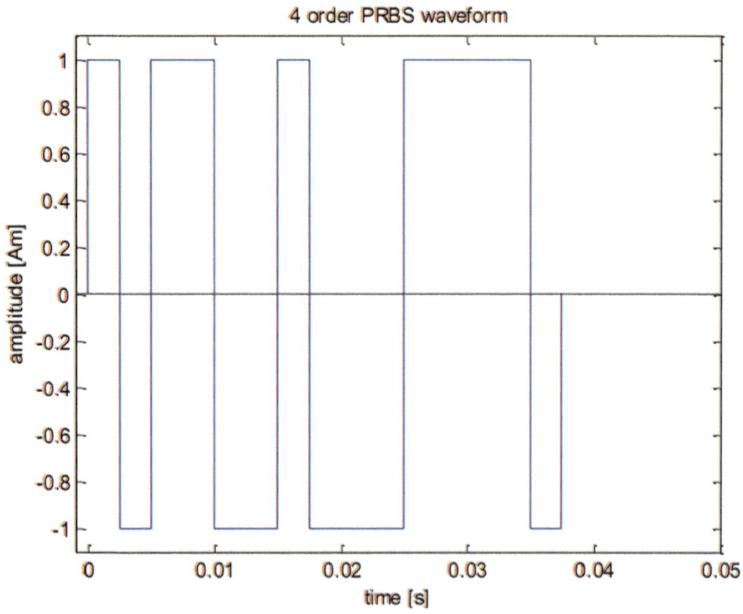

Fig. 7.6 4th order m-sequence coding

Using the m-sequence as the transmission waveform, the transmission process must be segmented according to different offset distances to the receiver position, and calculate the appropriate code element width for each segment. Because the choice of code element frequency is related to the offset distance, different offset distances require different code element frequencies. In this experiment, the distance between the transmitting electrodes was 400 m, the amplitude of the transmitting current was 500 A, and the distance between the receiving electrodes was 200 m. For observation points with a transmission and reception offset distance of 2000–3500 m, the maximum code element frequency is 40 Hz; for observation points with a transmission and reception offset distance of 3500–4900 m, the maximum code element frequency is 10 Hz; for observation points with a transmission and reception offset distance of 4900–6400 m, the maximum code element frequency is 5 Hz.

This choice of code element frequency depends to a certain extent on prior knowledge. If there is a lack of prior knowledge, it is necessary to expand the range and try multiple times. For example, for the above-mentioned observation points from 2000 to 3500 m, in the absence of prior knowledge, you will have to try multiple times within 100Hz, and then confirm the best set through data analysis. This characteristic to some extent reduces the efficiency of the MTEM method.

After obtaining the data, the MTEM method uses the m-sequence as the identification input signal to identify the earth system. The data processing process is divided into two steps: first, the earth impulse response is extracted through the correlation method, and then the earth impulse response is processed to extract the distribution information of the earth's electrical structure.

$$y(t) = z(t) + n(t) \tag{7.4}$$

$$z(t) - g(t)*u(t) \tag{7.5}$$

Figure 7.7 shows the measured response signal and the transmission waveform (segment). Using u(t) to represent the transmission waveform (identification input signal), g(t) to represent the earth impulse response, n(t) to represent noise, and y(t) to represent the response observation signal (identification input signal), the detection process satisfies the following relationship:

The correlation processing process records $u(t)$ and $y(t)$ separately, and calculates the correlation between the two:

$$CR(y, u) = CR(z, u) + CR(n, u) = g * AR(u) + CR(n, u) \tag{7.6}$$

where *CR* () represents cross-correlation calculation, *AR* () represents auto-correlation calculation. Since the *m* sequence is uncorrelated with noise, then *CR* (n, u) = 0. Therefore, Eq. (7.6) further equals:

$$CR(y, u) = g * AR(u) \tag{7.7}$$

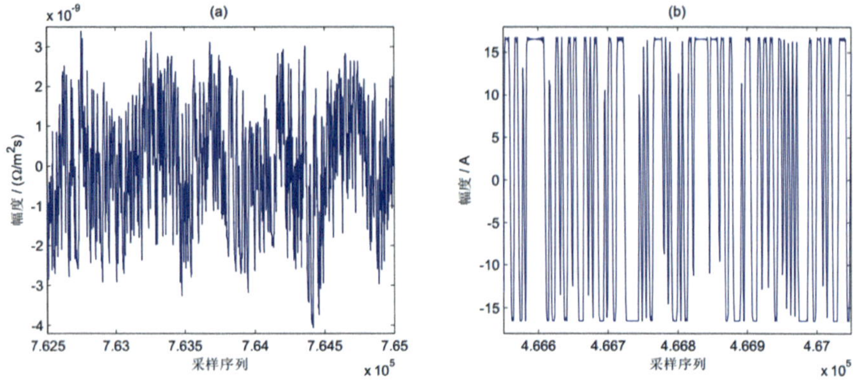

Fig. 7.7 Measured response signal and transmission waveform (segment) **a** Response signal y(t); **b** Transmission waveform u(t)

When the requirement for the fineness of the processing result is not too high (such as quick data quality assessment in the field), the characteristic of the m-sequence auto-correlation being approximately equal to the δ signal can be used to consider $CR\ (y,\ u) = g^*\ AR(u) \approx g^*\delta = g$, thereby obtaining a quick estimate of the earth impulse response g. However, in actual work, due to the distortion effect of the earth on the transmission current and the impact of the receiver response characteristics on the response signal, the actual $AR(u)$ The side lobes will have complex oscillation characteristics. Therefore, based on Eq. (7.7), the MTEM method uses deconvolution to utilize the cross-correlation between transmission and reception and the autocorrelation of transmission to achieve fine extraction of the earth's impulse response.

After obtaining the earth's impulse response, Ziolkowski et al. proposed a method to estimate the earth's resistivity based on the peak time of the earth's impulse response:

$$t_{peak} = \frac{\mu_0 r^2}{10\rho} \tag{7.8}$$

For a uniform half-space with resistivity ρ, t_{peak} is the arrival time of the peak of the earth's impulse response. Based on Eq. (7.8), the apparent resistivity of the earth under different transmission and reception combinations can be calculated. It should be noted that the aforementioned deconvolution method based on m-sequence is not the only method to extract the earth's impulse response. The processing method based on square wave can also extract the earth's impulse response. In fact, when MTEM proposed its apparent resistivity formula, it had not yet carried out detection work based on *m* coded transmission waveform, but still used the traditional square wave transmission waveform and completed all subsequent processing work.

Based on the apparent resistivity values of each transmission and reception combination, a common midpoint gather is formed according to the traditional seismic data

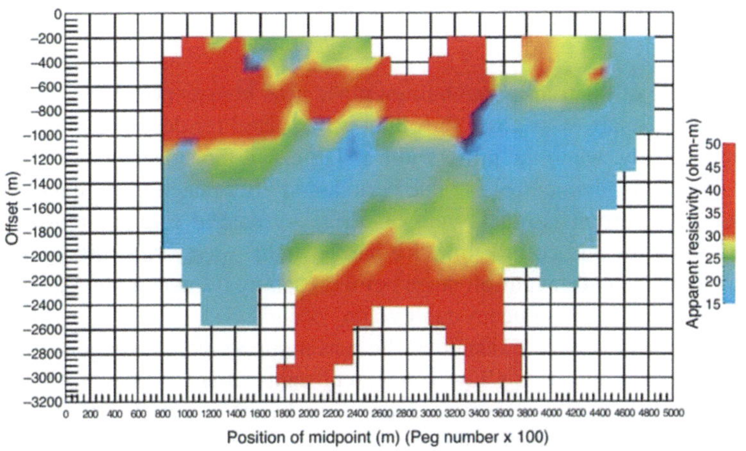

Fig. 7.8 Travel time-resistivity imaging under common midpoint coordinates

imaging method. As shown in Fig. 7.8, the pseudo-seismic imaging result obtained by MTEM in the experiment in southern France clearly shows a high-resistance structure under the low-resistance layer from a horizontal distance of 1800 m to 4000 m. Verification indicates that this high-resistance structure has a high correspondence with the pseudo-observation target.

The success of the MTEM method proves the feasibility of short-offset electromagnetic detection. After the MTEM method, the near-source electromagnetic method gradually became a hot topic in the international electromagnetic method research field. Inspired by this, Chinese researchers independently proposed the Short Offset Transient Electromagnetic Method (SOTEM) for metal ore and engineering environment detection, significantly promoting research development in this field in China.

7.3 Key Technologies of Short Offset Transient Electromagnetic Detection System

Simply studying near-source methods or equipment parameters will have limited practical effects. Only when the equipment and methods are matched can the detection performance be improved. In order to increase the detection depth, it requires a larger transmission current amplitude and a lower sensor noise level; to improve the detection accuracy, it requires a larger transmission signal bandwidth and a larger sensor working bandwidth (i.e., linear bandwidth). How to achieve large current fast cut-off transmission and low noise large bandwidth sensor becomes a key issue.

7.3.1 Large Current Fast Cut-off Transmitter

The amplitude of the transmitted current pulse is as high as possible, and the pulse width can be as narrow as possible, which is beneficial for deep and high-precision detection. Ideally, when the pulse width is large, the transmission current rises to the set amplitude, then enters a stable period, and after reaching the cut-off trigger condition, the current is quickly cut off (Fig. 7.9 dashed line); in reality, due to the long rise time and cut-off time, for narrow pulse waveforms, the transmission current enters the current cut-off stage before reaching the set amplitude value. The phenomenon of "high frequency can't go up" occurs.

For a magnetic source transmitter, the power supply voltage is generally at the level of tens of volts. Using low-voltage constant-voltage clamp technology can compress the rise and fall edge time, and the bandwidth can easily be expanded. For an electrical source transmitter, the bus voltage is generally 500–1000 V. Due to the harsh parasitic parameter conditions of the device, it is easy to introduce vibrations, spikes, etc. High-voltage constant-voltage clamp technology makes the control of the transmission waveform precise and reliable. This article proposes a super high-voltage constant-voltage clamp technology based on fine control of current feedback, which increases the clamp voltage from the current 1000–2500 V, and achieves fine control of the transmission waveform.

After the transmitter starts, it first completes the charging of capacitor C1 under the drive of a 1000 V bus voltage, and outputs a converter waveform (Fig. 7.10). Record the peak value of this current waveform as a reference value I_{ref}; then turn on K_5, so that the capacitor C1, the bus voltage quickly rises to 2500 V (the speed is 2.5 times that of 1000 V voltage); when the current pulse current rises to I_{ref}, the comparator flips and closes K_5, to restore the bus voltage to 1000 V, at this time the current continues to rise, the new peak value reached by the current at the falling edge trigger moment is recorded as the new reference value I_{ref}. This process is repeated until the emission current reaches the preset value. This strategy uses a combination of U_2 and C, which is similar to giving the original rising edge an extra "boost". Since the power of U_2 is much lower than the main power source U_1, its essence is to support a higher instantaneous power consumption with a lower average power consumption, without making additional demands on the main power device.

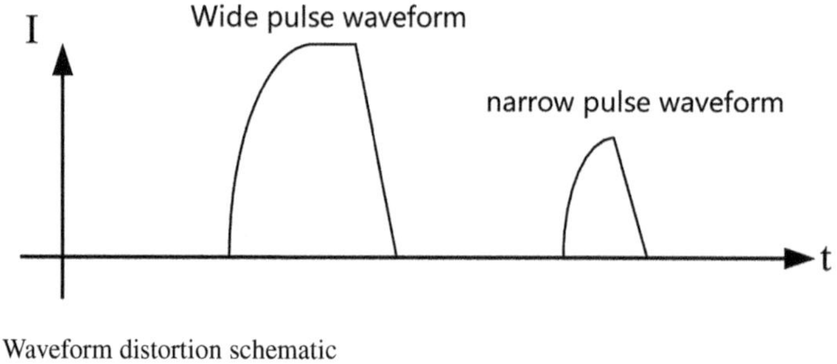

Fig. 7.9 Waveform distortion schematic

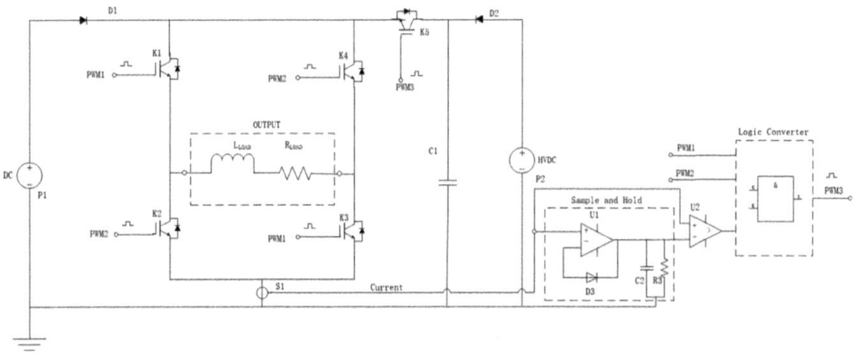

Fig. 7.10 Schematic diagram of fine control strategy for current feedback

7.3.2 Low Noise Wide Bandwidth Sensor

SOTEM uses an inductive magnetic field sensor (hollow coil). The noise sources of the hollow coil include the coil itself and the amplification circuit. The main work of low noise optimization is to develop a low noise amplification circuit. To balance low noise and wide bandwidth, we adopt a design and development strategy of first focusing on low noise and then on wide bandwidth.

A hybrid amplification circuit based on tube-level parallel connection is designed (as shown in Fig. 7.11), using 6 junction field effect transistors (J1 ~ J6) in parallel, and together with J7 and J8, they form a bootstrap common source common gate circuit. R11 and R12 provide bias voltage, and with the ± 15 V power supply voltage, the Miller effect is effectively suppressed, and common mode distortion is reduced. In addition, the overall voltage noise at the low frequency end (10 Hz) is no more than 0.8 nV/sqrt(Hz), and the current noise at the high frequency end (10 kHz) is no more than 0.5 pA/sqrt(Hz). The sensor noise performance optimization design is thus completed, and the measured system time domain noise is 0.1 nT/s.

The larger the damping coefficient (Damping Coefficient, DC), the smaller the linear segment bandwidth on the sensor sensitivity curve. However, if the sensor

Fig. 7.11 Hybrid amplification circuit based on tube-level parallel connection

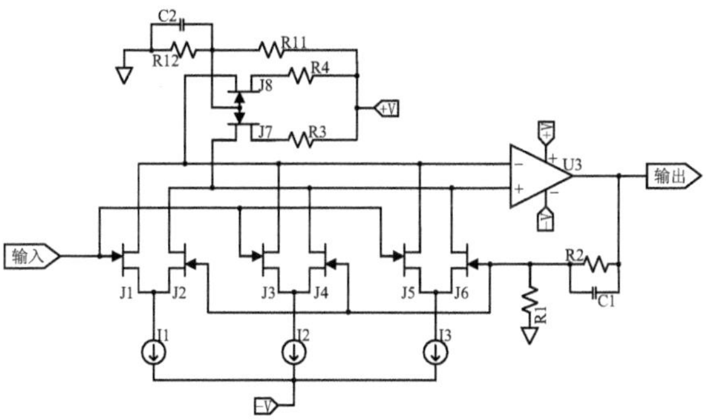

linear bandwidth is simply increased by reducing the DC, it will reduce the applicability of the sensor. We use adaptive variable damping technology to expand the bandwidth, as shown in Fig. 7.12. Through automatic control methods, feedback mediation based on actual field conditions and real-time programmable high-precision logic circuits are introduced to achieve precise matching of resistance values, thus obtaining more reliable wide-bandwidth observation signals. The actual operation process is similar to the calibration process after traditional equipment enters a new measurement area. The optimal damping value for the current measurement area is obtained through testing, and the parameters are kept stable throughout the measurement area. The adaptive variable damping module mainly consists of a feedback circuit, a microcontroller (MCU), and programmable resistors. The feedback circuit collects the output voltage V_{out}, amplifies and filters it through a conditioning circuit, and then inputs it into the MCU. The MCU analyzes the early features of the data, predicts the degree of distortion of amplitude and phase based on the analysis results, and generates a matching resistance adjustment value based on this. The MCU then transmits the matching resistance adjustment scheme to the programmable resistor through two data lines, serial data (SDA) and serial clock (SCL), to achieve adaptive adjustment of the damping resistance.

The programmable resistor used is the core component in the circuit. It is a new type of CMOS digital, analog hybrid signal processing integrated circuit that replaces traditional mechanical potentiometers (analog potentiometers). It can achieve fine adjustment of the damping resistance value R_t between 7000 Ω and 12000 Ω, thus achieving adaptive adjustment of the sensor DC corresponding to between 1.06 and 0.6. As shown in Fig. 6, the amplitude frequency response of the sensor was measured in the laboratory. It can be seen that without matching impedance (red curve), the curve has a main peak (the spike at the low frequency end is environmental interference in the laboratory), which corresponds to the resonance frequency. The measured curve shows that the resonance frequency of the sensor is about 100 kHz. The blue curve describes the situation when the damping coefficient is 0.9. It can be seen that the frequency deviating from the linear reference 1% (as a linear bandwidth condition) is about 30 kHz. When the damping coefficient increases, the corresponding curve will deviate from the linear reference earlier, resulting in a decrease in linear bandwidth. At the critical damping (damping coefficient = 1.0), the measured linear bandwidth is about 25 kHz. When the damping coefficient is further reduced, the

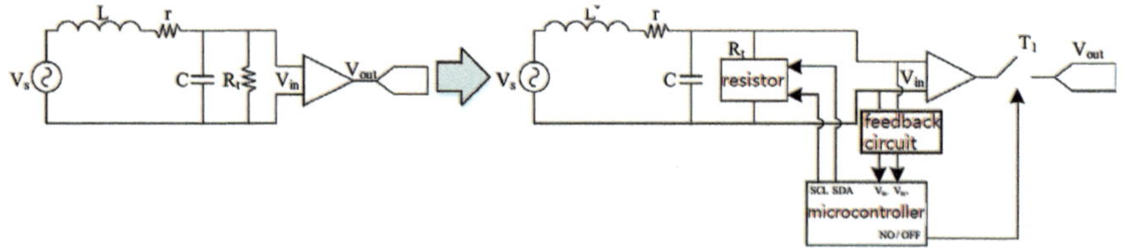

Fig. 7.12 Schematic diagram of adaptive variable damping technology scheme

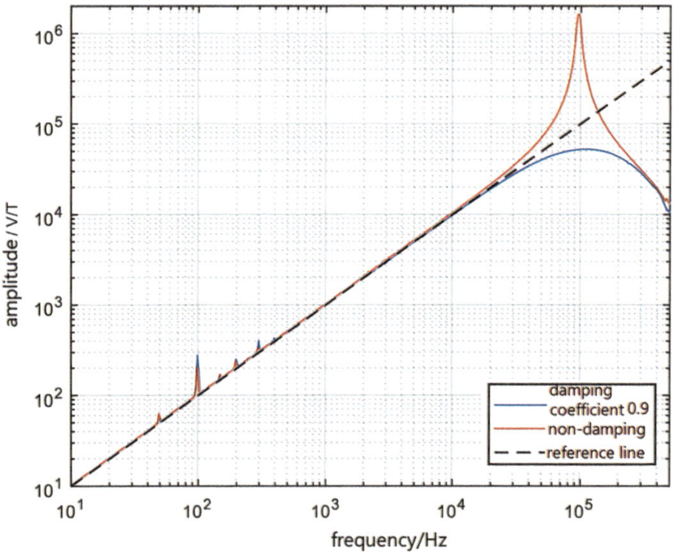

Fig. 7.13 Measured amplitude frequency response of the sensor

linear bandwidth can be further increased, which is beneficial for achieving more accurate detection (Fig. 7.13).

7.4 Field Application Test

Based on the breakthrough of the above key technologies, the development of SOTEM detection equipment has been completed, as shown in Table 7.2. The transmitter and its accessories are at the transmitting end, which includes the transmitter host at the bottom layer, the transmitter accessories and cable storage box at the middle layer, and the all-digital transmission control box at the top layer. The receiving end comprises the receiver box and sensor (hollow coil). The sensor and receiver are small in size, light in weight, large in bandwidth, and low in noise, which can not only be used for ground observation but also be carried by drones to conduct ground-air electromagnetic detection (Wu et al. 2019).

The developed transmitter was tested and compared under field conditions. The length of the transmission cable is 1.0 km, and the grounding resistance is 15 Ω. The maximum performance of the international mainstream transmitter is: transmission current 33 A, shutdown time is 135 μs, transmission power 16 kW (Fig. 7.14a); if the self-developed transmitter uses the same transmission current, its shutdown time is only 35 μs (Fig. 7.14b). Just comparing the above parameters, using the Bostick depth as the benchmark to describe the shallow blind area, under the common resistivity condition of 100 Ωm near the ground surface, the blind area depth of the international

Table 7.2 Short offset transient electromagnetic detection system parameter table

Equipment	Parameter	Indicator
	Maximum output power	40 kW
	Transmission waveform	Bipolar square wave\pseudo-random coding
	Shutdown time	53 μs
	Total weight	48kg
	Synchronization mode	GPS disciplined oven controlled crystal oscillator
Transmitter	Synchronization accuracy	200 ns
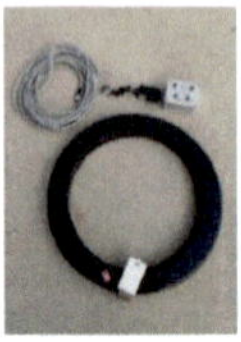	Sensor noise level	0.1 nT/s
	Linear bandwidth	25 kHz
	Effective area	20,000 m^2
	Receiver sampling rate	100 kHz
	Dynamic range	140 dB
Receiver	ADC bits	32 bits

mainstream transmitter is 41.3 m, and the self-developed transmitter is 21.1 m, which greatly reduces the blind area to about 1/2 of the international mainstream transmitter. If there is no limit on the current, the maximum performance of the self-developed transmitter under this condition is shown in Fig. 7.14c, with a transmission current of 51 A, a shutdown time of 53 μs, and a transmission power of 39.5 kW. It can be seen that under the same emission current conditions, the shutdown time of the self-developed transmitter has increased nearly 4 times, and the actual emission power of the transmitter has increased nearly 2.5 times. It can achieve shorter shutdown time under larger emission current conditions.

A comprehensive detection effect comparison was carried out in a coal mine area in Ningxia using self-developed equipment and international mainstream equipment (V8 system) (as shown in Fig. 7.15). The comparison line length is 1000 m, the measurement point distance is 50 m, the emission source length is 1060 m, the measurement line is parallel to the emission source, and the offset distance is 400 m. Both sets of equipment use the same working parameters: the emission base frequency is 25 Hz, the emission current is 18 A, the vertical induced electromotive force dBz/dt component is observed, and the single point measurement time is 2 min.

Figure 7.16 is the resistivity-depth cross-section obtained by one-dimensional inversion of the two sets of actual measurement data using SOTEMsoft software, where Fig. 9(a) is the result of the international mainstream system, and Fig. 9(b) is the result of the self-developed system. The inversion models and parameter settings of the two sets of data are exactly the same, the initial model is a uniform half-space of 100 Ωm, the maximum depth of the model is 600 m, the thickness of the first layer is 10 m, and the thickness of each layer below is increased by 1.1 times, with a total of 24 electrical layers, using the minimum gradient model constraint, and 15 iterations.

Receiver ADC Bits 32 bits

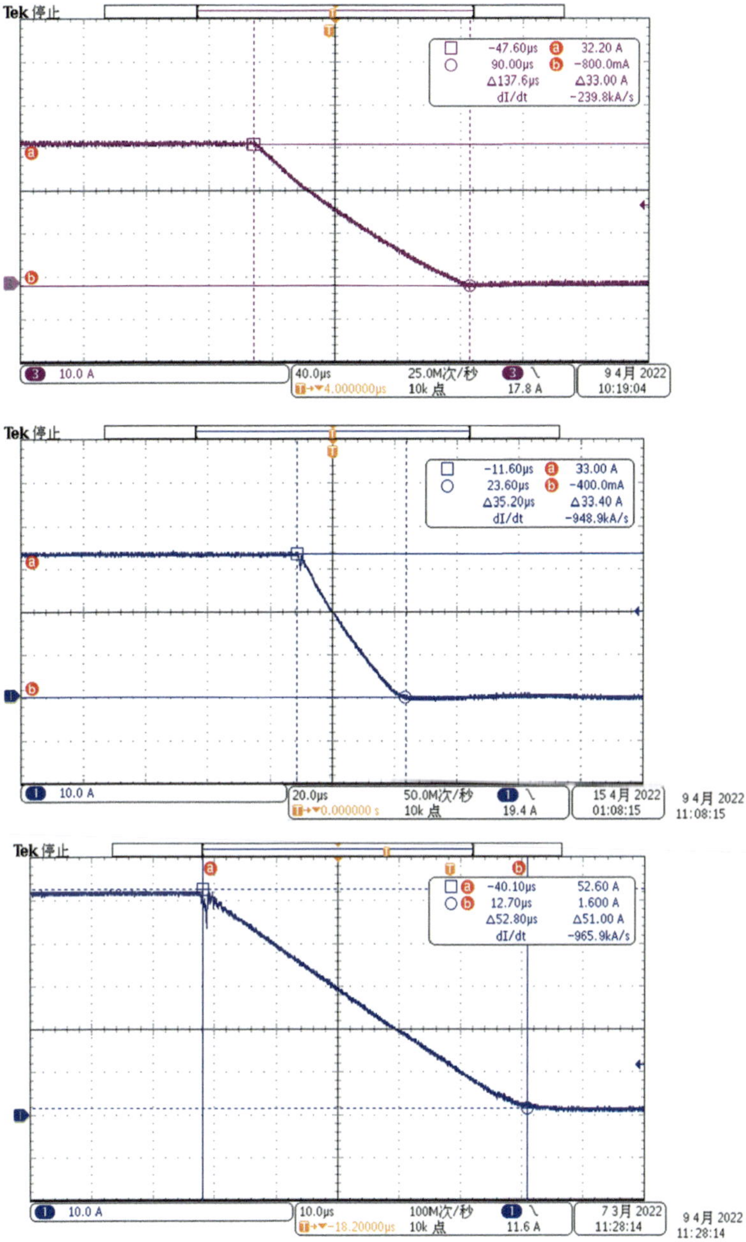

Fig. 7.14 Measured results of transmission waveform: **a** International mainstream transmitter, current 33 A; **b** Self-developed transmitter, current 33 A; **c** Self-developed transmitter, current 51 A

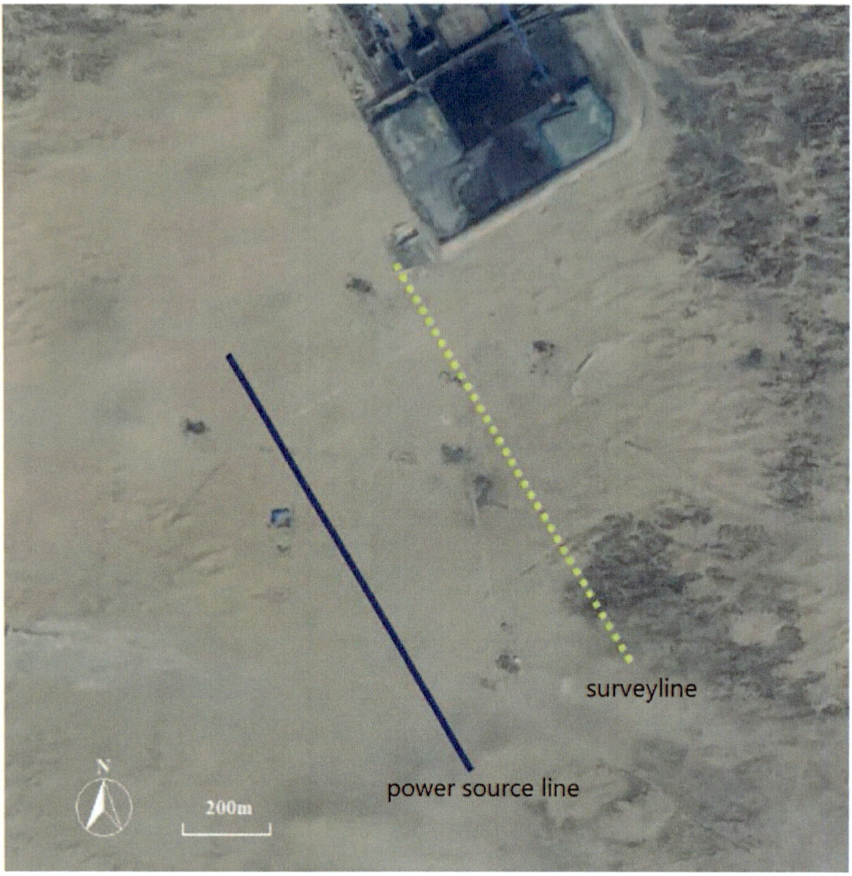

Fig. 7.15 Schematic diagram of the field device layout for the comprehensive detection effect comparison test

According to the known geology and drilling data in the area, the comparison line is located on one wing of the anticline, the main strata are the Quaternary (Q), the Middle Jurassic Zhiluo Group (J2z) and the Middle Jurassic Yan'an Group (J2y), and the main coal seam is the 12 coal of the Yan'an Group, among which the Zhiluo Group strata are water-rich sandstone strata. Figure 7.16 shows that the inversion results of both sets of data depict three electrical layers and the shape of the anticline. However, by comparing with the strata revealed by drilling, it can be reflected that the results of the self-developed system are more accurate in terms of the depth, thickness and change shape of each electrical layer, and the display of the fault near the 200 point number is also more obvious. At the same time, the results of the self-developed system show that the core of the anticline is located near the measurement point 900, and the detection results are consistent with the actual geological conditions.

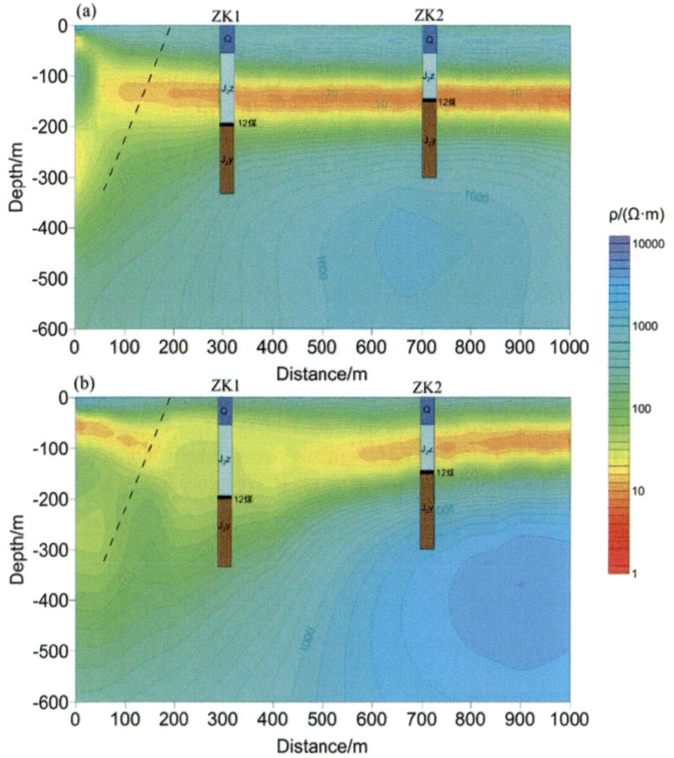

Fig. 7.16 Comparison of actual measurement data cross-sections. **a** International mainstream system results; **b** Self-developed system results

7.5 Conclusion

SOTEM deep high-precision detection is a systematic problem covering emission-observation-processing-solving, and equipment that matches the method needs to be developed to take advantage of the actual detection advantages of the new method. Therefore, equipment development from system design to key technology path selection needs to consider the whole process. In the process of SOTEM equipment development, the system design requirements are fully demonstrated first, and on this basis, the specific quantitative indicators of large current fast shutdown emission and low noise wideband sensor are refined. By overcoming key technologies such as ultra-high pressure constant pressure clamp, low noise hybrid amplification circuit, and adaptive variable damping, SOTEM method-specific equipment with large current, fast shutdown, low noise, and wideband characteristics is developed, which achieves good results in test comparison and practical application. In addition, because the developed system achieves large power emission and low noise observation under large bandwidth conditions, it can also cover the equipment requirements of traditional far-source area detection, so it can take into account the far-source area method. The system performance will be further improved in the future, and it is expected to contribute to the fine detection of deep resources in China.

References

Xue Guoqiang, Li Xiu, Di Qingyun. Progress in the theory and application of transient electromagnetic method[J], Progress in Geophysics, 2008, 22(4): 1195-1200

Park Hwarong. Principles of electromagnetic sounding method [M]. Beijing: Geological Publishing House, 1990

Xue Guoqiang, Yan Shu, Chen Weiyin. Electric source transient electromagnetic short offset detection method[J]. Journal of Nonferrous Metals, 2013, 23(9): 2365-2370

Wu H, Zhang B, Lin T F, Cao D P, Lou Y H. Semiautomated seismic horizon interpretation using the encoder-decoder convolutional neural network[J]. Geophysics, 2019, 84(6): B403–B417

Chapter 8
SOTEM Application Examples

This chapter elaborates on how the author's research team uses SOTEM to solve practical problems in the field of resource environment, such as metal ore and coal mine prevention and control, and provides nine examples, including: Henan Ye County SOTEM test and salt dissolution detection, Shandong Huafeng coal mine deep water-rich investigation, Qiaoshan mining area gold polymetallic ore detection, Anhui Taoyuan coal mine water body detection, Hebei Shunping city active fault detection, Shaanxi Hancheng deep coal seam water content, Qinghai area wild horse spring iron lead zinc polymetallic ore, Henan coal under bauxite, Anhui Wuwei lead zinc mine, etc. This chapter provides a detailed introduction to the geological problems, key technologies, and detection results in these practical detection problems, which can provide ideas and references for readers to understand SOTEM detection issues.

8.1 Henan Ye County SOTEM Test and Salt Dissolution Detection

8.1.1 Test Introduction

In order to verify the theoretical analysis results, compare the response curves of SOTEM under different parameters, and test the apparent resistivity algorithm and inversion technology, an actual test observation was conducted in Ye County, Henan Province, which has a flat terrain and simple stratum structure. This area is a salt mine enrichment area.

The stratum is relatively simple and stable, and the deep salt mine forms a cavity after mining, filling with water to create a very low resistance brine cavity. The strata in the survey area mainly include: Quaternary sand and clay interbeds, Tertiary sandstone and mudstone interbeds, and mudstone and salt rock interbeds. The salt

G. Xue et al., *Exploration with Short-Offset Grounded-Wire Transient Electromagnetic Method*, https://doi.org/10.1007/978-981-96-8846-3_8

mine in this area is approximately 1100–1400 m deep, and after the salt mine is artificially water-dissolved and fractured, it develops into a brine cavity, which is an ideal medium for underground deep energy storage. In addition, several boreholes up to 2 km deep have been completed in this area, and the known geological data is abundant, providing a guarantee for the reliability of the results. Figure 8.1 is a geological cross-section of the test area, and the corresponding rock electrical parameters are shown in Table 8.1.

In the test, the receiving point is located on the vertical line of the transmitter, and the position remains unchanged to ensure the consistency of the noise environment and the geological environment. By changing the position of the transmitter and the base frequency of the transmitting current, the electromagnetic field response under different offset distances and observation durations is studied. The geometric layout of the test transmission and reception is shown in Fig. 8.2. The parameters of the three transmitters are fixed at a transmitter length of 1 km and a transmitting current of 16 A; the observed components include three magnetic field components ($V_x(t)$, V_y (t), $V_z(t)$) and two electric field components (Ex, Ey); non-polarized electrodes

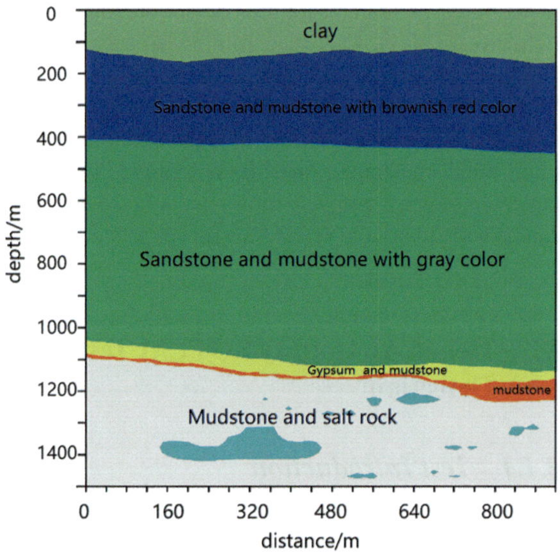

Fig. 8.1 Schematic diagram of the strata in the test area

Table 8.1 Electrical parameters of the strata

Lithology	Average thickness (m)	Resistivity range (Ω.m)
Sandstone, clay	110	100~200
Sandstone, mudstone	1000	20~80
Gypsum-containing mudstone	50	>500
Mudstone, rock salt	>500	>500
Rock salt cavity	1100~1400	<10

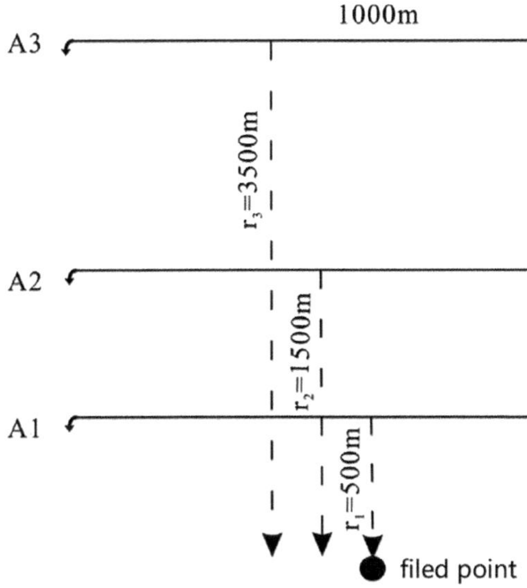

Fig. 8.2 Schematic diagram of the geometric layout of the test transmission and reception

are used to observe the electric field, the electrode distance is fixed at 50 m, and a 7K-SB model probe with an effective area of 40,000 m^2 is used to receive the induced voltage, and the measurement time for each point is 7 min.

8.1.2 Test Content and Results

(1) **The transmission and reception distance is fixed, and the transmission base frequency is different**

Firstly, with a fixed transmission distance of 500 m, the electromagnetic field responses were observed when the fundamental frequency of the transmission current was 5 Hz, 2.5 Hz, 1 Hz, and 0.5 Hz respectively. Figure 8.3 shows the observation results. It can be seen from the figure that when the observation delay is short, the signal strength of the five electromagnetic field components is relatively high. The signals of different transmission fundamental frequencies overlap well within the range of overlapping observation time, indicating that factors other than the transmission fundamental frequency have not significantly affected the signal. As the transmission fundamental frequency decreases, the signal attenuation delay becomes longer, and the signal quality in the late stage of each component of the electromagnetic field becomes worse. Overall, after 10 ms, all curves show different degrees of oscillation, indicating that the signal is now interfered with by external noise. It should be noted that when the transmission fundamental frequency is 5 Hz, the signal attenuation curve is relatively smooth throughout the observation time, while the curves of other transmission fundamental frequencies have already oscillated in the time period that overlaps with it. This is because the lower the transmission fundamental frequency, the fewer times the signal is superimposed within the same observation time, resulting in a decrease in the signal-to-noise ratio of the signal attenuation curve under the same moment, the same signal strength, and the same noise environment.

(2) **Fixed transmission fundamental frequency, different offset distances**

Next, with the transmission fundamental frequency fixed at 5 Hz, the responses of each component of the electromagnetic field were observed when the offset distances were 500 m, 1500 m, and 3500 m, respectively. The observation results are shown in Fig. 8.4. It is clear that as the offset distance increases, the signal strength of each component decreases significantly, and the smoothness of the signal curve deteriorates. At the same time, as the offset distance increases, the amplitude range of the signal becomes narrower, and the attenuation becomes slower.

According to the theory of electromagnetic wave propagation diffusion, high-frequency components of electromagnetic waves have a smaller wavelength and faster attenuation, so they are mainly concentrated at small offset distances. Low-frequency components of electromagnetic waves have a larger wavelength and slower attenuation, so they are mainly concentrated at large offset distances. Therefore, as the offset distance increases, the effect of the high-frequency electromagnetic field becomes weaker, leading to the early signals dominated by the high-frequency electromagnetic field not only decreasing in intensity but also being easily interfered with by geological noise and external noise.

8.1.3 Full-Period Apparent Resistivity and Inversion Results

Considering the overall signal quality, full-period apparent resistivity calculations and one-dimensional equivalent source inversions were performed using data from an offset distance of 500 m and transmission fundamental frequencies of 5 Hz, 2.5 Hz, and 1 Hz. The results are shown in Fig. 8.5. When calculating, the induced voltage needs to be converted into a vertical magnetic field (Chen et al. 2016). From the apparent resistivity curve, it can be seen that the calculation results of the three fundamental frequency data all indicate that the electrical properties of the strata in the survey area are H-type structures, and the calculation results of each frequency fit very well within the overlapping time range. However, due to the low quality of the late data for 2.5–1 Hz, there is a slight jitter in the apparent resistivity curve.

The one-dimensional equivalent source inversion results show different outcomes. The 5 Hz inversion result indicates that the stratum's electrical properties are H-type structures, while the 2.5–1 Hz inversion results suggest that the stratum is a G-type structure. This indicates that the earliest time signals corresponding to the two lower fundamental frequencies have penetrated the high-resistance layer on the

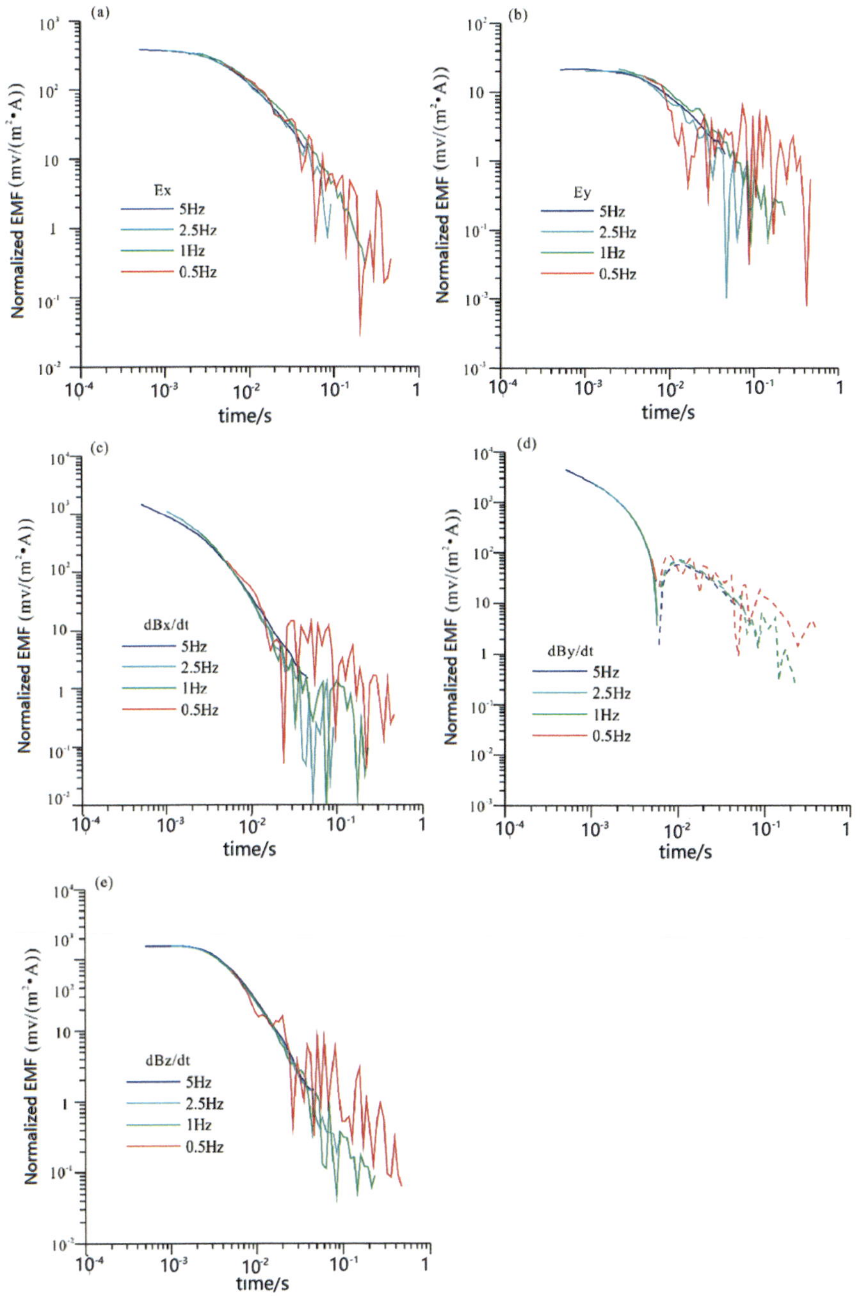

Fig. 8.3 Electromagnetic field response curves with fixed transmission distance and different transmission fundamental frequencies

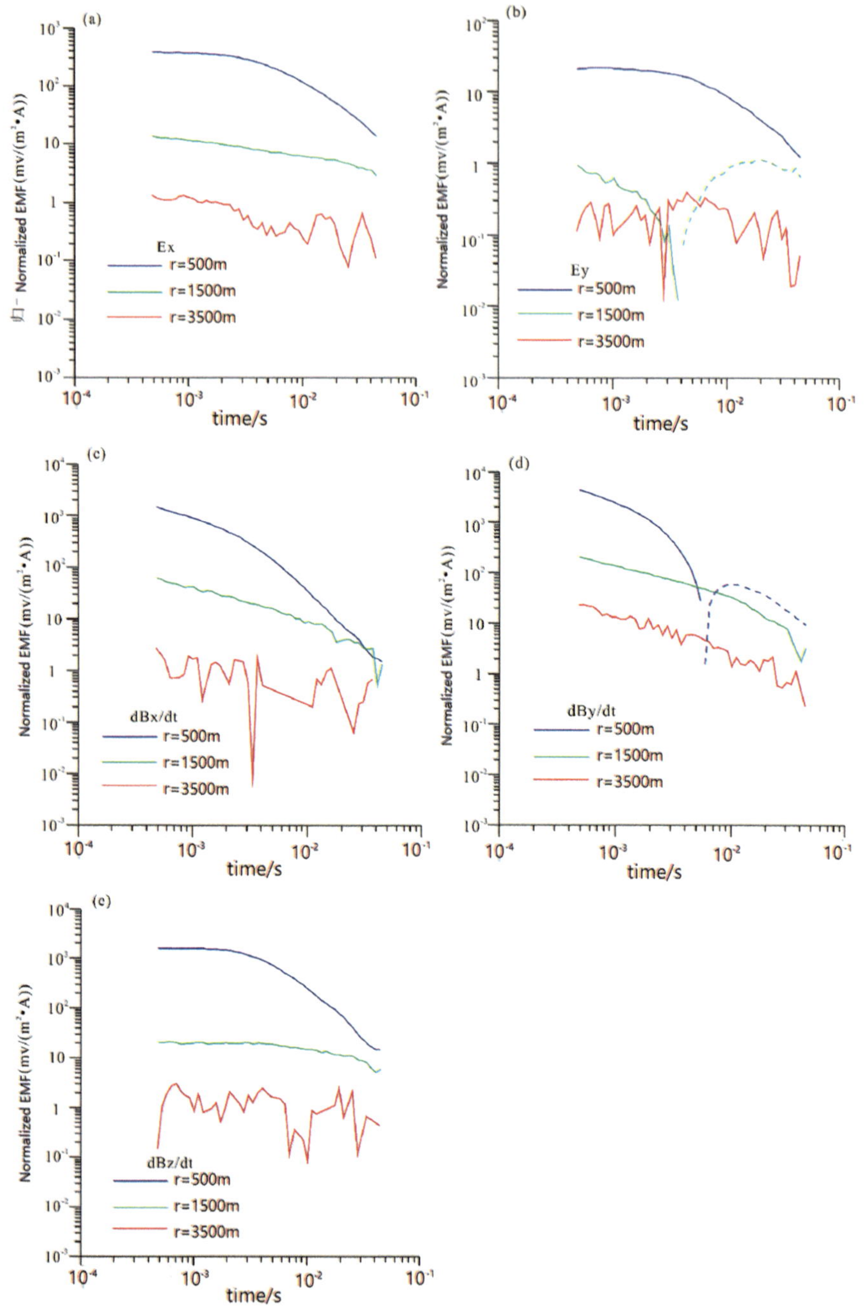

Fig. 8.4 Electromagnetic field response curves with fixed transmission fundamental frequency and different offset distances

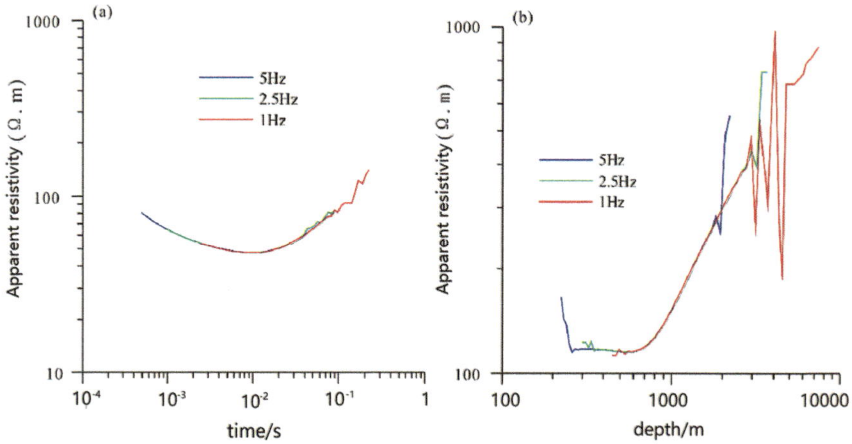

Fig. 8.5 Full-period apparent resistivity and coil inversion results **a** Full-period apparent resistivity **b** One-dimensional coil inversion results

ground surface, making the strata above this depth a blind area for observation. In addition, the inversion curve at the late stage (great depth) corresponding to 1 Hz has become very oscillatory and has lost its utility. Overall, both full-period apparent resistivity and one-dimensional equivalent source inversion can accurately reflect the real changes in the electrical properties of the strata below the measurement point.

8.1.4 Detection of Salt Caverns

Finally, a typical profile revealing deep salt dissolution cavities was selected in the test area for deep salt dissolution cavity SOTEM detection. According to the test situation, we chose the E_x component of the horizontal electric field, which has strong anti-interference ability, for observation (Guoqiang et al. 2013). The working parameters are: the base frequency of the transmitting current is 1 Hz, the intensity is 18 A, the electrode distance MN is 40 m, and the single-point observation time is 10 min. A total of 6 survey lines were arranged, as shown in Fig. 8.6, and the information for each survey line and the transmitting source is shown in Table 8.2.

Due to the use of shorter offset distance observations, coupled with the lower noise environment near the survey line, the original observed electric field data has a high signal-to-noise ratio. Most of the data from the measurement points exhibit a lower error bar level within the full delay range. The original observation decay curve of the typical measurement point is shown in Fig. 8.7.

After smoothing and filtering the data from the 6 survey lines, a one-dimensional inversion was performed to obtain the resistivity-depth cross-section diagrams, as shown in Figs. 8.8, 8.9 and 8.10. According to existing data, the depth of the salt mine in the survey area is within the range of 1100–1400 m. Before the salt-bearing strata

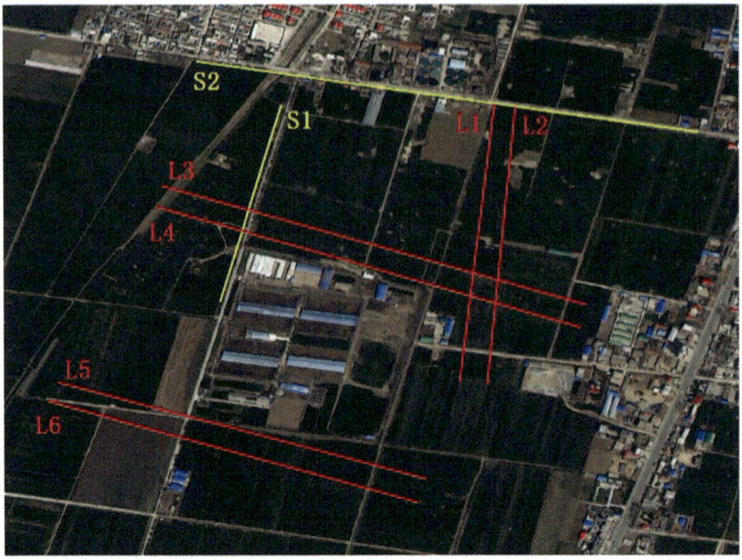

Fig. 8.6 SOTEM receiver layout

Table 8.2 Survey line and transmitting source information table

Survey Line/source	Length (m)	Azimuth (°)	Offset distance (m)	Control source
L1	680	186	480	S1
L2	680	186	485	S1
L3	920	108	400	S2
L4	920	108	450	S2
L5	800	108	890	S2
L6	800	108	940	S2
S1	500	195	–	–
S2	1060	100	–	–

were mined, they showed high resistance. After the dissolution water was extracted, water-filled cavities formed. Since brine is a highly conductive medium, it exhibits low resistance. Therefore, the low-resistance anomaly in the inversion resistivity-depth cross-section within this depth range can be qualitatively interpreted as a dissolution cavity.

First, looking at the results of lines L1 and L2 (Fig. 8.8), the inversion results of the two lines show a basically consistent electrical distribution. Based on the resistivity values, the strata can be divided into three electrical layers: the first layer is the high-resistance layer from the surface to a depth of 200 m, which consists of an interlayer of gravel and sticky clay covered by the Quaternary; the second layer is the low-resistance layer within the depth range of 200–900 m, which consists of an interlayer

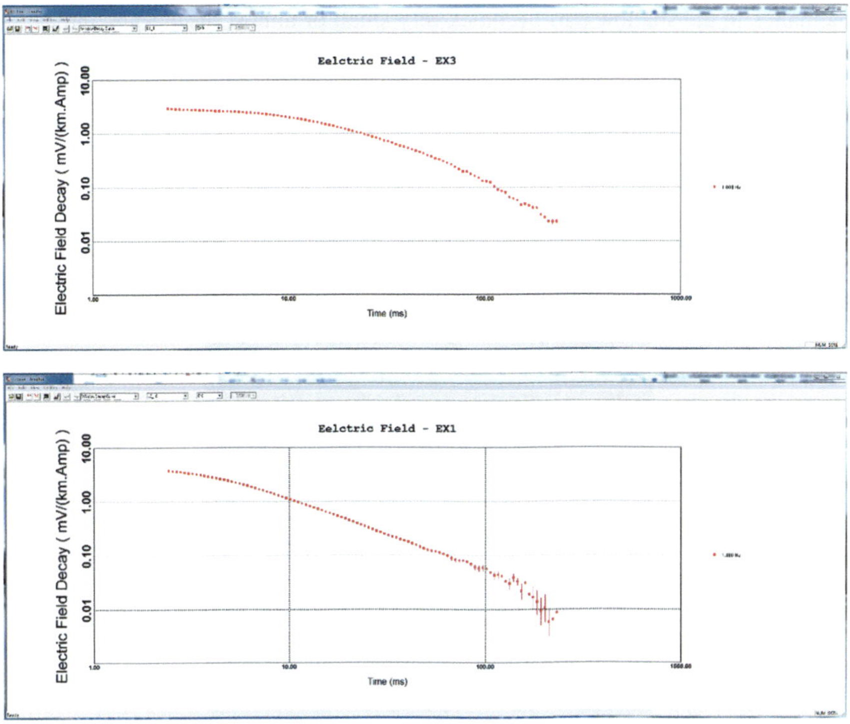

Fig. 8.7 Decay curve of original observation at some measurement points

of sandstone and mudstone of the Tertiary; the third layer is the high-resistance layer within the depth range of 900–1500 m, which is a layered salt-containing paste shale.

It can be observed that within the depth range of 1000–1400 m, there is a lateral abrupt change in resistivity, and three low-resistance anomalies exist under the overall high-resistance background. According to the aforementioned electrical characteristics analysis, it can be inferred that the low-resistance anomaly within this depth range is indicative of the salt-containing rock salt dissolution cavity, and the three dissolution cavities are designated as C1, C2, and C3, respectively. In addition, within the depth range of 100–900 m in the northern section of the two survey lines, there is also a certain degree of lateral discontinuity in resistivity, which is caused by the existence of a nearly east–west fault (Fig. 8.9).

The inversion results of lines L3 and L4 (Fig. 8.10) reveal the same electrical structure as L1 and L2, and there are also two low-resistance anomalies caused by dissolution cavities in the deeper part. Notably, the low-resistance anomaly at the 720 m point corresponds to the C2 anomaly revealed by the L1 and L2 lines. The consistent depiction of the same target body by different survey lines further demonstrates the reliability of the measurement data and inversion results.

The inversion results of lines L5 and L6 (Fig. 8.11) show a certain degree of difference from the previous four survey lines. First, the range of high-resistance anomalies in the shallow part becomes very small or almost disappears, indicating that the Quaternary strata in this area are thin or have good water content. The resistivity distribution of the deep salt-containing ore strata is relatively uniform, and there is no obvious lateral change in resistivity, suggesting that the strata here are relatively complete and that there is no water-filled salt dissolution cavity.

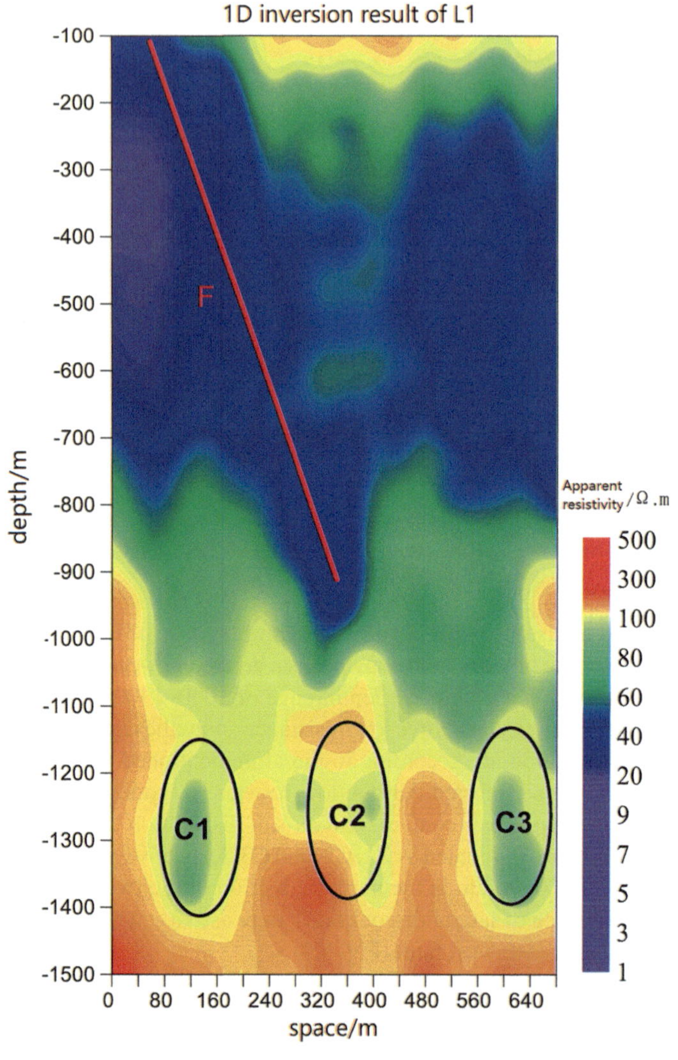

Fig. 8.8 L1, L2 line one-dimensional inversion resistivity-depth cross-section

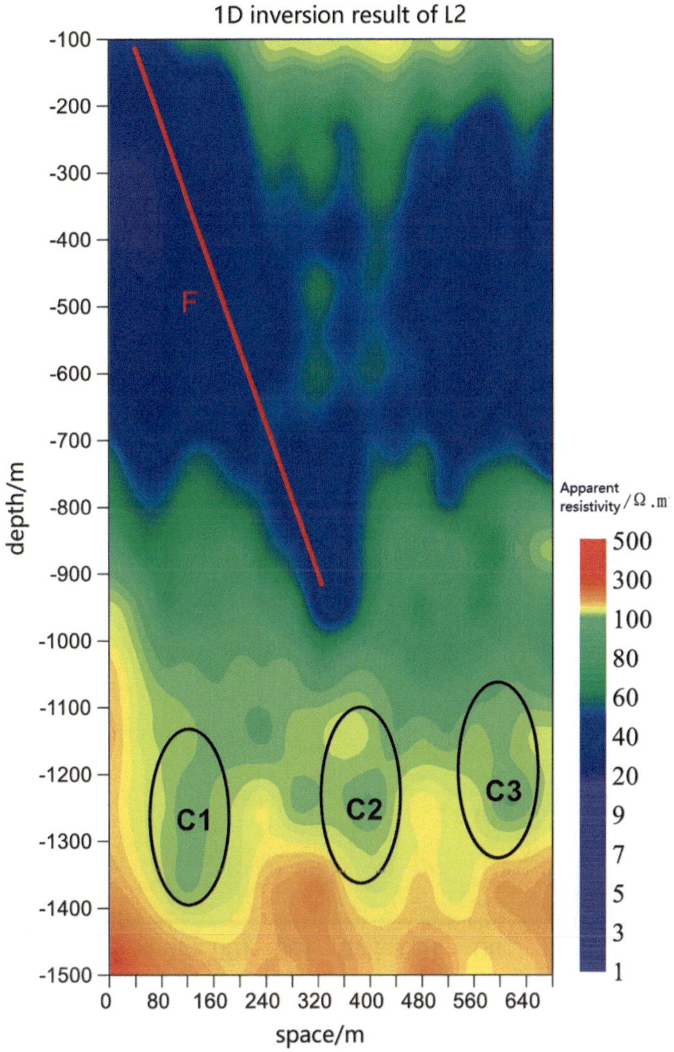

Fig. 8.8 (continued)

8.1.5 Test Conclusion

This test is an early systematic SOTEM observation experiment. The main purpose is to obtain the actual SOTEM response under different receiving and transmitting devices and various working parameters, and to compare and analyze these results with theoretical research findings to evaluate the effectiveness and superiority of the method. Through this test, it was found that the response characteristics of each

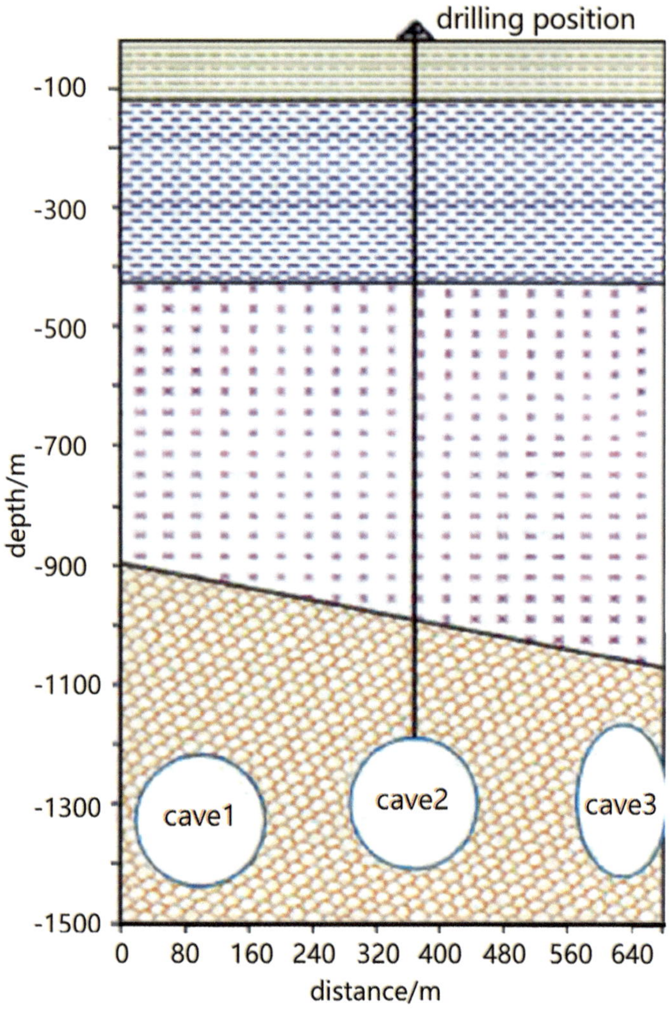

Fig. 8.9 Geological interpretation results of L1 line

component of the actual observed electromagnetic field largely coincide with the theoretical analysis results, verifying the accuracy of the theoretical work.

Through the actual observation of 6 typical cross-section data and subsequent processing and interpretation, multiple salt dissolution cavities within the depth range of 1100–1400 m underground were delineated. The interpretation results were consistent with the real situation revealed by known drilling, verifying SOTEM's ability to detect at great depths and the reliability of the corresponding data processing technology. In general, this test holds significant importance for the understanding, mastery, and improvement of the SOTEM method.

8.2 Investigation of Deep Water Richness in Shandong Huafeng Coal Mine

The transient electromagnetic method is often used for the exploration of coal mine goafs due to its sensitivity to low-resistance bodies. Huafeng Coal Mine has long been troubled by water from the third series of gravel rock on the top during the mining of coal seam 4. To determine the distribution of water-rich areas at the − 1100 level of the three mining areas of Huafeng Coal Mine, the mine commissioned the author's team to undertake the transient electromagnetic exploration task at this level. Since the mining area belongs to the North China type coal field, it is covered by a thick low-resistance layer, making it difficult for traditional ground loop source transient electromagnetic devices to achieve the target detection depth of 1500 m. Therefore, the SOTEM method, which offers a larger detection depth, was applied in this measurement.

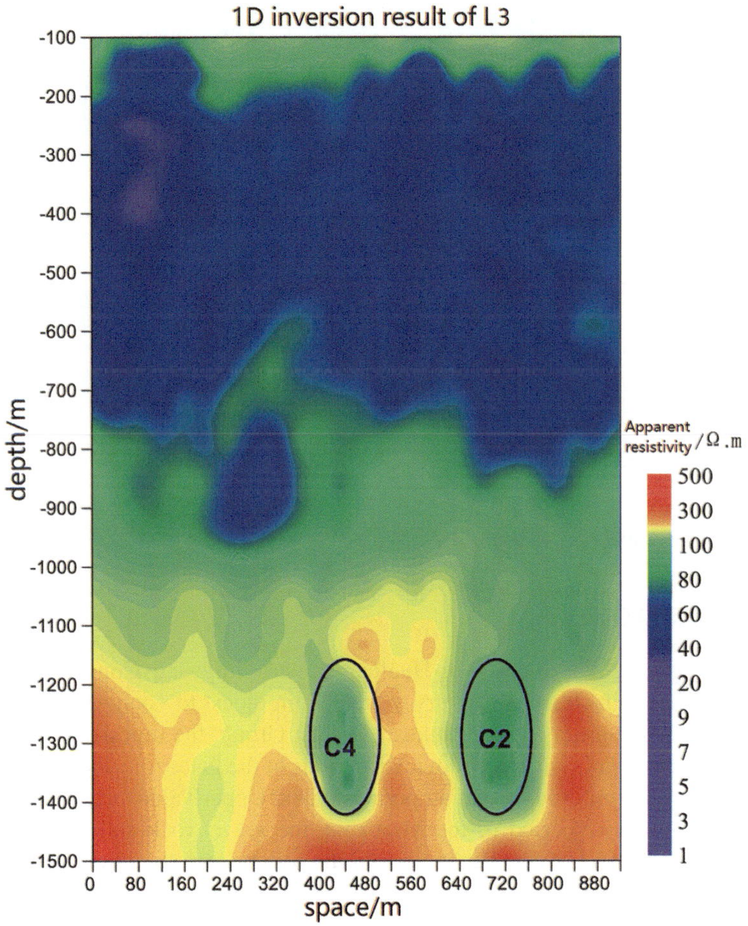

Fig. 8.10 L3, L4 line one-dimensional inversion resistivity-depth cross-section

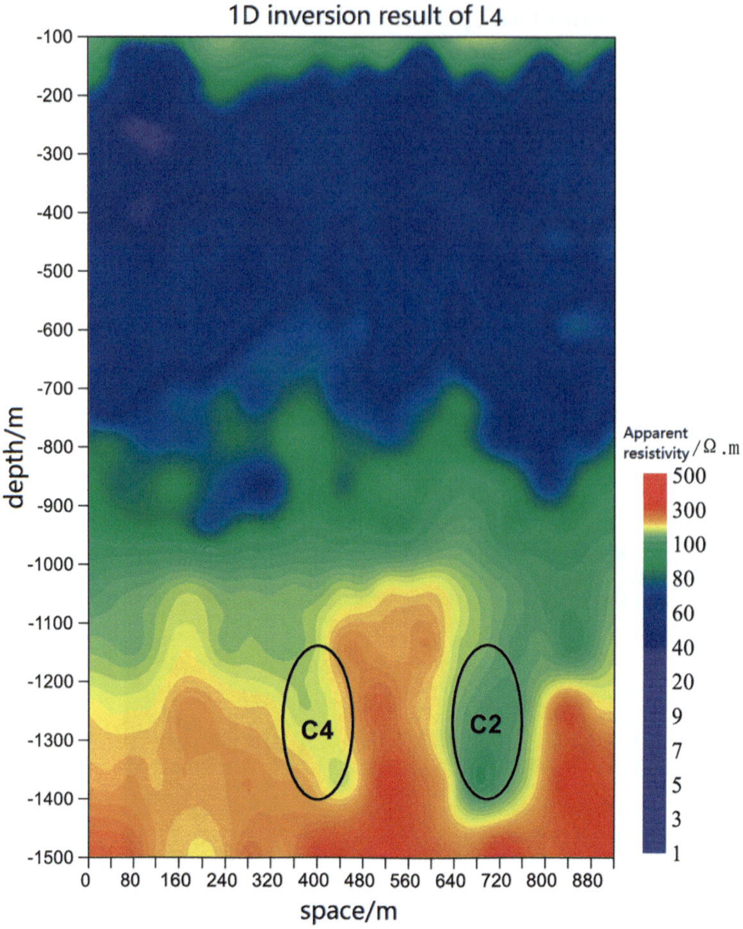

Fig. 8.10 (continued)

8.2.1 Survey Area Overview

(1) **Geological Overview**

Huafeng Well Field is located in Huafeng Town, Ningyang County, Shandong Province. The main strata in the survey area, from oldest to youngest, include the gneiss and granite gneiss of the Archean Taishan Group, the limestone, mudstone, and shale of the Cambrian and Ordovician periods of the Paleozoic, the coal-bearing strata of the Carboniferous–Permian periods of the Upper Paleozoic, the red sandstone of the Jurassic period of the Mesozoic, the gravel rock and red sandstone of the Tertiary period of the Cenozoic, and the topsoil and quicksand layer of the Quaternary. The stratigraphic column of the survey area is shown in Fig. 8.12.

The shallow structure in the well field is relatively simple, while the overall structure of the deep part of the well field (below −750 m) is a simple syncline tilting to the northeast, with sparse faults having a drop of less than 30 m, which indicates a simple structure. The coal-bearing strata are from the Carboniferous–Permian periods. The upper part is developed in the Mesozoic and Cenozoic, while the lower part is based on the Ordovician greywacke.

The main aquifers related to coal mining, from top to bottom, are the Quaternary gravel layer, the Tertiary gravel layer, the sandstone roof of the Permian Shanxi Formation 4 coal seam, the first and fourth limestone layers of the Carboniferous Taiyuan Formation, the limestone layer of the Benxi Formation, and the Ordovician limestone layer. Together, they constitute the northern type of multi-aquifer structure with karst fracture water-filled deposits. In the well field, the clayey siltstone (i.e., "red sandstone") at the bottom of the Tertiary, the variegated claystone of the Stone Box Group, and the siltstone, claystone, and mudstone between the aquifers of the

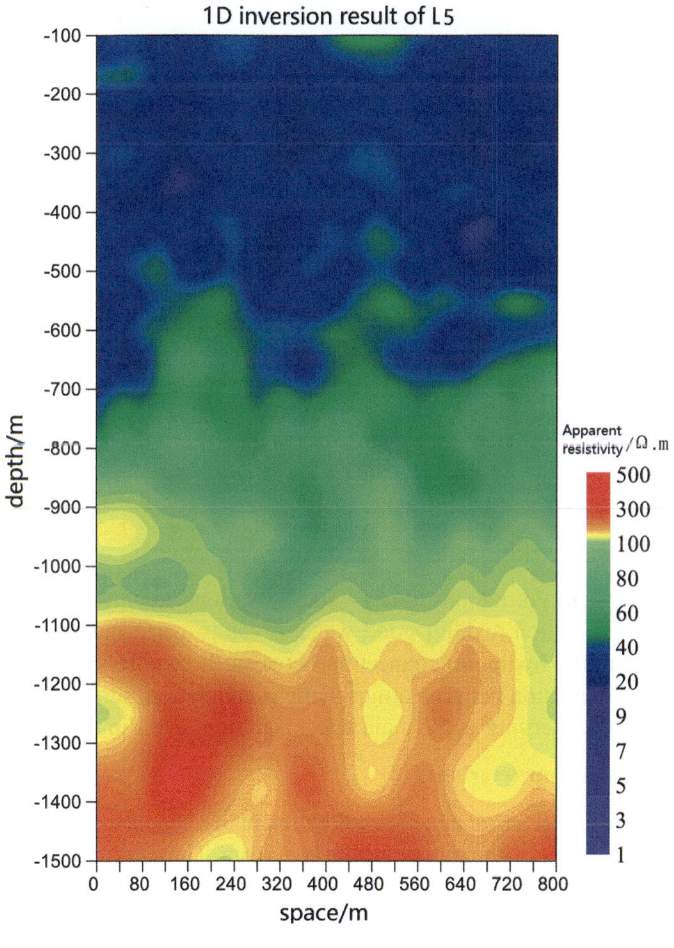

Fig. 8.11 L5, L6 line one-dimensional inversion resistivity-depth cross-section

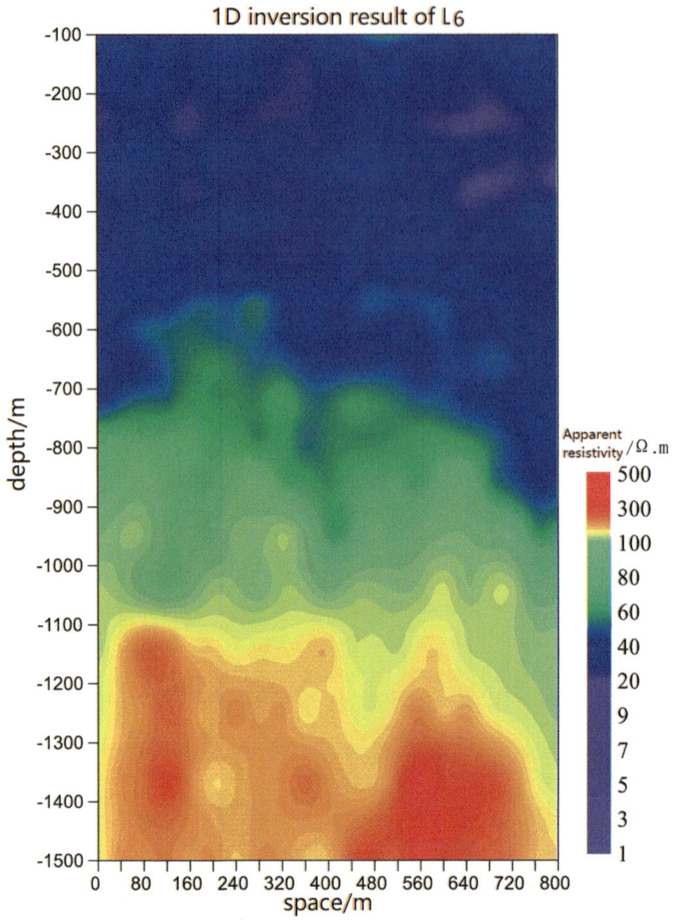

Fig. 8.11 (continued)

Shanxi and Taiyuan Formations, along with the variegated claystone and siltstone of the Benxi Formation, exhibit good water isolation performance. These layers normally isolate the hydraulic connection between the aquifers and also separate the hydraulic connection between atmospheric precipitation, surface water, and the coal-bearing strata, serving as the main water isolation layers in this area.

(2) **Geophysical characteristics**

The task of transient electromagnetic exploration in this area is to detect the distribution of water-rich areas in the roof of the 4th coal seam. The aquifers that significantly impact the mining of the 4th coal seam are the Tertiary Guanzhuang Group gravel aquifer and the sandstone aquifer at the roof of the 4th coal seam. The gravel in the gravel rock is primarily limestone, along with dolomitic limestone, flint, quartz, and other types of gravel. Its electrical characteristics exhibit high resistance. When the

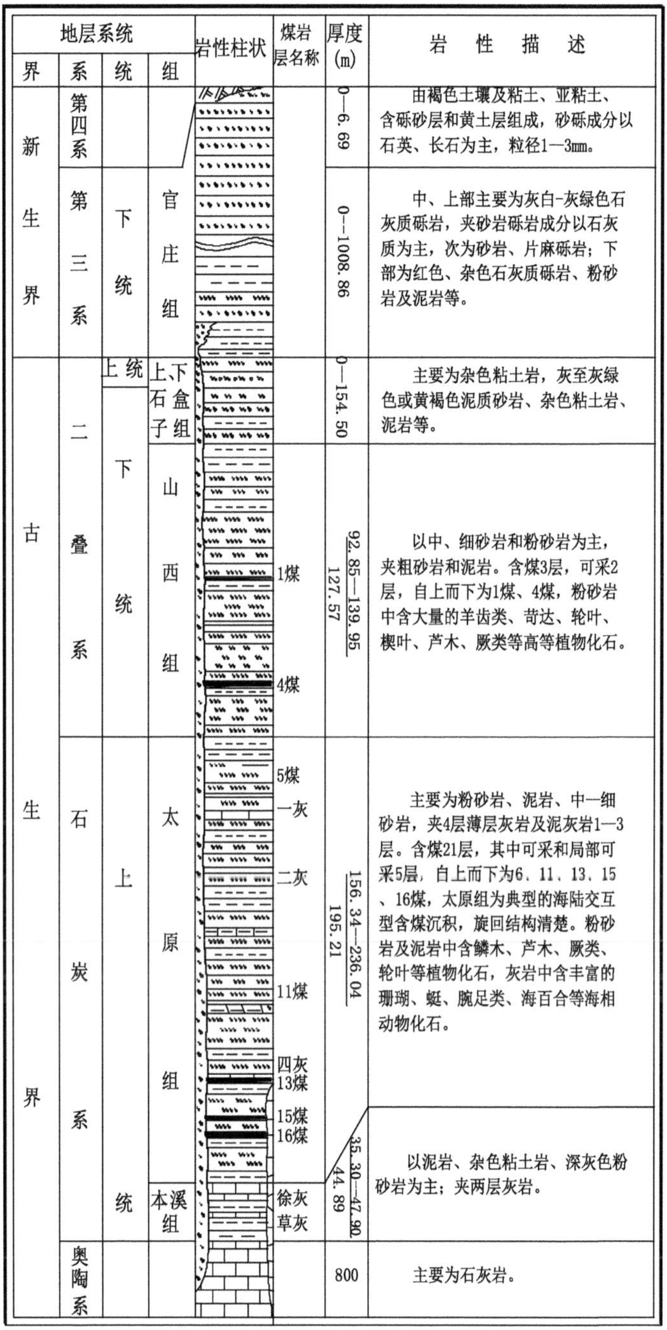

Fig. 8.12 Comprehensive stratigraphic column of Huafeng Well Field

proportion of gravel in the gravel layer is large and the mud and sand content is low, its resistivity will also be higher overall. While the resistivity of sandstone itself is high, it can significantly decrease when fractures are developed and filled with water.

In this area, due to the influence of the thick gravel layer of the Tertiary, the resistivity of the Cenozoic strata is high, whereas the resistivity of the Permian and Carboniferous strata does not change much. There is a significant electrical difference between the Tertiary and Quaternary strata and the underlying strata (see Table 8.3).

Figure 8.13 is the resistivity curve of the 700th measurement point of Line B300. The horizontal axis represents depth (in meters), while the vertical axis represents resistivity (in Ω·m). The basic shape of the curve is HK type. The resistivity curve shows a noticeable decrease at 700 m shallow, reflecting the Tertiary and Quaternary strata. Between 700 and 900 m, the resistivity change is minimal, indicating the Permian and Carboniferous strata. In the deeper sections, the curve reflects the Ordovician limestone strata, where the resistivity gradually increases. This indicates a significant electrical difference between the Tertiary and Quaternary strata and the underlying strata.

Table 8.3 Electrical properties of strata in the survey area

Strata	Lithology	Thickness(m)	Resistivity(Ω.m)
Tertiary and quaternary (Q + R)	Loess, sub-clay, sandy clay, gravel Gravel rock, siltstone, mudstone	0~1009	> 70
Permian (P)	Mudstone, sandstone and coal seam	93~294	30~200
Carboniferous (C)	Sandstone, mudstone, limestone and coal seam	240	30~200

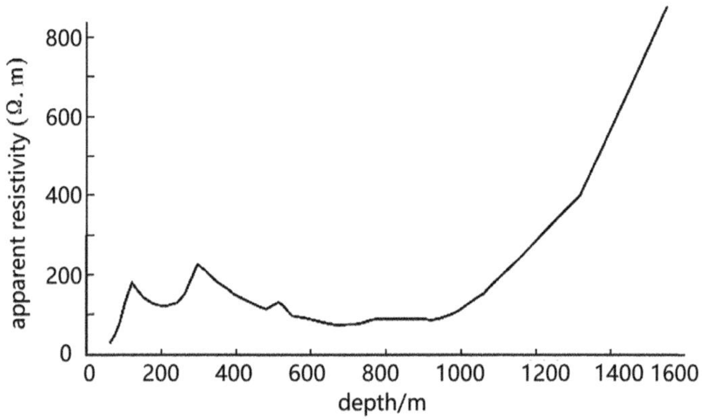

Fig. 8.13 Resistivity curve of strata in the survey area

8.2.2 Data Collection

The SOTEM exploration area for this survey is located in the northwest of Huafeng Well Field, with a total of 31 survey lines arranged. Among these, 29 are oriented along the dip direction, with a basic line spacing of 100 m. The line spacing is reduced to 50 m for densification between lines B100 and B1300, and the measurement point distance is 100 m. For comparison verification, two survey lines named A800 and A400 were taken along the strike direction, with a point spacing of 50 m, as shown in Fig. 8.14.

Through the experiment, the selected working parameters are as follows: transmitter length AB = 1000 m, transmitter–receiver distance r = 1000 m, transmission current base frequency 2.5 Hz, current intensity I = 20 A, and observing the vertical induced voltage component. The effective receiving area of the probe is 40,000 m^2, and the single point measurement time is 7 min.

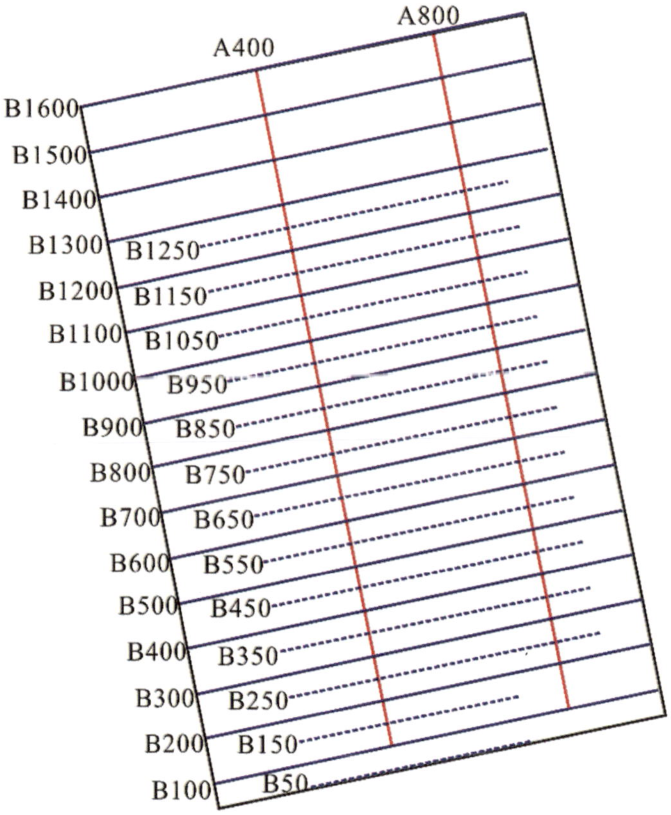

Fig. 8.14 Schematic diagram of survey line layout

8.2.3 *Data Interpretation*

First, based on the induced voltage multi-measurement section map and the inverted resistivity-depth section map of each survey line, the possible geological anomaly areas in the survey area are interpreted. Next, the resistivity slice maps for each horizontal and along-layer are analyzed, focusing on the distribution pattern of the geological anomaly areas. Finally, using the resistivity pseudo-section map and plan map, combined with geological data results, the distribution pattern and extent of water-rich areas are determined. A distribution map of water-rich areas on each layer of the **4th** coal roof in the survey area is then created.

(1) **Resistivity-Depth Section Map Electrical Characteristics**

A total of 31 survey lines were measured in this SOTEM exploration. After data processing for each survey line, the bottom interface of the third series and the position of the 4th coal seam were marked on the electrical section according to the drilling data and geological section data in the survey area. Hydrogeological interpretation was conducted for the resistivity section map of each survey line. Here, we choose the detection results of line B1000 for description.

Figure 8.15 is the inversion depth-resistivity section map of line B1000. The horizontal coordinate in the figure represents the measurement point number, while the vertical coordinate represents the detection depth. The black line in the figure indicates the 4th coal seam (where the dashed line denotes the mined-out area), the red dashed line represents the bottom interface of the third series, and the red solid line represents the fault.

The resistivity of the entire section generally exhibits a low–high-low–high electrical characteristic from shallow to deep. The resistivity value in the uppermost part of the figure (less than 100 m) is relatively low, showing a gradually increasing trend from shallow to deep, with values between 20 and 100 Ω·m, reflecting the strata of the upper group of the third series, which is a weak water-containing layer. The middle part of the third series (100–400 m section) corresponds to a larger range of resistivity changes, with values between 140 and 450 Ω·m, indicating it is water-free. The lower part of the third series (deeper than 400 m) also corresponds to a larger range of resistivity changes, with values between 80 and 400 Ω·m, which is either water-free or a weak water-containing layer. The resistivity near the 0 m pile number is slightly lower, suggesting it is weakly water-containing.

The resistivity change of the coal series strata is not large; the 4th coal is absent in the 0–300 m pile number section, but the rock layer shows a low resistance characteristic, indicating that the coal series strata at this position are weakly water-containing. The base of the coal series, the Au limestone layer, corresponds to high resistivity, suggesting that the Au limestone is water-free or weakly water-containing.

(2) **Different Horizontal Resistivity Characteristics**

In this exploration area, the thickness of the third series ranges from 600 m to 1008.86 m, which is relatively large. The dip angle of the underlying Permian strata is also

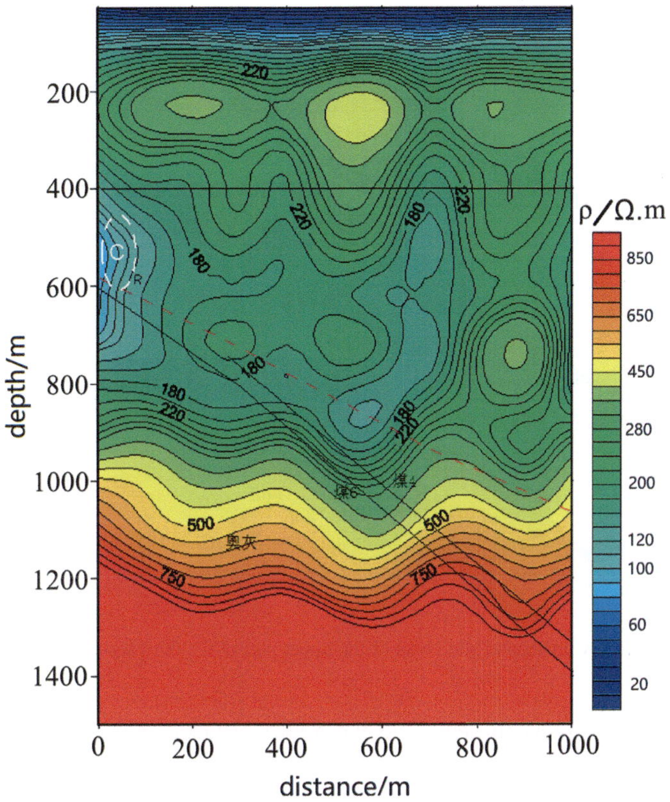

Fig. 8.15 B1000 line resistivity section map

significant, ranging from 32° to 43°. Therefore, to study the distribution of water-rich areas in the upper strata of the third series, we can utilize the resistivity horizontal slice map at different depths to analyze the water-richness of the upper gravel. For the lower strata of the third series and the Permian sandstone water-bearing layer, we can use the along-layer slice map to assess the water-richness of the strata at different layers.

a. **Electrical characteristics of the upper strata of the third series**

Figure 8.16 shows the resistivity horizontal slice maps at + 75, + 50, and + 25 m elevations, respectively, in the first section of the third series gravel (within a vertical depth of 100 m), which is the karst groundwater section.

Analyzing the three horizontal slice maps in this section, the resistivity change pattern from shallow to deep can be divided into three areas. The first area is south of the 800 m survey line pile number. In this area, from shallow to deep, the low resistance area gradually decreases, and the resistivity gradually increases, changing from relatively strong water-containing to weak water-containing or water-free. The second area is in the middle of the survey area, where the resistivity gradually

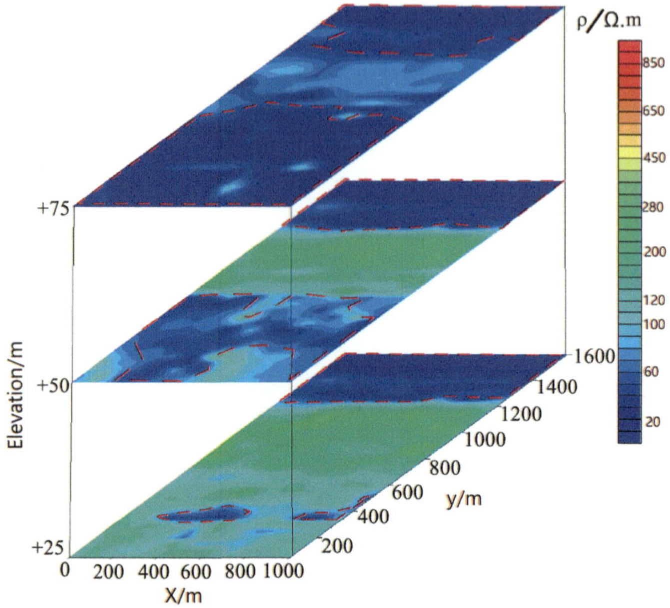

Fig. 8.16 Resistivity horizontal slice map of the upper part of the third series

increases from shallow to deep, and the high resistance area gradually expands, indicating that the water content of the rock layer changes from relatively weakly water-containing to water-free. The third area is north of the 1200 m survey line pile number, where the resistivity does not change much from shallow to deep, and it remains a low resistance area. The overall relative water content of the rock layer is strong, likely due to the development of fractures that increase the water content.

In the resistivity horizontal slice map at + 75 m elevation, the overall range of resistivity values is between 20 and 120 Ω·m. The resistivity values range between 20 and 240 Ω·m. From the figure, it can be seen that the resistivity values on both sides of the entire survey area are relatively low, indicating that the overall water content in this area is relatively abundant. The resistivity value near the middle of the survey area is slightly higher, with values between 60 and 120 Ω·m, which suggests relatively weak water content. Near the 1200 m survey line pile number, the red dashed line area shows a significant change in resistivity, indicating a large difference in water content between the strata on both sides of the red dashed line.

In the resistivity horizontal slice map at + 50 m elevation, the overall change range of resistivity values is between 20 and 240 Ω·m. From the figure, it can be observed that the resistivity values in the area circled by the red dashed line on both sides of the entire survey area are relatively low, indicating relative water content or weak water content. The resistivity in the area south of the 800 m survey line pile number is relatively low, with values between 30 and 80 Ω·m, which is assessed as relatively medium water content, with some areas showing relatively strong water content. The resistivity in the middle of the survey area is higher, with values between 80 and 240 Ω·m, which is judged to be relatively weak in water content or devoid of water. The resistivity in the area north of the 1200 m survey line pile number is lower, with values between 20 and 60 Ω·m, indicating relatively strong water content. Near the 1200 m survey line pile number, the red dashed line area exhibits a significant change in resistivity, indicating a large difference in water content between the strata on both sides of the red dashed line.

In the resistivity horizontal slice map at the + 25 m elevation, the overall range of resistivity values is between 20 and 280 Ω·m. From the figure, it can be seen that the resistivity in the area south of the 800 m survey line pile number is slightly low, with values between 50 and 140 Ω·m, which indicates relatively weak water content and locally medium water content. The resistivity in the middle of the survey area is higher, with values between 140 and 280 Ω·m, suggesting little to no water content. The resistivity in the area north of the 1200 m survey line pile number is low, with values between 20 and 60 Ω·m, indicating relatively strong water content. Near the 1200 m survey line pile number, the red dashed line area shows a significant change in resistivity, indicating a large difference in water content between the strata on both sides of the red dashed line.

b. **Electrical characteristics of the middle strata of the third series**

Figure 8.17 shows the resistivity horizontal slice maps of the entire survey area at −25, −75, −125, −175, −225, and −275 m elevations, located in the second section of the third series gravel (vertical depth between 100 and 400 m). Many factors affect the resistivity of the strata, with the most important being the lithology of the rock, i.e., the composition of the minerals contained in the rock. This is followed by the degree of fracture development, the mineralization of pore water, temperature, and other factors. For the third series, the most significant influencing factor is the degree of fracture development. Developed fractures are characterized by relatively abundant water, resulting in low resistance; conversely, relatively undeveloped fractures possess poorer water content, leading to higher resistance.

Examining the six horizontal slice maps in this section, from shallow to deep, the relatively high resistance area is mainly distributed in the middle of the survey area. The resistivity value in the north of the survey area is low, while the red circle area in the south indicates a low resistance area, with resistivity values showing a trend from low to high to low as one moves from south to north. From −25 m to −125 m, the high resistance area gradually expands, with local high resistance zones appearing, suggesting that the development of rock fractures weakens from shallow to deep. Between −175 m and −275 m, the change in the high resistance area is minimal, indicating that the degree of development of rock fractures in the upper and lower strata does not vary significantly.

In the area near the B200 line to the south of the B600 line, there is consistently a local zone with relatively low resistivity values, which can be interpreted as weak water content or no water content. In the region north of the B1200 line to the south of the B1600 line, the resistivity value exhibits a trend from low to high to low, with a local area of high resistivity appearing in the deeper sections. Based on the

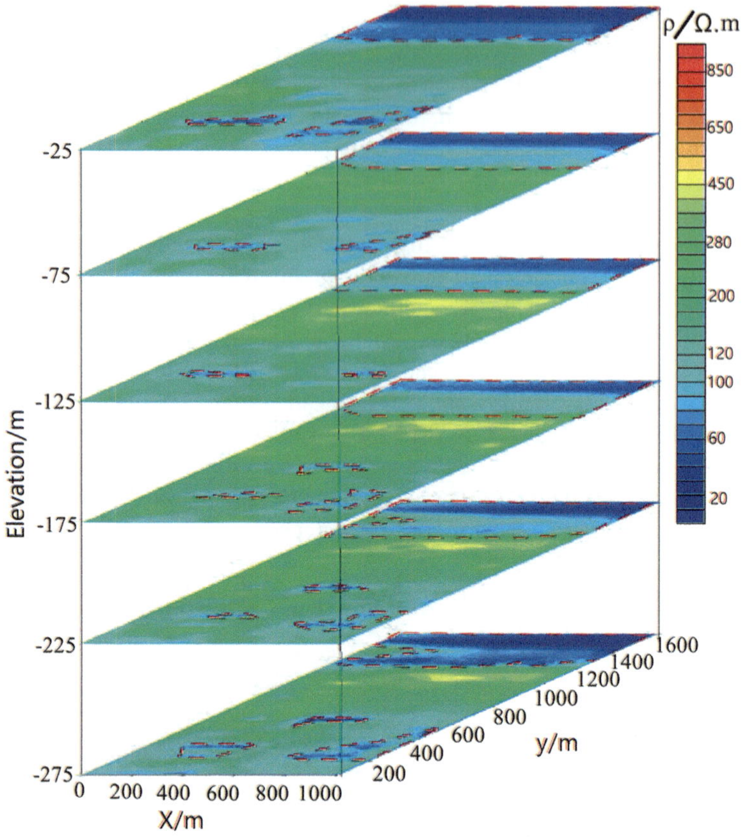

Fig. 8.17 −25, −75, −125, −175, −225, −275 m elevation resistivity horizontal slice map

electrical characteristics of each slice, it can be concluded that the connectivity of the upper and lower strata in the area north of the B1200 line is good. Near the B1200 survey line, the resistivity value shows significant variation, indicating a substantial electrical difference between the strata on both sides of this area.

c. **Electrical characteristics of the lower strata of the third series**

Figure 8.18 shows the resistivity horizontal slice maps of the entire survey area at −375, −475, −575, −675, −775, −875, −975, −1075, and −1175 m, including the third section of the third series gravel (vertical depth 400 m and below). The red solid line represents the bottom interface of the third series, the black solid line indicates the 4th coal stratum, the black dashed line marks the mined-out area, and the yellow solid line denotes the top interface of the Ordovician. The area east of the red solid line in the survey area comprises the third series strata, while the area west of the yellow solid line consists of the Ordovician strata, with the coal series strata located between the two.

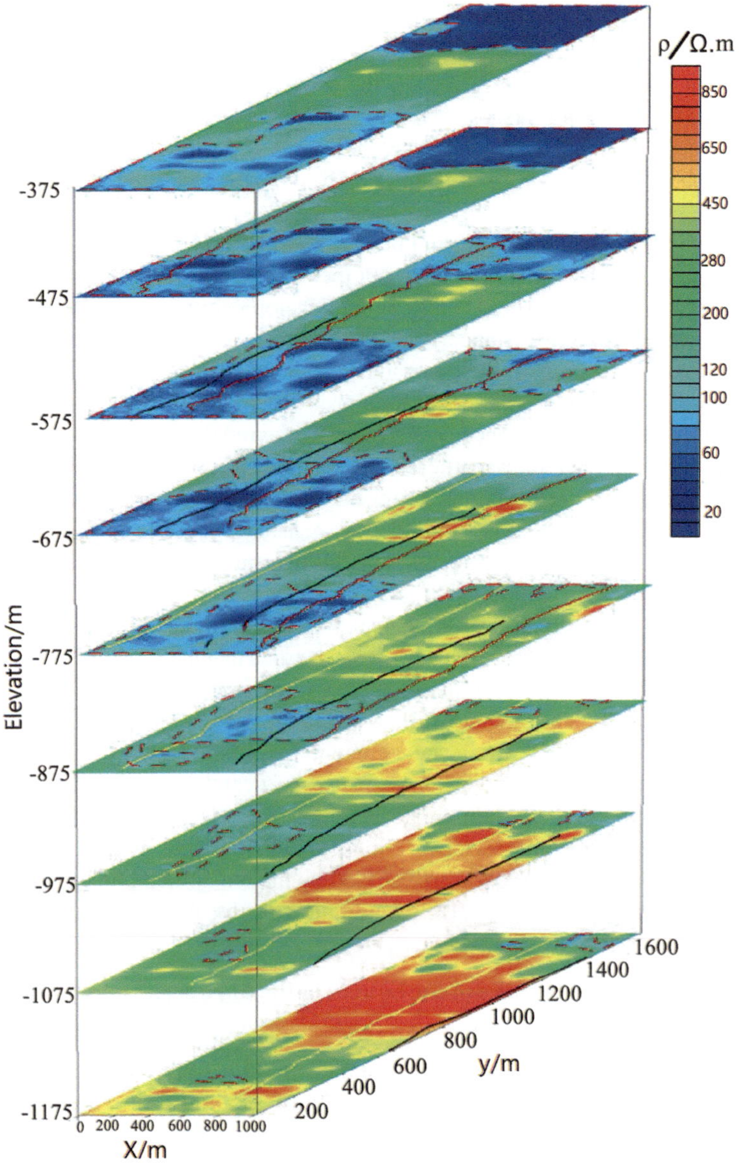

Fig. 8.18 Resistivity level slice map of −375 m to −1175 m elevation

Analyzing the nine resistivity slice maps in this section, the resistivity values throughout the survey area display a low-to-high-to-low trend from north to south. The resistivity values in the middle of the survey area are relatively high, indicating little to no water content or weak water content. The areas circled by the red dashed lines in the north and south exhibit relatively low resistivity values, suggesting well-developed fractures and relatively water-rich regions. In the northern area, as the depth of the horizontal slice increases, the low-resistance area circled by the red dashed line gradually diminishes. After the appearance of the Ordovician gray bottom interface at the −775 m horizontal slice, the extent of the low-resistance area in the northern part decreases with increasing depth, indicating that the water content in this region gradually weakens.

The high-resistance area in the middle of the survey area remains relatively stable in the range of −375 to −675 m depth, but the resistivity value gradually increases. At depths of −775 m and below, due to the development of conglomerate and the transition into the high-resistance layer of Ordovician limestone, the range of the high-resistance area in the middle expands, covering most of the survey area, indicating that the water content is decreasing or is absent. In the depth range of −375 m to −675 m, the low-resistance area circled by the red dashed line in the southern part of the survey area shows little change in range; however, the low-resistance anomaly becomes more prominent, suggesting that as depth increases, the water content strengthens. Below −775 m, the low-resistance area in the south gradually decreases until most of the area becomes high-resistance, indicating that the water content weakens.

In the −375 m resistivity level slice map, the overall range of resistivity values is between 20 and 600 Ω·m. From the map, it can be seen that the resistivity values from north to south in the entire slice display a low-to-high-to-low trend. The resistivity value of the area circled by the red dashed line in the northern part of the survey area is relatively low, indicating strong water content. Near the B1200 pile number survey line on both sides, the resistivity value changes significantly, suggesting a substantial electrical difference between the strata on either side. The resistivity value in the middle of the survey area is relatively high, indicating weak water content or no water. The resistivity value of the area circled by the red dashed line in the southern part of the survey area is also relatively low, indicating that it is a relatively water-rich area.

(3) **Electrical characteristics of the bottom of the third series and the top of the coal series**

In order to determine the water richness of the bottom rock layer of the third series, horizontal slices were created along the bottom of the third series, 50 m above the bottom, and 100 m above the bottom, respectively, for interpretation.

Figure 8.19 presents a horizontal slice map near 100 m above the bottom interface of the third series. The resistivity across the entire slice map from B100 to B1600 generally exhibits a low-to-high-to-low electrical characteristic. The range of resistivity changes in the map is significant, between 20 and 700 Ω·m. There is a considerable electrical difference between the middle and the sides of the entire slice map, reflecting variations in lithology and water content. The resistivity value in the area marked as II on the map changes significantly, ranging from 120 to 700 Ω·m, indicating that the water richness is relatively weak. In contrast, the resistivity

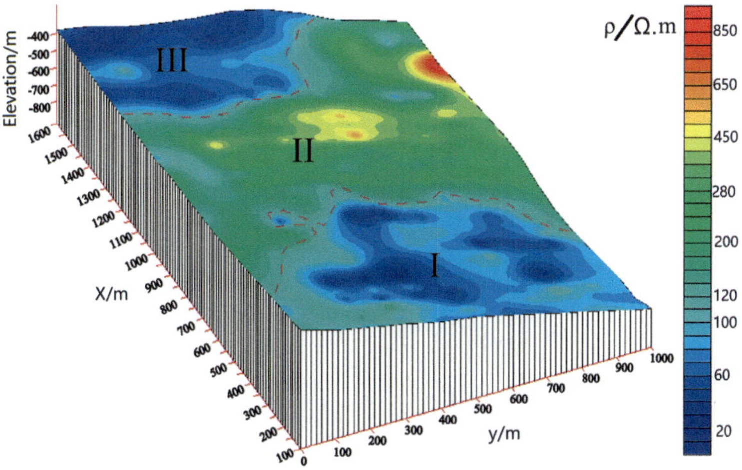

Fig. 8.19 Horizontal resistivity slice map 100 m above the bottom interface of the third series

values within the areas marked as I and III are relatively low, ranging from 20 to 140 Ω·m, suggesting that the water richness in these areas is relatively strong.

Figure 8.20 shows a horizontal slice map near 50 m above the bottom interface of the third series. The resistivity across the entire slice map from B100 to B1600 generally exhibits a low-to-high-to-low electrical characteristic. The range of resistivity changes in the map is significant, between 20 and 550 Ω·m. There is a notable electrical difference between the middle and the sides of the entire slice map, reflecting variations in lithology and water content. The resistivity of the area marked as II on the map is relatively high, ranging from 120 to 550 Ω·m, indicating that the water richness is relatively weak. In contrast, the resistivity values within the areas marked as I and III are relatively low, ranging from 20 to 120 Ω·m, suggesting that the water richness in these areas is slightly stronger than in the middle of the survey area. Notably, there is a relatively low-resistance area in the western part of the area corresponding to II, which may indicate some water conductivity, and it extends through the middle to connect the water-rich areas on both sides.

Figure 8.21 shows a horizontal slice map near the bottom interface of the third series. The resistivity across the entire slice map from B100 to B1600 generally exhibits a low-to-high-to-low electrical characteristic. The range of resistivity changes in the map is significant, between 20 and 500 Ω·m. There is a notable electrical difference between the middle and the sides of the entire slice map, reflecting variations in lithology and water content. The resistivity of the area marked as II on the map is relatively high, ranging from 120 to 500 Ω·m, indicating that the water richness is relatively weak. The resistivity values in the areas marked as I and III are relatively low, ranging from 20 to 150 Ω·m, suggesting that the water richness in these areas is relatively stronger than in the middle. Notably, there is a relatively low-resistance area in the western part of the area corresponding to II, which may

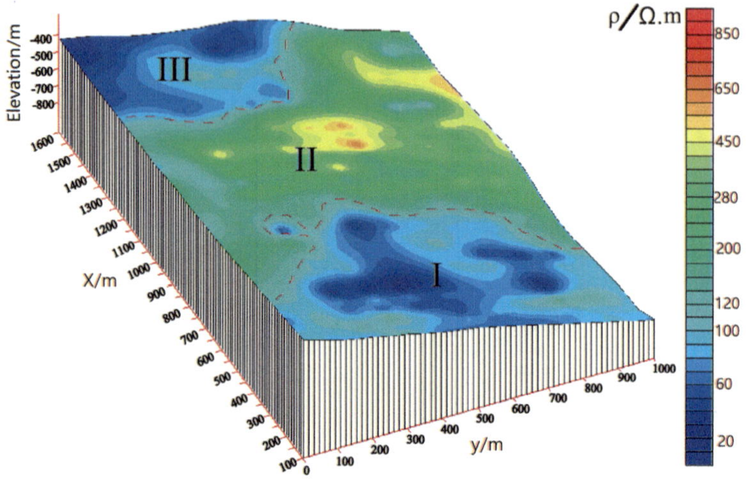

Fig. 8.20 Horizontal resistivity slice map 50 m above the bottom interface of the third series

indicate some water conductivity and connects the water-rich areas on both sides through the middle.

In order to determine the water richness near the top of the 4th coal series, a stratigraphic slice was made in the area near 50 m below the bottom of the third series. Figure 8.22 shows a stratigraphic slice map near 50 m below the bottom interface of the third series and close to the 4th coal seam. The resistivity across the entire slice map from B100 to B1600 generally exhibits a low–high-low electrical characteristic. The range of resistivity changes in the figure is significant, between

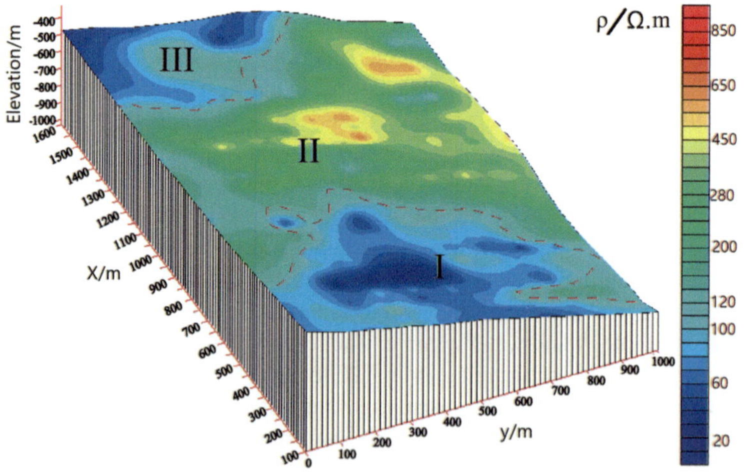

Fig. 8.21 Resistivity slice map of the bottom interface of the third series

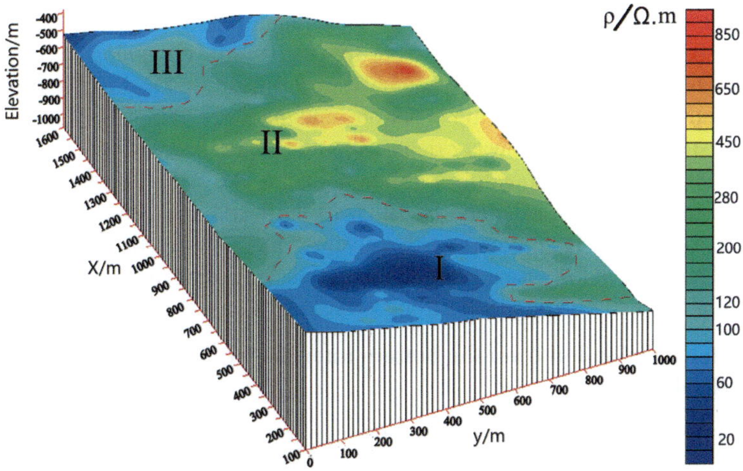

Fig. 8.22 Resistivity slice map 50 m below the bottom interface of the third series

20 and 600 Ω·m. It can be observed that there is a considerable electrical difference between the middle and the sides of the strata, reflecting variations in lithology and water content. The resistivity of the area marked as II in the figure is relatively high, ranging from 120 to 600 Ω·m, indicating that the water richness is relatively weak. In contrast, the resistivity values of the areas marked as I and III are relatively low, ranging from 20 to 160 Ω·m, suggesting that the water richness in these areas is relatively stronger than in the middle. Notably, there is a relatively low-resistance area in the western part of the area corresponding to II, which may indicate some water conductivity and connects the water-rich areas on both sides through the middle.

In the four sets of resistivity stratigraphic slice maps at 100 m above the bottom interface of the third series, 50 m above the bottom interface, at the bottom interface, and 50 m below the bottom interface, the overall range of resistivity changes is between 20 and 700 Ω·m. In these slice maps, the resistivity values exhibit a low–high-low trend along the B survey line direction. The areas with relatively high resistivity values are concentrated in the middle of the slice maps, while the areas with relatively low resistivity values are found on the sides. Moreover, from top to bottom, the resistivity shows a trend of changing from high to low. Overall, these partitions strongly correspond in plane position at different depths, indicating that the water bodies above and below the water-rich area have strong connectivity.

In these four stratigraphic slice maps, the overall range of resistivity values in the nearby area defined by B1400 to the north and A650 to the west is between 20 and 120 Ω·m, with resistivity values gradually increasing from top to bottom. The range of low resistance values narrows, indicating a decrease in the proportion of gravel rock, with water richness trending from strong to weak.

In the middle area corresponding to the slice maps at different depths, the overall range of resistivity values changes from top to bottom, remaining between 120 and 700 Ω·m, with high-resistance anomaly areas appearing locally. The resistivity values gradually increase from top to bottom, and the range of high resistance values expands, suggesting that the water richness in this area is decreasing.

In the slice maps corresponding to different depths, the overall range of resistivity values in the nearby area defined by B900 to the south and A900 to the west is between 20 and 150 Ω·m, with resistivity values gradually increasing from top to bottom. The range of low resistance values narrows, indicating that the water richness in this range is also trending from strong to weak.

(4) **Water distribution from the bottom interface of the third series to the top of the 4th coal**

The overall resistivity change from the bottom interface of the third series to the 4th coal is significant, with values ranging from 20 to 700 Ω·m. The resistivity values exhibit a characteristic of being low on both sides and high in the middle along the B survey line direction. Overall, this can be interpreted as weakly water-bearing or non-water-bearing, while the local areas with low resistivity values are relatively medium water-bearing or relatively strong water-bearing.

Figure 8.23 shows a stratigraphic slice map near 60 m above the top of the 4th coal, and Fig. 8.24 presents a stratigraphic slice map near 30 m above the top of the 4th coal. From these figures, we can analyze the water richness of these two layers and roughly judge the connectivity of the strata. Since the water richness of the layers represented by Figs. 8.23 and 8.24 is essentially the same, the water richness of the 60 m layer above the top of the 4th coal and the 30 m layer above the top of the 4th coal can be analyzed as follows:

The overall resistivity value in this area changes significantly, ranging from 20 to 650 Ω·m, which can be interpreted as weakly water-bearing or non-water-bearing, with local areas being relatively medium water-bearing. The entire survey area generally indicates that the resistivity value is relatively low in the small number survey line segment, which is weakly water-bearing or locally relatively medium water-bearing. In contrast, the resistivity value near the fault in the large number survey line segment is high, indicating non-water-bearing or locally weakly water-bearing conditions.

Among these findings, the resistivity in the segment from B100 to B300 shows a low value characteristic, which is judged to be weakly water-bearing. The resistivity between B300 to the north and B500 to the south shows a low value characteristic, indicating relatively medium water-bearing conditions. Notably, at the B350 line, the resistivity value in the 200–550 m pile number section is relatively lower, suggesting relatively strong water content. The entire area north of the B500 line and south of the B1600 line exhibits relatively high resistivity values, indicating no water content or weak water content in local areas. Only near the B550 line in the 400 m pile number and the B600 line in the 200–400 m pile number section is the resistivity value relatively low, indicating weak water content.

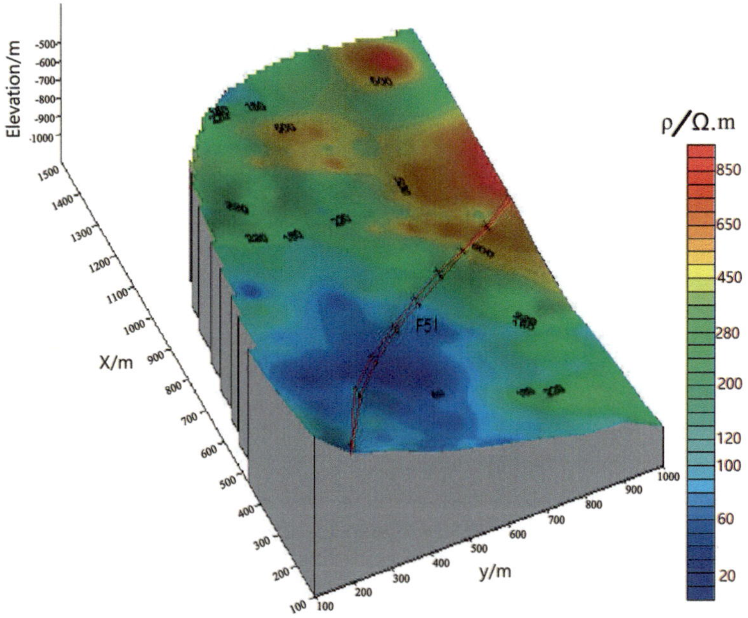

Fig. 8.23 Resistivity slice map of 60 m stratigraphic above the 4th coal

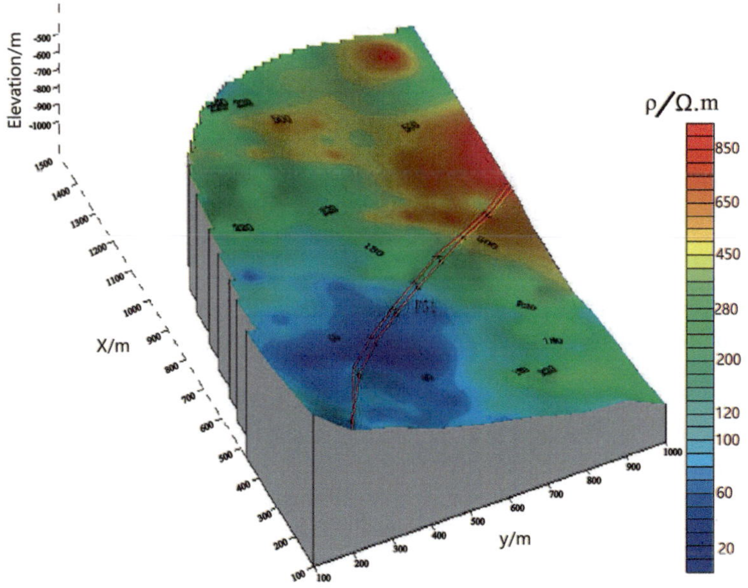

Fig. 8.24 Resistivity slice map of 30 m stratigraphic above the 4th coal

Based on the above water-rich analysis, it can be roughly concluded that the connectivity between the stratum near 60 m above the top of the 4th coal and the stratum near 30 m above the top of the 4th coal is good.

8.3 Detection of Magmatic Copper–Nickel Ore in Xinjiang

Global magmatic copper–nickel sulfide deposits are primarily found in cratonic areas, while China's magmatic copper–nickel sulfide deposits are mainly distributed in orogenic belts. The Kalatongke copper–nickel mine in Xinjiang is situated on the northern edge of the Central Asian orogenic belt and is the largest magmatic copper–nickel sulfide deposit in this region. It is characterized by blocky and disseminated sulfide mineralization and falls under the conduit-type ore-forming model for magmatic copper–nickel sulfide deposits.

The Kalatongke mafic rock mass group developed during the late Carboniferous (Y5 rock mass, 320 Ma) and early Permian (Y1, Y3, Y9, and G21 rock mass, 290–281 Ma) periods. The 13 basic rock masses that have been discovered are mostly concealed underground, and the overall basic degree of these rock masses is low. Unlike other typical copper–nickel deposits globally, the ore-forming rock masses in Kalatongke are small and grouped, characterized by large deposits within small rock masses. Additionally, the Kalatongke magmatic copper–nickel deposit generally develops within carbonaceous strata, which may contribute to sulfide saturation and mineralization.

Previous geophysical detection research has primarily focused on blocky ore bodies, while the exploration of non-blocky ore bodies (such as disseminated ore bodies) still presents significant challenges. The disseminated ore body does not exhibit obvious differences from the surrounding rock in terms of resistivity, polarization rate, density, or seismic velocity, making it difficult to locate these bodies directly through geophysical methods. In the Kalatongke deposit, the extensively developed carbonaceous layer may play a role in the formation of the copper–nickel deposit. The carbon provided by this layer acts as a reductant during the sulfide saturation process, suggesting a potential spatial contact relationship between the carbonaceous layer and the ore body. Detecting this copper–nickel deposit, with its unique geological features such as small rock masses forming large ore bodies and the widespread development of carbonaceous strata, has remained a significant area of research.

Based on the determination of the physical properties of the ore and surrounding rocks of the magmatic copper–nickel deposit, the surface position of the Kalatongke ore-forming rock mass is delineated using the gravity-magnetic method. The top and bottom interfaces, along with the storage depth of the ore-forming rock mass, are defined using metal ore seismic exploration methods. For the low-resistance and high-polarization characteristics of the carbonaceous layer, the newly developed SOTEM employs its enhanced sensitivity to the electric effects of the carbonaceous layer for extraction and utilization of its electrical information. The depth and shape of the disseminated ore body are analyzed comprehensively using the spatial distribution relationship between the ore-forming rock mass and the carbonaceous layer, as well as the storage relationship between the ore body and the ore-forming rock mass. Through the analysis of measured data and drilling verification, this research lays a theoretical foundation and provides technical support for further exploration of similar copper–nickel deposits.

8.3.1 *Geological Characteristics and Rock and Mineral Physical Property Analysis of Kalatongke Copper–Nickel Deposit*

The Kalatongke deposit is located in the western part of the Central Asian orogenic belt (Fig. 8.25). The mafic rock mass group consists of multiple small rock masses, with 11 identified in the mining area, numbered Y1-Y11 (Fig. 8.26). The rock masses are mainly controlled by NW-directed faults and intrude into the South Ming Water Group of the Lower Carboniferous. The northwest ends of the Y1 rock mass and the Y4, Y6-Y11 rock masses are exposed on the surface, while the rest are hidden. Current mining activities are concentrated in the Y1 and Y2 rock masses. The scale, morphology, petrographic characteristics, and ore-bearing properties of these rock masses have been extensively studied.

The ore body is located in the middle and lower parts of the Y1 rock mass, primarily composed of disseminated and dense blocky ores. Sulfides are observed to be nearly oriented within the disseminated ore, while dense blocky ores are distributed inside the disseminated ore body and between the Y1 and Y2 rock masses. The deep part of the eastern section of the Y2 rock mass, from lines 47 to 53 (650–740 m below the surface), has yielded mineral-bearing olivine and pyroxenite, with locally visible ore bodies up to 10 m thick. However, there is limited research on the characteristics and

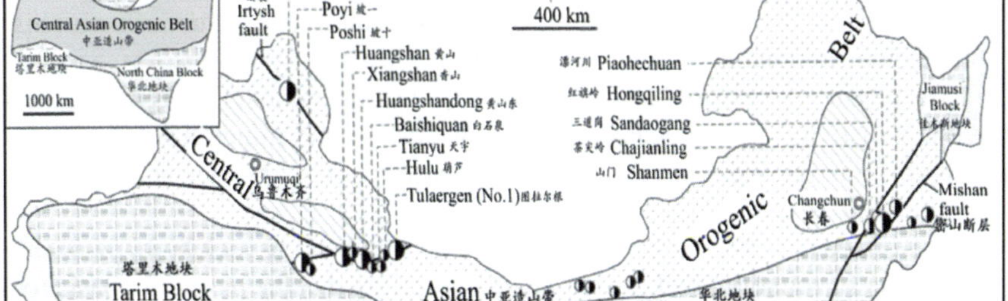

Fig. 8.25 Location of Kalatongke copper–nickel deposit in the Central Asian orogenic belt

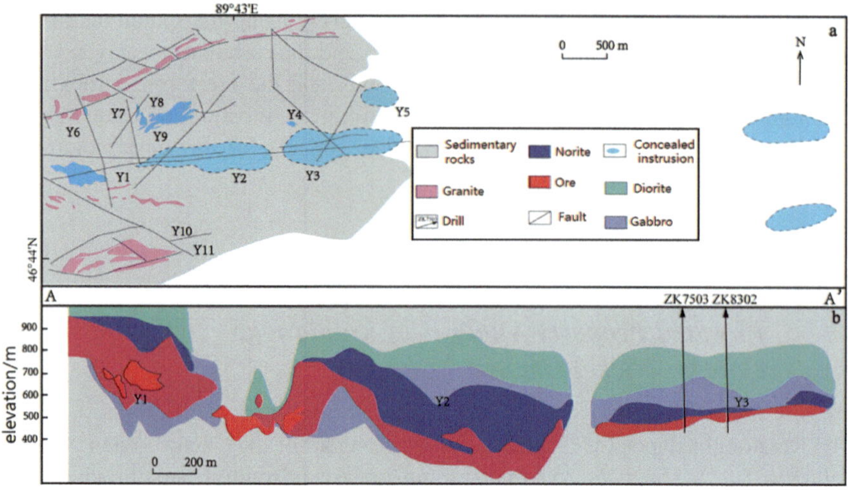

Fig. 8.26 Geological sketch of the Karatongke rock body group

lithology of rock bodies and ore bodies such as Y3 and Y5, so this study primarily focuses on these two rock bodies for geophysical research.

The physical properties of rocks and ores serve as the bridge connecting geological models and geophysical exploration. The main types of rocks and ores in the mining area have been selected for physical property testing. The results of these physical property tests are presented in Table 8.4.

Analysis from Table 8.4 shows:

(1) The density of the main ore-bearing rock, pyroxenite, is 0.1–0.2g/cm^3 higher than the surrounding rock, and the density will significantly increase with the increase of sulfide content.
(2) Carbonaceous condensed limestone contains almost no magnetism, while well-mineralized rock bodies have a high magnetization rate, and the magnetism of rock ore is mainly controlled by the contained pyrrhotite. The magnetization rate of ore-bearing magnesia-ultramafic rock is significantly higher than that of surrounding diorite and condensed limestone, which can produce effective high magnetic anomalies.
(3) The resistivity value of blocky, angular gravel copper–nickel ore is the smallest, the resistivity of impregnated copper–nickel ore and mineralized diorite and pyroxenite is the largest, and the resistivity of carbonaceous condensed limestone is medium.
(4) The impedance difference between the main ore-bearing rock body pyroxenite and the surrounding rock is significant, indicating that the basic-ultrabasic rock in the mining area and the ore with a high degree of mineralization can form an effective impedance difference with the surrounding rock, providing a good physical basis for shallow seismic exploration.

Table 8.4 Physical characteristics of rocks and ores in the Karatongke deposit

Rock name	Density (g/cm^3)		Magnetization (10^{-5} SI)		ηs (%)		ρs (Ω·m)	Z (km/s)(g/cm^3)	
Massive ore	> 4.0		2500–4000		22.9–84.5		2.61–18.5	12	
Angular gravel ore	3.40–3.8								
Impregnated ore	3.0–3.5		Star Point	200–660	1.5–85.5		273–845	Star Point	16.2–18.4
			Sparse	1200–1600					
			Medium	2200–3000				Medium	17.9–19.4
								Dense	18.5–20.7
Diorite	2.61–2.78		20		0.64–51.6		1034–11,286	13–14	
Gabbro	2.81–2.93		500–1000		1.04–35.3		685–6976	18	
Tuff	Carbonaceous	2.65–2.75	20–69		Carbonaceous	1.72–79.8	330–6676	16	
	Siliceous	2.75–2.8	–		Siliceous	0.24–53.2	–	15	
Granite	2.69		125				–	–	

8.3.2 Locating Deep Ore Bodies Based on SOTEM Data

The layout of the survey lines is shown in Fig. 8.27, with a total of 3 survey lines arranged. The length of survey line L2 is 5000 m, spanning two rock bodies Y3 and Y5.

Affected by the Induced Polarization (IP) effect, the transient electromagnetic response appears distorted, sometimes even exhibiting a reverse sign phenomenon. To better extract the electrical information contained in the transient electromagnetic response influenced by IP, we chose the Short Offset Transient Electromagnetic Method (SOTEM) for electromagnetic field collection. The measured electric field data at points 2080 m and 2520 m are shown in Fig. 8.28a. Using the adaptive regularization algorithm, we aimed to find a model that can fit the measured data to the maximum extent while meeting the requirements of the model objective function, as illustrated in Figs. 8.28b and 8.28c.

As shown in Fig. 8.28a, the measured electric field data exhibits a clear sign reversal phenomenon, which is a distortion of the transient electromagnetic response caused by the electric shock effect. Through data inversion, resistivity and polarization rate results were obtained simultaneously, as illustrated in Figs. 8.28b, c. Within the depth range of 0–200 m, the models recovered by the inversion of the two measurement point data are nearly identical. The low charge rate region is interpreted as the Quaternary low-resistance cover layer at the surface, while the low-resistance,

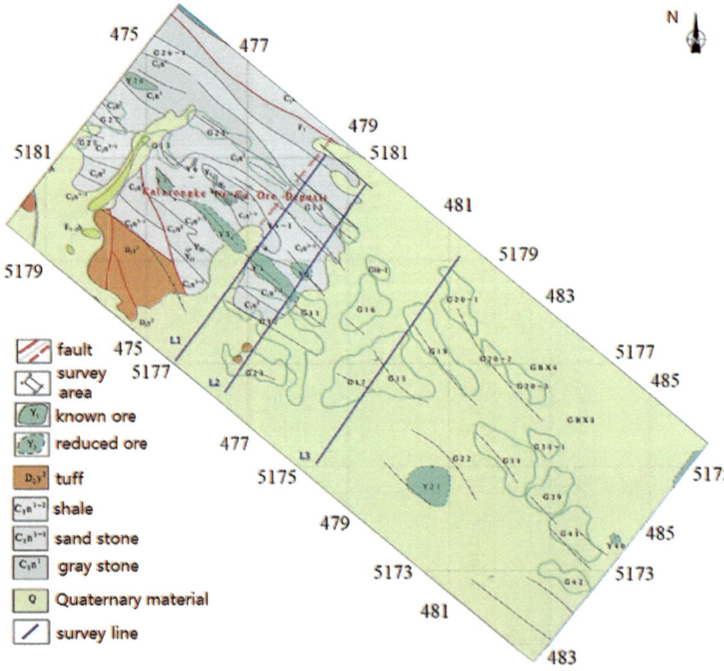

Fig. 8.27 Seismic and electromagnetic survey line map of Karatongke metal mine

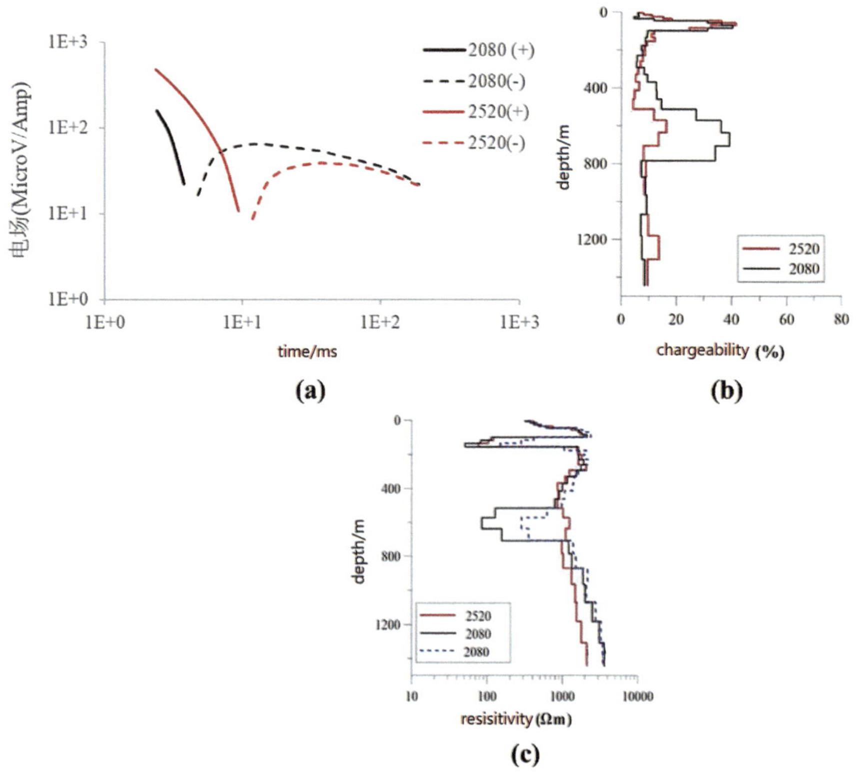

Fig. 8.28 shows the SOTEM measured electric field curve and charge rate, resistivity inversion results at points 2080 m and 2520 m. **a** Measured electric field; **b** Charge rate inversion result; **c** Resistivity inversion result, with the dashed line representing the inversion result when the IP effect is not considered

high-polarization area at 50–150 m is interpreted as the widely developed shallow carbonaceous layer. However, in the deeper range of 200–700 m, the models recovered by the inversion of the two measurement point data are completely different. A low-resistance, high-polarization anomaly appears at point 2080 m (depth about 600 m), with the lowest resistivity being less than 100 Ωm and the maximum charge rate greater than 0.3, approaching 0.4. It is speculated that this anomaly represents carbonaceous tuff. The inversion result obtained without considering the IP effect at point 2080 m is also provided, showing a false low-resistance anomaly in the shallow part and a false high-resistance anomaly in the middle and deep parts. This resistivity anomaly is attributed to the neglect of the IP effect, a conclusion supported by drilling information.

Figure 8.29a displays the resistivity inversion profile in the range of 1200–3400 m on line L2. The vertical and horizontal axes represent elevation and horizontal distance, respectively. Resistivity is displayed in logarithmic intervals. Red and yellow indicate low and medium–low resistivity, while green and blue represent medium–high and high resistivity. The high resistivity in the model (deep blue) is interpreted as diorite, while the green area is interpreted as tuff. Pyroxenite and gabbro are located in the blue area, appearing light blue.

A low resistivity layer is present at the measurement points 1500–2300 m, with an elevation of 600–300 m, marked as C1. The low-resistance area C1 in survey line L2 (Fig. 8.29a) is displayed in yellow and red, inferred to be carbonaceous tuff. The high-resistance area, represented by deep blue, is inferred to be diorite. Gabbro and diorite often coexist, and the light blue area (resistivity less than $10^{2.6}$–$10^{3.2}$ Ωm) appearing within the dark blue stratum (resistivity greater than $10^{3.2}$ Ωm) at an elevation of 800–600 m is inferred to be gabbro, which is a basic ore-forming rock

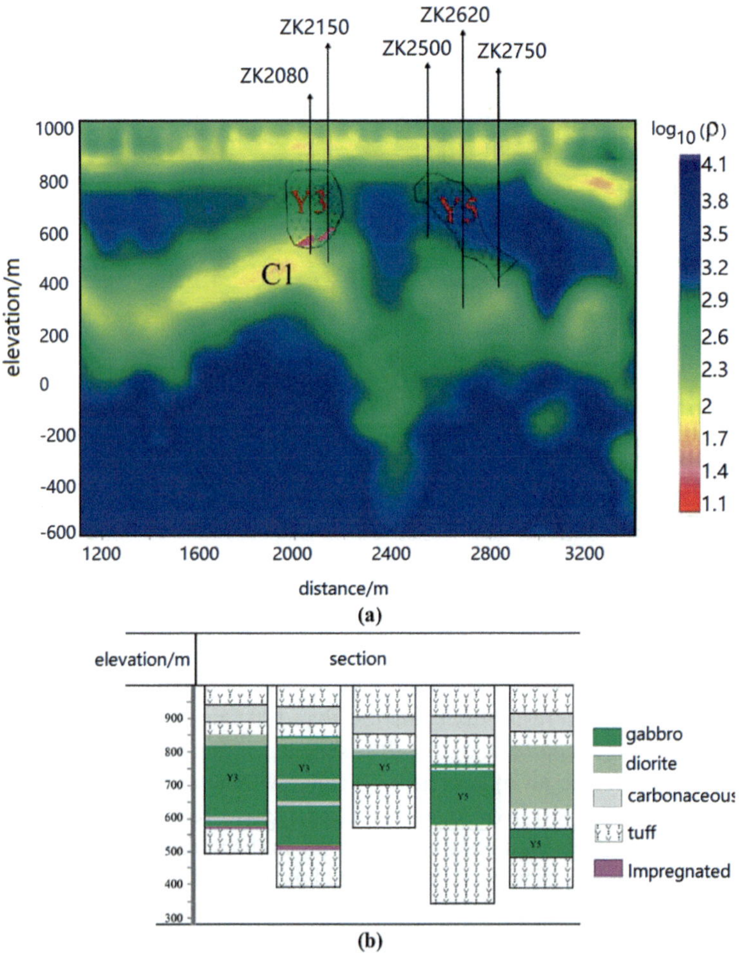

Fig. 8.29 Shows the L2 survey line resistivity inversion profile and borehole profiles at 2080, 2150, 2500, 2620, and 2750 m. Among these, red and yellow indicate resistivity less than $10^{2.3}$ Ωm, while blue indicates resistivity greater than $10^{3.2}$ Ωm. ZK-indicates the location of the borehole, while Y3 and Y5 indicate the rock bodies. **a** Resistivity inversion profile; **b** Borehole profile at 2080, 2150, 2500, 2620, and 2750 m

body, marked as Y3 and Y5 rock bodies in Fig. 8.29a. The Y3 intrusion is close to the carbonaceous tuff C1, and the disseminated ore body (pink in Fig. 8.29a) is located at the bottom of the Y3 intrusion. In contrast, the Y5 intrusive rock body has no spatial contact with C1, so the absence of a disseminated ore body at the bottom of Y5 is attributed to the lack of a carbonaceous layer.

Drilling verification was conducted at the L2 survey line points of 2080, 2150, 2500, 2620, and 2750 m. According to the drilling results (Fig. 8.29b), disseminated ores were found at 2080–2150 m. Gabbro exists between elevations of 850–500 m, and disseminated sulfides were located between 550 and 500 m. Within the gabbro, carbonaceous tuff, as a large trap body, can be found in the rock core that passes through the Y3 intrusion. However, no ore was found at the points of 2500 m, 2620 m, and 2750 m, and no carbonaceous tuff was identified in the rock core of the Y5 intrusion at 2500, 2620, and 2750 m.

8.4 Nickel Ore Detection in Nanyang Basin, Henan

Zhou'an in Henan is the largest nickel and platinum group element deposit in the Qinling-Dabie orogenic belt (Fig. 8.30a). Early exploration primarily relied on drilling, and the ore-bearing rock body was identified as ultrabasic rock, hidden beneath the Neogene strata, intruded into the Dajue Mountain Group of the Zhujia Mountain Group. The surrounding rocks are affected by mild contact metamorphism. The top of the ore-bearing rock body is buried at a depth of about 300–400 m, while the bottom is buried at a depth of about 700–900 m. Due to the influence of the Archean metamorphic rocks and the thick Neogene strata in the area, exploring deep hidden ore bodies is challenging, and deep prospecting geophysical exploration has not yet achieved a breakthrough.

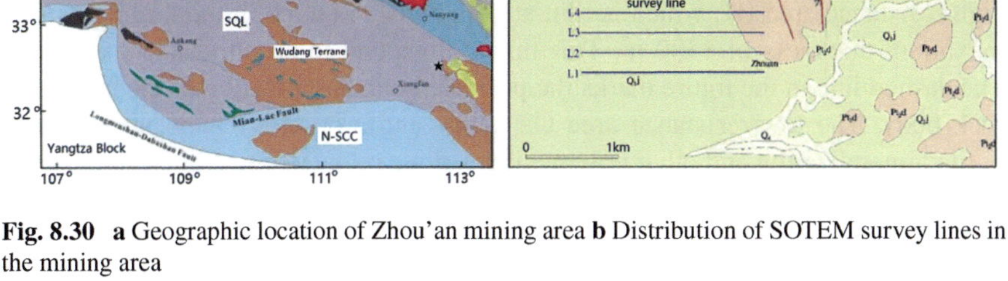

Fig. 8.30 **a** Geographic location of Zhou'an mining area **b** Distribution of SOTEM survey lines in the mining area

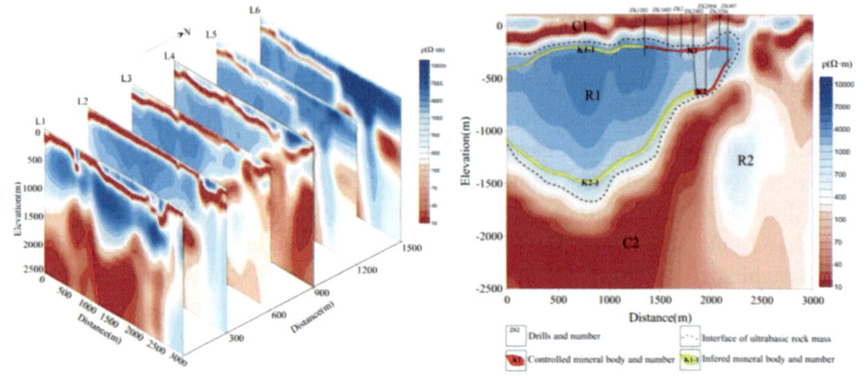

(a) **(b)**

Fig. 8.31 **a** Electric results of 6 side line profiles **b** Resistivity profile of Line L3 and possible geological interpretation

In order to detect deep ore bodies, we arranged six electrical source short-offset transient electromagnetic method (SOTEM) survey lines in the mining area, as shown in Fig. 8.30b. The survey lines run east–west, each line is about 3 km long, with a line spacing of 300 m and a point spacing of 50 m. Data collection was carried out using self-developed wide-band equipment, with the main working parameters being: transmission current 30 A, transmission base frequency 5 Hz, time window range 0.475–47.5 ms, and the effective area of the receiving sensor is about 10,000 m2.

Using the dual-mode drive inversion algorithm described earlier, we obtained a high-resolution three-dimensional resistivity structure about 2.5 km deep under the survey line, revealing that the strata in the survey area exhibit a low-resistance-high-resistance-low-resistance change feature from shallow to deep (Fig. 8.31). In Fig. 8.31a, the low-resistance layer at the shallow surface of each survey line mainly reflects the Neogene cover layer. The large high-resistance body below it may be an ultrabasic rock body, and the low-resistance feature of the deep strata may be related to rock alteration.

Taking Line L3 as an example (Fig. 8.31b), there is a low-resistance area C1 at the shallow surface, approximately 100 m thick, corresponding to the Neogene cover layer. Below it, at a measurement depth of about 200–1600 m, there is a significant high-resistance area R1, which is thin at both ends and thick in the middle, with a maximum thickness of 1500 m. This may correspond to an ultrabasic rock body. The dashed line in the figure marks the possible boundary position of the ultrabasic rock body. The low-resistance area C2 below and around R1 may be related to pyrite mineralization and other polymetallic sulfide mineralization alterations in the surrounding rock of the ultramafic rock body. The relatively high-resistance area R2 on its right may correspond to the metamorphic rock series of the Daque Mountain Group, which is consistent with the inclination (260–290°) and dip angle (30–42°) of this group of strata.

The results indicate that the SOTEM electromagnetic method can effectively delineate the boundary between the ore body and the rock body, as well as the strata in this area, with high imaging accuracy for the ore body and the high- and low-resistance interfaces. The delineated rock body interface is consistent with the drilling detection results. By combining the ore-forming belt identified by drilling with the high-resolution imaging results, we speculate on the possible mineralization belt along the periphery of the ultrabasic rock body, providing a solid geophysical basis for further prospecting and exploration of the Zhou'an copper–nickel mine.

8.5 Xiaoshan Mining Area Gold Polymetallic Ore Detection

8.5.1 Overview of the Survey Area

(1) **Geological Overview**

The Xiaoshan area is located in Sanmenxia City, western Henan Province, at the junction of the East Qinling orogenic belt and the southern margin of the North China Craton (Fig. 8.32). It is a typical area for the development of metamorphic core complexes. The region has superior ore-forming geological conditions and is rich in gold-polymetallic geochemical anomalies, with medium-small deposits and mineralization points widely distributed. As a result, it has long been considered an area with significant prospecting potential.

Despite extensive mineral exploration work in the Xiaoshan area in recent years, only a few medium-small gold-polymetallic deposits, such as Ban Kuan, Shen Jiayao, Da Fangshan, and Si Jiagou, have been discovered. The area is still characterized by the phrase "seeing stars but not the moon," with no major prospecting breakthroughs achieved. In contrast, the neighboring Xiao Qinling and Xiong Ershan areas, which have extremely similar ore-forming geological conditions, have made significant resource discoveries, achieving high-profile prospecting breakthroughs. This has led to western Henan becoming the second largest gold-producing area in China, after Jiaodong.

During the Late Jurassic-Early Cretaceous period, the subduction of the Pacific Plate and the Eurasian Plate caused a change in the regional stress field, leading to a transition of the southern margin of the North China landmass from compression to extension. The overlying lower strata were intruded by upwelling asthenosphere, and under the influence of mantle fluid, the thickened lower crust partially melted and rose to form a crust-mantle mixed source granite body. During its formation, the mantle fluid brought a significant amount of silver, gold, and other polymetallic elements.

Simultaneously, due to the high heat flow environment of the metamorphic core complex structural system, the rocks in the lower part of the crust were mixed and remelted, providing energy for the migration of ore-forming elements, resulting in the further enrichment of silver, gold, and other polymetallic elements. As the

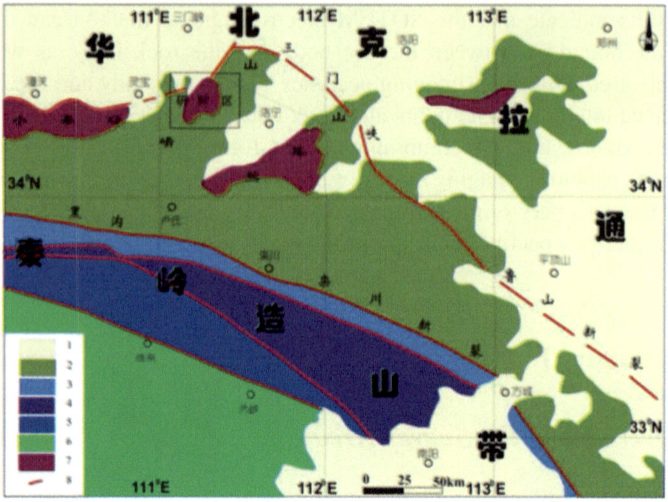

Fig. 8.32 Simplified tectonic map of Western Henan, 1-Quaternary; 2-North China Craton cover; 3-Kuanping Group; 4-Erlangping Group; 5-Qinling Group; 6-South Qinling Series; 7-Archean metamorphic rocks; 8-Regional deep major faults

metamorphic core complex continued to uplift, it further extended and, due to gravity, collapsed into the surroundings, leading to the incision of the detachment normal fault. Concurrently, the brittle-ductile shear zone of the metamorphic core complex moved upward, and subsequent ductile–brittle and brittle faults overlapped. The hydrothermal fluids generated in this process enriched and formed ores in the reformed steep brittle-ductile shear zone and the later gentle fracture zone, as shown in Fig. 8.33.

Studies on the occurrence of gold-polymetallic mineralization in this area indicate that most of the veins discovered so far are mainly controlled by brittle-ductile shear zones of varying directions, attitudes, and scales developed in the deep metamorphic rock series of the Xiaoshan metamorphic core complex basement. These ore-controlling brittle-ductile shear zones, whether the main detachment zone of the metamorphic core complex or secondary shear zones nearly parallel or oblique to the main detachment zone, have a close genetic relationship with the formation, evolution, and tectonic emplacement of the metamorphic core complex.

In the existing mining areas of Xiaoshan, the current exploration depth is mostly in the near-surface range of 100–300 m; some sporadic mining depths are even shallower, mostly within the 100–200 m range. Only a few drill holes in individual mining areas exceed 800 m in depth. Such a shallow average exploration and mining depth is insufficient for the gold (polymetallic) mineralization system in the ductile (brittle) shear zone of the Archean metamorphic nucleus. It is well known that the neighboring Xiaoqinling and Xiong'ershan areas, which have similar geological conditions and mineralization types, have reached exploration depths of 500–800 m,

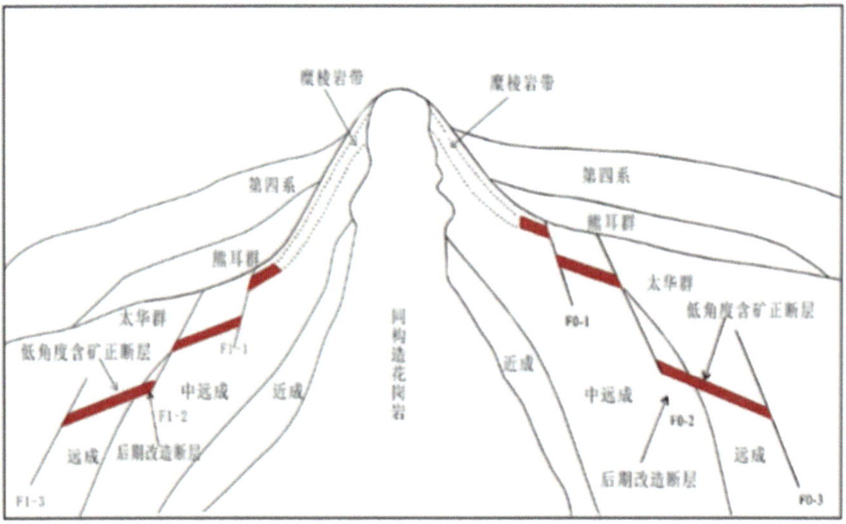

Fig. 8.33 Ore-forming geological model

where industrial ore bodies still exist. In contrast, exploration depths in other high-grade metamorphic areas worldwide, such as Western Australia and North America, are even greater.

In some major mining areas of Xiaoshan, the surface exposure scale of the main ore-controlling ductile–brittle shear zone is considerable, with strike lengths mostly close to or greater than 1000 m. For example, the K4 vein of the Kuanping deposit is 960 m, the 9–1 vein of the Dafangshan deposit is 1000 m, the 3 vein of the Hulu Valley deposit is 970 m, the K1 vein of the Sijiagou deposit is 1000 m, and the 9 vein of the Banwidth deposit is 1130 m. Notably, the K1 vein of the Shenjiayao deposit even exceeds 5000 m. Clearly, considering the relatively shallow overall erosion depth of this area, the current exploration depth does not align with the known exposure length of the ore veins.

For the mineralization system controlled by ductile–brittle shear structures, it is theoretically expected that its development depth will not be less than the strike length. There are often multiple sections of varying mineralization strength in both strike and dip directions, typically manifesting as alternating ore-bearing and non-ore-bearing sections, or strong and weak mineralization sections from shallow to deep. Given that shallow structural phenomena of ore-controlling faults can still be observed near the surface in this area, it is reasonable to believe that what is currently seen at the exploration and mining depth represents only the upper or shallow part of the mineralization system, with considerable gold-polymetallic mineralization likely existing deeper within the ore-controlling structural system.

In terms of neighboring geological comparisons, the three geographically adjacent metamorphic nucleus complex bodies in western Henan—Xiaoqinling, Xiaoshan, and Xiong'ershan—show that two of the more deeply eroded bodies (Xiaoqinling and

Table 8.5 Stratigraphic rock types and resistivity characteristics in the survey area

Stratum	Rock type	Resistivity range	Average value
Taihua group	Gneiss and mixed rock	1128 ~ 124,887 Ω·m	8591 Ω·m
Xiong'er group	Andesite	1230 ~ 84,255 Ω·m	11,321 Ω·m
fourth series	Loose overburden	30 ~ 120 Ω·m	70Ω·m
Ore	Massive galena ore	11 ~ 1550 Ω·m	164 Ω·m
	Fine vein infiltration ore	5848 ~ 12,227,572 Ω·m	89,147 Ω·m

Xiong'ershan) have already achieved significant prospecting breakthroughs, while the less eroded Xiaoshan has not yet made a breakthrough. This reflects the issue of the hidden burial depth of the main mineralization depth zone: the deeper the mineralization system is buried, the more difficult it is to discover. Therefore, for a major prospecting breakthrough in the Xiaoshan area, a deeper exploration is essential.

(2) **Stratigraphic resistivity**

The rocks and minerals in the work area are generally similar, and their resistivity follows the same general rule, with obvious resistivity differences, that can produce significant apparent chargeability and resistivity anomalies. According to the stratigraphic and logging data in the area and neighboring areas, the comprehensive stratigraphic resistivity is as follows: The rock type of the Taihua Group deep metamorphic rock is mainly gneiss and mixed rock, with a resistivity range of 1128 ~ 124,887 Ω·m, and the mode is 8591 Ω·m. The rock type of the Xiong'er Group is andesite, with a resistivity range of 1230 ~ 84,255 Ω·m, and the mode is 11321 Ω·m. The resistivity of the shallow Quaternary cover layer is 30 ~ 120 Ω·m. The ore generally shows a low resistance anomaly feature, but due to its composition and structural structure, the resistivity of deep blocky galena ore is lower, with a range of 11 ~ 1550 Ω·m, the mode is 164 Ω·m, showing a low resistance anomaly; fine vein-like infiltrated ore, because its main component is quartz or calcite, the metal mineral particles have poor connectivity and high resistivity, with a range of 5848 ~ 12,227,572 Ω·m, the mode is 8 generally similar 9147 Ω·m, showing a high resistance anomaly feature with its surrounding rock Archean gneiss and mixed rock. The main stratigraphic rock types and resistivity ranges in the survey area are shown in Table 8.5.

8.5.2 Data Collection and Processing

(1) **Data Collection**

To determine the location and shape of the electrical structure within a depth range of 1000 m in this area and to delineate the location of the anomaly, SOTEM exploration is conducted in the prospective mineralization area based on existing geological data. In this study, four survey lines were arranged, and a 1 km long transmitter was positioned along the mountain path. The information regarding the survey lines and the transmitter is shown in Fig. 8.34. The V8 electromagnetic system produced by Phoenix Company of Canada is utilized for signal transmission and reception, with a transmission current of 16 A, a base frequency of 2.5 Hz, and a vertically induced electromotive force measured by a magnetic probe with an effective area of 40,000 m^2.

(2) **Data Processing and Interpretation**

After denoising and inversion of the measured data using SOTEMsoft software, the depth-resistivity profile of each survey line was obtained. Here, the L3 survey line is used as an example (Fig. 8.35) to illustrate the SOTEM detection effect. From Fig. 8.35, it can be observed that the resistivity of the shallow part of the strata in this area is generally low, while the resistivity of the deep part is high. This is consistent with the electrical characteristics of the shallow Quaternary sedimentary layer and the mixed granite of the Taihua Group.

In the range of 300–600 m, the high-resistance stratum extends from the deep part to the shallow part, indicating that the Taihua Group in this area has been uplifted to the surface and that the top Xiong'er Group has been eroded. Additionally, the presence of F1 and F2 faults between the two strata is the main reason for the low-resistance reflection. Therefore, combined with the relevant mineralization and

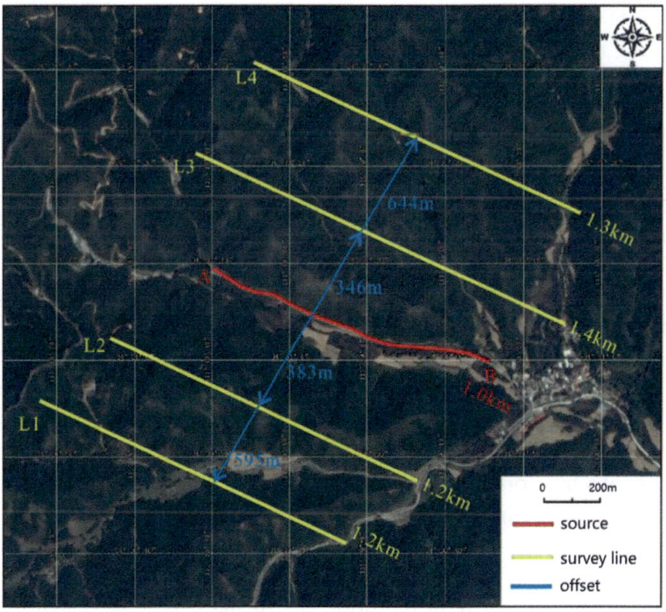

Fig. 8.34 Layout of transmitter and survey line

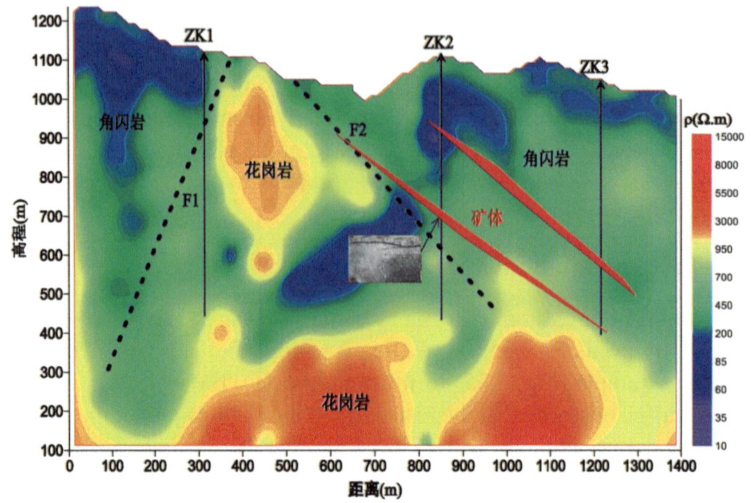

Fig. 8.35 Elevation-resistivity profile and geological interpretation of L3 survey line

ore-controlling data in this area, it is speculated that there may be a significant range of mineralization anomalies near the F2 fault.

Subsequently, three verification boreholes (No. 1–No. 3) were drilled along the L3 line. Among them, boreholes 2 and 3 revealed two layers of metal mineralization at a depth of 800 m.

8.6 Investigation of Deep Coal Seam Water Content in Hancheng, Shaanxi

8.6.1 Overview of the Survey Area

The survey area is located in Longmen Town, north of Hancheng City, Shaanxi Province, and borders the Yellow River to the east (Fig. 8.36). This area is characterized as a typical low mountain and hilly region. The bedrock is widely exposed in the valleys, while the mountain tops are covered with extensive loess. Due to erosion and long-term scouring by surface water, a crisscrossing network of valleys and winding ridges and peaks has formed. The valleys are predominantly V-shaped, with steep sides. In the thick loess areas, gullies are well developed, leading to the formation of loess cliffs, loess pillars, loess funnels, and other geomorphological features.

The terrain in the survey area is rugged, exhibiting a significant difference in altitude. The highest point is in the northeast of the survey area, with an altitude of 916 m, while the lowest point is located in the valley in the southeast, with an

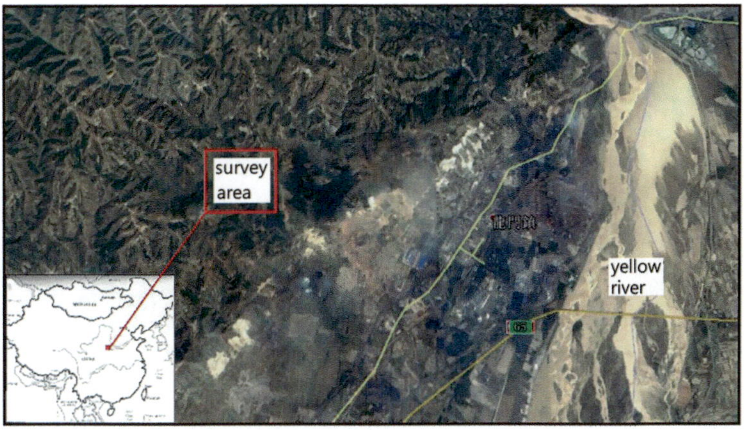

Fig. 8.36 Schematic diagram of the survey area location

altitude of 585 m. The terrain and geomorphology of the survey area are illustrated in Figs. 8.37 and 8.38.

The stratigraphy of the survey area is organized as follows, from newest to oldest: Quaternary, Triassic, Permian, Carboniferous, Ordovician, Cambrian, etc.

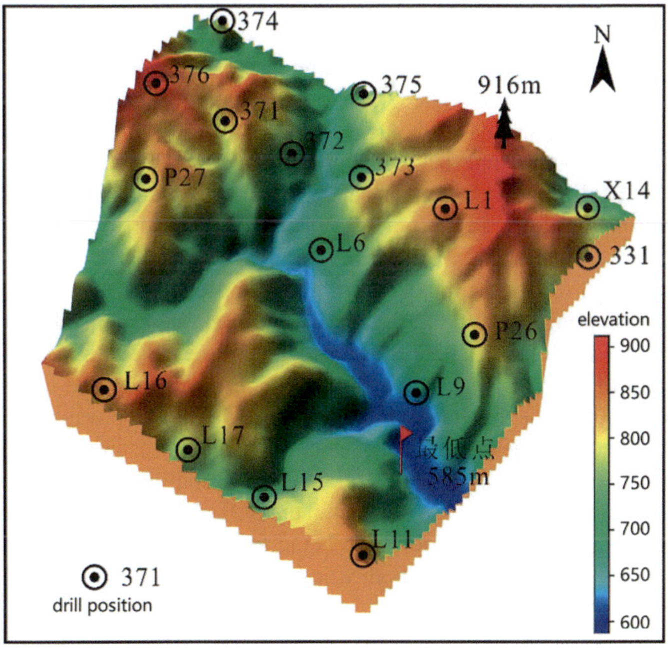

Fig. 8.37 Topographic map of the survey area

Fig. 8.38 Typical geomorphology of the survey area

Many units have conducted extensive drilling in the survey area, performing resistivity logging, long-source-distance sonic logging, natural gamma logging, and other measurement activities. This work has resulted in a comprehensive summary of the stratigraphic distribution and rock geophysical properties of the area.

Based on the drilling and geophysical measurements that have been carried out, the distribution, lithology, water content, and electrical properties of the strata in the survey area are summarized in Table 8.6.

The purpose of this measurement work is to determine the water-rich conditions of each water-bearing layer within the range of 2# coal (approximately 700 m deep) to 100 m below the top interface of the Ordovician limestone (approximately 1100 m deep), and to analyze the water-bearing structure as well as the horizontal and vertical connectivity in the area. The Shaanxi Coalfield Geophysical Exploration Team has conducted large loop transient electromagnetic measurements over an area of 6.1 square kilometers, obtaining the electrical distribution within a depth of one kilometer underground in the survey area. This SOTEM detection focused on two typical sections, with the main objective being to verify the capability of SOTEM detection in identifying deep water bodies and to compare the results with those from the large loop detection.

Table 8.6 Main lithological parameters of the strata in the survey area

Stratum	Thickness (m)	Lithology	Water content	Resistivity (Ω.m)
Quaternary	0 ~ 56	Loess and alluvial layer	Sandstone pore water	40 ~ 80
Triassic	150 ~ 220	Sandstone, mudstone	Basically no water	200 ~ 250
Permian	507 ~ 773	Sandstone, sandy mudstone, coal	Sandstone fracture water	35 ~ 300
Carboniferous	35 ~ 112	Sandstone, siltstone, mudstone, limestone, coal	Sandstone and limestone fracture water	45 ~ 300
Ordovician		Limestone	Limestone karst fracture dissolution water	> 700

8.6.2 *Experiment and Data Collection*

Before the section measurement, we first conducted an observation experiment near borehole 373 to determine the noise level of the survey area, establish the best working parameters, and verify the inversion results based on borehole data. According to Table 8.6, the maximum burial depth of the Ordovician limestone in this area is about 1200 m, so the depth of this detection must exceed this value.

In the experiment, a long wire source was deployed along the low-lying mountain road in the southwest of the survey line, measuring 723 m in length, with the observation point located 588 m away from the center of the emission source. The emission current waveform is a bipolar rectangular wave with a duty cycle of 50%, and two sets of emission frequencies, 2.5–1 Hz, were tested, with a current intensity of 16 A. The observation instrument utilized the V8 system from Phoenix Geophysics Company of Canada, and the effective receiving area of the magnetic probe is 40,000 m^2.

The large loop TEM detection implemented by the Shaanxi Coalfield Geophysical Exploration Team in this area has the following working parameters: loop size 800 × 800 m, emission current base frequency 5 Hz, and current intensity 4.5 A.

Figure 8.39a shows the measurement attenuation curves of 50 time channels for both SOTEM and loop source TEM at the same measurement point. Analysis of this figure reveals that the signal from the loop source TEM decays faster than that from SOTEM, and after a delay time of more than 70 ms, the signal falls below the noise level. Although the 1 Hz signal has a longer delay time, signals below the noise level are difficult to utilize. Therefore, the 1 Hz signal does not have an advantage compared to the 2.5 Hz signal; on the contrary, the 2.5 Hz signal can provide more shallow geological information.

Figure 8.39b presents the one-dimensional inversion result of the measured data. According to the estimation formula of the transient electromagnetic detection depth, the maximum inversion depth of the two devices is determined to be 1500 m. From Fig. 8.48b, it can be seen that at depths shallower than 1200 m, the inversion results of the loop source TEM and SOTEM are generally consistent, but there is a significant

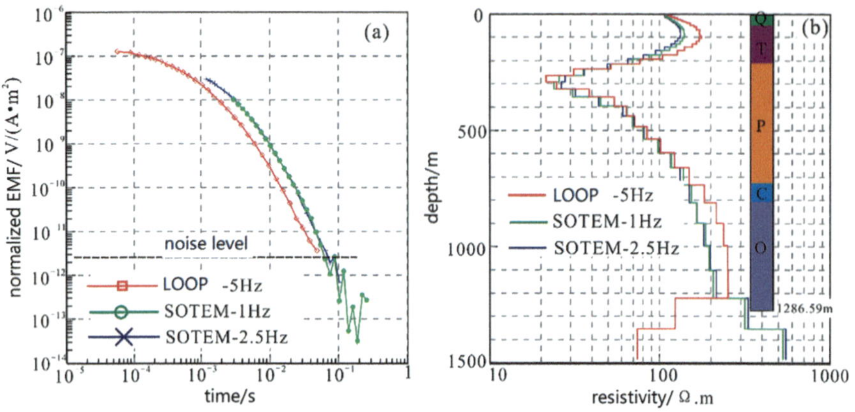

Fig. 8.39 Measured response and inversion results of different TEM devices at borehole 373

difference below this depth. The inversion resistivity of the loop source drops sharply, while the resistivity of SOTEM continues to increase. According to drilling data, at this test point, the top interface depth of the Ordovician limestone formation is about 800 m deep. The stratum below this depth should have a higher resistivity; therefore, the inversion result of SOTEM data more accurately reflects the resistivity of the deep stratum.

To compare with the loop source detection results, we selected two lines with minimal interference for SOTEM observations. The lengths of both lines are 600 m, designated as S1 and S2, respectively, with the corresponding loop source lines being L8 (from 5600 to 6200 m) and L10 (from 5600 to 6200 m).

We continued to use the transmitter set up during the test work, with the distance from the center of the transmitter to the center of line L1 being 588 m, and the distance to L2 being 488 m. The measurement point distance is 20 m, the transmission base frequency is 2.5 Hz, and the current intensity is 16 A. The single-point measurement time is 2 min, with approximately 1200 superimpositions. The working receiver layout is shown in Fig. 8.40.

8.6.3 Data Processing and Result Evaluation

Using SOTEMsoft software, the measured data is sorted, preprocessed, and inverted to obtain the inversion results for all measurement points. Given the favorable sedimentation of the coal-bearing strata in this area, we employed the quasi-two-dimensional inversion technique with lateral resistivity constraints. This approach overcomes the abrupt changes in resistivity between adjacent measurement points, suppresses noise interference, and improves inversion speed.

Figure 8.41 presents the inversion results for several measurement points along the two lines. It is evident that the trend of electrical changes in the strata is relatively stable and distinct, exhibiting a "high-low–high" pattern from shallow to deep. This

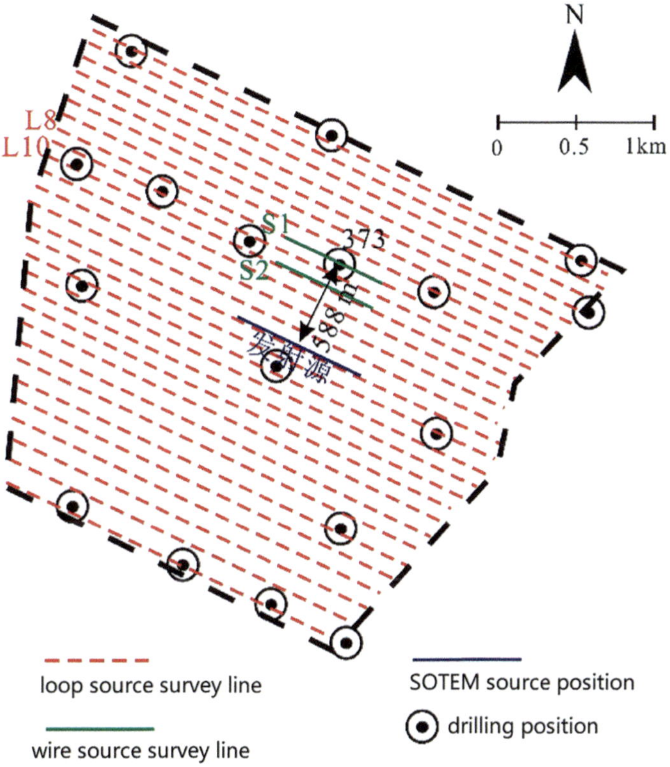

Fig. 8.40 Schematic diagram of transient electromagnetic working receiver layout

pattern is consistent with the inversion results and the strata distribution revealed by drilling. The high resistivity layer within the upper 200 m corresponds to the loess and sand mudstone of the Quaternary and Triassic periods. As the burial depth increases, the resistivity gradually decreases, revealing a clear low-resistance layer in the depth range of 300–500 m. Combined with the drilling data in this area, this indicates a rich water content in the sandstone pore water layer of the Shanxi Group. As depth continues to increase, the resistivity begins to rise again, and no low-resistance layer is observed.

Finally, the SOTEM and large loop TEM inversion results are presented in resistivity-elevation cross-section diagrams, as shown in Figs. 8.42 and 8.43. Comparing the inversion results reveals that within the elevation range of −200 to 800 m, the trend of stratum resistivity changes indicated by the two devices is generally consistent. This clearly reflects the shallow Quaternary and Triassic high-resistance strata, as well as the water-bearing low-resistance Permian sand mudstone strata.

However, at greater depths, there are notable differences in the inversion results from the two methods. As depth increases, the resistivity in the SOTEM inversion results shows a consistent upward trend, indicating the presence of deep Ordovician limestone strata. In contrast, the large loop TEM results show that the resistivity of

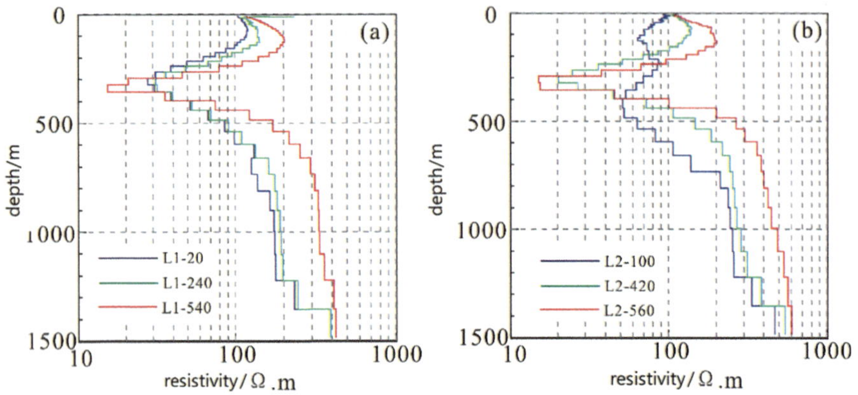

Fig. 8.41 Typical measurement point inversion results

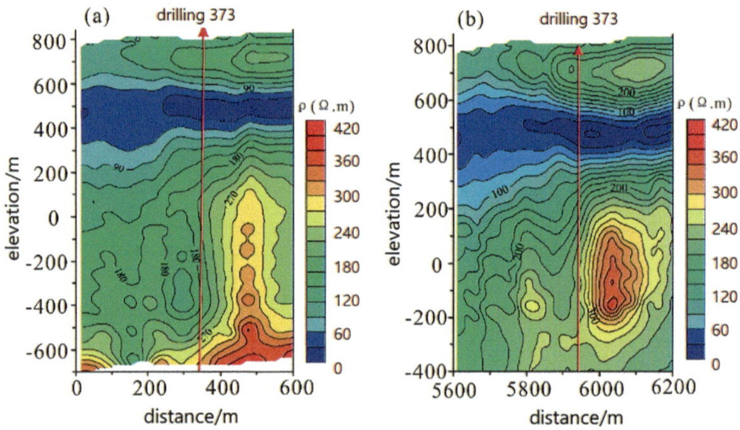

Fig. 8.42 Resistivity-elevation cross-section diagram of S1 and L8 lines **a** S1 line SOTEM results **b** L8 line large loop TEM results

some measurement points initially increases and then decreases, which is inconsistent with the actual geological conditions, suggesting that the large loop TEM results may be distorted in this depth range.

In summary, due to the relatively shallow detection depth of the loop source transient electromagnetic method, only the inversion results between elevations of −200 to 800 m are considered reliable. Based on the SOTEM detection results, it is speculated that there may be a water-bearing layer at the top of the Ordovician limestone, located at 0–400 m on line S1, and 0–200 m and 340–440 m on line S2. This inference aligns with the water-bearing characteristics of the deep strata revealed by existing boreholes.

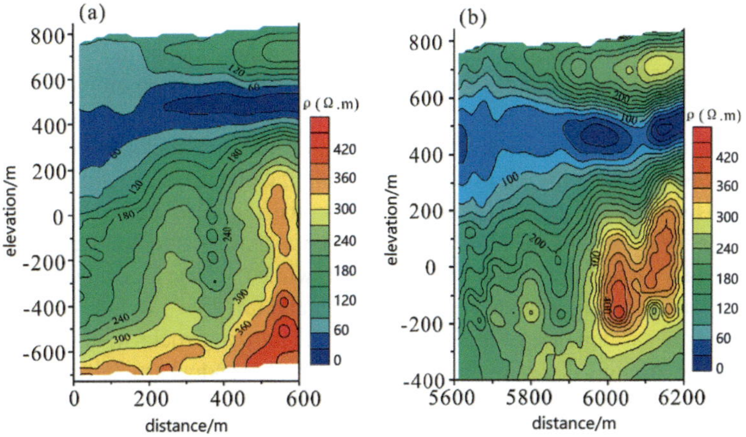

Fig. 8.43 Resistivity-elevation cross-section diagram of S2 and L10 lines **a** S2 line SOTEM results **b** L10 line large loop TEM results

8.7 Exploration of Multi-metal Iron-Lead-Zinc Mine in Qinghai's Wild Horse Spring Area

The Wild Horse Spring mining area is located on the southwestern edge of the Qaidam Basin, extending 11.4 km from east to west and about 6 km from north to south. It belongs to the western section of the East Kunlun Mountains (see Fig. 8.44). The area is rich in iron, copper, lead, zinc, and other beneficial metal minerals, making it a large-scale copper–lead–zinc deposit and a key exploration target for metal mineral resources in Qinghai Province.

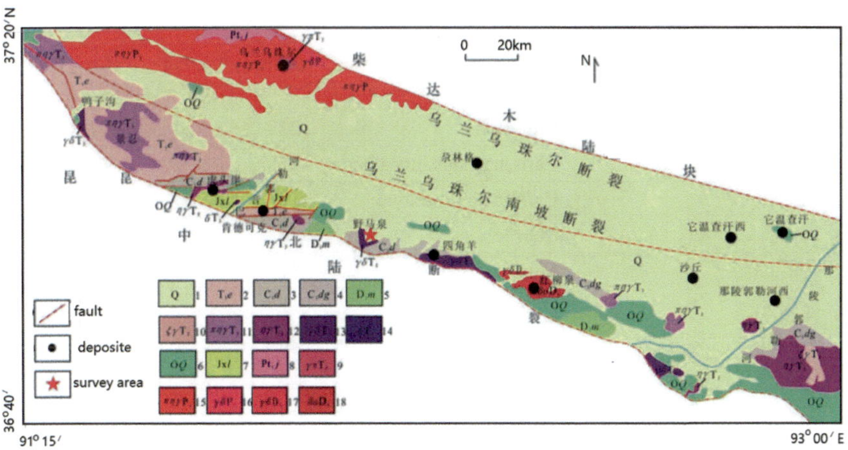

Fig. 8.44 Geological map of the Wild Horse Spring deposit area

In recent years, extensive geological, geochemical, and drilling explorations have revealed that the formation of the Wild Horse Spring deposit has undergone multiple periods of intense tectonic events. The area has a history of significant magmatic activity, featuring many complex intrusive rocks, extrusive rocks, and vein rocks (main references or previous work). The types of intrusive rocks primarily include granodiorite, biotite quartz diorite, and monzonite, with a smaller amount of diorite. The extrusive rocks in the mining area mainly consist of Early Paleozoic rift fissure-type extrusive rocks and Late Devonian continental extrusive rocks. The Early Paleozoic extrusive rocks are predominantly basalt, while the Late Devonian continental extrusive rocks mainly include andesite, dacite, and tuff.

There exists a sequence of rock types from the intrusive rocks to the surrounding rocks in the mining area: unaltered intrusive rocks–endoskarn (altered intrusive rocks)–skarnized carbonate rocks–skarn–skarnized carbonate rocks (hornfels metamorphic clastic rocks)–carbonate rocks. The ore bodies are primarily located in the skarn at the contact zone between the intrusive rocks and the carbonate rocks, with some also found in the skarnized carbonate rocks at the top. The former mainly produces iron-copper–zinc minerals, while the latter primarily yields lead–zinc minerals. The ore-bearing skarn is mainly iron-magnesium skarn, followed by calcium-magnesium skarn and manganese skarn, all strictly controlled by the fault structures in the area. The deeper parts of the mining area show promising prospects for exploration.

However, due to the harsh natural conditions of the mining area, the relatively low level of basic geological work, and other factors, the understanding of the stratigraphic interface, fault structure, and the current deep distribution pattern of the ore-bearing skarn in the area remains significantly insufficient. This limitation hinders the progress of deep prospecting in the mining area (Table 8.7).

The rock physics measurements of the ore samples in the mining area reveal significant differences among various lithologies. The resistivity of monzonitic granite in

Table 8.7 Main rock and ore resistivity statistics of the Wild Horse Spring mining area

Rock type	Sample number	Resistivity /Ωm
Limestone	23	719
Carbonaceous limestone	25	99
Carbonate limestone	18	48
Marble	19	1057
Monzonite granite	31	1420
Granodiorite	17	913
Mineralized skarn	20	367.6
Galena magnetite ore	13	179
Sphalerite magnetite ore	19	13.7
Brass magnetite	21	12.7
Magnetite	23	10.3

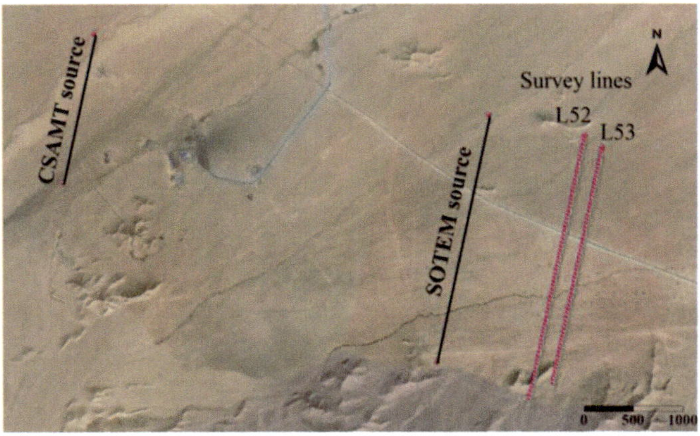

Fig. 8.45 Distribution of electromagnetic detection measurement line positions

the mining area is the highest, with an average resistivity of 1400 Ωm. The resistivity of limestone, marble, and granodiorite is above 700 Ωm, exhibiting clear high-resistance characteristics. The resistivity of mineralized siliceous rock is around 400 Ωm, while the resistivity of galena magnetite is approximately 200 Ωm. The resistivity of carbonaceous limestone is around 100 Ωm, and the resistivity of sphalerite magnetite, brass magnetite, and magnetite is around 10 Ωm.

Given the significant resistivity difference between the surrounding rock and the mineralized siliceous rock, we employed the short-offset transient electromagnetic method (with a transmit-receive distance of 1 km) and the controlled-source audio-frequency magnetotelluric method (with a transmit-receive distance of 10 km) to conduct resistivity depth measurement research. As shown in Fig. 8.45, there are 2 measurement lines and 112 measurement points, with a measurement point spacing of about 40 m.

Figure 8.46 shows the inversion results for the near-source and far-source of line L52. As depicted in Fig. 8.46a, the high-resistance area (red area) at the 800–1600 m measurement points corresponds to the granodiorite body of the Late Carboniferous Diaoersu Formation (C2d). The medium–low resistance area (green area) at the 1800–2600 m measurement points corresponds to the carbonaceous stratum. The transition area between the high-resistance area and the medium–low resistance area aligns with the known siliceous rock belt. The low-resistance area on the right side of the profile (blue area) corresponds to the Ordovician Qiman Tag Group stratum (OQ); due to severe carbonization of the carbonate rock, it exhibits obvious low-resistance characteristics.

Combined with drilling results (see Fig. 8.47), the lead-zinc ore is located at the 800–1600 m measurement points, within the elevation range of 3300–3100 m, corresponding to the relatively low-resistance green area in Fig. 8.46a. This ore is primarily found at the horizontal interface between the red high-resistance and the

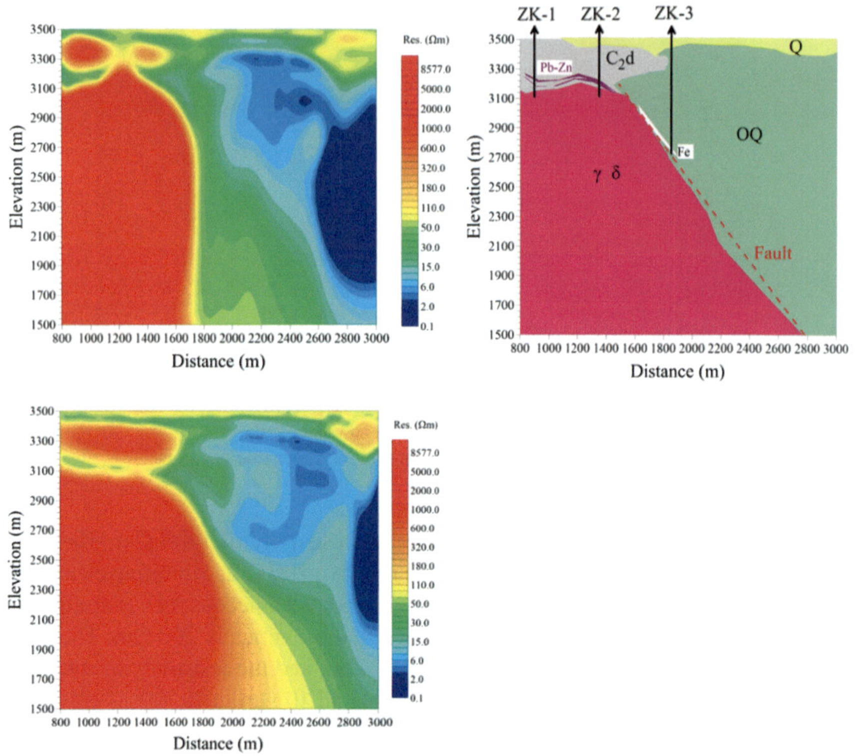

Fig. 8.46 Comparison of electrical profiles. **a** Near-source measurement results **b** Far-source measurement results. The black solid line represents the inferred lead–zinc ore body, and the red dashed line represents the inferred fault

green low-resistance areas, distributed along this interface. The shape of the lead-zinc ore body resembles the change in the green area in the near-source results. In the far-source results, the corresponding green area shows a single inclined distribution, which does not accurately represent the distribution shape of the ore body.

The iron ore is mainly located at the 1600–1900 m measurement points, within the elevation range of 3100–2700 m. It is produced at the steeply inclined interface between the red high-resistance and the green low-resistance areas on the right side of the near-source results (as indicated by the red dashed line in Fig. 8.46a). In contrast, the red high-resistance and green interface on the right side in the far-source results are much gentler, as shown by the red dashed line in Fig. 8.46b. The entire section of the near-source results clearly reflects the deep spatial characteristics of each geological body, with the rock body interface consistent with the actual drilling results.

Overall, compared to the CSAMT profile (Fig. 8.46b), SOTEM aligns more closely with the actual position of the ore body (the low-resistance area within the elevation range of 3300–3100 m), and the depiction of details is more accurate (Fig. 8.46b). This indicates that the SOTEM electromagnetic method can effectively

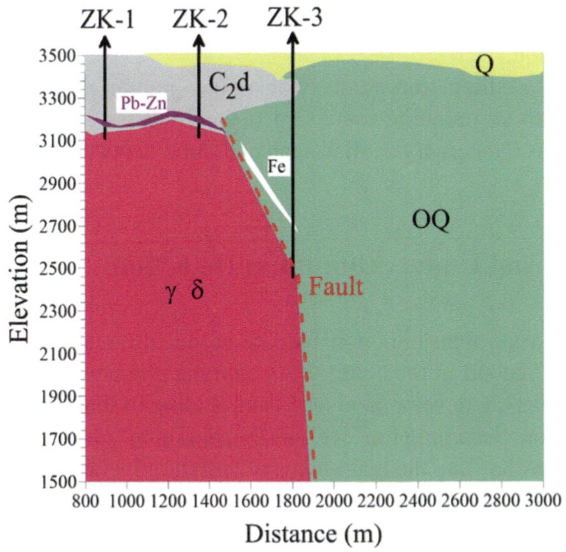

Fig. 8.47 Geological interpretation result map

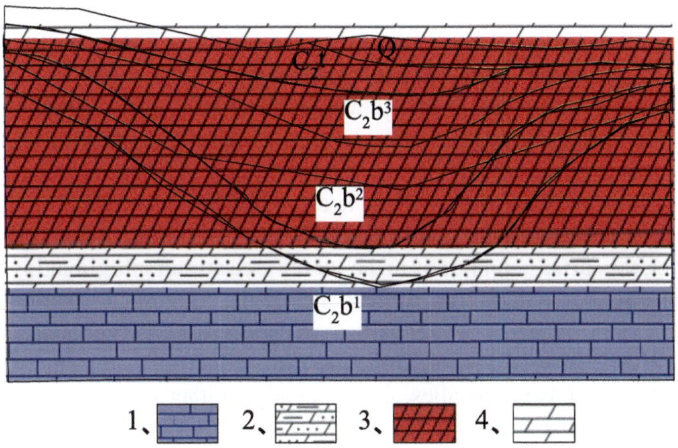

1. Limestone; 2. Iron-rich claystone; 3. Bauxite, coal seam; 4. Clayey dolomite

Fig. 8.48 Cross-sectional diagram of typical karst sinkhole ore formation in Henan Province

delineate the ore body and the contact boundary between the rock body and the strata in this area. However, the accuracy of the far-source detection results in depicting the ore body and the horizontal interface between high and low resistivity areas still requires improvement. For fault shape depiction, the SOTEM results are more accurate, showing a larger dip angle of the fault consistent with drilling results, while

CSAMT still needs to enhance the depiction of steeply inclined faults due to a lack of lateral resolution.

Zk–represents drilling, represents monzonitic granite, OQ is the Ordovician Qiman Tag Group surrounding rock, C_2d is the Diaoersu Formation surrounding rock, Q represents Quaternary aeolian sand and flood deposits.

8.8 Henan Coal Under Aluminum Detection

Henan Province is one of the important bauxite-producing areas in China, with proven reserves ranking second in the country. As alumina production capacity increases, the intensity of bauxite development has risen, leading to shallow bauxite resources nearing exhaustion. Searching for "coal-under-aluminum" resources less than 500 m deep or deeper has become the main focus of current mineral exploration.

The bauxite in Henan Province primarily deposits in the Carboniferous Benxi Group strata on the weathered erosion surface of the Ordovician (or Cambrian) system, with the upper Permian strata containing coal seams. Due to past disorganized mining, the bauxite layer and the upper coal seams have resulted in numerous mined-out areas and old tunnels. The lack of data complicates the planning of drilling projects in the exploration process for "coal-under-aluminum."

The transient electromagnetic method using magnetic sources has been widely applied in investigating water-filled mined-out areas in coal mines due to its sensitivity to low-resistance bodies and high work efficiency, resulting in many successful cases. Audio-frequency magnetotellurics and controlled-source audio-frequency magnetotellurics effectively detect the location of bauxite-bearing rock layers, the changes in the trend of ore-bearing rock layers, and the morphological characteristics of the Ordovician weathering crust.

However, the surface conditions in the survey area are complex, making the layout of the magnetic source line frame difficult and costly. Additionally, audio-frequency magnetotellurics in adjacent residential areas face severe interference, and controlled-source audio-frequency magnetotellurics are easily affected by the field source effect, which can lead to false anomalies. Consequently, neither of these two methods is suitable for use in this area.

8.8.1 Geological and Geophysical Features

The bauxite distribution area in Henan Province is primarily located in the central Henan region, encompassing Zhengzhou, Pingdingshan, Sanmenxia, Jiyuan, and Jiaozuo, with a total ore-bearing area exceeding 20,000 square kilometers. It is divided into the Ruyang-Ruzhou-Baofeng ore-forming area, the Songji ore-forming area, the Shaanxi-Mianchi-Xin'an ore-forming area, and the Jiyuan-Jiaozuo ore-forming area north of the Yellow River. The survey area is situated in the Songji ore-forming area.

The bauxite resources in Henan are found at the discontinuous contact interface between the Early Paleozoic and Late Paleozoic strata. Due to long-term uneven crustal activity and intermittent sedimentation interruptions, the bottom surface of the bauxite deposit has developed various undulating ancient weathered landforms and karst topographic features. According to the analysis of data obtained from the exploration of the bauxite area in Henan, the bauxite deposit is closely related to ancient karst activity. The sinkhole is the primary ore-controlling karst type of high-quality thick bauxite in Henan, as illustrated in Fig. 8.48.

The Santian Bauxite Mine in Xinmi City is located in Pingmo Town, Henan Province. The terrain in the area is high in the north and low in the south, with a relative height difference of 224 m. The surface vegetation is dense, and there are north–south gullies, characterizing the region as a low mountain and hilly area.

The strata in the mining area are divided from bottom to top into the Ordovician, Carboniferous, Permian, and Quaternary systems. The Ordovician consists of the Middle Ordovician Majiagou Formation, where the lower part includes silicified dolomite, gravel dolomite limestone, and angular limestone. The upper part comprises dolomite crystalline limestone, crystalline limestone, and dolomite mudstone (leopard limestone), with a thickness of about 88 m. The Carboniferous includes the Upper Carboniferous Benxi Group (C_2b) and Taiyuan Group (C_2t). The Benxi Group is the bauxite-bearing layer, divided into four lithological sections, of which three sections (C_2b_3) are the main industrial ore layers in the mining area. These are composed of gray-white, gray-green, and dark gray bauxite, bauxite rock, and clay ore. This section is generally a single layer of ore, locally varying to 2–3 layers, with thicknesses ranging from 0.41 to 40.80 m, and an average thickness of 8.72 m.

The lower part of the Taiyuan Group consists of gray and gray-white bioclastic crystalline limestone, while the middle part includes gray-white, light brown–red claystone, siltstone, and carbonaceous claystone with thin coal seams. The upper part is composed of gray and dark gray mudstone bioclastic limestone along with claystone or carbonaceous claystone. The Permian strata are represented by the Lower Permian Shanxi Group (P_1s), primarily consisting of dark gray and gray-black claystone, carbonaceous claystone, carbonaceous siltstone, quartz sandstone, sandy mudstone, and mudstone. This group contains the main mineable coal seam in the area, known as the second 1 coal seam, which has a stable layer position and an average thickness of 4.31 m.

The statistical results of the rock and mineral specimens are shown in Table 8.8. The resistivity of each stratum can be categorized into three groups based on numerical values: the Permian and Carboniferous strata exhibit medium to high resistance characteristics, with resistivity generally ranging from 300 to 5000 Ω·m; the Ordovician strata show high resistance characteristics, with resistivity ranging from 3000 to 12,000 Ω·m; and the bauxite and coal-bearing strata present medium to low resistance characteristics, with resistivity ranging from 100 to 1000 Ω·m.

Table 8.8 Electrical characteristics of rock and mineral specimens in the mining area

Stratum	Main rock type	Resistivity/(Ω·m)
Quaternary	Loess, gravel	15–120
Neogene	Claystone, sandstone	113–2421
Permian, carboniferous	Sandstone, claystone, muddy limestone	302–5013
Ordovician	Limestone	3142–12,352
Coal Bauxite		102–843 121–1970

There is a significant electrical difference between the Ordovician and other strata. The electromagnetic method can be used to detect the physical properties of Ordovician limestone, allowing for assessment of the spatial distribution and enrichment degree of bauxite by understanding the upper surface position and undulation characteristics of the Ordovician strata. Additionally, the electrical characteristics of sedimentary strata are relatively continuous. When there is a mined-out area, it disrupts the continuity of the electrical layer, causing contour lines to appear twisted, depressed, or stepped anomalies. Depending on the water content of the mined-out area, the anomaly position will exhibit high and low resistance characteristics. Thus, the electromagnetic method can be employed to investigate mined-out areas or old tunnels.

8.8.2 SOTEM Interpretation

(1) Work Arrangement

According to the requirements of the "Coal Under Aluminum" exploration task, a total of 6 survey lines were established for this work, covering a total length of 4800 m, with a point distance of 20 m, resulting in 246 points. The azimuth is set at 0 degrees, with the specific locations shown in Fig. 4.18.

As indicated in the layout diagram, there are numerous pits left by open-pit mining in the middle of the area, along with many terraced fields featuring steep slopes, making traffic conditions poor. If a traditional return line device is used, it becomes difficult to set up the line frame, leading to significant construction challenges.

To address this, the flexible layout characteristics of the short-offset transient electromagnetic emission field source were utilized. The emission source is arranged near the road at the east end of the work area, which not only ensures good electrical conditions for optimal emission power but also significantly enhances the speed of field source layout by utilizing the existing road. Additionally, the emission source is set up once for all area observations, greatly improving project work efficiency (Fig. 8.49).

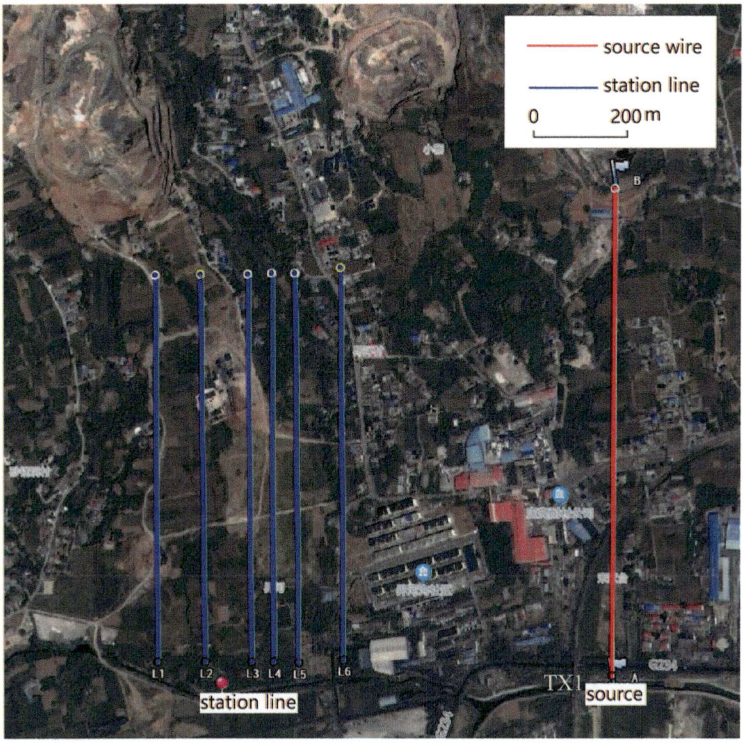

Fig. 8.49 Schematic diagram of work arrangement

(2) Data Collection and Processing

This work selected the 30 kW high-power GDP-32 II multifunctional electrical method observation system from Zonge Company in the United States for construction. The system utilizes a short-offset device, with an offset distance (transmission distance) ranging from 0.6 to 1 km, and an AB length of 1025 m (Table 8.9).

The signal source for transmission uses a base frequency of 16 Hz, a duty cycle of 1:1 square wave, and a power supply current of 15 A. The collection probe employs the Zonge TEM/3 magnetic rod, with an equivalent area of 10,000 m^2, observing the vertical induced electromotive force component (Hz). The collection time window utilizes a 30.52 μs equidistant time window, totaling 512 windows. The collection time window range is 0.03–15.63 ms, and a single measurement is superimposed 1024 times.

The collected data undergoes initial jump point removal followed by five-point smoothing processing. Based on data quality, the window data after the last 10 ms is discarded. As shown in Figs. 8.50 and 8.51, after preprocessing, the measurement data is used for OCCAM one-dimensional inversion, with the fitting error for the

Table 8.9 Survey line information table

Survey Line number	Survey line length/m	Transmitter	Offset distance/m
L1	800	Tx1	953–1016
L2	800	Tx1	851–991
L3	800	Tx1	764–905
L4	800	Tx1	709–864
L5	800	Tx1	657–827
L6	800	Tx1	567–748

entire area controlled within 10%. The resistivity inversion cross-section diagram for each profile is then obtained.

(3) Analysis and interpretation of abnormal features

Through one-dimensional inversion of the data, the resistivity cross-section diagrams for each measurement line were obtained. Subsequently, the horizontal distribution characteristics of the Carboniferous, Permian, and Ordovician were analyzed. The morphology of the Ordovician top plate was outlined, and the favorable ore-forming positions were inferred. Additionally, the suspected mined-out areas were delineated based on the resistivity mutation characteristics, combined with surface exploration data.

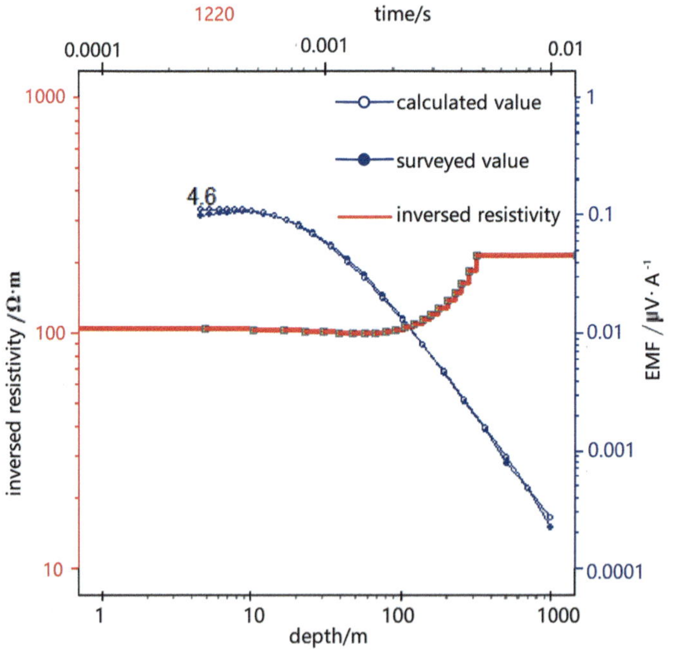

Fig. 8.50 Inversion diagram of point 1220 on line L4

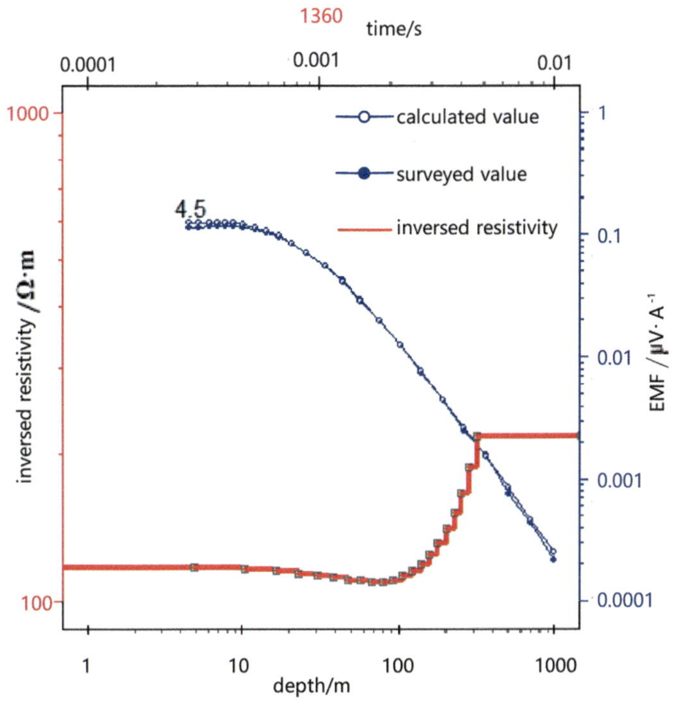

Fig. 8.51 Inversion diagram of point 1360 on line L4

As shown in Fig. 8.52, the resistivity of line L3 exhibits horizontal stratification characteristics. The shallow medium and low resistance area is inferred to be the Permian and Carboniferous, while the deep high resistance area is inferred to be the Ordovician. The depth of the Ordovician gradually increases from north to south. It is observed that between 300 and 500 m, the top plate of the Ordovician at an elevation of about 60 to 150 m shows distinct characteristics of changing from steep to gentle. This section is inferred to represent a favorable ore-forming position for bauxite. At the same time, there are two high-resistance anomalies in this area, which are inferred to be old mining caves. The high-resistance characteristics suggest that there is no significant water filling. Later, it was discovered that there are two abandoned vertical shafts from civilian mining in this area, corroborating the inferred results.

In order to understand the distribution morphology of the two mined-out areas on line L3 at depth, line L4 was established 40 m east of line L3. As shown in Fig. 8.53, the horizontal stratification characteristics of L4 resistivity are consistent with those of line L3, and the deep high resistance area is inferred to be the Ordovician. The Ordovician gradually deepens from north to south.

It is observed that the top plate of the Ordovician at points 250 to 550 also exhibits a trend of changing from steep to gentle, which is inferred to represent a favorable ore-forming position for bauxite. The two high-resistance anomalies observed on line

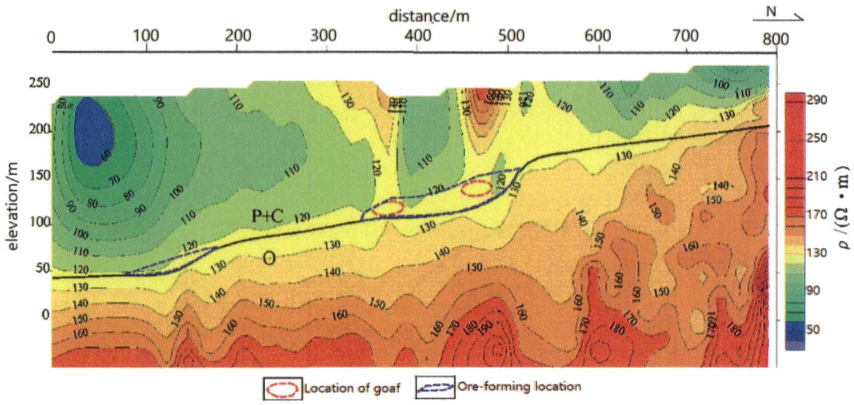

Fig. 8.52 Transient electromagnetic depth inversion cross-section diagram of line L3

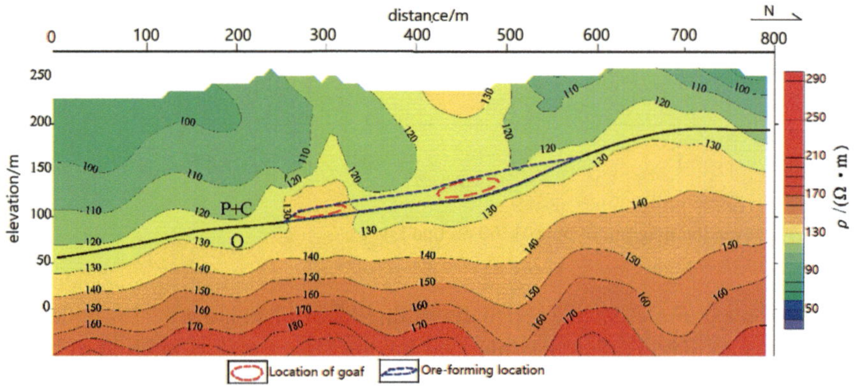

Fig. 8.53 Transient electromagnetic depth inversion cross-section diagram of line L4

L3 are still present on line L4, though their positions have shifted slightly toward the lower number point, and the range is larger. This anomaly is inferred to be an extension of the two mined-out areas on line L3, and its high-resistance characteristics suggest that there is no significant water filling.

This work employs SOTEM technology to stratify the strata within the working range, depict the morphology of the Ordovician top plate, and infer favorable ore-forming positions based on the characteristics of karst funnels conducive to the enrichment of bauxite ore bodies. Notably, two layers of bauxite bodies were identified in line L3 through drilling verification, with a total thickness of 7.48 m, thereby confirming the inferred results. During the work, two mined-out areas were inferred, and it was speculated that there was no significant water filling, providing a reference for later mine safety production.

8.9 Application Case Detection of Anhui Wuwei Lead-Zinc Mine

The three-dimensional inversion algorithm proposed in this study is utilized to image the resistivity structure of the SOTEM data collected in the Xiwan lead-zinc mine in Wuwei County, Anhui Province. This approach delineates the position of the lead-zinc ore body, determines the distribution of the electrical structure surface, and provides an interpretation of the ore-controlling structure and ore-forming model of the survey area.

8.9.1 Geological Background

The lead-zinc mine is located in the middle section of the mineralization belt in the middle and lower reaches of the Yangtze River. Three northeastward faults (F1, F2, F3) and one northwestward fault (F4) are primarily developed in the mining area (Fig. 8.54). The discovered lead-zinc ore bodies are mainly situated on the southeast side of the F2 fault. The spatial distribution of the ore bodies is controlled by the interlayer fracture zone.

The lead-zinc ore bodies are primarily deposited in the contact zone between volcanic rocks and conglomerate limestone, as well as in the fracture zone of conglomerate limestone. According to the rock resistivity measurement data, the resistivity of volcanic rocks is relatively low, ranging from tens to hundreds of ohm-meters. In contrast, the resistivity of hard gypsum and conglomerate limestone is higher, with values ranging from hundreds to thousands of ohm-meters.

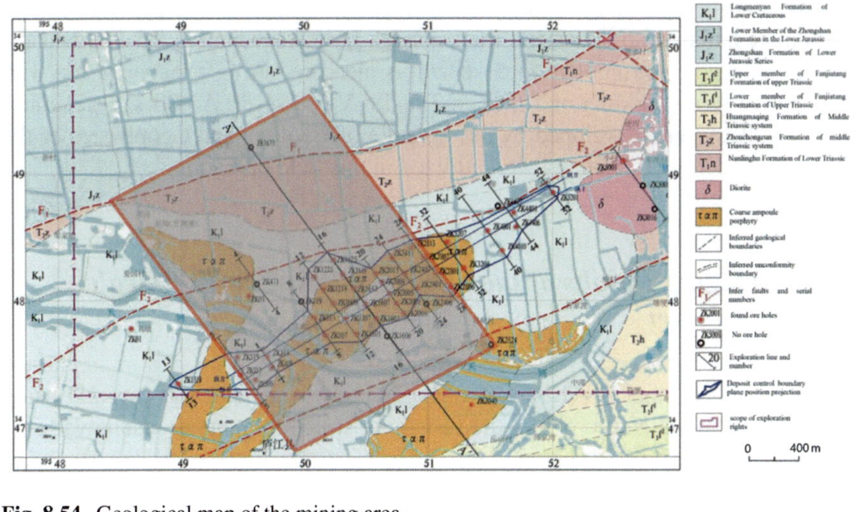

Fig. 8.54 Geological map of the mining area

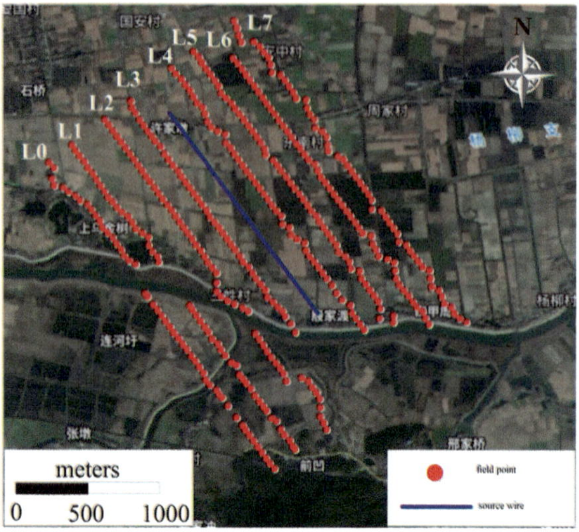

Fig. 8.55 Schematic diagram of the receiver placement in the measurement area

While these conditions are favorable for geophysical exploration, the thick Quaternary cover layer in this area results in a lack of outcrops for field observation, making prospecting challenging. Therefore, high-precision geophysical exploration methods need to be employed to detect the location of deep hidden ore bodies.

This time, the SOTEM exploration method was employed to detect the Wuwei Xiwang lead-zinc deposit. The gray shadow area in Fig. 8.55 indicates the measurement range of this SOTEM exploration. A total of 8 measurement lines were established, running from west to east: L0 to L7, with offsets of 900 m, 700 m, 400 m, 200 m, 200 m, 400 m, and 600 m, respectively. The length of each measurement line ranges from 2000 to 2700 m, with a point distance of 50 m. The transmitter is positioned between the L3 and L4 lines. The transmission current is 15 A, the transmitter length is 2000 m, the transmission base frequency is 2.5 Hz, and the vertical induced electromotive force dBz/dt is measured using a magnetic rod with an effective receiving area of 10,000 m2 and a V8 receiver. A 100 kW generator and a Tx-40 system, developed by the Institute of Geology and Geophysics of the Chinese Academy of Sciences, were utilized in this field operation.

8.9.2 Imaging Results of the Structure Based on Full-Scale 3D U-Net

The three-dimensional SOTEM data of the measurement area was inverted using the full-scale 3D U-Net to obtain the resistivity structure imaging results within a depth range of 1100 m, as shown in Fig. 8.56a. The subsurface of the measurement area is predominantly in a high resistivity state (greater than 200 Ω•m). There is a distinct high-resistance anomaly area (greater than 800 Ω•m) in the range from point 500 to point 2000, which is speculated to be the contact zone between volcanic rocks and Dongmasanshan conglomerate limestone, as well as the fracture zone of the Dongmasanshan conglomerate limestone.

Three notable resistivity lateral discontinuities are observed near points 500, 2000, and 2600, with these three anomaly areas inferred to be fault zones F3, F2, and F1, respectively. The structure with a resistivity greater than 800 Ω•m is imaged, as shown in Fig. 8.56b. Within the depth range of 0–800 m underground, there are shallow (less than 250 m) high-resistance anomalies, speculated to be the contact zone of mountain rocks and conglomerate limestone, as well as mid-deep (250–450 m) high-resistance anomalies, which are inferred to be conglomerate limestone bodies. The high-resistance anomaly at the bottom (below 500 m) is likely the underlying Dongmasanshan gypsum ore body.

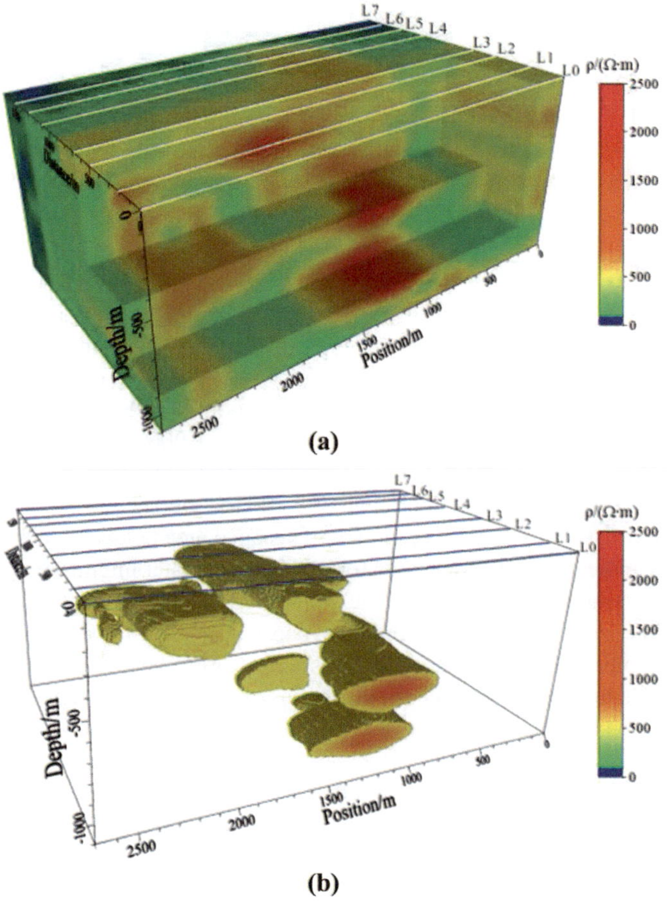

Fig. 8.56 **a** Full-scale 3D U-Net inversion result **b** Resistivity structure below 800 m depth

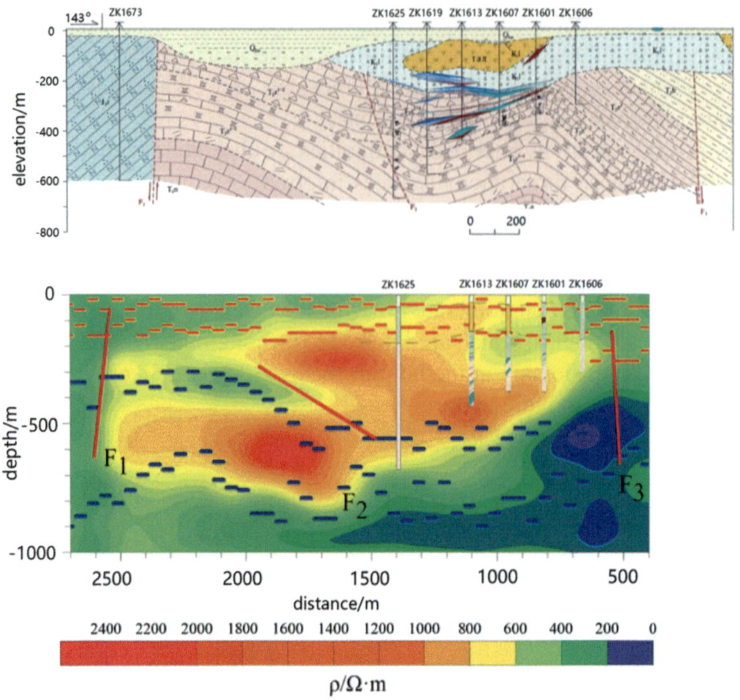

Fig. 8.57 Comparison of geological cross-section and imaging results

The resistivity structure imaging results of the Anhui Xiwang lead-zinc mine provide the resistivity structure distribution within 1100 m underground in the measurement area, as shown in Fig. 8.57. Near point 1400 of line L4, the high-resistance anomaly area within the elevation range of −200 to −700 m corresponds to the hard gypsum ore body. Between points 600 and 1100, a lead-zinc ore structure was identified within the elevation range of −100 to −500 m.

In the southeast of point 1500, the low-resistance area below the elevation of −500 m may indicate a water-rich zone formed by river water migrating underground along the interlayer fracture zone. The lead-zinc ore body is primarily deposited in the contact zone between the Longmenyuan volcanic rock and the Dongmasanshan conglomerate limestone, as well as in the fracture zone of the Dongmasanshan conglomerate limestone, between the elevations of −150 and −700 m. The ore body is thin and lens-shaped, and its spatial distribution is controlled by the interlayer fracture zone.

Based on the undulation of the interface, the positions of faults F1, F2, and F3 are determined. These three fault zones collectively control the direction of the mineralization zone and local uplift.

References

Chen Bench, Zhou Fengtong, Li Jinming. Study on wave field transformation of transient electromagnetic field[J]. Geophysical and Geochemical Exploration, 1999, 3: 36–42

Chen W Y, Xue G Q, Khan M Y. Quasi MT inversion of short-offset transient electromagnetic data[J]. Pure and Applied Geophysics, 2016, 173(7): pp.2413–2422

Xue Guoqiang, Chen Weiyin, Zhou Nannan, et al. Ground source transient electromagnetic short offset deep detection technology[J]. Chinese Journal of Geophysics, 2013, 56(1): 255–261

Ma Jie, Li Xiu, Qi Zhipeng, et al. Transient electromagnetic virtual wave field layered Grenn function Born approximation imaging method. Chinese Journal of Geophysics, 2023, 66(6): 2631-2645

Outlook

SOTEM leverages the advantages of the small volume effect from the loop-source transient electromagnetic method while overcoming its limitation of shallow exploration depth. Simultaneously, it incorporates the strong field source signals from the grounded-source long-offset transient electromagnetic method, discarding its drawbacks of low interpretation accuracy and work efficiency. This has resulted in the development of a new grounded-source all–zone detection method, along with advanced data preprocessing and inversion technologies, and a comprehensive SOTEM detection equipment system. These innovations have been successfully applied in the fields of mineral resources, geothermal energy, and engineering exploration.

Looking ahead, SOTEM will continue to focus on enhancing detection accuracy, depth, and speed. This includes the development of multi-source transient electromagnetic detection devices, higher precision three-dimensional inversion techniques, and refined extraction technologies for multi-electrical information. In recent years, to achieve greater detection accuracy, short-offset transient electromagnetics has introduced various new device configurations, such as double-line sources, multi-line sources, and vertical line sources. These advancements significantly enrich the application potential of transient electromagnetic methods. By integrating advanced technologies, these new devices have improved both detection accuracy and efficiency, thereby promoting the development of resource exploration, environmental monitoring, and engineering surveying.

In the realm of three-dimensional inversion, convolutional layers used as feature extraction structures in neural networks excel at local feature extraction but often struggle with global feature extraction. Given that three-dimensional inversion is a holistic process with strongly correlated data elements, the next step involves incorporating Transformer architectures to enhance the U-Net's capability for global feature extraction from three-dimensional SOTEM data. Additionally, to improve the accuracy of structural surface and body imaging, it is essential to utilize three-dimensional forward modeling methods to construct training datasets that more accurately reflect

G. Xue et al., *Exploration with Short-Offset Grounded-Wire Transient Electromagnetic Method*, https://doi.org/10.1007/978-981-96-8846-3

actual geological models. Key technologies will focus on the precise depiction of undulating layer interfaces to achieve high-precision three-dimensional imaging.

Regarding the synchronous extraction of electrical information, such as resistivity and polarization rate, existing methods typically utilize the magnetic field component for information extraction. However, due to the complex terrain and geological conditions in the field, significant electrical variations can occur between adjacent measurement points along the measurement line. Poor grounding conditions can hinder the acquisition of high-quality electric field data, leading to suboptimal inversion results. Therefore, it is crucial to employ SOTEM electric field data for induced polarization information extraction, conduct research on the full parameter inversion algorithm of the Cole–Cole induced polarization model, and carry out inversion studies on the induced polarization effects in two-dimensional and three-dimensional SOTEM contexts that align more closely with actual geological conditions.

Bibliography

Andersen W L. A hybrid fast hankel transform algorithm for electromagnetic modeling[J]. Geophysics, 1989, 54(2): 263–266.

Araya-Polo M, Jennings J, Adler A, Dahlke T. Deep-learning tomography[J]. Lead. Edge, 2018, 37 (1): 58–66.

Bai P, Vignoli G, Viezzoli A, Nevalainen J, Vacca G. (Quasi-) real-time inversion of airborne time-domain electromagnetic data via artificial neural network[J]. Rem. Sens., 2020, 12 (20): 3440

Byrd R H, Lu P, Nocedal J, et al. A Limited Memory Algorithm for Bound Constrained Optimization[J]. Siam Journal on Scientific Computing, 1995, 16(5): 1190-1208.

Cai Wei, Song Xianhai, Yuan Shichuan, et al. Inversion of Rayleigh wave dispersion curve based on firefly and bat swarm intelligence algorithm[J]. Chinese Journal of Geophysics, 2018, 61(06): 239-250.

Cao XinZhi, Zhang Wangsheng, Sun Huashan. Review of the progress in deep mineral exploration in China[J]. Geological Science and Technology Information, 2009, 28(2): 104-109.

Castillo J E, Miranda G F. Mimetic discretization methods[M]. 2013, CRC Press.

Chang J H, Xue G Q, Malekian R. A comparison of surface-to-coal mine roadway TEM and surface TEM responses to water-enriched bodiesa[J]. IEEE Access, 2019, 7: 167320-167328.

Chen Weiyin, Xue Guoqiang. Application of multi-device detection technology of transient electromagnetic method in coal mine goaf survey[J]. Progress in Geophysics , 2013, 28(5): 2709-2717.

Chen Weiyin, Xue Guoqiang. Effective skin depth of grounding wire source electromagnetic field[J]. Chinese Journal of Geophysics, 2014, 57(7): 2314-2320

Chen Weiyin, Xue Guoqiang, Li Meifang. Characteristics of transient electromagnetic response in the near field of electric source[J]. Geophysical and Geochemical Exploration Computing Technology, 2011, 33(Supplement): 122-126.

Chen Weiyin, Xue Guoqiang, Cui Jiangwei. Analysis of the influence of electric source transient electromagnetic emission source deformation on observation results[J]. Progress in Geophysics, 2015, 30(1): 0126-0132.

Chen Mingsheng. Research on depth measurement of electric dipole source transient electromagnetic (I)[J]. Coal Geology & Exploration, 1999a, 27(1): 55-59.

Chen Mingsheng. Research on depth measurement of electric dipole source transient electromagnetic (II)[J]. Coal Geology & Exploration, 1999b, 27(2): 54-57.

Chen Mingsheng. Research on depth measurement of electric dipole source transient electromagnetic (III)[J]. Coal Geology & Exploration, 1999c, 27(3): 58-61.

Chen Mingsheng. Research on depth measurement of electric dipole source transient electromagnetic (IV)[J]. Coal Geology & Exploration, 1999d, 27(4): 52-55.

G. Xue et al., *Exploration with Short-Offset Grounded-Wire Transient Electromagnetic Method*, https://doi.org/10.1007/978-981-96-8846-3

Chen Mingsheng. Research on depth measurement of electric dipole source transient electromagnetic (V)[J]. Coal Geology & Exploration, 1999e, 27(5): 63-65.

Chen Mingsheng. Research on depth measurement of electric dipole source transient electromagnetic (VI)[J]. Coal Geology & Exploration, 2000, 28(1): 46-49.

Chen Guisheng. Discussion on the application effect and existing problems of transient electromagnetic method in metal mineral exploration[J]. Mineral and Geology , 2006, 20(4-5): 543-547.

Cheng Jiulong, Li Mingxing, Xiao Yanli, et al. Particle swarm optimization inversion of mine transient electromagnetic method under full space condition[J]. Chinese Journal of Geophysics, 2014, 057(010):3478-3484

Cheng J L, Li, F, Peng S, et al. Joint inversion of TEM and DC in roadway advanced detection based on particle swarm optimization[J]. Journal of Applied Geophysics, 2015, 123: 30-35.

Cockett R, Kang S, Heagy L J, et al. SimPEG: An open source framework for simulation and gradient based parameter estimation in geophysical applications[J]. Computers & Geosciences, 2015, 85(DEC.PT.A):142–154.

Colombo D, Turkoglu E, Li W C, Sandoval-Curiel E, Rovetta D. Physics-driven deep-learning inversion with application to transient electromagnetics. Geophysics, 2021, 86 (3): E209–E224.

Di Qingyun, Xue Guoqiang, Yin Changchun, Li Xiu. New methods of artificial source electromagnetic detection in China[J]. Science China: Earth Sciences, 2020, 50(09):1219-1227.

Doyoro Y G, Chang P Y, Puntu J M, et al. A review of open software resources in python for electrical resistivity modelling[J]. Geoscience Letters, 2022, 9 (1).

Du Ruiqing. Research on geophysical prospecting model for deep iron ore exploration—A case study of a certain iron ore area in Henan Province[D]. Beijing: China University of Geosciences (Beijing), 2013.

Fan Yanan, Li Xiu, Qi Zhipeng, Sun Dali. Research on Transient Electromagnetic Virtual Wave Field Born Approximation Algorithm[J]. Progress in Geophysics, 2019, 34(2): 529-536.

Fu H H, Wang Y Q, Um E S, et al. A parallel finite-element time-domain method for transient electromagnetic simulation[J]. Geophysics, 2015, 80(4): E213-E224.

Haber E, Ascher U, Oldenburg D W. 3D forward modelling of time domain electromagnetic data[C]. SEG Technical Program Expanded Abstract. Salt Lake City: Society of Exploration Geophysicists, 2002, 641–644.

Haber E, Ascher U M, Oldenburg D W. Inversion of 3D electromagnetic data in frequency and time domain using an inexact all-at-once approach[J]. Geophysics, 2004, 69(5):1216-1228.

Hamoda M, Mamat M, Rivaie M,et al. A Conjugate Gradient Method with Strong Wolfe-Powell Line Search for Unconstrained Optimization[J]. Applied Mathematical Sciences, 2016, 10: 721-734.

Han Ziqiang, Liu Tao, Ouyang Jin, et al. Application of Full Area Apparent Resistivity of Rectangular Large Fixed Source Loop TEM Method in Coal Mine Goaf Exploration[J]. Progress in Geophysics, 2015, 30(1): 343-349.

He Q, Wang Y. Reparameterized full waveform inversion using deep neural networks[J]. Geophysics, 2020, 86(1): 1-71.

He Jishan, Xue Guoqiang. Overview of Short Offset Electromagnetic Detection Technology[J]. Journal of Geophysics, 2018, 061(001):1-8

Heagy L J, Kang S, Cockett R, et al. Open source software for simulations and inversions of airborne electromagnetic data[J]. Exploration Geophysics, 2019, 51(1): 38-44.

Hinton G E, Osindero S, Teh Y W. A fast learning algorithm for deep belief nets, Neural Computation[J]. 2006, 18, 1527–1554.

Hou Yanwei. Transient Electromagnetic Detection of Deep Water Sources in Inner Mongolia Region[J]. Progress in Geophysics, 2012, 27(6): 2698-2703.

Hua Qian, Liang Chuntao, Yang Yihai, et al. Research on Noise Tomography Imaging of the Eastern Margin of the Qinghai-Tibet Plateau Based on Simulated Annealing Method[J]. Journal of Geophysics, 2020, 63(05): 69-83.

Huang Lijun, Lu Guifu, Liu Ruide. Application of Electric Source Transient Electromagnetic Method in Coalfield Hydrogeological Survey[J]. Journal of Engineering Geophysics, 2004, 1(2): 174-77

Huang Lijun, Lu Guifu, Liu Ruide. Research on the Application of Electric Source Transient Electromagnetic Method in Oilfield[J]. Geophysical and Geochemical Exploration, 2005, 29(4): 316-318

Hyman J M, Shashkov M. Mimetic Discretizations for Maxwell's Equations[J]. Journal of Computational Physics[J]. 1999, 151(2):881–909.

Ito H, Kaieda H, Suzuki K, Kiho K, Mogi T, Yuuki Y, Jomori A. A semi-airborne TEM (GREATEM) survey and its applicability to geo-hazards[C]. 10th Asian Regional Conference of IAEG, 2017.

Ji Yanju, Lin Jun, Zhu Kaiguang, et al. Groundwater Resources Exploration Using Transient Electromagnetic Technology[J]. Progress in Geophysics, 2005, 20(3): 828-833

Jiang Bangyuan. Practical Near Zone Magnetic Source Transient Electromagnetic Exploration[M]. Beijing: Geological Publishing House, 1998.

Kazei V, Ovcharenko O, Plotnitskii P, Peter D, Zhang X L, Alkhalifah T. Mapping full seismic waveforms to vertical velocity profiles by deep learning[J]. Geophysics, 2021, 86 (5): R711–R721.

Kim Y, Nakata N. Geophysical inversion versus machine learning in inverse problems[J]. Lead. Edge, 2018, 37 (12): 894–901.

Kong Xiangru, Wang Qianshen. Comprehensive Geophysical Profile and Lithospheric Structure and Dynamics in the Western Qinghai-Tibet Plateau[J]. Science Bulletin, 1999, 44(12): 1257-1265.

Kunetz G. Processing and interpretation of magnetotelluric soundings[J]. Geophysics, 1972, 37,: 1005–1021.

Lee K H, Liu G, Morrison H F. A new approach to modeling the electromagnetic response of conductive media[J]. Geophysics, 1989, 54: 1180-1192.

Lei Kangxin, Xue Guoqiang, Chen Weiyin, Zhou Nannan. Relationship between Resolution of Transient Electromagnetic Detection of Thin Layers and Offset Distance[J], Journal of Earth Science and Environment, 2020, 42(6): 731-736.

Leong Z X, Zhu T Y. Direct velocity inversion of ground penetrating radar data using GPRNet[J]. Journal of Geophysical Research: Solid Earth, 2021, 126: e2020JB021047.

Levy S, Oldenburg D, Wang J Y. Subsurface imaging using magnetotelluric data[J]. Geophysics, 1988, 53: 104-117.

Li Xiu, Xue Guoqiang, Song Jianping, Guo Wenbo, Wu Junjie. Optimized Algorithm from Transient Electromagnetic Field to Wave Field[J]. Journal of Geophysics, 2005, 5: 1185-1190

Li Jianhui, Liu Shucai, Zhu Ziqiang, et al. Research on the Symmetry Relationship between Electromagnetic Field and Magnetic Field Excited by Rectangular Loop[J]. Journal of Central South University (Science and Technology), 2010, 41(2): 638-642

Li J H, Farquharson C G, Hu X Y. Three effective inverse Laplace transform algorithms for computing time-domain electromagnetic responses[J]. Geophysics, 2016, 81(2): E113-E128.

Li D, Tian Z, Ma Y, Gu J, Li S. Application of grounded electrical source airborne transient electromagnetic (GREATEM) system in goaf water detection[J]. Journal of Environmental & Engineering Geophysics, 2019a, 24(3).

Li H, Xue G Q, He Y M. Decoupling induced polarization effect from time domain electromagnetic data in Bayesian framework[J]. Geophysics, 2019b, 84(6):1-17.

Li Wenhan, Liu Bin, Li Shucai, Li He, Lu Kailiang, Li Xiu. Research on Multi-resolution Imaging Method of Urban Underground Space Based on High Performance Transient Electromagnetic Radiation Source[J]. Journal of Geophysics, 2020a, 63(12): 4553-4564

Li J F, Liu Y H, Yin C C, Ren X Y, Su Y. Fast imaging of time-domain airborne EM data using deep learning technology[J]. Geophysics, 2020b, 85(5): E163–E170.

Li R, Yu N, Wang X, et al. Model-Based Synthetic geoelectric sampling for magnetotelluric inversion with deep neural networks[J]. IEEE Transactions on Geoscience and Remote Sensing, 2020c, PP(99):1–14.

Li Zhuang, Cheng Peisheng, Zhang Jianming, Miao Xuhuang. Discovery of the Large-scale Lead-zinc Ore in the North Margin of the Luzong Basin and Its Prospecting Indicators[J]. Mineral Exploration, 2020d, 11 (10): 2163-2169

Li Jinming. Geoelectric Field and Electric Prospecting[M]. Beijing: Geological Publishing House, 2005

Liang Guanghe, Xu Xingwang, Xiao Qibin, et al. Application of Magnetotelluric Sounding Method in Copper-Nickel Ore Prospecting - A Case Study of Tulaergen Copper-Nickel Ore Related to Ultramafic Rock in Xinjiang[J]. Mineral Deposits, 2007, 26(1): 120-127

Liang Shengxian, Wang Yonghua, Liao Guozhong, et al. Research and Application of Seismic Interpretation Method of Magnetotelluric Impedance Tensor Mode[J]. Progress in Geophysics, 2015, 30(1): 153-158.

Liao Zhong, Li Wenyao. The application effect of line source TEM in searching for geothermal in Tengchong hot water pond [J]. Engineering Survey, 2002, 1: 67-69

Lin Jun, Xue Guoqiang, Li Xiu. Technical innovation thinking of semi-aerial electromagnetic detection method [J]. Chinese Journal of Geophysics, 2021, 64(9): 2995-3004

Liu Guangding, Hao Tianyao. Application of Geophysical Methods in Searching for Hidden Ore Deposits[J]. Journal of Geophysics, 1995, 38(6): 850-854.

Liu Changsheng, Lin Jun. Modeling and seawater impact analysis of seafloor surface magnetic source transient response [J]. Chinese Journal of Geophysics, 2006, 49(6): 1891-1898.

Liu Xiaoshan, Yang Ling, Fang Sheng, Wang Jiaying. A New Magnetotelluric Inversion Method - Vertical Time-Distance Curve Method[J]. Petroleum Geophysical Prospecting, 1987, 4: 445-453.

Liu Jianxin, He Jishan, Zhang Zongling, et al. Dual-frequency excitation method and its application in the demonstration area [J]. Geology of China, 2001, 28(3):9

Liu B, Guo Q, Li S, et al. Deep learning inversion of electrical resistivity data[J]. IEEE Transactions on Geoscience and Remote Sensing, 2020, PP(99):1–14.

Loke M H, Dahlin T. A comparison of the Gauss–Newton and quasi-Newton methods in resistivity imaging inversion[J]. Journal of Applied Geophysics, 2002, 49(3): 149-162.

Lu Kailiang, Li Xiu, Qi Zhipeng, Fan Yanan, Zhou Jianmei, Li Wenhan, Li He, Zhang Mingjing, Wang Yangzhou. Fine integral transformation algorithm from transient electromagnetic diffusion field to virtual wave field. Chinese Journal of Geophysics, 2021, 64(9): 3379-3390

Luo Yanzhong, Chang Yanjun. Fast algorithm of G-S transformation [J]. Chinese Journal of Geophysics, 2000, 43(5): 684-690.

Marina R C, Niklas L, Thomas K,et al. Two-dimensional probabilistic inversion of plane-wave electromagnetic data: Methodology, model constraints and joint inversion with electrical resistivity data[J]. Geophysical Journal International, 2014, 3: 1508-1524.

McCormack M D, Zaucha D E, Dushek D. First-break refraction event picking and seismic data trace editing using neural networks[J]. Geophysics, 1993, 58: 67–78.

Mohsen A A, Hashish E A. The fast Hankel transform[J]. Geophysical Prospecting, 1994, 42(2): 131–139.

Mu R, Zeng X. A review of deep learning research[J]. KSII Transactions on Internet and Information Systems: 2019, 13: 1738-1764.

Mulder W A, Wirianto M, Slob E C. Time-domain modeling of electromagnetic diffusion with a frequency-domain code[J]. Geophysics, 2007, 73(1): F1-F8.

Nabighian M N. Electromagnetic methods in applied geophysics[M]. Tulsa: Society of Exploration Geophysicists, 1988.

Newman G A, Hohmann G W, Anderson W L. Transient electromagnetic response of a Three-dimensional body in a layered earth[J]. Geophysics, 1986, 51(8): 1608–1627.

Niu Zhilian. Principles of time domain electromagnetic method [M]. Changsha: Central South University Press, 1992.

Oh S, Noh K, Seol S J, Byun J. Cooperative deep learning inversion of controlled-source electromagnetic data for salt delineation[J]. Geophysics, 2020, 85: E121–E137.

Oristaglio M. Conductivity imaging of transient electromagnetic data by conjugate-gradient method[J]. Seg Technical Program Expanded Abstracts, 1993, 12(1): 1396.

Pace F, Santilano A, Godio A. Particle Swarm Optimization of 2D Magnetotelluric data[J]. Geophysics, 2019, 84: 1-59.

Park Hwarong, Yin Changchun. Calculation of transient sounding forward problem using G-S inverse Laplace transformation method [J]. Physical and Chemical Exploration Calculation Calculation, 1987, 9(4): 295-302.

Qi Zhipeng, Li Xiu, Wu Qiong, Sun Huaifeng, Yang Zenglin. Full time domain reverse transformation algorithm from transient electromagnetic diffusion field to pseudo-seismic wave field [J]. Chinese Journal of Geophysics, 2013, 56(10): 3581-3595

Rumelhart D, Hinton G, Williams R. Learning representations by back-propagating errors[J]. Nature, 1986, 323: 533–536. Scales J A. Tomographic inversion via the Conjugate Gradient method[J]. Geophysics, 1987, 52, 179-185.

Song T, Hu X, Du W, Cheng L, Xiao T, Li Q. Lp-Norm inversion of gravity data using adaptive differential evolution. Applied Sciences-Basel, 2021, 11(14): 64-85.

Sun J, Niu Z, Innanen K A, et al. A theory-guided deep learning formulation and optimization of seismic waveform inversion[J]. Geophysics, 2019, 85(2): 1-63.

Sun Huaifeng, Liu Shangbin, Yang Yang. Transient electromagnetic Crank-Nicolson FDTD three-dimensional forward [J]. Chinese Journal of Geophysics Report, 2021, 64(01): 343-354.

Tang Xingong, Hu Wenbao, Yan Liangjun. Study on the influence of rift terrain on long-offset transient electromagnetic sounding [J]. Journal of Engineering Geophysics, 2004, 1(4): 313-317

Teng Jiwen. Strengthening the geophysical prospecting, exploration and development of metal mineral resources in the second depth space of the crust [J]. Geological Bulletin, 2006, 25(7): 767-771.

Wang Ruo, Wang Miaoyue, Lu Yuanlin. Preliminary study on three-dimensional three-component CSAMT method finite element forward simulation [J]. Progress in Geophysics, 2007, 22(2): 579-585

Wang Dayong, Li Tonglin, Gao Yuan, et al. Application of CSAMT method and TEM method in the exploration of concealed ore in Longhushan area, Tongling [J]. Journal of Jilin University (Earth Science Edition), 2009, 39(6): 1134-1140

Wang Zhishui, Miu Xuhuang, Zhang Jianming, et al. Application of comprehensive electrical method in the exploration of Xiwan lead-zinc mine in Anhui [J]. Journal of Geology, 2023, 47(02): 208-215.

Weng Aihua, Wang Xueqiu. Numerical calculation of long-offset transient electromagnetic sounding late response and apparent resistivity [J]. Seismic Geology, 2003, 25(4): 664-670.

Wu X, Xue G Q, Fang G Y, et al. The development and applications of the semi-airborne electromagnetic system in China[J]. IEEE Access, 2019b, 7: 104956-104966.

Xu Shizhe. Finite Element Method in Geophysics [M]. Beijing: Science Press, 1994

Xue Guoqiang, Li Xiu, Song Jianping, Guo Wenbo. Theoretical analysis and numerical calculation of loop source transient electromagnetic imaging [J]. Chinese Journal of Geophysics, 2004, 2: 338-343

Xue G Q, Yan Y J, Li X. Control of the waveform dispersion effect and applications in a TEM imaging technique for identifying underground objects[J]. Journal of Geophysics and Engineering, 2011a, 8(2): 195.

Xue Guoqiang, Wang Heyuan, Yan Shu, et al. Solution of transient electromagnetic field time domain Green's function[J]. Chinese Journal of Geophysics. 2014, 57(2): 671-678.

Xue G Q, Chen W Y, Yan S. Research study on the short offset time-domain electromagnetic method for deep exploration[J]. Journal of Applied Geophysics, 2018, 155: 131-137.

Xue G Q, Zhang L B, Zhou N N, Chen WY. Developments measurements of TEM sounding in China[J]. Geological Journal, 2019, 55(5).

Xue G Q, Li H, He Y M, et al. Development of the inversion method for transient electromagnetic data[J]. IEEE Access, 2020, PP(99), 1-1.

Xue Guoqiang, Chen Weiyin, Wu Xin. Theory and application of electric source short offset transient electromagnetic method[M]. Beijing: Science Press, 2022

Xue J J, Cheng J L, Gelius, Wu X, Zhao Y. Full waveform inversion of transient electromagnetic data in the time domain[J]. IEEE Transactions on Geoscience and Remote Sensing, 2022, 60: pp. 1-8.

Yan Shu, Shi Xianxin, Chen Mingsheng. The problem of detection depth of transient electromagnetic method[J]. Chinese Journal of Geophysics, 2009, 52(6): 1583-1591.

Yan Jiayong, Teng Jiwen, Lv Qingtian. Geophysical exploration and application of deep metal mineral resources[J]. Progress in Geophysics, 2009, 23(3): 871-891

Yan Shu, Xue Guoqiang, Chen Mingsheng. Review and prospect of theoretical research on large loop source transient electromagnetic response[J]. Progress in Geophysics, 2011, 26(3): 941-947.

Yan Shuai, Yin Changchun, Su Yang, et al. Research on fast imaging algorithm of transient electromagnetic data based on convolutional neural network[J]. Chinese Journal of Geophysics, 2023, 66(10): 4290-4300

Yang D, Oldenburg D W. Three-dimensional inversion of airborne time-domain electromagnetic data with applications to a porphyry deposit[J]. Geophysics, 2012, 77(2): 23.

Yang Q, Hu X, Liu S, Jie Q, Wang H, Chen Q. 3-D gravity inversion based on deep convolution neural networks[J]. IEEE Geoscience and Remote Sensing Letters, 2022, 19: pp. 1-5.

Ye Yixin, Deng Juzhi, Li Man, Yang Haiyan. Current status and prospect of electromagnetic method in deep mineral exploration[J]. Progress in Geophysics, 2011, 26(1): 327-334

Yu Dongkai, Song Xianhai, Jiang Dongwei, et al. Improved swarm algorithm and its application in surface wave dispersion curve inversion[J]. Chinese Journal of Geophysics, 2018, 61(4): 1482-1495

Zeng Zhiwen, Chen Xiao, Yang Haiyan, et al. Joint inversion of geomagnetic depth sounding and gravity wide range physical property constraint based on improved differential evolution algorithm[J]. Chinese Journal of Geophysics, 2020, 63(12): 281-293.

Zeng Guo. Numerical simulation of two-dimensional finite element forward modeling of geomagnetic field[D]. Changsha: Central South University, 2008.

Zhai Yusheng, Deng Jun, Wang Jianping, et al. Problems in deep mineral exploration[J]. Mineral Deposits, 2004, 23(2): 142-149.

Zhang S, Zhou N N. Extraction of multiple electrical parameters from IP-affected transient electromagnetic data based on LSTM-ResNet[J]. IEEE Transactions on Geoscience and Remote Sensing, 2023, 61:1-14.

Zhang S, Yin C C, Cao X Y, Sun S Y, Liu Y H, Ren X Y. DecNet: Decomposition network for 3D gravity inversion[J]. Geophysics, 2022, 87(5): G103–G114.

Zhao Guoze, Tang Ji, et al. Using geomagnetic network to detect the conductivity structure of the upper mantle in Changchun and other places[J]. Seismology and Geology, 2001, 23(2): 143-153.

Zhao Y, Wu X, Chen W Y. U-net-based pseudoseismic imaging for the short-offset transient electromagnetic method[J]. Journal of Geophysics and Engineering, 2023, 20(2): 312–332.

Zheng Liping, Yan Liangjun, Xie Xingbing, et al. LOTEM and CSAMT joint exploration test[J]. Journal of Engineering Geophysics, 2009, 6(6): 694-697.

Zhou Nannan, Xue Guoqiang, Chen Weiyin, et al. Geophysical detection of iron ore mining area[J]. Progress in Geophysics, 2011, 26(2): 669-674.

GPSR Compliance

The European Union's (EU) General Product Safety Regulation (GPSR) is a set of rules that requires consumer products to be safe and our obligations to ensure this.

If you have any concerns about our products, you can contact us on ProductSafety@springernature.com

In case Publisher is established outside the EU, the EU authorized representative is:

Springer Nature Customer Service Center GmbH
Europaplatz 3
69115 Heidelberg, Germany

Batch number: 10704305

Printed by Printforce, the Netherlands